Emergence, Complexity and Computation

Volume 30

The Emergence, Complexity and Computation (ECC) series publishes new developments, advancements and selected topics in the fields of complexity, computation and emergence. The series focuses on all aspects of reality-based computation approaches from an interdisciplinary point of view especially from applied sciences, biology, physics, or chemistry. It presents new ideas and interdisciplinary insight on the mutual intersection of subareas of computation, complexity and emergence and its impact and limits to any computing based on physical limits (thermodynamic and quantum limits, Bremermann's limit, Seth Lloyd limits...) as well as algorithmic limits (Gödel's proof and its impact on calculation, algorithmic complexity, the Chaitin's Omega number and Kolmogorov complexity, non-traditional calculations like Turing machine process and its consequences,...) and limitations arising in artificial intelligence field. The topics are (but not limited to) membrane computing, DNA computing, immune computing, quantum computing, swarm computing, analogic computing, chaos computing and computing on the edge of chaos, computational aspects of dynamics of complex systems (systems with self-organization, multiagent systems, cellular automata, artificial life,...), emergence of complex systems and its computational aspects, and agent based computation. The main aim of this series it to discuss the above mentioned topics from an interdisciplinary point of view and present new ideas coming from mutual intersection of classical as well as modern methods of computation. Within the scope of the series are monographs, lecture notes, selected contributions from specialized conferences and workshops, special contribution from international experts.

More information about this series at http://www.springer.com/series/10624

Andrew Adamatzky

Editor

Reversibility and Universality

Essays Presented to Kenichi Morita
on the Occasion of his 70th Birthday

 Springer

Editor
Andrew Adamatzky
Unconventional Computing Centre
University of the West of England
Bristol
UK

ISSN 2194-7287   ISSN 2194-7295 (electronic)
Emergence, Complexity and Computation
ISBN 978-3-030-10334-7   ISBN 978-3-319-73216-9 (eBook)
https://doi.org/10.1007/978-3-319-73216-9

Kenichi Morita

Preface

Kenichi Morita entered the world of automata and universality with his first Japanese language paper in 1972 [21] (Fig. 1), where Kazuhiro Sugata and Morita designed a cellular automaton computer based on interactions of signals propagating in two-dimensional automaton arrays, and English language paper in 1977 [22], where Kazuhiro Sugata, Hiroshi Umeo and Morita developed a rebound automaton—a two-dimensional automaton which accepts several classes of irregular languages. The latter paper inspired a series of further research threads on non-deterministic rebound automata [20], fooling rebound automata [19], probabilistic rebound automata [26], rebound Turing machines [25], and inspired findings on complexity of multi-head automata [2] and existence of a context-free language not accepted by any deterministic rebound automaton [4]. In 1978, Kenichi wrote an influential paper on lower bounds on tape complexity of two-dimensional tape Turing machines [8]. 'Universality', or computational equivalence, component emerged in Kenichi's research in 1982, when, together with Umeo and Sugata, he demonstrated that any deterministic two-way real-time cellular automaton can be simulated by a deterministic one-way cellular automaton in twice real-time [23]. In parallel, Kenichi and co-authors produced impressible results on isometric context-free array grammar generating upright rectangles [24] and shown that membership problems for regular array grammars and context-free array grammars are NP-complete, and their emptiness problems are undecidable [9]. Kenichi established himself in the field of reversible computation universality when he showed, in 1989, that any irreversible Turing machine can be converted to a single-tape two-symbol reversible Turing machine [11]. This reversible Turing machine was used by Kenichi Morita and Masateru Harao to solve a problem proposed by Tommaso Toffoli—whether a one-dimensional reversible cellular automaton is computationally universal. Morita and Harao designed a one-dimensional partitioned cellular automaton that simulates the computationally universal reversible Turing machine [10]. The automaton was further extended in computationally universal one-way reversible cellular automaton [13]. In 1992, Kenichi Morita and Satoshi Ueno freed spatial and temporal non-uniformity of Norman Margolus model by designing a two-dimensional reversible partitioned cellular automaton with 16 states [14]. They proved

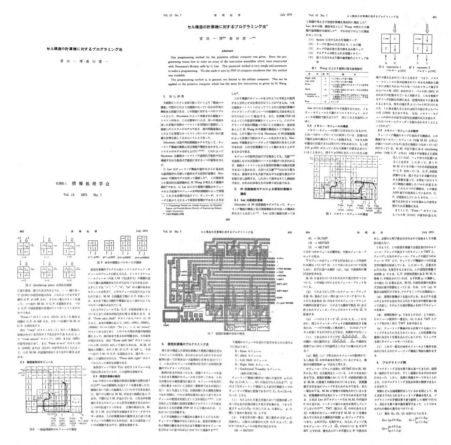

Fig. 1 Pages of the first paper authored by Kenichi Morita, in collaboration with Kazuhiro Sugata on computing via interacting signals in cellular automata [21]. Copyright (c) 1972 by the Information Processing Society of Japan

Fig. 1 (continued)

computational universality of this automaton by implementing Fredkin gate in its evolution. Next quarter of century of Kenichi Morita research was full of remarkable results in automata and grammars, universality and reversibility, theory of computation and unconventional computing. Most high impact results include parallel generation and parsing of array languages by reversible automata [15], construction of a reversible automaton from Fredkin gates [12], solving a firing squad synchronisation problem in reversible cellular automata [3], self-reproduction in reversible cellular spaces [17], universal reversible two-counter machine [16], solution of NP problems in hyperbolic cellular automata [6, 7], reversible P-systems [1], a new universal reversible logic element with memory [18] and reversibility in asynchronous cellular automata [5]. Kenichi Morita's achievements in reversibility, universality and theory of computation are celebrated by over twenty high-profile contributions from Kenichi's colleagues, collaborators, students and friends. Theoretical constructs presented in this book are amazing in their diversity and depths of intellectual insights: queue automata, hyperbolic cellular automata, Abelian invertible automata, number-conserving cellular automata, Brownian circuits, chemical automata, logical gates implemented via glider collisions, computation in swarm networks, picture arrays, universal reversible counter machines, input-position-restricted models of language acceptors, descriptional complexity and persistence of cellular automata, partitioned cellular automata, firing squad synchronisation algorithms, reversible asynchronous automata, reversible simulations of ranking trees, Shor's factorisation algorithms and power consumption of cellular automata. The book is a unique source of knowledge in universality and reversibility in computation, an indispensable textbook for computer scientists, mathematicians, physicists and engineers.

Bristol, UK   Andrew Adamatzky
February 2018

References

1. Alhazov, A., Morita, K.: On reversibility and determinism in P systems. In: Workshop on Membrane Computing, pp. 158–168. Springer (2009)
2. Holzer, M., Kutrib, M., Malcher, A.: Complexity of multi-head finite automata: Origins and directions. Theor. Comput. Sci. **412**(1–2), 83–96 (2011)
3. Imai, K., Morita, K.: Firing squad synchronization problem in reversible cellular automata. Theor. Comput. Sci. **165**(2), 475–482 (1996)
4. Inoue, K., Takanami, I., Taniguchi, H.: A note on rebound automata. Inf. Sci. **26**(1), 87–93 (1982)
5. Lee, J., Peper, F., Adachi, S., Morita, K., Mashiko, S.: Reversible computation in asynchronous cellular automata. In: Unconventional Models of Computation, pp. 220–229. Springer (2002)
6. Margenstern, M., Morita, K.: A polynomial solution for 3-SAT in the space of cellular automata in the hyperbolic plane. J. Univ. Comput. Sci. **5**(9), 563–573 (1999)
7. Margenstern, M., Morita, K.: NP problems are tractable in the space of cellular automata in the hyperbolic plane. Theor. Comput. Sci. **259**(1), 99–128 (2001)
8. Morita, K., Umeo, H., Ebi, H., Sugata, K.: Lower bounds on tape complexity of two-dimensional tape Turing machines. IECE Jpn. D **61**(6), 381–386 (1978)
9. Morita, K., Yamamoto, Y., Sugata, K.: The complexity of some decision problems about two-dimensional array grammars. Inf. Sci. **30**(3), 241–262 (1983)
10. Morita, K., Harao, M.: Computation universality of one-dimensional reversible (injective) cellular automata. IEICE Trans. **72**(6), 758–762 (1989)
11. Morita, K., Shirasaki, A., Gono, Y.: A 1-tape 2-symbol reversible Turing machine. IEICE Trans. **72**(3), 223–228 (1989)
12. Morita, K.: A simple construction method of a reversible finite automaton out of Fredkin gates, and its related problem. IEICE Trans. **73**(6), 978–984 (1990)
13. Morita, K.: Computation-universality of one-dimensional one-way reversible cellular automata. Inf. Process. Lett. **42**, 325–329 (1992)
14. Morita, K., Ueno, S.: Computation-universal models of two-dimensional 16-state reversible cellular automata. IEICE Trans. Inf. Syst. **75**(1), 141–147 (1992)
15. Morita, K., Ueno, S.: Parallel generation and parsing of array languages using reversible cellular automata. Int. J. Pattern Recognit. Artif. Intell. **8**(02), 543–561 (1994)
16. Morita, K.: Universality of a reversible two-counter machine. Theor. Comput. Sci. **168**(2), 303–320 (1996)
17. Morita, K., Imai, K.: Self-reproduction in a reversible cellular space. Theor. Comput. Sci. **168** (2), 337–366 (1996)
18. Morita, K.: A simple reversible logic element and cellular automata for reversible computing. In: Proc. MCU 2001, LNCS 2055, pp. 102–113. Springer (2001)
19. Petersen, H.: Fooling rebound automata. Math. Found. Comput. Sci. 1999, 241–250 (1999)
20. Sakamoto, M., Inoue, K., Takanami, I.: A two-way nondeterministic one-counter languages not accepted by nondeterministic rebound automata. IEICE Trans. **73**(6), 879–881 (1990)
21. Sugata, K. and Morita, K.: A programing method for cellular computer. J. Inf. Process. **13**(7), 460–466 (1972) (in Japanese)
22. Sugata, K., Umeo, H., Morita, K.: Language accepted by a rebound automaton and its computing abilities. Electron. Commun. Jpn. **60**(4), 11–18 (1977)
23. Umeo, H., Morita, K., Sugata, K.: Deterministic one-way simulation of two-way real-time cellular automata and its related problems. Inf. Process. Lett. **14**(4), 158–161 (1982)
24. Yamamoto, Y., Morita, K., Sugata, K.: An isometric context-free array grammar that generates rectangles. IEICE Trans. **65**(12), 754–755 (1982)
25. Zhang, L., Inoue, K., Ito, A., Wang, Y.: A leaf-size hierarchy of alternating rebound Turing machines. J. Automata Lang. Comb. **7**(3), 395–410 (2002)
26. Zhang, L., Okazaki, T., Inoue, K., Ito, A., Wang, Y.: A note on probabilistic rebound automata. IEICE Trans. Inf. Syst. **81**(10), 1045–1052 (1998)

Contents

A Snapshot of My Life

Kenichi Morita

I was born on 30 March 1949 in Osaka as a son of my father Masukichi and my mother Fusako. My parents influenced me largely, since I had neither brother nor sister. My father was an electrical engineer, and thus I also liked making or designing various functional things. My later research style may be affected by such an activity. When I was a high school student, I started to make electrical and electronic circuits as a hobby. As many hobbyists in those days did, I first assembled a radio from vacuum tubes and some other components. I also composed digital circuits, such as logic gates and flip-flops, out of transistors, diodes, resistors and capacitors. Around the end of 1960's, TTL (transistor-transistor logic) ICs became available. So, I made several logic circuits for digital apparatus, for example a digital clock, using ICs also as a hobby. At that time, I dreamed to assemble a small digital computer out of TTL ICs, but, of course, it was not possible for me.

In 1967, I entered Osaka University, Faculty of Engineering Science. I was in the Department of Biophysical Engineering, which was established in 1967 for the research and education in the interdisciplinary area of biology, neuroscience, physics, and information science. There, many classes were held for undergraduate and graduate students on basics of molecular biology, biophysics, neurophysiology, quantum mechanics, control theory, information theory, automata theory, and so on. Although only a few topics among them are directly concerned with my current study, a wide variety of topics in them stimulated me very much, and had indirect influences on my research.

In 1970, I started a graduation study on cellular automata (CAs) under the supervision of Professor Kazuhiro Sugata. I first learned the classical work of von Neumann on the self-reproducing CA, and then studied a realization method of a Turing machine in the von Neumann's 29-state CA. Since then I have been

K. Morita (✉)
Hiroshima University, Higashi Hiroshima, Japan
e-mail: km@hiroshima-u.ac.jp

© Springer International Publishing AG 2018
A. Adamatzky (ed.), *Reversibility and Universality*, Emergence, Complexity
and Computation 30, https://doi.org/10.1007/978-3-319-73216-9_1

studying CAs till now. One day around the end of 1970, a senior student taught me that an interesting CA was introduced in the latest issue of Scientific American. I looked at the issue, and found that "The fantastic combination of John Conway's new solitaire game *life*" is explained in the article "Mathematical Games" by Martin Gardner. The game *life* was again discussed in the issue of February 1971, where the surprising pattern of "glider gun" by William Gosper was introduced. In the same issue, Edward Fredkin's simple self-replicating CA was also described. These CAs, as well as interesting patterns of *life*, gave a great impression to me.

Around that year, a new mini-computer PDP-12 of DEC (Digital Equipment Corporation) was introduced to my department. Though its memory size was quite small (it had only 8 K words of magnetic-core memory, where 1 word consists of 12 bits), it was a very useful computer since a CRT (cathode ray tube) display was equipped. Note that, in those days, only a very few computers had such a graphic display. In the spring of 1971, after finishing the graduation study, I wrote simulation programs for the game *life*, and for the Fredkin's self-replicating CA. The programs were written by an assembly language based on a very naive algorithm. Although they were not good simulators and thus slow, it was very fascinating for me to watch evolution processes of the CAs through the CRT display. These simulator programs were indeed the first two I wrote. Since then, I have been much interested in creating simulators for CAs, Turing machines, and other computing models for assisting theoretical studies on them.

From 1974 to 1987, I was a research associate of Osaka University, Faculty of Engineering Science. In 1974, I started to study automata models of two-dimensional information processing. They are two-dimensional CAs, Turing machines with two-dimensional tapes, and some others. I submitted my PhD thesis entitled "Computational complexity in one- and two-dimensional tape automata" in 1978. The chairman of the assessment committee of my thesis was Professor Tadao Kasami. Later, I also studied several types of array grammars, e.g., isometric regular array grammars, to investigate two-dimensional formal languages.

In my private life, I married Kiyomi in 1975, and later I had two daughters Mayumi and Emiko, and a son Juichiro. Several months after the marriage, i.e., in 1976, I started a new hobby of running. Although I was not good at other sports, I continued running almost 40 years. My best record in full marathon is 4:05:34 (3 March 1985), but it is far from the record of Alan Turing. According to Wikipedia (see "List of non-professional marathon runners"), Turing ran a full marathon with the time 2:46:03 in 1947, a really good record in those days.

One day around the middle of 1980's, Professor Toshio Mitsui of Osaka University, who is an expert of biophysics and is also interested in computation theory, asked me a question how physical reversibility is related to computing processes, namely, how computing can be carried out in a physically reversible environment. This question gave me a new vista, and I was attracted in the topic of "reversible computing," which may play an important role in the future computing systems. I first looked for the past literature on this topic, and found several pioneering papers by Landauer, Feynman, Bennett, Fredkin and Toffoli, and some others. I thought there are still many problems on reversible computing to be studied from the viewpoint of automata theory. Since then, I have been studying reversible computing extensively.

In 1987, I moved to Yamagata University, Faculty of Engineering. There, I was an associate professor from 1987 to 1990, and a full professor from 1990 to 1993. The university is in Yonezawa city, which lies in the north-east region of Japan, and is surrounded by mountains. Since snow falls very heavily in winter, I had to remove a large pile of snow lying in front of my house almost every morning to make a path, but I enjoyed it. In summer, I often climbed up mountains, and sometimes ran there like a cross-country. In Yamagata University, I mainly studied reversible CAs. I also studied formal inference systems for knowledge processing with my colleagues. I proposed a framework of "partitioned CAs" for designing reversible CAs, and proved, e.g., computational universality of one-dimensional reversible CAs by it.

I then moved to Hiroshima University, Graduate School of Engineering, in 1993, where I was a full professor till 2013. I continued to study reversible computing with my colleagues and students. Besides reversible CAs, I investigated various models of reversible computing, such as reversible Turing machines, reversible counter machines, reversible logic elements with memory, and so on. I showed that even very simple instances of these reversible models have computational universality. When I was in Hiroshima, I made international research exchange actively; I visited many places in the world to attend conferences, and to pursue joint researches. Also many researchers visited Hiroshima. Thus, I had many friends in the 20 years in Hiroshima, which was indeed a happy period for me. In 2013, I retired from the university, and I am now a professor emeritus of Hiroshima University.

November, 2017   *Kenichi Morita*

FSSP Algorithms for 2D Rectangular Arrays. Recent Developments

Hiroshi Umeo

Abstract Synchronization of large-scale networks is an important and fundamental computing primitive in parallel and distributed systems. The synchronization in cellular automata, known as the Firing Squad Synchronization Problem (FSSP), has been studied extensively for more than fifty years, and a rich variety of synchronization algorithms has been proposed. In this article, we give a survey on a class of minimum-time FSSP algorithms for two-dimensional (2D) rectangular arrays. The algorithms discussed here are all based on an L-shaped mapping, where the synchronized configurations on 1D arrays are mapped efficiently onto a 2D array in an L-shaped form, yielding a 2D minimum-time FSSP algorithm.

1 Introduction

The synchronization in ultra-fine-grained parallel computational model of cellular automata (CA) has been known as the Firing Squad Synchronization Problem (FSSP) since its development, in which it was originally proposed by J. Myhill in the book edited by Moore [2] to synchronize all/some parts of self-reproducing cellular automata. The FSSP has been studied extensively for more than fifty years, and a rich variety of synchronization algorithms has been proposed [1–16].

In the present article, we focus our attention to a class of 2D minimum-time FSSP algorithms that is based on an L-shaped mapping, where the synchronized configurations on 1D arrays are mapped efficiently onto a 2D array in an L-shaped form, yielding a minimum-time FSSP algorithm and a smaller-state implementation. In Sect. 2, we give a description of the 2D FSSP and review some basic results on the 2D FSSP algorithms. In Sects. 3–6, we introduce a minimum-time FSSP algorithm in a new class of 2D FSSP algorithms, each based on the L-shaped mapping, and present several small-state implementations of those algorithm. In the last section, we give a summary of the paper.

H. Umeo (✉)
University of Osaka Electro-Communication, Neyagawa-shi,
Hastu-cho, 18-8, Osaka 572-8530, Japan
e-mail: umeo@cyt.osakac.ac.jp; umeo@osakac.ac.jp

© Springer International Publishing AG 2018
A. Adamatzky (ed.), *Reversibility and Universality*, Emergence, Complexity
and Computation 30, https://doi.org/10.1007/978-3-319-73216-9_2

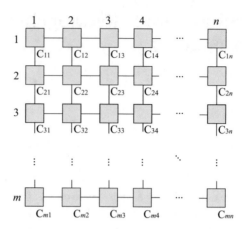

Fig. 1 A two-dimensional (2D) rectangular cellular automaton of size $m \times n$, arranged in m rows and n columns

2 Firing Squad Synchronization Problem

2.1 FSSP on 2D Cellular Arrays

Figure 1 shows a finite two-dimensional (2D) cellular array consisting of $m \times n$ cells. Each cell is an identical (except the border cells) finite-state automaton. The array operates in lock-step mode in such a way that the next state of each cell (except border cells) is determined by both its own present state and the present states of its north, south, east, and west neighbors. All cells (*soldiers*), except the north-west corner cell (*general*), are initially in the quiescent state at time $t = 0$ with the property that the next state of a quiescent cell with quiescent neighbors is the quiescent state again. At time $t = 0$, the north-west upper corner cell $C_{1,1}$ is in the *fire-when-ready* state, which is the initiation signal for the array. The FSSP is to determine a description (state set and next-state function) for cells that ensures all cells enter the *fire* state at exactly the same time and for the first time. The tricky part of the problem is that the same kind of soldier having a fixed number of states must be synchronized, regardless of the size $m \times n$ of the array. The set of states and transition rules must be independent of m and n.

A formal definition of the 2D FSSP is as follows: a cellular automaton $\mathcal{M}$ is a pair $\mathcal{M} = (\mathcal{Q}, \delta)$, where

1. $\mathcal{Q}$ is a finite set of states with three distinguished states G, Q, and F, each in $\mathcal{Q}$. G is an initial general state, Q is a quiescent state, and F is a firing state, respectively.
2. δ is a next state function such that $\delta : \mathcal{Q} \times (\mathcal{Q} \cup \{*\})^4 \rightarrow \mathcal{Q}$. The state $* \notin \mathcal{Q}$ is a pseudo state of the border of the array. Each tuple in the next state function δ means that:

$$S_{\text{itself}}^{t+1} = \delta(S_{\text{itself}}^t, S_{\text{north}}^t, S_{\text{south}}^t, S_{\text{east}}^t, S_{\text{west}}^t).$$

Here, we denote a state of $C_{i,j}$ at time (step) t by $S_{i,j}^t$, where $t \geq 0$, $1 \leq i \leq m, 1 \leq j \leq n$.

3. The quiescent state Q must satisfy the following conditions:

$$\delta(Q, Q, Q, Q, Q) = \delta(Q, *, Q, Q, *) = \delta(Q, *, Q, Q, Q) = \delta(Q, *, Q, *, Q) =$$
$$\delta(Q, Q, Q, *, Q) = \delta(Q, Q, *, *, Q) = \delta(Q, Q, *, Q, Q) = \delta(Q, Q, *, Q, *) =$$
$$\delta(Q, Q, Q, Q, *) = Q.$$

A 2D cellular automaton of size $m \times n$, $\mathcal{M}_{m \times n}$ consisting of $m \times n$ copies of $\mathcal{M}$, is a 2D array of $\mathcal{M}$. Each $\mathcal{M}$ is referred to as a cell and denoted by $C_{i,j}$, where $1 \leq i \leq m$ and $1 \leq j \leq n$.

A *configuration* of $\mathcal{M}_{m \times n}$ at time t is a function $\mathcal{C}^t : [1, m] \times [1, n] \to Q$ and it is denoted as:

$$\begin{matrix} S_{1,1}^t S_{1,2}^t \dots S_{1,n}^t \\ S_{2,1}^t S_{2,2}^t \dots S_{2,n}^t \\ S_{3,1}^t S_{3,2}^t \dots S_{3,n}^t \\ \cdot \\ \cdot \\ \cdot \\ S_{m,1}^t S_{m,2}^t \dots S_{m,n}^t. \end{matrix}$$

A *computation* of $\mathcal{M}_{m \times n}$ is a sequence of configurations of $\mathcal{M}_{m \times n}$, $\mathcal{C}^0$, $\mathcal{C}^1$, $\mathcal{C}^2$, ..., $\mathcal{C}^t$, ..., where $\mathcal{C}^0$ is a given initial configuration such that:

$$S_{i,j}^0 = \begin{cases} G & i = j = 1 \\ Q & \text{otherwise.} \end{cases} \tag{1}$$

A configuration at time $t + 1$, $\mathcal{C}^{t+1}$ is computed by synchronous applications of the next state function δ to each cell of $\mathcal{M}_{m \times n}$ in $\mathcal{C}^t$ such that:

$$S_{i,j}^{t+1} = \delta(S_{i,j}^t, S_{i-1,j}^t, S_{i+1,j}^t, S_{i,j+1}^t, S_{i,j-1}^t).$$

A *synchronized configuration* of $\mathcal{M}_{m \times n}$ at time t is a configuration $\mathcal{C}^t$: $S_{i,j}^t = F$, for any $1 \leq i \leq m$ and $1 \leq j \leq n$.

The FSSP is to obtain an $\mathcal{M}$ such that, for any $m, n \geq 2$,

1. A synchronized configuration at time $t = T(m, n)$, $\mathcal{C}^{T(m,n)} : S_{i,j}^{T(m,n)} = F$, for any $1 \leq i \leq m$ and $1 \leq j \leq n$, can be computed from an initial configuration $\mathcal{C}^0$ in Eq. (1).
2. For any t, i such that $1 \leq t \leq T(m, n) - 1, 1 \leq i \leq m, 1 \leq j \leq n, S_{i,j}^t \neq F$.

No cells fire before time $t = T(m, n)$. We say that the array $\mathcal{M}_{m \times n}$ is synchronized at time $t = T(m, n)$ and the function $T(m, n)$ is the time complexity for the synchronization.

2.2   Lower-Bound and Optimality in 2D FSSP Algorithms

Concerning the time optimality of the 2D FSSP algorithms, Beyer [1] and Shinahr [4] gave a lower bound of the algorithms and proposed a minimum-time FSSP algorithm.

Theorem 1 *There exists no cellular automaton that can synchronize any 2D array of size $m \times n$ in less than $m + n + \max(m, n) - 3$ steps, where the general is located at one corner of the array.*

Theorem 2 *There exists a cellular automaton that can synchronize any 2D array of size $m \times n$ at exactly $m + n + \max(m, n) - 3$ steps, where the general is located at one corner of the array.*

3   Beyer–Shinahr Algorithm

The first 2D minimum-time FSSP algorithm $\mathcal{A}_1$, developed independently by Beyer [1] and Shinar [4], is based on a rotated L-shaped mapping which maps configurations of generalized FSSP (GFSSP) solutions for 1D arrays onto a 2D array in an L-shaped fashion. The GFSSP is an extended FSSP version which allows the initial general to be located at any cell, where the solution with the same kind of soldier having a fixed number of states must be designed, regardless of the position of the general and the length n of the array.

A rectangular array of size $m \times n$ is regarded as $\min(m, n)$ rotated L-shaped 1D arrays, where each rotated L-shaped 1D array is synchronized independently by using the GFSSP algorithm.

Figure 2 (left) is a space-time diagram for the original FSSP on which most of the minimum-time FSSP algorithms have been developed. At time $t = 0$ the general on the left end emits an infinite number of signals which propagates at $1/(2^{\ell+1} - 1)$ speed, where ℓ is non-negative integer. These signals with $\ell = 1, 2, 3, \ldots$ meet with a reflected signal at half point, quarter points, ..., etc., denoted by • in Fig. 2. It is noted that these cells indicated by • are synchronized. By increasing the number of *pre-synchronized* cells (not in firing state) exponentially, eventually all of the cells are synchronized at the last stage for the first time.

A key idea behind the GFSSP algorithm proposed by Moore and Langdon [3] is to reconstruct the original FSSP algorithm as if an initial general had been at the left or right end with being in the general state at time $t = -(k - 1)$, where k is the number of cells between the general and the nearest end. Figure 2 (right) illustrates a space-time diagram for the GFSSP algorithm. The initial general emits a left- and right-going signal with $1/1$ speed and keeps its position by marking a special symbol. The propagated signals generate a new general at each end. On reaching the end, they generate the necessary signals assuming that the end is the far end. The special marking symbol tells the first $1/1$ signal generated by the left and right end generals that side was the just nearest end. At that point the slope $1/1$ signal is generated

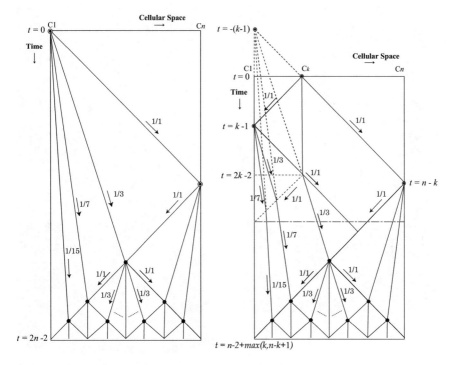

Fig. 2 Space-time diagram of the synchronization algorithms for the original FSSP with a general at one end (left figure) and the GFSSP with a general at an arbitrary position (right figure) in a 1D array of length n

and it changes the slope of all the preceding signals to the next higher one, that is, $1/(2^\ell - 1)$ becomes $1/(2^{\ell+1} - 1)$ for $\ell = 1, 2, \ldots,.$

Note that the original minimum-time solution is working below the dotted line in the Fig. 2 (right). Therefore the minimum-time complexity for the GFSSP is $\min(k - 1, n - k)$ steps smaller than the original FSSP with a general at one end. Thus, the time complexity is $2n - 2 - \min(k - 1, n - k) = n - 1 + \max(k - 1, n - k) = n - 2 + \max(k, n - k + 1)$. Most of the GFSSP algorithms presented in the past are based on the space-time diagram shown in Fig. 2 (right). Figure 3 presents some snapshots for the Moore and Langdon's [3] 17-state implementation on 21 cells with a general on 8th and 16th cells. A comprehensive survey on GFSSP algorithms and their implementations can be seen in Umeo et al. [9]. The minimum-time GFSSP algorithm [3, 9] is stated as follows:

Theorem 3 *There exists a cellular automaton that can synchronize any 1D array of length n in minimum $n + \max(k, n - k + 1) - 2$ steps, where the general is located on the kth cell from left end.*

We overview the algorithm $\mathcal{A}_1$ operating on an array of size $m \times n$. Configurations of the generalized synchronization processes on 1D array can be mapped on rotated

Fig. 3 Snapshots of the Moore and Langdon's [3] 17-state implementation on 21 cells with a general on 8th and 16th cells

L-shaped arrays. We refer the 1D array as L-array. See Fig. 4. At time $t = 0$, the north-west cell $C_{1,1}$ is in general state and all other cells are in quiescent state. For any i such that $1 \leq i \leq \min(m, n)$, the cell $C_{i,i}$ will be in the general state at time $t = 3i - 3$. A special signal which travels towards a diagonal direction is used to generate generals on the cells $\{C_{i,i} | 1 \leq i \leq \min(m, n) \}$. For each i such that $1 \leq i \leq \min(m, n)$, the cells $\{C_{i,j} | i \leq j \leq n \}$ and $\{C_{k,i} | i \leq k \leq m\}$ constitute the ith L-shaped array. Note that the ith general generated at time $t = 3i - 3$ is on the $(m - i + 1)$th cell from the left end of the ith L-array. The length of the ith L-array is $m + n - 2i + 1$. Thus, using Theorem 3, the ith L-array can be synchronized at exactly $t_i = 3i - 3 + m + n - 2i + 1 - 2 + \max(m - i + 1, n - i + 1) = m + n + \max(m, n) - 3$, which is independent of i. In this way, all of the L-arrays can be synchronized simultaneously.

Thus, an $m \times n$ array synchronization problem is reduced to independent $\min(m, n)$ 1D GFSSP problems such that:

$$\begin{cases} \mathcal{P}(m, m + n - 1), \mathcal{P}(m - 1, m + n - 3), ..., \\ \qquad ..., \mathcal{P}(1, n - m + 1) & m \leq n, \\ \mathcal{P}(m, m + n - 1), \mathcal{P}(m - 1, m + n - 3), ..., \\ \qquad ..., \mathcal{P}(m - n + 1, m - n + 1) & m > n. \end{cases}$$

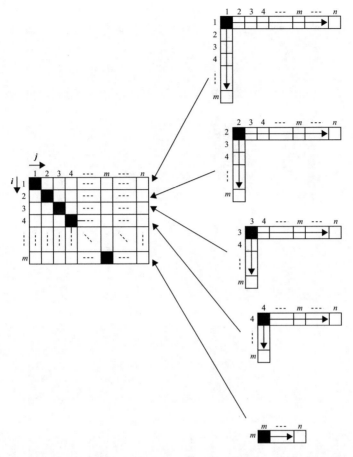

Fig. 4 A 2D synchronization scheme based on an L-shaped mapping developed in Beyer [1] and Shinahr [4]

Here, $\mathcal{P}(k, \ell)$ means the 1D GFSSP problem for ℓ cells with a general on the kth cell from left end.

Shinahr [4] presented a 28-state implementation of the algorithm, where most of (97%) the transition rules had *wild cards* which can match any state. In 2009, Umeo, Ishida, Tachibana, and Kamikawa [8] showed that the 28-state rule set consists of 12849 transition rules and it is valid for the synchronization for any rectangle arrays of size $m \times n$ such that $2 \leq m, n \leq 500$. Figures 5 and 6 illustrate snapshots of the configurations on an array of size 14×9 and 9×14 based on the new 28-state, 12849-rule implementation given in Umeo et al. [8]. Thus, we have:

Theorem 4 *The algorithm $\mathcal{A}_1$, implemented on a cellular automaton with 28 states and 12849 rules, can synchronize any $m \times n$ rectangular array in minimum $m + n + \max(m, n) - 3$ steps.*

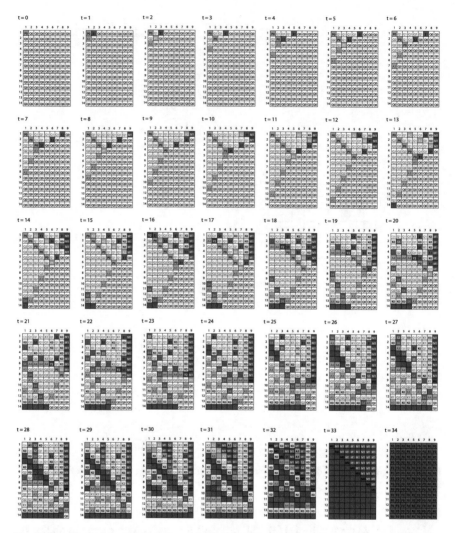

Fig. 5 Snapshots of the configurations of the Shinahr's 28-state synchronization algorithm on a rectangle array of size 14 × 9, implemented in Umeo et al. [8]

4 Umeo–Kubo–Nomura Zebra Implementations

In this section, we give three state-efficient implementations $\mathcal{A}_2$ of the FSSP algorithm $\mathcal{A}_1$. The implementation is based on a zebra mapping which was originally developed for realizing a 7-state square synchronizer in Umeo and Kubo [10]. We overview an implementation technique of the zebra mapping for square arrays.

The zebra mapping is basically similar to the rotated L-shaped mapping scheme presented in the previous section, however, the mapping onto square arrays consists

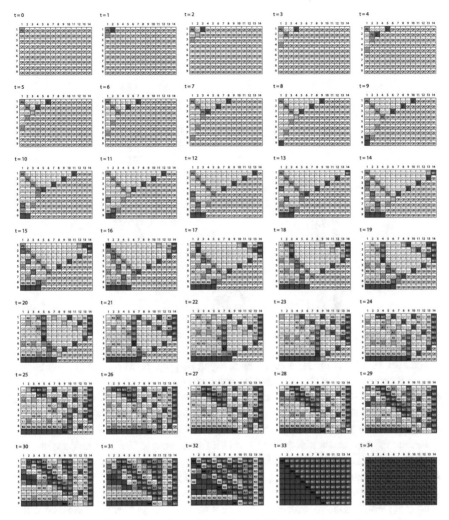

Fig. 6 Snapshots of the configurations of the Shinahr's 28-state synchronization algorithm on a rectangle array of size 9×14, implemented in Umeo et al. [8]

of two types of configurations: one is a synchronized configuration and the other is a filled-in configuration with a stationary state. The stationary state remains unchanged once filled-in by the time before the final synchronization. Each configuration is mapped alternatively onto a square array *in a zebra fashion*. The mapping is referred to as *zebra mapping*. Figure 7 illustrates the zebra mapping which consists of an embedded synchronization and a filled-in layer.

A key idea of the zebra implementation is:

- Alternative embedding of two types of configurations. A stationary layer separates two synchronization layers and it allows us to use an equal state set for the vertical

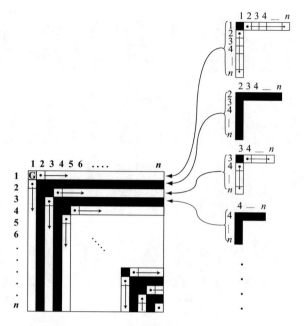

Fig. 7 A zebra mapping scheme for an $n \times n$ square cellular automaton

and horizontal synchronization on each layer, helping us to construct a small-state transition rule set for the synchronization layers.

- A one-cell smaller synchronization configuration than the usual L-shaped mapping is embedded, where we can save synchronization time by two steps.
- A single state X is used for an initial general state of the square synchronizer, the stationary state in stationary layers, and a firing state of the embedded one-cell-smaller synchronization algorithm. The state X itself acts as a pre-firing state.
- Any cell in state X, except $C_{n,n}$, enters the final synchronization state at the next step if all its neighbors are in state X or the boundary state of the square. The cell $C_{n,n}$ enters the synchronization state if and only if its north and west cells are in state X and its east and south cells are in the boundary state. A cell in state X that is adjacent to the cell $C_{n,n}$ is also an exception. This is an only condition that makes cells fire.

The zebra implementation for squares can be applied to rectangles with small modifications. As is shown in Fig. 4, a 1D GFSSP configuration is mapped on an L-shaped array, where the cells on the horizontal and vertical segments have to cooperate with each other for the synchronization. Thus, in contrast to the square implementation, two independent, small-size synchronization configurations cannot be implemented on the horizontal and vertical segment on a single synchronization layer in the rectangle case. Here we do not go to the details of the implementation. All the implementations given below are variants of the zebra-mapping.

The first ten-state implementation is a straightforward implementation of the zebra-mapping, which yields a non-minimum-time algorithm. The second 11-state

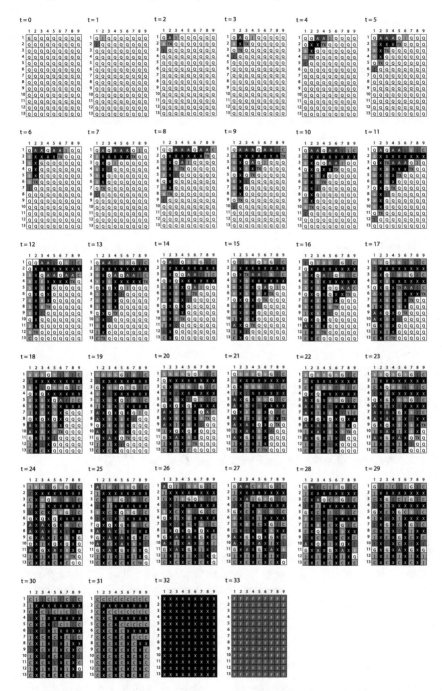

Fig. 8 Snapshots of the non-minimum-time ten-state synchronization process on a 13×9 rectangular array

implementation is a variant of the zebra-mapping, where the first synchronization layer L_1 and the thereafter layers L_i, $i \geq 3$ take a different set of synchronization rule set. The third one is a nine-state implementation which regards the marking symbol used in the recursive division as the pre-firing state, making the algorithm operate in minimum-steps. Those three implementations are stated in Theorems 5, 6 and 7. Some snapshots of the synchronization processes in those three implementations are given in Figs. 8, 9, and 10. See Umeo and Nomura [13] for details.

Theorem 5 *There exists a ten-state 2D CA that can synchronize any $m \times n$ rectangle arrays in $m + n + \max(m, n) - 2$ non-minimum steps.*

Theorem 6 *There exists an eleven-state 2D CA that can synchronize any $m \times n$ rectangle arrays in $m + n + \max(m, n) - 3$ minimum steps.*

Theorem 7 *There exists a nine-state 2D CA that can synchronize any $m \times n$ rectangle arrays in $m + n + \max(m, n) - 3$ minimum steps.*

5 Umeo–Yunès–Yamawaki Algorithm

In this section, we develop a minimum-time FSSP algorithm $\mathcal{A}_3$ which is also based on a new L-shaped mapping. First, we introduce a *freezing-thawing* technique that yields a delayed synchronization for 1D array.

5.1 Freezing-Thawing Technique for Delayed FSSP

The following technique was developed in Umeo [5] for designing several fault-tolerant FSSP algorithms for 1D arrays.

Theorem 8 *Let t_0, t_1, t_2 and Δt be any integer such that $t_0 \geq 0$, $t_1 = t_0 + n - 1$, $t_1 \leq t_2$ and $\Delta t = t_2 - t_1$. We assume that a usual minimum-time synchronization operation is started at time $t = t_0$ by generating a special signal at the left end of 1D array of length n. We also assume that the right end cell of the array receives another special signals from outside at time $t_1 = t_0 + n - 1$ and $t_2 = t_1 + \Delta t = t_0 + n - 1 + \Delta t$, respectively. Then, there exists a 1D cellular automaton that can synchronize the array at time $t = t_0 + 2n - 2 + \Delta t$.*

The array operates as follows:

1. Start a minimum-time FSSP algorithm at time $t = t_0$ at the left end of the array. A 1/1 speed, i.e., 1 cell per 1 step, signal is propagated towards the right direction to wake-up cells in quiescent state. We refer the signal as *wake-up signal*.

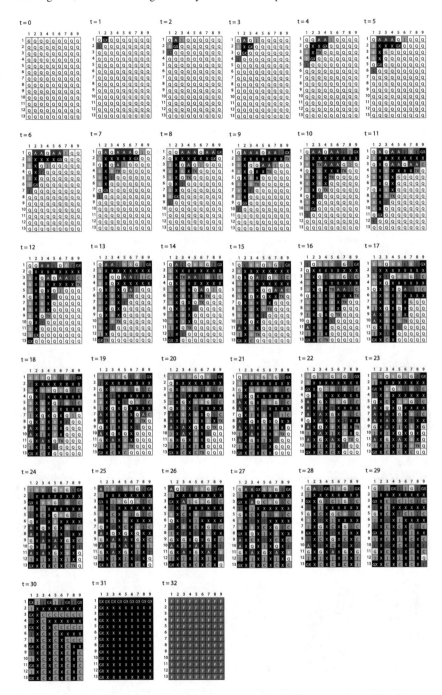

Fig. 9 Snapshots of the minimum-time eleven-state synchronization process on a 13 × 9 array

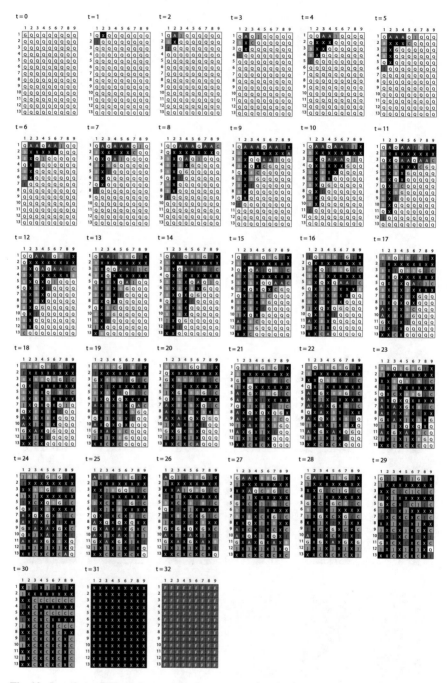

Fig. 10 Snapshots of the minimum-time nine-state synchronization process on a 13 × 9 array

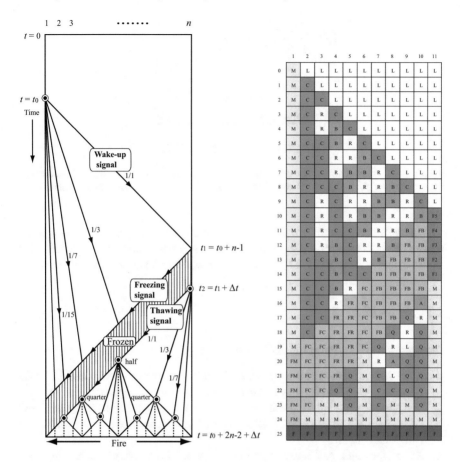

Fig. 11 Space-time diagram for delayed FSSP scheme based on the *freezing-thawing* technique (left) and synchronized configurations on 11 cells with delay $\Delta t = 5$ (right)

A *freezing signal* is given from outside at time $t_1 = t_0 + n - 1$ at the right end of the array. The signal is propagated in the left direction at its maximum speed, that is, 1 cell per 1 step, and freezes the configuration progressively. Any cell that receives the freezing signal from its right neighbor has to stop its state-change and transmits the freezing signal to its left neighbor. The frozen cell keeps its state as long as no thawing signal will arrive. Figure 11 (right) shows those frozen states in grey color, sandwitched between the freezing and thawing signals.

2. A special signal supplied with outside at time $t_2 = t_1 + \Delta t$ is used as a *thawing signal* that thaws the frozen configuration progressively. The thawing signal forces the frozen cell to resume its state-change procedures immediately. See Fig. 11 (left). The signal is also transmitted toward the left end at speed $1/1$.

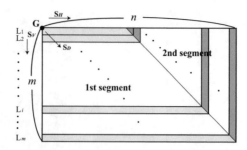

Fig. 12 A 2D array of size $m \times n$ ($m \leq n$) is regarded as consisting of m rotated (90° in counter-clockwise direction) L-shaped 1D arrays. Each L-shaped array is divided into two segments

It is easily seen that the array can be synchronized at time $t = t_0 + 2n - 2 + \Delta t$. We can freeze the entire configuration during Δt steps and delay the synchronization on the array for Δt steps. We refer the scheme as the *freezing-thawing technique*.

5.2 Algorithm $\mathcal{A}_3$

The overview of the algorithm $\mathcal{A}_3$ is as follows:

1. A 2D array of size $m \times n$ is regarded as $\min(m, n)$ rotated or mirrored *L-shaped* 1D arrays, each consisting of a horizontal and a vertical segment.
2. The shorter segment is synchronized by the freezing-thawing technique with $\Delta t = |m - n|$ steps delay. The longer one is synchronized in the usual way without any delay.
3. All of the *L-shaped* arrays fall into a special firing (synchronization) state simultaneously and for the first time.

First, we consider the case where $m \leq n$. We assume that the initial general G is on the north-west corner denoted by • in Fig. 12. We regard a 2D array of size $m \times n$ as consisting of m rotated (90° in counterclockwise direction) *L-shaped* 1D arrays. Each *L-shaped* array is denoted by L_i, $1 \leq i \leq m$, shown in Fig. 12. Each L_i is divided into two segments, that is, one horizontal and one vertical segment, each referred to as the 1st and 2nd segments. The length of each segment of L_i is $n - m + i$ and i, respectively.

5.2.1 Starting Synchronization

At time $t = 0$, a 2D array M has a general at $C_{1,1}$ and any other cells of the array are in quiescent state. The general G generates three signals s_V, s_D and s_H, simultaneously, each propagating at 1/1-speed in the vertical, diagonal and horizontal directions, respectively. See Fig. 12. The s_V- and s_H-signals work for generating wake-up signals for the 1st and 2nd segments on each *L-shaped* array. The s_D-signal is used for printing a special marker "∎" for generating a thawing signal that thaws frozen configurations on shorter segment. Their operations are as follows:

- **Signal s_V**: The s_V-signal travels along the 1st column and reaches $C_{m,1}$ at time $t = m - 1$. Then, it returns there and begins to travel again at $1/2$-speed along the 1st column towards $C_{1,1}$. On the return's way, the signal initiates the synchronization process for the 1st segment of each L_i. Thus a new general G_{i1} for the synchronization of the 1st segment of each L_i is generated, together with its wake-up signal, at time $t = 3m - 2i - 1$ for $1 \leq i \leq m$.
- **Signal s_D**: The s_D-signal travels along a principal diagonal line by repeating a zigzag movement: going one cell to the right, then going down one cell. Each time it visits cell $C_{i,i}$ on the diagonal, it marks a special symbol "∎" to inform the wake-up signal on the segment of the position where a thawing signal is generated for the neighboring shorter segment. The symbol on $C_{i,i}$ is marked at time $t = 2i - 2$ for any i, $1 \leq i \leq m$. Note that the wake-up signal of the 1st segment of L_m knows the right position by the arrival of the s_D-signal, where they meet at $C_{m,m}$ at the very time $t = 2m - 2$.
- **Signal s_H**: The s_H-signal travels along the 1st row at $1/1$-speed and reaches $C_{1,n}$ at time $t = n - 1$. Then it reflects there and returns the same route at $1/2$-speed. Each time it visits a cell of the 1st row on its return way, it generates a general G_{i2} at time $t = 2m + n - 2i - 1$ to initiate a synchronization for the 2nd segment on each L_i, $1 \leq i \leq m$.

The wake-up signals for the 1st and 2nd segments of L_i meet on $C_{i,n-m+i}$ at time $t = 2m + n - i - 2$. The collision of the two signals acts as a delimiter between the 1st and 2nd segments. Note that the synchronization operations on the 1st segment are started at the left end of the segment. On the other hand, the synchronization on the 2nd segment is started at the right (upper) end of the segment. The wake-up signal generated by G_{i1} reaches a cell where the special mark is printed at time $t = 3m - i - 2$ and generates a thawing signal which travels at $1/2$-speed along the 1st segment to thaw the configuration on the 2nd segment.

5.2.2 Synchronization

Now we consider the synchronization on L_m. Figure 13 (left) shows a space-time diagram for synchronizing L_m. As was mentioned in the previous subsection, the synchronization of the 1st and 2nd segments of L_m are started by the generals G_{m1} and G_{m2} at time $t = m - 1$ and $t = n - 1$, respectively. Each General generates a wake-up signal propagating at $1/1$-speed. The wake-up signal for the 1st and 2nd segments meets $C_{m,n}$ at time $t = m + n - 2$, where $C_{m,n}$ acts as an end of the both two segments. A freezing signal is generated simultaneously there for the 2nd segment at time $t = m + n - 2$. It propagates in upper (right in Fig. 13 (left)) direction at $1/1$-speed to freeze the synchronization operations on the 2nd segment. At time $t = 2m - 2$ the wake-up signal of the 1st segment reaches

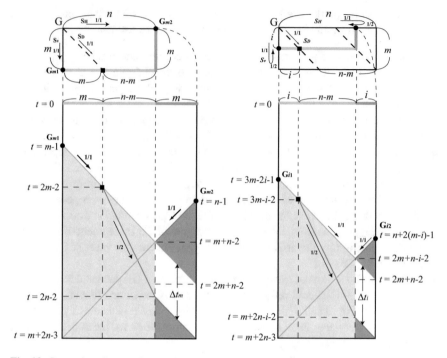

Fig. 13 Space-time diagram for synchronizing L_m (left) and L_i (right)

the symbol "■" and generates a thawing signal for the 2nd segment. The thawing signal starts to propagate from the cell at 1/2-speed in the same direction. Those two signals reach at the left end of the 2nd segment with time difference $n - m$ which is equal to the delay for the 2nd segment. The synchronization for the 1st segment is started at time $t = m - 1$ and it can be synchronized at time $t = m + 2n - 3 = m + n + \max(m, n) - 3$ by the usual way. On the other hand, the synchronization for the 2nd segment is started at time $t = n - 1$ and its operations are delayed for $\Delta t = \Delta t_m = n - m$ steps. Now letting $t_0 = n - 1$, $\Delta t = n - m$ in Theorem 8, the 2nd segment of length m on L_m can be synchronized at time $t = t_0 + 2m - 2 + \Delta t = m + 2n - 3 = m + n + \max(m, n) - 3$. Thus, L_m can be synchronized at time $t = m + n + \max(m, n) - 3$.

Now we discuss the synchronization for L_i, $1 \le i \le m - 1$. Figure 13 (right) shows a space-time diagram for synchronizing L_i. The wake-up signals for the two segments of L_i are generated at time $t = 3m - 2i - 1$ and $n + 2(m - i) - 1$, respectively. Generation of freezing and thawing signals is done in a similar way as employed in L_m. Synchronization operations on the 2nd segment are delayed for $\Delta t_i = n - m$ steps. The synchronization for the 1st segment of length $n - m + i$ is started at time $t = 3m - 2i - 1$ and it can be synchronized by a usual method at time $t = m + 2n - 3 = m + n + \max(m, n) - 3$. On the other hand, the synchronization for the 2nd segment is started at time $t = n + 2(m - i) - 1$ and its

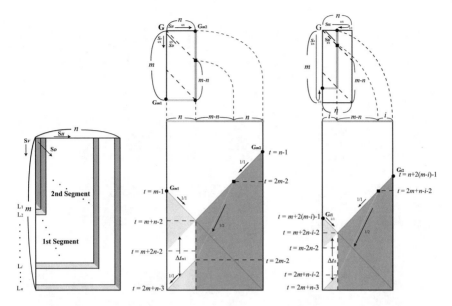

Fig. 14 Segmentation of a 2D array of size $m \times n$ ($m > n$) (right) and space-time diagram for synchronizing L_n (middle) and L_i, $1 \le i \le n - 1$ (left)

operations are delayed for $\Delta t = \Delta t_i = n - m$ steps. Now letting $t_0 = n + 2$ $(m - i) - 1$, $\Delta t = n - m$ in Theorem 8, the 2nd segment of length i of L_i can be synchronized at time $t = t_0 + 2i - 2 + \Delta t = m + 2n - 3 = m + n + \max(m, n) - 3$. Thus, L_i can be synchronized at time $t = m + n + \max(m, n) - 3$.

In the case where $m > n$, a 2D array of size $m \times n$ is regarded as consisting of n mirrored L-*shaped* arrays. See Fig. 14 (left). Segmentation and synchronization operations on each L-*shaped* array can be done almost in a similar way. It is noted that the thawing signal is generated on the 2nd segment to thaw frozen configurations on the 1st segment. Any rectangle of size $m \times n$ can be synchronized at time $t = 2m + n - 3 = m + n + \max(m, n) - 3$. Figure 14 (middle and right) shows the space-time diagram for the synchronization on L_n and L_i, $1 \le i \le n - 1$.

One notes that the algorithm needs no a priori knowledge on side length of a given rectangle, that is, whether wider than long or longer than wide. To check that our algorithm works correctly on a 2D cellular automaton, we have generated a set of local transition rules by a computer program, yielding that each cell has 84 internal states and 8979 transition rules in its realization. Our computer simulation shows that the rule set generated is valid for the synchronization on any rectangle of size $m \times n$ such that $2 \le m, n \le 500$. Now we can establish the next theorem.

Theorem 9 *The synchronization algorithm* $\mathcal{A}_3$, *implemented on an cellular automaton with 84 states and 8979 local rules, can synchronize any* $m \times n$ *rectangular array in minimum* $m + n + \max(m, n) - 3$ *steps.*

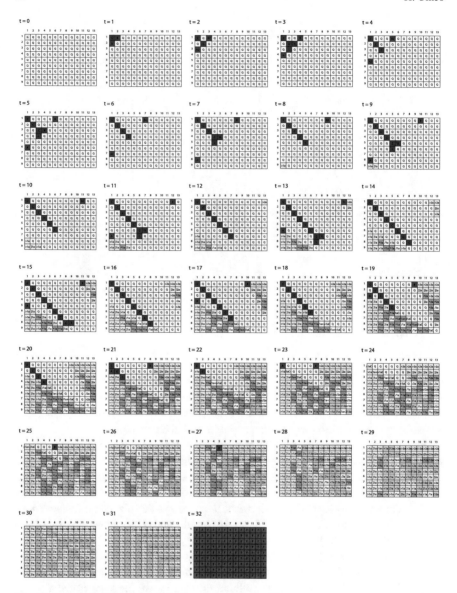

Fig. 15 Snapshots of the synchronization process on 9 × 13 array

Figures 15 and 16 show some snapshots of the synchronization process operating in minimum-steps on 9 × 13 and 13 × 9 arrays. See Umeo and Uchino [14] and Umeo et al. [16] for details.

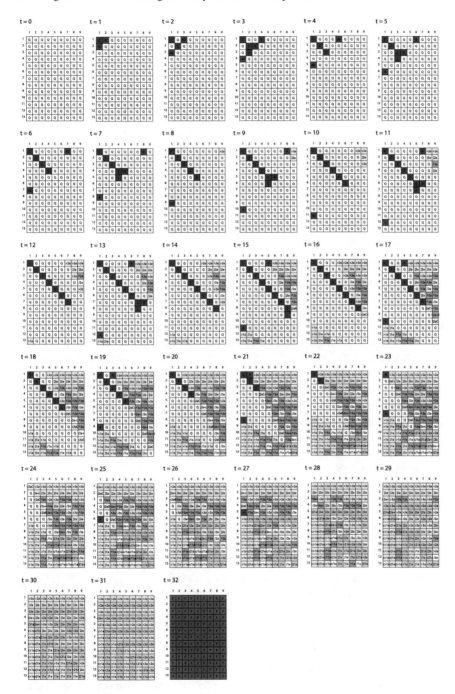

Fig. 16 Snapshots of the synchronization process on 13×9 array

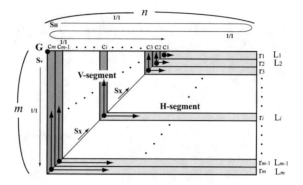

Fig. 17 A 2D array wider-than-long of size $m \times n$ ($m \leq n$) is regarded as m ($= \min(m, n)$) mirrored and rotated *L-shaped* 1D arrays, each consisting of a long, horizontal H-segment and a short, vertical V-segment, respectively

6 Umeo–Yamawaki–Nishide Algorithm

In this section, we develop a minimum-time FSSP algorithm $\mathcal{A}_4$ which is also based on a new L-shaped mapping. First, we consider the synchronization for rectangles which are wider-than-long, i.e., the case where $m \leq n$. The algorithm is based on the freezing-thawing technique presented in the previous section. We regard a 2D array of size $m \times n$ as m ($= \min(m, n)$) mirrored and rotated *L-shaped* 1D arrays. Each *L-shaped* array is denoted by L_i, $1 \leq i \leq m$, shown in Fig. 17. Each L_i is divided into two segments, that is, one vertical V-segment and one horizontal H-segment. The length of each segment of L_i is i and $n - m + i$, $1 \leq i \leq m$, respectively. Each vertical and horizontal segment is referred to as c_i and r_i, $1 \leq i \leq m$, respectively, as is shown in Fig. 17.

6.1 Algorithm $\mathcal{A}_4$

The overview of the algorithm $\mathcal{A}_4$ is as follows:

1. We assume that the initial general G is on the north-west corner $C_{1,1}$ denoted by
   • in Fig. 17 and the algorithm has a priori knowledge on side length of a given rectangle, that is, it is wider-than-long.
2. The general generates two signals s_H and s_V at time $t = 0$. Their operations are as follows:

   • **Signal s_V**: The s_V-signal travels along the 1st column and reaches $C_{m,1}$ at time $t = m - 1$. The s_V-signal generates a new signal, referred to as s_X-signal, on the south-west corner $C_{m,1}$ at time $t = m - 1$. It propagates toward the diagonal direction and generates a new general at each bending point of L_i, $1 \leq i \leq m$.

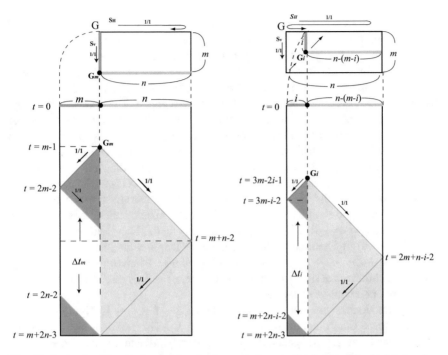

Fig. 18 Space-time diagram for synchronization operations on L_m (left) and L_i (right), $1 \leq i \leq m - 1$, in a rectangle wider-than-long of size $m \times n$

- **Signal s_H:** The s_H-signal works for generating a thawing signal that thaws frozen configurations on shorter V-segments. The s_H-signal travels along the 1st row at 1/1-speed and reaches $C_{1,n}$ at time $t = n - 1$. Then it reflects there and returns the same route at 1/1-speed, and reaches $C_{1,1}$ again at time $t = 2n - 2$. Then, it travels again along the 1st row at 1/1 speed with generating a thawing signal at each $C_{1,i}$, $1 \leq i \leq m$.

3. Synchronization operations on each L-shaped array are started by a newly generated general at each bending point of the array. The shorter V-segment is synchronized by the freezing-thawing technique with $\Delta t = 2(n - m)$ steps delay. The longer H-segment is synchronized by the usual way without any delay. All of the L-shaped arrays are synchronized at the same time and for the first time.

6.2 Synchronization

Now we consider the synchronization on L_m. Figure 18 (left) shows a space-time diagram for synchronizing L_m. As was mentioned in the previous subsection, the synchronization of the vertical and horizontal segments of L_m are started simultaneously by the generals G_m generated at the bending point of the L-shaped array at time $t = m - 1$. The general generates a wake-up signal propagating at 1/1-speed for each segment. The synchronization for the horizontal (longer) H-segment is started at time $t = m - 1$ and the segment of length n can be synchronized by the usual way at time $t = m - 1 + 2n - 2 = m + 2n - 3 = m + n + \max(m, n) - 3$. We focus our attention on the V-segment. The wake-up signal on the vertical segment arrives at $C_{1,1}$ at time $t = 2m - 2$ and a freezing signal for the segment is generated there simultaneously. It propagates in lower direction at 1/1-speed to freeze the synchronization operations on the V-segment. At time $t = 2n - 2$, the signal s_H returns to $C_{1,1}$ and generates a thawing signal for the segment. The thawing signal starts to propagate from the cell at 1/1-speed to thaw the configuration on the V-segment. In this way, the synchronization for the shorter V-segment is started at time $t = m - 1$ and its synchronization operations are delayed for $\Delta t = \Delta t_m = 2n - 2 - (2m - 2) = 2(n - m)$ steps. Now letting $t_0 = m - 1$, $\Delta t = 2(n - m)$ in Theorem 8, the V-segment of length m can be synchronized at time $t = t_0 + 2m - 2 + \Delta t_m = m + 2n - 3 = m + n + \max(m, n) - 3$. Thus, L_m can be synchronized at time $t = m + n + \max(m, n) - 3$. Figure 18 (left) illustrates how those operations work successfully.

Now we discuss the synchronization for L_i, $1 \leq i \leq m - 1$. Figure 18 (right) shows a space-time diagram for synchronizing L_i. The wake-up signals for the two segments of L_i are generated by a general initiated by the signal s_X at time $t = 3m - 2i - 1$. Generation of freezing and thawing signals is done in a similar way as employed in L_m. The synchronization for the longer H-segment of length $n - m + i$ is started at time $t = 3m - 2i - 1$ and it can be synchronized by a usual method at time $t = 3m - 2i - 1 + 2(n - m + i) - 2 = m + 2n - 3 = m + n + \max(m, n) - 3$. On the other hand, the synchronization for the shorter V-segment is started at time $t = 3m - 2i - 1$ and its operations are frozen progressively from time $t = 3m - i - 2$ when the wake-up signal reaches C_{1m-i}. At time $t = m + 2n - i - 2$, the s_H-signal arrives at the cell C_{1m-i} and the configuration on the segment is also thawed progressively. Thus, the synchronization operations on the V-segment are delayed for $\Delta t_i = m + 2n - i - 2 - (3m - i - 2) = 2(n - m)$ steps. Now letting $t_0 = 3m - 2i - 1$, $\Delta t = 2(n - m)$ in Theorem 8, the V-segment of length i of L_i can be synchronized at time $t = t_0 + 2i - 2 + \Delta t = m + 2n - 3 = m + n + \max(m, n) - 3$. Thus, for any i, $1 \leq i \leq m$, L_i can be synchronized at time $t = m + n + \max(m, n) - 3$.

In this way, the algorithm presented can synchronize any rectangle wider-than-long of size $m \times n$ in minimum $t = m + n + \max(m, n) - 3$ steps.

In the case of rectangles longer-than-wide, each horizontal segment, that is a shorter one in this case, falls into a firing state at time $t = 2n + m - 3$, which is smaller than the minimum-time $t = 2m + n - 3$. Moreover at time $t = 2n + m - 3$ some cells on some vertical segments are still in quiescent state. Thus the algorithm presented above fails to synchronize rectangles longer-than-wide. See Fig. 19. The figure illustrates a space-time diagram for unsuccessful synchronization on L_m (left) and L_i (right), when applying the wider-than-long synchronization algorithm to

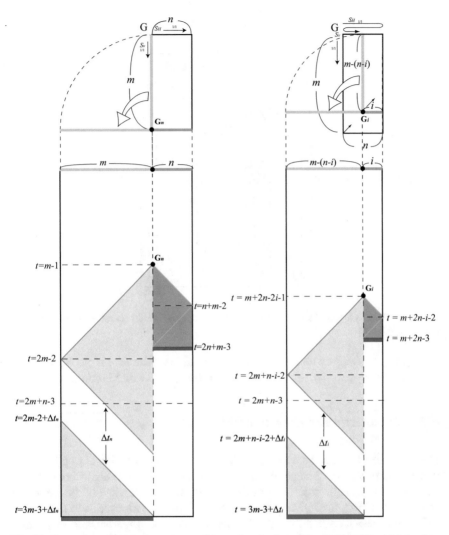

Fig. 19 Space-time diagram for unsuccessful synchronization on L_m (left) and L_i (right), when applying the wider-than-long synchronization algorithm to a rectangle longer-than-wide

rectangles longer-than-wide. In the next section we propose a synchronization algorithm for the rectangles longer-than-wide.

In the case where $m > n$, a 2D array of size $m \times n$ ($m \geq n$) is regarded as consisting of n mirrored *L-shaped* arrays. See Fig. 20. Functions of the signals s_H and s_V are interchanged with each other. Synchronization operations on each *L-shaped* array can be done almost in a similar way. It is noted that the thawing signal is generated on the 1st column to thaw the frozen configurations on the shorter H-segments. Figure 21 shows the space-time diagram for the synchronization on L_n

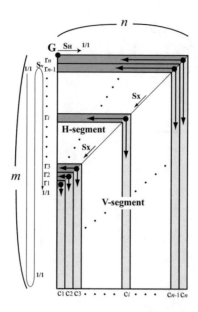

Fig. 20 A 2D array longer-than-wide of size $m \times n$ $(m > n)$ is regarded as consisting of n mirrored *L-shaped* 1D array

and L_i, $1 \leq i \leq n$. Any rectangles longer-than-wide of size $m \times n$ can be synchronized at time $t = 2m + n - 3 = m + n + \max(m, n) - 3$ which is minimum in time.

6.3   Construction of the Algorithm $\mathcal{A}_4$

In this section we develop a new minimum-time FSSP algorithm $\mathcal{A}_4$ operating in $m + n + \max(m, n) - 3$ steps for any array of size $m \times n$. The algorithm can synchronize any rectangle regardless of its side length. The overview of the algorithm is as follows: First, for a given array, the general gives an instruction to make each cell operate the wider-than-long and the longer-than-wide synchronization algorithm simultaneously. The general also checks the shape of the array, i.e., wider-than-long or longer-than-wide, by sending some signals on the array. Then, each cell inhibits undesirable synchronization operations after getting the shape information.

6.3.1   Inhibition of Undesirable Synchronization Operations

Generally speaking, an initial general has no a priori knowledge on side length of a given rectangle, that is, whether it is wider-than-long or longer-than-wide. The synchronization scheme presented in the previous sections can only synchronize either wider-than-long or longer-than-wide arrays. To synchronize any arrays successfully in minimum-time, the array has to prepare synchronization operations for both cases from time $t = 0$.

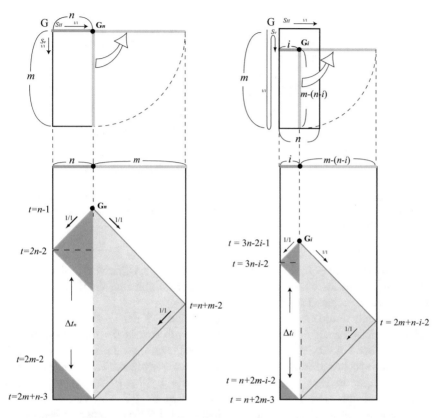

Fig. 21 Space-time diagram for synchronizing L_n (left) and L_i (right), $1 \le i \le n - 1$, of a rectangle longer-than-wide

We assume that a 2D array M has initially a general at $C_{1,1}$ and any other cells of the array are in quiescent state. At time $t = 0$, the general generates three signals s_V, s_H, and s_D simultaneously, each propagating at $1/1$-speed in the vertical, horizontal, and diagonal directions. The functions of s_V- and s_H-signals are extended in such a way that both of them generate the s_X-signal on the cells $C_{m,1}$ and $C_{1,n}$ at time $t = m - 1$ and $t = n - 1$, respectively, to prepare the synchronization operations for the wider-than-long and longer-than-wide cases, then return to $C_{1,1}$ at $1/1$ speed. In order to thaw the frozen configurations those s_V- and s_H-signals have to return their routes at $1/1$ speed as described in the previous section. In this way, the array begins its synchronization operations from two corner cells, $C_{m,1}$ and $C_{1,n}$, to prepare the synchronization in time for both cases. Note that one of them must be stopped in the future, depending on its side length of a given array.

For this purpose, at time $t = 0$, the general of the array introduces a new signal, referred to as s_D-signal, which travels along a principal diagonal line by repeating a zigzag movement: going one cell to the right, then going down one cell.

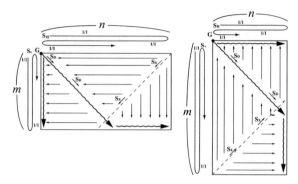

Fig. 22 A trajectory of signals generated by a general on a 2D array wider-than-long (left) and longer-than-wide (right) of size $m \times n$. Newly introduced signals (denoted by bold arrows in the figure) for stopping undesirable synchronization operations are also illustrated

In the case where $m \leq n$, the signal hits the lower edge. Otherwise it hits the right edge. The information where the signal hits helps to inhibit undesirable synchronization operations prepared so far.

We consider the case $m \leq n$. At time $t = 2m - 2$ the s_D-signal reaches the cell $C_{m,m}$, then travels at 1/1 speed along the mth row in the right direction toward $C_{m,n}$. See Fig. 22 (left). During its travel on the mth row it prints a special mark that inhibits synchronization operations for shorter V-segments. The signal prints the mark on C_{mi} at time $t = m + i - 2$ for any $i, m + 1 \leq i \leq n$. The wake-up signal for the V-segment arrives at $C_{m,i}$ at time $t = 2n + m - i - 2$, which is later than the marking at the right cell $C_{m,i}$, for any i such that $m + 1 \leq i \leq n$, thus enabling the inhibition of the operations for those V-segments. As for the H-segments, the inhibition for them is made in the following way. The s_V- and s_H-signal returns to C_{11} at time $t = 2m - 2$ and $t = 2n - 2$, respectively. The general on C_{11} gets the side length information at time $t = 2m - 2$ by the first arrival of the return of the s_V-signal. Then, a new signal generated there begins to travel along the 1st column and prints a similar inhibition mark on the 1st column which stops synchronization operations for the longer H-segment in the case where $m \leq n$. The wake-up signal for the H-segment arrives at $C_{1,j}$ at time $t = 2m + j - 3$, which is later than the marking at the cell $C_{1,j}$, for any j such that $1 \leq j \leq m$, thus enabling the inhibition of the operations for the H-segment. In this way, both of the synchronization operations can be stopped before the correct synchronization in the case where $m \leq n$. A similar inhibitions can be made for the case $m > n$, where the s_D-signal hits the right edge. See Fig. 22 (right).

6.3.2   Final Algorithm

One notes that the algorithm proposed needs no a priori knowledge on side length of a given rectangle, that is, whether it is wider-than-long or longer-than-wide. To check that our algorithm works correctly on a 2D cellular automaton, we have generated a

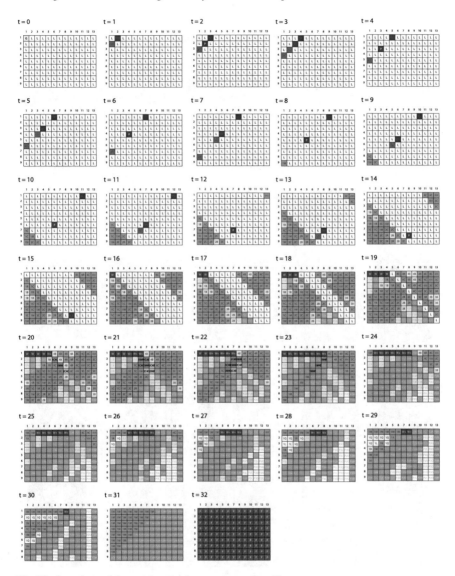

Fig. 23 Snapshots of the synchronization process on 9×13 array

set of internal states and a set of local transition rules by a computer program, yielding that each cell has 124 internal states and 45128 transition rules in its realization. Our computer simulation shows that the rule set generated is valid for the synchronization on any rectangle of size $m \times n$ such that $2 \leq m, n \leq 253$. Figures 23 and 24 show some snapshots of the synchronization process operating in minimum-steps on 9×13 and 13×9 arrays. Now we can establish the next theorem. See Umeo et al. [15] for details.

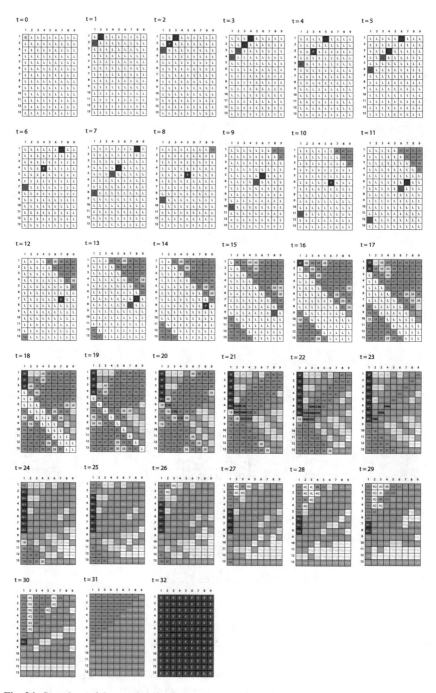

Fig. 24 Snapshots of the synchronization process on 13×9 array

Table 1 A quantitative comparison of 2D FSSP algorithms based on L-shaped mapping

Algorithm	Time complexity	# of states	# of transition rules	Reference
$\mathcal{A}_1$	$m + n + \max(m, n) - 3$	–	–	Beyer [1]
$\mathcal{A}_1$	$m + n + \max(m, n) - 3$	28	–	Shinahr [4]
$\mathcal{A}_1$	$m + n + \max(m, n) - 3$	28	12849	Umeo et al. [8]
$\mathcal{A}_{2-1}$	$m + n + \max(m, n) - 2$	10	1629	Umeo and Nomura [13]
$\mathcal{A}_{2-2}$	$m + n + \max(m, n) - 3$	11	4044	Umeo and Nomura [13]
$\mathcal{A}_{2-3}$	$m + n + \max(m, n) - 3$	9	2561	Umeo and Nomura [13]
$\mathcal{A}_3$	$m + n + \max(m, n) - 3$	84	8979	Umeo et al. [16]
$\mathcal{A}_4$	$m + n + \max(m, n) - 3$	124	45128	Umeo et al. [15]

Theorem 10 *The synchronization algorithm $\mathcal{A}_4$ can synchronize any $m \times n$ rectangular array in minimum $m + n + \max(m, n) - 3$ steps.*

7 Summary

In the present paper, we gave a survey on recent developments in FSSP algorithms for 2D cellular arrays. We focused our attention on a new class of the 2D minimum-time FSSP algorithms based on L-shaped mapping. It is seen that the L-shaped mapping presents a rich variety of minimum-time FSSP algorithms and their implementations. Most of the 2D algorithm proposed are isotropic with respect to shape of a given rectangle array, i.e. no need to control the FSSP algorithm for longer-than-wide and wider-than-long input rectangles, however the underlying algorithm used in $\mathcal{A}_4$ is not isotropic. The non-isotropic property led to the increase of the number of states required in its implementation. The isotropic property plays an important role in the design of higher dimensional minimum-time FSSP algorithms. A class of isotropic multi-dimensional minimum-time FSSP algorithms is given in Umeo et al. [11]. Here, we present Table 1 based on a quantitative comparison of those 2D FSSP algorithms and their transition tables discussed in this chapter.

Acknowledgements The author would like to thank reviewers for many helpful comments and suggestions to improve the paper.

References

1. Beyer, W.T.: Recognition of topological invariants by iterative arrays. Ph.D. Thesis, MIT, pp. 144 (1969)
2. Moore, E.F.: The firing squad synchronization problem. In: Moore, E.F. (ed) Sequential Machines, Selected Papers, pp. 213–214. Addison-Wesley, Reading MA (1964)
3. Moore, F.R., Langdon, G.G.: A generalized firing squad problem. Inf. Control 12, 212–220 (1968)
4. Shinahr, I.: Two- and three-dimensional firing squad synchronization problems. Inf. Control 24, 163–180 (1974)
5. Umeo, H.: A simple design of time-efficient firing squad synchronization algorithms with fault-tolerance. In: IEICE Transactions on Information and Systems, vol. E87-D (3), pp. 733–739 (2004)
6. Umeo, H.: Firing squad synchronization problem in cellular automata. In: Meyers, R.A. (ed.) Encyclopedia of Complexity and System Science, vol. 4, pp. 3537–3574. Springer, Berlin (2009). A new edition will appear in 2020
7. Umeo, H.: Synchronizing square arrays in optimum-time. Int. J. Gen. Syst. 41(6), 617–631 (2012)
8. Umeo, H., Ishida, K., Tachibana, K., Kamikawa, N.: A transition rule set for the first 2-D optimum-time synchronization algorithm. In: Proceedings of the 4th International Workshop on Natural Computing, PICT, vol. 2, pp. 333–341. Springer, Berlin (2009)
9. Umeo, H., Kamikawa, N., Nishioka, K., Akiguchi, S.: Generalized firing squad synchronization protocols for one-dimensional cellular automata - a survey. In: Acta Physica Polonica B, Proceedings Supplement. vol. 3, pp.267–289 (2010)
10. Umeo, H., Kubo, K.: A seven-state time-optimum square synchronizer. In: Proceedings of International Conference on Cellular Automata for Research and Industry, ACRI 2010, LNCS, vol. 6350, pp. 219–230 (2010)
11. Umeo, H., Kubo, K., Nishide, K.: A class of FSSP algorithms for multi-dimensional cellular arrays. Commun. Nonlinear Sci. Numer. Simul. 21, 200–209 (2015)
12. Umeo, H. Kubo, K., Takahashi, Y.: An isotropic optimum-time FSSP algorithm for two-dimensional cellular automata. In: Proceedings of the 12th International Conference on Parallel Computing Technologies, PaCT 2013, LNCS, vol. 7979, pp. 381–393 (2013)
13. Umeo, H., Nomura, A.: A state-efficient zebra-like implementation of synchronization algorithms for 2D rectangular cellular arrays. In: BIOMATH, vol. 1, pp. 1–6 (2012)
14. Umeo, H., Uchino, H.: A new time-optimum synchronization algorithm for rectangle arrays. Fundam. Inform. 87(2), 155–164 (2008)
15. Umeo, H., Yamawaki, T., Nishide, K.: An optimum-time firing squad synchronization algorithm for two-dimensional rectangle arrays –freezing-thawing technique based. J. Cell. Autom. 7, 31–46 (2012)
16. Umeo, H. Yunès, J.-B., Yamawaki, T.: A simple-optimum-time firing squad synchronization algorithms for two-dimensional arrays. In: Proceedings of 2009 International Conference on Computational Intelligence, Modelling and Simulation, CSSim 2009, IEEE Computer Society, pp. 120–125 (2009)

Abelian Invertible Automata

Klaus Sutner

Abstract Invertible transducers are particular Mealy automata that define so-called automata groups, subgroups of the full automorphism group of the infinite binary tree. In the recent past, automata groups have become a major source of interesting and challenging constructions in group theory. While this research typically focuses on properties of the associated groups, we describe the topological structure of the associated automata in the special case where the group in question is free Abelian. As it turns out, there are connections between these automata and the theory of algebraic number fields as well as the theory of tiles. We conclude with a conjecture about the connectivity properties of the canonical invertible automata generating free Abelian groups.

1 Motivation

Historically, the idea of reversible computation had its roots in physics rather than logic: at the fundamental level, the laws of physics are reversible. Since computing devices can obviously be realized within the context of these laws, it is plausible that computation itself should be amenable to reversibility: there ought to be a way to make the requisite logical operations reversible [4, 16, 17]. Perhaps surprisingly, this idea turns out to be of some practical importance, since reversible computation can be carried out without any thermodynamical cost [5], at least as a matter of principle. Morita has given many ingenious examples of reversible computation in the context of discrete dynamical systems, and in particular cellular automata [18–21]. As these examples show, and contrary to what was tacitly assumed till the 1960s, reversible computation is a rich and endlessly challenging area of computability theory. We now even have the beginnings of a more structural approach to reversible computation [1], following roughly Girard's Geometry of Interaction.

It can be argued that in the realm of algebra it is the concept of a group that best captures the notion of reversibility: any action by a group element g can be undone by

K. Sutner (✉)
Carnegie Mellon University, Pittsburgh, PA, USA
e-mail: sutner@cs.cmu.edu

© Springer International Publishing AG 2018
A. Adamatzky (ed.), *Reversibility and Universality*, Emergence, Complexity
and Computation 30, https://doi.org/10.1007/978-3-319-73216-9_3

the action associated with g^{-1}. It is thus natural to ask whether reversible computation might have any direct connections with group theory. A first step in this direction was taken by Serre, albeit in a different context: he suggested to study subgroups of the full automorphism group $\mathsf{Aut}(2^\star)$ of the infinite binary tree $2^\star$ [28]. The topological group $\mathsf{Aut}(2^\star)$ is profinite, and thus Hausdorff, compact and totally disconnected. In some interesting cases, subgroups can be described by certain finite state machines that are naturally reversible. More precisely, there are Mealy automata $\mathcal{A}$ over a binary alphabet that have associated inverse automata $\mathcal{A}'$ such that the composition of their respective transductions is the identity. The transductions in question are length-preserving and thus do not admit universal computation. Yet their associated groups are surprisingly complicated and there are many challenging questions associated with these automata. In fact, groups defined by invertible automata have become a standard source of examples and counterexamples in group theory [22, 29]. A case in point is a 5-state automaton due to Grigorchuk [10] that defines a subgroup of $\mathsf{Aut}(2^\star)$ with intermediate growth, answering a question by Milnor. Grigorchuk goes so far as to describe the discovery of very small automata associated with complicated groups as one of the "wonderful phenomena in modern mathematics." If one considers semigroups rather than groups, there is even a 2-state non-invertible Mealy automaton that exhibits intermediate growth [2].

We are here interested in a slightly different perspective: the computational complexity of the discrete dynamical systems defined by invertible Mealy automata. Given an automorphism f defined by some automaton $\mathcal{A}$, one would like to understand the orbits $x\, f^\star$ of words $x \in 2^\star$ under f. In particular, one would like to analyze the computational complexity of the question whether a word appears in the orbit of another (Orbit Problem) and where in the orbit it appears (Timestamp Problem). These questions are quite difficult in general, so it makes sense to restrict one's attention to a scenario that is of little interest from the group theory perspective: all the groups in question are free Abelian. In this limited setting, one can give a description of the automorphism f in terms of the algebraic integers in an algebraic number field associated with the automaton. Nonetheless, it requires some amount of effort to characterize the associated automata. The existence of numeration systems for algebraic integers then produces a convenient normal form for the automorphisms and is helpful in classifying the relevant automata and tackling the orbit problems just mentioned. For example, somewhat surprisingly it turns out that, in some cases, the Orbit Problem can be solved by a finite state machine, despite the fact that the orbits have exponential length. Alas, little is known about the general situation.

We will here refrain from giving detailed proofs and confine ourselves to simple sketches. We refer the reader to the literature for details, in particular [10, 22, 24, 29, 30, 32, 33].

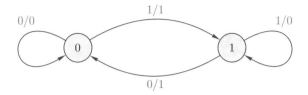

Fig. 1 The smallest interesting invertible automaton

2 Transducers and Automorphisms of the Binary Tree

For our purposes, an invertible transducer is a type of Mealy automaton
$\mathcal{A} = \langle Q, 2, \tau \rangle$ where $\tau : Q \times 2 \to 2 \times Q$. In the customary arrow notation, all
transitions are of the form $p \xrightarrow{a/\pi_p(a)} q$; here π_p is a permutation of the alphabet **2**
depending on the source state p; for background see [6, 8, 27]. Our automata have
no initial and final states; Eilenberg referred to these devices as output modules [7].
In order to obtain a transduction $\mathcal{A}(p)$ from $\mathbf{2}^\star$ to $\mathbf{2}^\star$ we select an arbitrary initial
state p in $\mathcal{A}$. To lighten notation, we write $\underline{p}$ for this map whenever the automaton is clear from context. Note that all our automorphisms naturally extend to maps
$\mathbf{2}^\omega \to \mathbf{2}^\omega$. We refer to states such that $\pi_p = I$ as copy states, and as toggle states
otherwise. We will write application of our automorphisms as a right action on finite
or infinite words, $x f$.

Any interesting invertible automaton must have at least one copy state and
one toggle state. Surprisingly, the 2-state machine in Fig. 1, with one toggle and
copy state each, already generates the lamplighter group, in perfect keeping with
Grigorchuk's observation from above.

The transduction semigroup generated by all the $\mathcal{A}(p)$, $p \in Q$, under composition
will be written $S(\mathcal{A})$; and $\mathcal{G}(\mathcal{A})$ the corresponding group. While it is convenient to
admit infinite automata, the situation where the Mealy automaton has finite state
set is by far the most interesting. As it turns out, many interesting groups admit
such a representation: free groups, free Abelian groups, certain nilpotent groups,
the lamplighter group, and in particular Grigorchuk's group. Computationally, it is
straightforward to construct automata that represent the elements of $S(\mathcal{A})$ using the
standard product construction for the composition of rational transductions described
in [8]. Of course, the size of these machines grows exponentially, so the construction
is only feasible in rather limited circumstances. To handle the group $\mathcal{G}(\mathcal{A})$, one
usually also needs the inverse automaton $\mathcal{A}'$ that is obtained by interchanging the
input/output bits for all toggle states. Note that we may have $S(\mathcal{A}) = \mathcal{G}(\mathcal{A})$, in which
case there is no need for the inverse automaton; we call $\mathcal{A}$ group-like in this situation.
It is easy to see that $S(\mathcal{A})$ is Abelian if, and only if, $\mathcal{G}(\mathcal{A})$ is Abelian.

We will transfer standard notions from semigroup and group theory to the corresponding automata. For example, we may refer to an automaton as being torsion-free
Abelian.

Here are some simple examples. The automaton Suc_2 in Fig. 2, the successor
automaton of rank 2, generates the monoid $\mathbb{N}^2$. If the loop has length n rather than 2

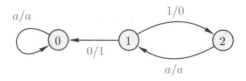

Fig. 2 The successor automaton Suc_2 of rank 2, generating the free monoid $\mathbb{N}^2$

as in Fig. 2 we obtain the general successor automaton of rank n, see below for an algebraic definition. In the case where $n = 1$ these machines are also referred to as adding machines or odometers; regrettably, for $n \geq 2$ the notion of sausage automaton appears in the literature. It is not hard to see that all orbits of our invertible binary automata have length a power of 2. For Suc_2, the orbit of a word x of length $2k$ under transduction $\underline{1}$ has length 2^k. To see why, let $u, v \in \mathbf{2}^k$ and write shf for the perfect shuffle operation. Then

$$\mathsf{shf}(u, v)\,\underline{1} = \mathsf{shf}(u\,f, v)$$

where f is the truncated successor function in reverse binary (another automorphism of $\mathbf{2}^\star$, defined by the analogous successor automaton where state 1 has a self-loop under $1/0$) and $\underline{1} = \mathsf{Suc}_2(1)$ as per our notational convention.

By contrast, the machine $\mathcal{A}$ in Fig. 3 generates $\mathbb{Z}^2$ as a semigroup. To see why, note that $\underline{1}^{-1} = \underline{2}$, and similarly $\underline{3}^{-1} = \underline{4}$. This is an example of an automaton that admits a skew-symmetry φ: a transition $p \xrightarrow{a/b} q$ is mapped to $\varphi(p) \xrightarrow{\overline{a}/\overline{b}} \varphi(q)$. Thus, state $\varphi(p)$ defines the inverse function of state p and the automaton is obviously group-like. As in the previous example, the orbit of a word of length $2k$ under $\underline{1}$ has length 2^k and has a similar description in terms of shuffle. Alas, this time there is no simple description for the associated map f. Note that $\underline{1}^k$ similarly produces orbits of length 2^k for odd k, whereas even powers of $\underline{1}$ produce shorter orbits.

The example also shows that full edge labels in the diagrams lead to visual clutter. It is preferable to relabel transitions as follows; this convention will also be useful in

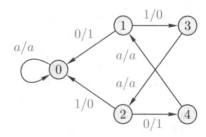

Fig. 3 An automaton that uses a successor-like function to generate the free Abelian group of rank 2

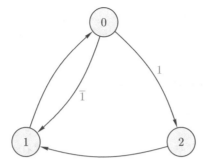

Fig. 4 Another automaton generating the free Abelian group of rank 2, albeit it in a less obvious manner

Sect. 4. Henceforth, a toggle transition $p \xrightarrow{0/1} q$ will be written as $p \xrightarrow{1} q$, a toggle transition $p \xrightarrow{1/0} q$ as $p \xrightarrow{\bar{1}} q$ and a copy transition $p \xrightarrow{a/a} q$ as $p \xrightarrow{0} q$ or even simply as $p \longrightarrow q$. The labels $\mathbf{2}_s = \{\bar{1}, 0, 1\}$ will be referred to as trits to emphasize similarity to balanced ternary numeration systems. Numerically we will interpret $\bar{1}$ as -1.

Here is our last example of an automaton, using this convention. The 3-state machine is group-like and indeed generates $\mathbb{Z}^2$, but this time there is no obvious reason for this. As it turns out, the identity $\underline{0}\,^2\underline{1}\,^2\underline{2} = I$ holds, from which observation our claim can easily be derived (to avoid confusion, we always write I for the identity automorphism of $\mathbf{2}^\star$). Again, the orbit of words of length $2k$ under $\underline{0}$ has length 2^k, but this time the situation is a bit more complicated: one can show that for any fixed orbit under $\underline{0}$ and any $u \in \mathbf{2}^k$ there is precisely one $v \in \mathbf{2}^k$ such that $\mathsf{shf}(u, v)$ lies in the orbit. Hence the orbits of even length words under $\underline{0}$ have the form $\{\,\mathsf{shf}(u, v\,f) \mid u \in \mathbf{2}^k\,\}$ for some finite transduction f.

The description of automorphisms of $\mathbf{2}^\star$ in terms of Mealy automata is convenient since standard algorithms from automata theory can be exploited in the study of these automorphisms. For example, we may safely assume that the Mealy automata are minimal in the sense that no two states have the same behavior, if we consider them as acceptors over $\mathbf{2} \times \mathbf{2}$ in the obvious manner. All the sample automata we have seen so far are indeed minimal. Thus we may safely assume that all the basic maps $\underline{k}$, $k \in Q$, are distinct as, for example, in Fig. 4.

Of course, it is also important to have a more algebraic description available. To begin with, note that any automorphism f of $\mathbf{2}^\star$ can be written in the recursive form $f = (f_0, f_1)s$ where $s \in \mathfrak{S}_2$, the symmetric group on two letters: s describes the action of f on $\mathbf{2}$, construed as the first level of $\mathbf{2}^\star$, and f_0 and f_1 are the automorphisms induced by f on the two subtrees of the root (which are naturally isomorphic to the whole tree). Thus there are residuation maps $\partial_a : \mathsf{Aut}(\mathbf{2}^\star) \to \mathsf{Aut}(\mathbf{2}^\star)$, $a \in \mathbf{2}$, and a parity map par : $\mathsf{Aut}(\mathbf{2}^\star) \to \mathfrak{S}_2$ that produce the corresponding decomposition. Here we write σ for the transposition in $\mathfrak{S}_2$ and suppress the identity in this context. We refer to an automorphism of the form $f = (f_0, f_1)\sigma$ as odd, all others $f = (f_0, f_1)$

as even. In other words, f is even if $a\,f = a$ for $a \in \mathbf{2}$, and odd otherwise. Clearly, $\underline{p}$ is odd if, and only if, p is a toggle state. We can describe the full automorphism group as a wreath product:

$$\mathsf{Aut}(\mathbf{2}^\star) \simeq (\mathsf{Aut}(\mathbf{2}^\star) \times \mathsf{Aut}(\mathbf{2}^\star)) \rtimes \mathfrak{S}_2.$$

The group operation in the wreath product has the form

$$(f_0, f_1)s\,(g_0, g_1)t = (f_0 g_{s(0)}, f_1 g_{s(1)})\,st.$$

In the context of sequential functions, residuals were first introduced by Raney [26] and correspond exactly to the recursive components in the wreath decomposition. Note that a subgroup G of $\mathsf{Aut}(\mathbf{2}^\star)$ may not be closed under residuation; if it is, we call G self-similar or state-closed. In this case, the wreath characterization of the full automorphism group carries over: $G \simeq (G \times G) \rtimes \mathfrak{S}_2$.

For legibility, we will occasionally write k^- rather than $k-1$, and k^+ rather than $k+1$. As an example, using wreath notation, the successor automaton Suc_n of rank n, with a loop of length n rather than just 2 as in Fig. 2, has the form

$$\underline{0} = (\underline{0}, \underline{0}) \qquad \underline{1} = (\underline{0}, \underline{n})\,\sigma \qquad \underline{k} = (\underline{k^-}, \underline{k^-}), \quad 2 \le k \le n.$$

Using the shuffle characterization from above, it is not hard to show that Suc_n generates the free Abelian monoid $\mathbb{N}^n$, but not a group. The automaton in Fig. 4 can be generalized like so. A cycle-cum-chord transducer is given by

$$\underline{0} = (\underline{n^-}, \underline{m^-})\,\sigma \qquad \underline{k} = (\underline{k^-}, \underline{k^-}), \quad 1 \le k < n.$$

where $1 \le m \le n$. We will write CC^n_m for this transducer. The diagram of CC^6_4 is shown in Fig. 5. One can show that CC^n_m generates the free Abelian group of rank $n - \gcd(n, m)$. As we will see, these are in a sense the most basic invertible automata generating free Abelian groups.

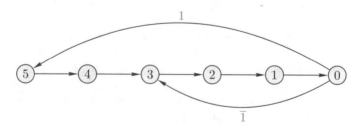

Fig. 5 A cycle-cum-chord automaton that generates the free Abelian group of rank 4

Self-similar subgroups of $\mathsf{Aut}(2^\star)$ can always be translated into Mealy automata, though not necessarily finite ones. For suppose G is self-similar. We can construct the complete group automaton for G, in symbols $\mathfrak{C}_G$, as follows: the automaton has G as state set and the transitions $f \xrightarrow{a/af} \partial_a f$. In general, this invertible transducer will be infinite, but certainly $\mathcal{S}(\mathfrak{C}_G)$ is a group and isomorphic to G. Note that the complete automaton always admits the skew-symmetry mentioned above. As already mentioned, more interesting is a representation of G in terms of a finite automaton. To this end, call G is finite-state if for all $f \in G$ the number of residuals $\partial_x f$ is finite. If G is self-similar, finite-state and finitely generated, we can construct the group automaton $\mathfrak{A}_G$, a subautomaton of $\mathfrak{C}_G$, just like the complete group automaton, but with state set restricted to the collection of all residuals of the generators of G. The group generated by $\mathfrak{A}_G$ is isomorphic to G, but the semigroup may be different as is the case for successor automata. Note that $\mathfrak{A}_G$ is minimal by construction. To recover parts of $\mathfrak{C}(\mathcal{A})$ computationally from $\mathfrak{A}_G$ we can use the standard product machine construction combined with minimization to obtain a machine for each automorphism $f \in G$. Unless $\mathfrak{A}_G$ is group-like, some of the components in these products will be copies of the inverse automaton $\mathfrak{A}'_G$. The complete automaton is then the limit of these automata. Note that the product machine construction combined with minimization is directly related to questions of growth, so one should in general expect no simple descriptions of the resulting automata [3].

3 Abelian Automata

Given a nontrivial, state-closed group G acting on $2^\star$, it is clear that the collection of even elements forms a subgroup H of index 2. Moreover, restricted to H, the residuation maps are group homomorphisms. Correspondingly, one can define an action of a group G on $2^\star$, given a subgroup $H \leq G$ of index 2 and a homomorphism $\Phi : H \to G$. Fix coset representatives $h_0 = 1$ and $h_1 \in G - H$ and define the action via

$$ax\,f = b\,(x\,\Phi(h_b^{-1}fh_a))$$

where b is determined by the condition that $h_b^{-1}fh_a \in H$. Unfortunately, this action may fail to be faithful and the conditions under which it is are slightly complicated, see [22, 23]. In the Abelian case, no problems arise and one can rewrite this characterization using additive notation in the form

$$ax\,f = \begin{cases} a\,(x\,\Phi(f)) & \text{if } f \in H, \\ \overline{a}\,(x\,\Phi(f + (-1)^a g)) & \text{if } f \notin H \end{cases}$$

where g is a suitably chosen coset representative, $g \in G - H$. As an example, consider $G = \mathbb{Z}^2$ with generators $\mathbf{e}_1$ and $\mathbf{e}_2$ and $H = \langle 2\mathbf{e}_1, \mathbf{e}_2 \rangle$. We can set

$\Phi(2a, b) = (b, a)$ and let $g = \mathbf{e}_1$. Then, for example, $0^\omega(4, 3) = 01001^\omega$ and $1^\omega(4, 3) = 101001^\omega$.

It is fairly easy to check whether a given invertible Mealy automaton generates an Abelian group. Suppose $G \leq \mathsf{Aut}(2^\star)$ is self-similar. For any automorphism $f \in G$ define its gap to be $\gamma_f = (\partial_0 f)(\partial_1 f)^{-1} \in G$, so that $\partial_0 f = \gamma_f \partial_1 f$. An easy induction using wreath representation shows the following.

Lemma 1 *A self-similar group $G \leq \mathsf{Aut}(2^\star)$ is Abelian if, and only if, all even elements of G have gap value I, and all odd elements have the same gap value.*

The conditions of the lemma naturally carry over to an automaton generating the group G. So suppose $\mathcal{A}$ is an invertible automaton that satisfies the gap conditions so that $\mathcal{S}(\mathcal{A})$ is Abelian. To avoid tedious special cases, let us note that the transduction monoid and group are Boolean precisely when every toggle state has gap I. By minimality, this means that every such state has out-degree 1 (we consider the transition diagram to be a simple graph rather than a multi-graph). In the other case we obtain a free Abelian monoid and group. From now on, we will always assume that the we are in the second case. We refer to $\gamma(\underline{p})$ for any toggle state p as the gap value of $\mathcal{A}$. It follows easily from minimality that any state p has at most one predecessor copy state. For toggle predecessors note that for any $a \in 2$ there can be at most one q such that $q \xrightarrow{a/\bar{a}} p$. However, it may well happen that there are distinct predecessors q_0 and q_1 such that $q_a \xrightarrow{a/\bar{a}} p$. Thus, every state has indegree at most 3 in a minimal invertible automaton, and it is not hard to show that this bound is tight.

So suppose $\mathcal{A}$ is a minimal invertible automaton with n states and n_0 toggle states. We may safely assume $n_0 < n$. It is clear that one can test whether $\mathcal{A}$ is Abelian in polynomial time by checking directly that all the maps $\underline{p}$ commute. This requires a product machine construction $\mathcal{A}_p \otimes \mathcal{A}_q$ as described in [8] and a test that $\mathcal{A}_p \otimes \mathcal{A}_q$ and $\mathcal{A}_q \otimes \mathcal{A}_p$ are behaviorally equivalent. The latter property can be handled in time $\widetilde{O}(n^2)$ using the standard algorithm in [11]. Using the gap characterization, we can instead check that all the copy states have out-degree 1 and check that all product automata $\mathcal{A}_p \otimes \mathcal{A}'_p$ where p is a toggle state have the same behavior.

Lemma 2 *Given a minimal invertible automaton on n states, one can test commutativity in $\widetilde{O}(n_0 n^2)$ steps.*

Note that the test is trivial for an invertible automaton that contains just a single toggle state p. In this case, the automaton will be Abelian if, and only if, it consists of a copy chain, a directed path of copy states, ending in a toggle state of the form

$$q_k \xrightarrow{0} q_{k-} \xrightarrow{0} \ldots \xrightarrow{0} q_1 \xrightarrow{0} q_0$$

plus two back transitions starting at the toggle state. We refer to this part of the complete automaton as the copy chain at q_0 of length k. As an example, consider the cycle-cum-chord automaton in Fig. 5. Note that copy chains of arbitrary length always exist; in fact, we can construct an infinite copy chain at any toggle state, see Sect. 5 for an application of this idea.

For the two transitions emanating from the toggle state q_0, there are two possibilities. First, they may both end at two copy states in the chain. We may safely assume that one of the transitions leads back to q_k, so we are dealing with a cycle-cum-chord transducer. The other possibility is that one of these transitions leads to the identity state (recall that we assume minimality). In this case we obtain a successor automaton Suc_n, see [22].

Consider a group $G \leq \mathsf{Aut}(2^\star)$. Following Nekrashevych and Sidki [23], we will refer to G as an m-lattice if G is state-closed and free Abelian of finite rank m. Suppose $\mathcal{A}$ generates an m-lattice. As we have seen, the complete automaton associated with $\mathcal{A}$ is a computable structure. In particular, we can effectively construct a finite subautomaton for any automorphism $f \in \mathcal{G}(\mathcal{A})$. The reference shows that a computationally preferable representation of the complete automaton can be obtained by using $\mathbb{Z}^m$ directly as state set. Call $u \in \mathbb{Z}^m$ even if its first component u_1 is even, and odd otherwise. Then the transition function τ of the complete automaton can be described in terms of a residuation matrix $\mathsf{A} \in 1/2\,\mathbb{Z}^{m \times m}$ and an odd residuation vector $e \in \mathbb{Z}^m$ by

$$\tau(u, d) = \mathsf{A}(u + de) \tag{1}$$

where $d = 0$ whenever u is even, and $d = \pm 1$ otherwise (recall our labeling convention from above). Writing c_i for the ith column of A, we have $e = \mathsf{A}^{-1}(c_1 + v)$ where v is integral. The matrices A in question are non-singular and 1/2-integral: $\mathsf{A}^{-1}(\mathbb{Z}^m) \cap \mathbb{Z}^m$ is a sublattice of $\mathbb{Z}^m$ of index 2. As a matter of fact, using similarity with respect to $\mathsf{GL}_m(\mathbb{Z})$, we can safely assume that the matrix A has the form

$$\mathsf{A} = \begin{pmatrix} \frac{a_{11}}{2} & a_{12} & \dots & a_{1m} \\ \frac{a_{21}}{2} & a_{22} & \dots & a_{2m} \\ \vdots & \vdots & \ddots & \vdots \\ \frac{a_{m1}}{2} & a_{m2} & \dots & a_{mm} \end{pmatrix} \tag{2}$$

where all the coefficients a_{ij} are integral. One can verify that the characteristic polynomials of these matrices have the form

$$\chi(z) = z^m + 1/2\left(g_{m-1}z^{m-1} + \dots + g_1 z + g_0\right) \tag{3}$$

where all the coefficients g_i are integral; in particular $g_0 = \pm 1$. In the case of interest to us when the action induced by A is faithful, it is shown in the reference that $\chi(z)$ is irreducible, a property we will tacitly assume from now on. Computational evidence suggests that most matrices A have but one $\mathsf{GL}_m(\mathbb{Z})$ class [23, 24]. In this case we can further assume that the matrix A has the form of a companion matrix with fractional components again only appearing in the first column, and all other columns being unit vectors. One property of these characteristic polynomials $\chi(z)$ that is crucial for our purposes is the fact that all their roots have modulus strictly less than 1. Thus any residuation matrix A has spectral radius strictly less than 1, and A is a contraction. As elements of the corresponding algebraic number field, all roots have denominator 2.

The complete automaton $\mathfrak{C}(A, e)$ now takes the following simple form and is obviously computable: the state set is $\mathbb{Z}^m$ and the transition function is given by Eq. (1). Unlike the product automata mentioned earlier, this infinite Mealy automaton is always reduced in the sense that any two distinct states have distinct behavior. In the following we will always interpret the complete automaton in this manner.

3.1 Canonical and Principal Automata

As in the last section, consider the complete automaton $\mathfrak{C}(A, e)$ for some m-lattice G. We are interested in finite subautomata of $\mathfrak{C}(A, e)$ that generate the same lattice. Following [22], define the nucleus $\mathcal{N}$ of the action as the following subset of G:

$$\mathcal{N} = \bigcup_{g \in G} \bigcap_{n \in \mathbb{N}} \{ \partial_x g \mid |x| \geq n \}$$

Thus, $\mathcal{N}$ consists of all states of the complete automaton that are reachable from a cycle. Note that $\mathcal{N}$ naturally defines a subautomaton of $\mathfrak{C}(A, e)$ and it is shown in [22] that this automaton generates the lattice. The action is called contracting if $\mathcal{N}$ is finite and one can show that all m-lattices are contracting in this sense. As a consequence, it is decidable whether two Abelian automata generate the same lattice. The algorithm given in the reference relies on the claim that the nucleus can be effectively generated; see below for a plausible method. An image of the nucleus of the successor automaton Suc_2 of rank 2 is shown in Fig. 6. Note the extraneous strongly connected components that could be removed without changing the generated group (the subautomaton colored red already generates the group).

Select some anchor point $u \in \mathbb{Z}^m$, $u \neq 0$, and form the closure under the transition function defined by Eq. (1). The elements of the closure will have the form

$$A^k \cdot u + \left(a_k A^k + a_{k-1} A^{k-1} + \ldots + a_1 A \right) \cdot e \tag{4}$$

where all the coefficients a_i are trits, $a_i \in 2_s$. Since A is a contraction, the closure is always finite. We will refer to the resulting automaton as a canonical automa-

Fig. 6 The nucleus automaton associated with the sausage automaton of rank 2

ton (for A) and write $\mathfrak{A}(\mathsf{A}, e, u)$. We may safely assume that u is odd, otherwise $\mathfrak{A}(\mathsf{A}, e, u)$ has the form $\mathfrak{A}(\mathsf{A}, e, u_0)$ plus a copy chain ending at u_0, the first toggle state obtained by repeated residuation from u. The description of points in the closure becomes somewhat simpler if the anchor point u is equal to $\mathbf{e}_1$; correspondingly we write $\mathfrak{A}(\mathsf{A}, e)$ and even $\mathfrak{A}(\mathsf{A})$ if in addition $e = \mathbf{e}_1$. The latter automaton will be called the principal automaton (for A) and is entirely determined by the residuation matrix A. In either case, we are dealing with subautomata of the finite nucleus. Note that the principal automaton always contains the sink 0 since there is a transition $\mathbf{e}_1 \xrightarrow{-1} 0$. Computational evidence suggests that the principal automaton almost always contains $-\mathbf{e}_1$ and is skew-symmetric. The only exception to this rule appears to be the successor automata based on the characteristic polynomial $\chi(z) = z^m - 1/2$ as in Fig. 2: here the principal automaton has three strongly connected components, the sink plus two parts that are skew-symmetric to each other. In this case, either one of these components alone produces just a monoid, not a group. To avoid special cases, in this situation we will here refer to the automaton comprised of all three strongly connected components as the principal automaton. Note that the nucleus automaton is strictly larger than the principal one for all $m \geq 2$.

Now consider the condensation graph of $\mathfrak{C}(\mathsf{A}, e)$, i.e., the graph whose nodes are the strongly connected components of $\mathfrak{C}(\mathsf{A}, e)$ and whose edges are inherited. Then remove all nodes that fail to be both reachable and co-reachable from non-trivial strongly connected components; call this the strict condensation graph. It is easy to see from Eq. (4) that there are only finitely many non-trivial strongly connected components, and they are all finite. A component induces a subautomaton if it is terminal in the strict condensation graph: it has no out-edges. Again, we will think of the terminal component 0 as being part of the principal automaton.

An algorithm to compute the nucleus is implicit in Okano [24]. Let V be the Vandermonde matrix given by the m roots of the characteristic polynomial $\chi(z)$, define the vector norm $\|x\| = \|V \cdot x\|_\infty$ in terms of the Chebyshev norm and let $\lambda < 1$ be the spectral radius of A. Then $\|c_1\| = \lambda^m$ and, for the induced matrix norm, we have $\|\mathsf{A}\| = \lambda$. For any point u on a cycle one can then show that $\|u\| \leq \lambda^m/(1 - \lambda)$. Thus the search for strongly connected components can be limited to a finite region of the complete automaton. Note, though, that the region becomes quite large when λ is close to 1.

As an example, consider the residuation matrix

$$\mathsf{A} = \begin{pmatrix} -1 & 1 \\ -1/2 & 0 \end{pmatrix} \qquad \chi(z) = z^2 + z + 1/2$$

This is the matrix associated with the cycle-cum-chord transducer CC_2^3 in Fig. 4 and has spectral radius $\lambda = 1/\sqrt{2}$. The roots of $\chi(z)$ have absolute norm 2. The corresponding principal automaton, a 7 state machine, is shown in Fig. 7; the state labels will be explained in Sect. 4. In this case, the strongly connected component of the anchor point $\mathbf{e}_1$ admits a skew-symmetry. The principal automaton here coincides with the nucleus. On the other hand, if we select the residuation vector to be

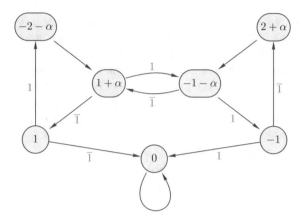

Fig. 7 The principal automaton associated with $\chi(z) = z^2 + z + 1/2$. States are labeled by algebraic integers. Note the skew-symmetry

$e = (3, 2)$, the canonical automaton $\mathfrak{A}(A, e)$ has but 3 states and is isomorphic to CC_2^3. Clearly $\mathfrak{A}(A, e)$ fails to admit any skew-symmetry, yet it still generates the free Abelian group of rank 2 [33]. In this case, the nucleus has 5 subautomata: the principal automaton, plus two pairs of skew-symmetric ones. These automata are generated by the powers of the transduction $CC_2^3(0)$.

Consider some subautomaton $\mathcal{A}$ of $\mathfrak{C}(A, e)$ and a transition $p \xrightarrow{d} q$ whose target q lies in $\mathcal{A}$, but whose source p does not. Let $\mathcal{A}_+$ denote the smallest subautomaton of $\mathfrak{C}(A, e)$ that contains $\mathcal{A}$ and p.

Lemma 3 $\mathcal{A}_+$ *generates the same group as* $\mathcal{A}$.

To see why, first assume that p is a copy state and consider the copy chain at $q = q_0$ of length m^-. As a consequence of Eq. (3) and using additive notation we have

$$2q_0 + g_{m-1}q_1 + \ldots \pm q_{m^-} = I. \tag{5}$$

By repeated residuation it follows that $p = q_1 \in \mathbb{Z}[Q]$ where Q is the state set of $\mathcal{A}$. If p is a toggle state, we may safely assume that the new transition has the form $p \xrightarrow{\bar{1}} q$. Let $p \xrightarrow{1} q'$, so that $q' = q + \gamma \in \mathbb{Z}[Q]$ where γ is the gap value from above. Hence we can apply the same residuation argument as in the first case to show that $p \in \mathbb{Z}[Q]$.

Reading the lemma in the opposite direction, we see that there are two types of interesting subautomata of the nucleus of $\mathfrak{C}(A, e)$

- the principal automaton $\mathfrak{A}(A)$, and
- terminal strongly connected automata.

To see why, consider a terminal strongly connected component S in a subautomaton $\mathcal{A}$ of the complete automaton. If S consists only of 0 we are dealing with the principal automaton; otherwise S itself defines a subautomaton that generates the same group as $\mathcal{A}$. The principal automata all seem to have a non-trivial skew automorphism, all the others do not (but recall our convention regarding successor automata as in Fig. 2).

3.2 Self-affine Tiles

Let us digress briefly to comment on the connection between Abelian invertible automata and questions related to tilings and iterated function systems. Since A is a contraction, the representation of the elements of a subautomaton of $\mathfrak{C}(A, e)$ in Eq. (4) naturally gives rise to a so-called tile, a compact subset of $\mathbb{R}^m$ of positive Lebesgue measure. More precisely, fix a set $\mathcal{D} \subseteq \mathbb{R}^m$ of generalized digits. We are only interested in the case $|\mathcal{D}| = 2$; moreover, we may assume without loss of generality that one of the digits is 0 so that $\mathcal{D} = \{0, d\}$. We can think of the tile as being determined by an iterated function system given by A and $\mathcal{D}$: we are interested in the compact set $T \subseteq \mathbb{R}^m$ such that

$$T = A(T + \mathcal{D}) \tag{6}$$

It is easy to see that T has the explicit representation

$$T = \left\{ \sum_{i=1}^{\infty} A^{-i} d_i \,\middle|\, d_i \in \mathcal{D} \right\} \subseteq \mathbb{R}^m \tag{7}$$

The tile T is called self-affine if it has positive Lebesgue measure, a property that is somewhat rare. To develop a test for positive measure, consider all real vectors that admit a description in terms of a k-digit expansion of the form

$$\mathcal{D}_k = \left\{ \sum_{i=0}^{k-1} A^{-i} d_i \,\middle|\, d_i \in \mathcal{D} \right\} \tag{8}$$

It was shown by Lagarias and Wang [14, 15] that T has positive measure if, and only if, the cardinality of $\mathcal{D}_k$ is 2^k for all k. It follows from the results in Sect. 5 that this condition is satisfied for our residuation matrices and digit sets described below.

Now define the containment lattice of A^{-1} and $\mathcal{D}$ to be the $\mathbb{Z}$-module generated by the pre-images of $\mathcal{D}$ under A^{-1}:

$$\mathbb{Z}[A^{-1}, \mathcal{D}] = \mathbb{Z}[\mathcal{D}, A^{-1}\mathcal{D}, \dots, A^{-n+1}\mathcal{D}] \tag{9}$$

Clearly, the containment lattice contains the symmetric digit set $\Delta\mathcal{D} = \{-d, 0, +d\}$ and is closed under A^{-1}. A digit set is said to be primitive if $\mathbb{Z}[A^{-1}, \mathcal{D}] = \mathbb{Z}^m$. Now consider the digit $d = A^{-1}(c_1 + v)$. The containment lattice is none other than the $\mathbb{Z}$-linear closure of the set of points co-reachable from the origin in $\mathfrak{C}(A, d)$. Indeed,

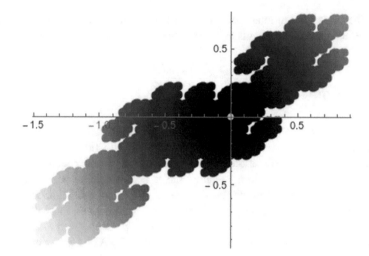

Fig. 8 The integral self-affine tile associated with $\chi(z) = z^2 + z + 1/2$, using the standard digit $(1, 0)$. The tile has Lebesgue measure 1 and lattice tiles $\mathbb{R}^2$ with tiling set $\mathbb{Z}^2$. It is known as the twin-dragon

in conjugation with the conjecture in Sect. 6, it consists precisely of these points; to wit, the collection of all tame automorphisms, see Sect. 6 for definitions. At any rate, in particular for $d = \mathbf{e}_1$ we obtain a primitive digit set. A primitive digit set is standard if the digits form a complete residue system of $\mathbb{Z}^n / \mathsf{A}^{-1} \mathbb{Z}^n$. In other words, for a non-zero digit d we must have $\mathsf{A}\, d$ non-integral. This is, of course, precisely the choice of the residuation vector. Hence we obtain self-affine tiles, and for $d = \mathbf{e}_1$ the tiling set can be chosen to be the full lattice $\mathbb{Z}^m$. Figure 8 shows an example of such a tile. Note that the origin is the only integral point in the tile.

4 Path Polynomials

We will now develop yet another representation of the complete automaton that is helpful in studying path existence problems. As a starting point, consider the question of how the nuclei of $\mathfrak{C}(\mathsf{A}, e)$ and their subautomata compare for different values of the residuation vector e. Fix some residuation matrix A and vector $e = \mathsf{A}^{-1}(c_1 + v)$ where c_1 is the first column of A and v is integral. Our first result shows that as far as the gap value is concerned, only the residuation matrix matters.

Theorem 1 *The principal automaton is isomorphic to a subautomaton of $\mathfrak{C}(\mathsf{A}, e)$ for all e. As a consequence, the gap value of $\mathfrak{C}(\mathsf{A}, e)$ depends only on the residuation matrix A.*

There are two ways to establish this result. The first is essentially taken from [24] and directly constructs a linear map that defines the embedding. To this end, define an $m \times m$ integral matrix with column vectors

$$E = \left(e, A^{-1}e, \ldots, A^{-m^{-}}e\right) \tag{10}$$

One can check that E is non-singular and it commutes with A. As a consequence, E induces a monomorphism from $\mathfrak{C}(A)$ to $\mathfrak{C}(A, e)$, so that $\mathfrak{C}(A)$ is always (isomorphic to) a subautomaton of $\mathfrak{C}(A, e)$. In particular, E shows that $\mathfrak{A}(A, e, e)$ is an isomorphic copy of the principal automaton $\mathfrak{A}(A)$.

A rather different approach to Theorem 1 focuses on a description of paths in the complete automaton $\mathfrak{C}(A, e)$, and in particular in the principal automaton. Consider a path

$$\pi : e \xrightarrow{d_0} q_1 \xrightarrow{d_1} \ldots \xrightarrow{d_{k-2}} q_{k-} \xrightarrow{d_{k-}} q_k \tag{11}$$

Only paths such that $q_i \neq 0$ for $i < k$ are of interest. We will interpret the label $\mathsf{lab}(\pi) = d = d_0 d_1 \ldots d_{k-}$ as a word over the three-letter alphabet $\mathbf{2}_s = \{-1, 0, 1\}$. For reasons of legibility, we will often write $\bar{1}$ instead of -1 in this context. Define the path polynomial $P_d^e(z) \in \mathbb{Z}[z]$, for any word d over $\mathbf{2}_s$, as follows: $P_\varepsilon^e(z) = 1$ and

$$P_{d\delta}^e(z) = z \cdot (P_d^e(z) + \delta) \tag{12}$$

Then $P_d^e(A)$ is a linear map from $\mathbb{Q}^m$ to $\mathbb{Q}^m$ and $P_d^e(A)(e)$ is the state of $\mathfrak{A}(A, e, e)$ after scanning $d \in \mathbf{2}_s{}^*$, starting at initial state e. We write $P_d(A)$ for the path polynomial for the principal automaton $\mathfrak{A}(A)$, in which case the initial state is $\mathbf{e}_1 = A^{-1}c_1$. Note that all coefficients of a path polynomial encode the corresponding path in an entirely straightforward manner.

The extension $d0$ is valid if, and only if, $P_d^e(A)(e)$ is even; otherwise the valid extensions are $d1$ and $d\bar{1}$. An induction on the length of a path then shows that P_d is defined whenever P_d^e is and that $P_d(A) = P_d^e(A)$. This provides another proof of the claim that the principal automaton $\mathfrak{A}(A)$ is embedded in all complete automata $\mathfrak{C}(A, e)$.

As we will see shortly, path polynomials define algebraic integers in a natural way. To see why, consider the algebraic number field $\mathbb{F} = \mathbb{Q}[z]/\chi(z)$ of degree m. We write α for the representative of $1/z$ of the inverse of a root of $\chi(z)$, and $\chi^\star(z)$ for the minimal polynomial of α, the reciprocal of χ. We have $\mathbb{F} = \mathbb{Q}(\alpha)$, but α is an algebraic integer and more useful for our purposes. Now suppose we have two directed paths $\mathbf{e}_1 \longrightarrow p$, with corresponding path polynomials P_1 and P_2. Then $P_1 = P_2 \pmod{\chi(z)}$. The see this, observe that $P_1(A)(\mathbf{e}_1) = P_2(A)(\mathbf{e}_1)$ implies that $\mathbf{e}_1$ is in the null space of $P = P_1 - P_2$. By Cayley-Hamilton, the remainder operation with respect to $\chi(z)$ does not affect the corresponding linear operators. Hence the null space of $P \pmod{\chi(z)}$ is non-zero. If P is not zero, it has degree less than m and is thus coprime with $\chi(z)$. Hence there are cofactors Q_1 and Q_2 such that $Q_1 P + Q_2 \chi(z) = 1$. But then Q_1 is the inverse of P, contradiction. By slight abuse of notation, we will refer to $P \bmod \chi(z) \in \mathbb{F}$ also as the path polynomial for p. Hence we can label states p in the principal automaton by algebraic integers. Figure 7 shows an example of such a labeling.

Path polynomials suggest a generalization where we label a node p in $\mathfrak{C}(A)$ by an algebraic integer $\Phi(p) \in \mathbb{Z}_{\mathbb{F}}$ subject to the constraint

$$p \xrightarrow{d} q \iff \Phi(p) = \alpha \Phi(q) - d\beta \tag{13}$$

where $\beta = e \circ (\alpha^i)_{i<m}$. An easy induction shows the analogue of Eq. (4):

Lemma 4

$$\alpha^k \Phi(q_k) = \Phi(q_0) + \beta \sum_{i<k} d_i \alpha^i.$$

Considering the self-loop at 0 it follows that $\Phi(0) = 0$. In particular for $e = \mathbf{e}_1$ we have $\beta = 1$. In this case for source $q_0 = \mathbf{e}_1$ and target $q_k = 0$ we have

$$1 + \sum_{i<k} d_i \alpha^i = 0.$$

Similarly, if there is a cycle of length k at a point q we have

$$(\alpha^k - 1) \Phi(q) = \sum_{i<k} d_i \alpha^i.$$

The string d is nothing but the reverse base α expansion of the algebraic integer $\sum_{i<k} d_i \alpha^i$ on the right hand side. It is easy to check that

$$P_d(z) = z^k \left(1 + \sum_{i<k} d_i z^{-1}\right).$$

Thus, path polynomials and labels are closely connected:

Lemma 5 *For any path from $\mathbf{e}_1$ to q labeled d: $P_d(1/\alpha) = \Phi(q)$.*

The polynomial representation is useful because it allows us to use polynomial arithmetic to search for paths. As an example, consider again the characteristic polynomial $\chi(z) = z^2 + z + 1/2$ with principal automaton shown in Fig. 7. There is a trivial path from 0 to $\mathbf{e}_1$ labeled $\bar{1}$, corresponding to $\rho = -1$. This is obviously the shortest path, but we can try to find others by writing

$$\rho = \beta\chi^* + \sigma \qquad \sigma = \sum s_i \alpha^i \tag{14}$$

where the digits s_i are again trits. We have $2 + 2\alpha + \alpha^2 = 0$ in $\mathbb{Z}_{\mathbb{F}}$. Hence we can rewrite the digits string -1 as follows:

1	α	α^2	α^3	α^4
-1				
2	2	1		
	-2	-2	-1	
		2	2	1
1	0	1	1	1

Thus $\beta = 1 - \alpha + \alpha^2$ and $\sigma = 1 + \alpha^2 + \alpha^3 + \alpha^4$. One can easily check in Fig. 7 that this corresponds indeed to the shortest path from 0 to $\mathbf{e}_1$ that passes through $-\mathbf{e}_1$. This approach is similar in spirit to Gilbert's clearing algorithm [9].

Now consider a copy chain of the form

$$q_k \xrightarrow{0} q_{k-} \xrightarrow{0} \dots \xrightarrow{0} q_1 \xrightarrow{0} q_0$$

where q_0 belongs to some subautomaton $\mathcal{A}$ with state set Q, while some or all of the other states in the chain lie outside of $\mathcal{A}$. As we have seen in Lemma 3, adjoining these states does not change the group generated by $\mathcal{A}$. In the special case where $q_0 = \mathbf{e}_1$ is the generator of the principal automaton, and there are at least $m - 1$ copy states in the chain, we even have equality: $\mathbb{Z}[q_{m-}, \dots, q_0] = \mathbb{Z}[Q]$. This follows easily by induction on the length of a path from the generator to a state in the automaton. In this case, the label of q_i is simply α^i. To generalize Lemma 5 one needs to admit Laurent polynomials as path polynomials. Since $\mathbf{A}$ is invertible this causes no difficulties. At any rate, the transduction group generated by $\mathcal{A}$ can thus be represented by a lattice of algebraic integers.

Lemma 6 *Let $\mathcal{A}$ be a canonical automaton generating an m-lattice G. The G is isomorphic to a lattice of algebraic integers of the form $\sum_{i<m} a_i \alpha^i$, $a_i \in \mathbb{Z}$.*

For example, in a cycle-cum-chord automaton the length of the backbone copy chain is always larger than the rank of the lattice. Thus the group is generated by the first m states on the chain.

5 Knuth Normal Form

The last lemma suggests that one consider numeration systems for the algebraic integers in an algebraic number field, a subject first breached by Knuth in 1960 [13] in the case of the Gaussian integers $\mathbb{Z}[\mathbf{i}]$. As it turns out, every Gaussian integer ρ can be written uniquely in the form $\rho = \sum_{i<k} b_i \alpha^i$ where $\alpha = -1 + \mathbf{i}$ using only binary digits 0 and 1. Since $1/\alpha$ is a root of $\chi(z) = z^2 + z + 1/2$, we can translate this result into the world of subautomata as follows [12]. Attach an infinite copy chain to the generator $\mathbf{e}_1$ of $\mathbf{CC}_2^3$ and refer to the resulting automaton as $\mathcal{A}_+$, see Fig. 9. By slight abuse of notation, let us refer to these states by their labels as α^k, with $\alpha^0 = 1$ representing the generator $\mathbf{e}_1$. As we have seen, $\mathcal{A}_+$ still generates the same group.

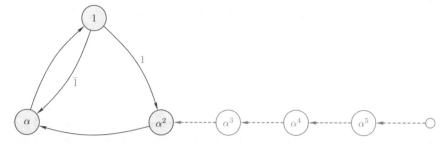

Fig. 9 Adding an infinite copy chain to a strongly connected canonical automaton

More importantly, every element of the associated 2-lattice can be written uniquely in Knuth normal form (KNF) $\sum_{i<k} b_i \alpha^i$, where $b_i \in \mathbf{2}$, according to [13, 34].

As an example, consider $f = \underline{1}^{\,5}$ in the automaton from Fig. 7. Written as a bit-vector, the Knuth normal form of f is 100010111. The next table shows the corresponding rewrite process.

1	α	α^2	α^4	α^4	α^5	α^6	α^7	α^8
5								
-4	-4	-2						
	4	4	2					
		-2	-2	-1				
				2	2	1		
					-2	-2	-1	
						2	2	1
1	0	0	0	1	0	1	1	1

Knuth normal form is extremely helpful in exploring properties of the automorphisms generated by $\mathcal{A}$, rather than just their group structure. For example, one can show that $\underline{1}^{\,2^{4k}}$ has normal form α^{8k}, a map that copies the first $8k$ input bits and then behaves like the odd transduction $\underline{1}$ on the remainder of the input. This can be used to show that Knuth normal form for this particular automaton can be computed by a suitable finite state transducer [31]. Similarly one can show that the group of automorphisms acts transitively on each level set $\mathbf{2}^k$ and a little more work makes it possible to identify the levels where the group acts simply transitively. To construct Schreier graphs like the one in Fig. 10, one can use for $\mathbf{2}^k$ one can use the fact that in Knuth normal form the generators look like $0^i 10^j \in \mathbf{2}^k$.

Uniqueness of Knuth normal form is not hard to see: assume $\sum_{i<k} b_i \alpha^i = \sum_{i<k} b_i' \alpha^i$ for digits $b_i, b_i' \in \mathbf{2}$. Then $b_i = b_i'$ for all $i < \ell$, but, say, $0 = b_\ell \neq b_\ell' = 1$ and we have $\sum_{i>\ell} b_i \alpha^i = \alpha^\ell + \sum_{i>\ell} b_i' \alpha^i$. But the first automorphism copies at least ℓ bits, whereas the second changes the bit in position ℓ.

Existence is much harder to deal with. For example, it is known that for $m = 2$ there are only six characteristic polynomials that give rise to residuation matrices;

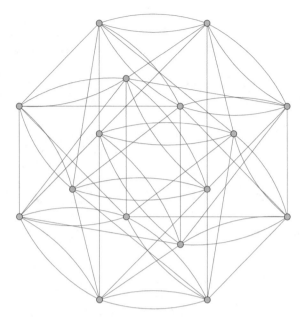

Fig. 10 A Schreier graph associated with CC_2^3 and the subtree 2^4

all of these are unique up to $GL_m(\mathbb{Z})$ similarity. The matrices together with the corresponding α values and their minimal polynomials are shown below. The first matrix, which gives rise to the successor automaton, is the only one that produces a real field. The forth matrix is associated with the automaton in Fig. 3. As the table shows, only 4 out of the 6 admit a Knuth normal form.

A	α	min. pol.	KNF
$\begin{pmatrix} 0 & 1 \\ 1/2 & 0 \end{pmatrix}$	$-\sqrt{2}$	$-2 + z^2$	no
$\begin{pmatrix} 1 & 1 \\ -1/2 & 0 \end{pmatrix}$	$1 + i$	$2 - 2z + z^2$	no
$\begin{pmatrix} 1/2 & 1 \\ -1/2 & 0 \end{pmatrix}$	$\left(1 + i\sqrt{7}\right)/2$	$2 - z + z^2$	yes
$\begin{pmatrix} 0 & 1 \\ -1/2 & 0 \end{pmatrix}$	$i\sqrt{2}$	$2 + z^2$	yes
$\begin{pmatrix} -1/2 & 1 \\ -1/2 & 0 \end{pmatrix}$	$\left(-1 + i\sqrt{7}\right)/2$	$2 + z + z^2$	yes
$\begin{pmatrix} -1 & 1 \\ -1/2 & 0 \end{pmatrix}$	$-1 + i$	$2 + 2z + z^2$	yes

In the case where all coefficients of the minimal polynomial of α are nonnegative, one can show that the Gilbert-style rewrite process always terminates. More precisely, we can interpret Eq. (5) as a cancellation rule that simplifies some expressions. Applying Eq. (5) twice we obtain the shift rule

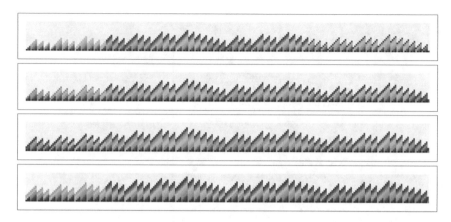

Fig. 11 The Knuth normal forms for some automorphisms defined by a cycle-cum-chord transducer

$$2\alpha \mapsto (2 - g_{m-1})\alpha^2 + (g_{m-1} - g_{m-2})\alpha^3 + \ldots + g_0\alpha^{m+2} \qquad (15)$$

that can be used to eliminate coefficients other than 0 and 1. Given an algebraic integer $\rho = \sum_{i<k} a_i\alpha^i$, we may assume that all coefficients are nonnegative. Let us refer to $\sum_{i<k} a_i$ as the weight of the integer. Clearly, application of the cancellation rule reduces weight. But note that the sum of exponents in the shift rule is telescoping, so that its application does not affect weight. Now consider a rewrite system that tries to remove all coefficients other than 0 and 1 by first applying the cancellation rule, and then the shift rule. Assume that the rewrite process fails to terminate on some input ρ. By deleting an initial segment, we may safely assume that the weight of the expression remains constant throughout the process, i.e., we only apply the shift rule. Using the notation from Eq. (14) we see that after a transient phase, only the last $m + 1$ coefficients of σ will be non-zero. Since the weight does not change, this block of coefficients must ultimately repeat in some later term σ'. But that means that $\sigma' = \alpha^k\sigma$, a contradiction.

The copy chain extension to the principal automaton is perhaps the most natural way to define Knuth normal form, but there are other options. For example, we can use the generator of one of the strongly connected automata in $\mathfrak{C}(A, e)$ that generate the lattice as an anchor for the chain. For CC_2^3 this is indicated in Fig. 9. The Knuth normal forms of the automorphisms $\underline{1}^k$ for $0 \le k < 2^{10}$ are shown in Fig. 11. The height of each column indicates the number of terms, and the actual terms are color coded.

In cases where Knuth normal form fails to exist one can try allow for larger digit sets, and in particular for trits $\{-1, 0, 1\}$. Define the weak Knuth normal form to be an expansion of the form $\sum_{i<k} d_i\alpha^i$ where $d_i \in \mathbf{2}_s$. This generalization corresponds to the step from canonical numeration systems for algebraic number fields, where the digit set is of the form $\{0, 1, \ldots, N^-\}$ to symmetric canonical number systems where the digits are chosen from $\{0, \pm 1, \ldots, \pm N^-\}$. Here N is typically the absolute

value of the absolute norm of the generator α and thus $N^- = 1$ in our case, There is large body of literature on these numeration systems, see [25] and the references therein. Call an algebraic integer expanding if all its conjugates have modulus larger than 1. The following result follows from the work in [14, 15].

Theorem 2 (Lagarias, Wang 1997) *Let α be an expanding algebraic integer of norm 2. Then all algebraic integers ρ in $\mathbb{F} = \mathbb{Q}(\alpha)$ have an expansion $\rho = \sum_{i<k} d_i \alpha^i$ where $d \in \{-1, 0, 1\}$.*

Corollary 1 *All m-lattices admit a weak Knuth normal form.*

Of course, we no longer have uniqueness. For example, for the 2-lattice associated with $\alpha = 1 + \mathbf{i}$ the non-trivial weak normal forms of $\overline{1} = -1$ are $1011 \ldots 11\overline{1}1$. Another possibility is to move to the completion of the group and consider infinite normal forms $\sum_i d_i \alpha^i$: the automorphism corresponding to digit d_k leaves the first k input bits unchanged, so this formal infinite sum indeed defines an automorphism of $\mathbf{2}^\star$ [22]. For example, for the successor automaton from Fig. 2 with characteristic polynomial $\chi(z) = z^2 + 1/2$, the automorphism $\underline{1}^{-1}$ produces the infinite digit sequence $(1, 0, 1, 0, 1, \ldots)$.

6   Open Problems

It is shown in [22] that the nucleus automaton is a natural finite subautomaton of the complete automaton $\mathfrak{C}(A, e)$. Alas, from the perspective of automata theory, the nucleus hides a lot of interesting fine structure. As we have seen, there may be smaller subautomata that also generate the same m-lattice; the cycle-cum-chord transducer CC_2^3 being a case in point.

Question: Is there a reasonable description of the smallest subautomaton of $\mathfrak{C}(A, e)$ that generates the full lattice? Is its state complexity computable in polynomial time?

Note that the algorithm to compute the nucleus outlined above is clearly not polynomial in A. On the other hand, the principal automaton might be computable in polynomial time.

The question arises at to what the essential differences between the principal automaton and strongly connected subautomata of the nucleus might be. Let us call an automorphism f tame if there is some word w such that $\partial_w f = I$, and strongly tame if all its residuals are tame. Thus f is tame if, and only if, the corresponding state p in $\mathfrak{C}(A, e)$ is co-reachable from 0. Similarly, f is strongly tame if, and only if, every state reachable from the corresponding state p in $\mathfrak{C}(A, e)$ is co-reachable from 0. If a principal automaton $\mathcal{A}$ has the two component structure just described, then all its basic transductions are obviously strongly tame. But the same holds true for the whole group generated by $\mathcal{A}$: any composition of basic transductions can be residuated to I by systematic removal of components in the representation of f. By contrast, all the basic transductions defined by canonical automata of the single component type such as CC_m^n fail to be tame. Note, though, that this classification depends crucially on the fact that there are final strongly connected components in the principal automaton other than the trivial one. Otherwise, an automorphism defined by a state in such a putative final component obviously fails to be tame. Computational evidence supports the following conjecture.

Conjecture: Every principal automaton $\mathfrak{A}(A)$ other than a successor automaton is skew-symmetric and consists of exactly two strongly connected components, one of them being the sink 0. For successor automata, the principal automaton has 3 strongly connected components.

Note that it suffices to show that in all these principal automata there is a path from e_1 to $-e_1$: this would suffice to rewrite any improper path polynomial as a proper one. Equivalently, one needs to rewrite the trit representation of $\bar{1}$ to a form $(1, d_2, d_3, \ldots, d_\ell, 1)$. In case the conjecture fails in this strong form, one might wish to consider the situation where the underlying residuation matrix has only one similarity type.

Acknowledgements It is a pleasure to acknowledge many helpful conversations with Tsutomo Okano and Tim Becker.

References

1. Abramsky, S.: A structural approach to reversible computation. Theor. Comput. Sci. **347**, 441–464 (2005)
2. Bartholdi, L., Reznykov, I.I., Sushchansky, V.I.: The smallest mealy automaton of intermediate growth. J. Algebr. **295**, 387–414 (2005)
3. Bartholdi, L., Silva, P.V.: Groups defined by automata. CoRR (2012). arXiv:1012.1531
4. Bennett, C.H.: Logical reversibility of computation. IBM J. Res. Dev. **17**, 525–532 (1973)
5. Bennett, C.H.: The thermodynamics of computation–a review. IJTP **21**(12), 905–940 (1982)
6. Berstel, J.: Transductions and context-free languages (2009). http://www-igm.univ-mlv.fr/berstel/LivreTransductions/LivreTransductions.html
7. Eilenberg, S.: Automata, Languages and Machines, vol. A. Academic Press, Dublin (1974)
8. Elgot, C.C., Mezei, J.E.: On relations defined by generalized finite automata. IBM J. Res. Dev. **9**, 47–68 (1965). January
9. Gilbert, W.J.: Radix representations of quadratic fields. J. Math. Anal. Appl. **83**, 264–274 (1981)
10. Grigorchuk, R.R., Nekrashevich, V.V., Sushchanski, V.I.: Automata, dynamical systems and groups. Proc. Steklov Inst. Math. **231**, 128–203 (2000)
11. Hopcroft, J.E., Ullman, J.D.: Introduction to Automata Theory, Languages and Computation. Addison-Wesley, Reading (1979)
12. Knuth, D.: Private communication, 2010
13. Knuth, D.E.: Commun. ACM. An imaginary number system **3**, 245–247 (1960)
14. Lagarias, J.C., Wang, Y.: Self-affine tiles in $\mathbb{R}^n$. Adv. Math. **121**, 21–49 (1996)
15. Lagarias, J.C., Wang, Y.: Integral self-affine tiles in $\mathbb{R}^n$ II. Lattice tilings. J. Fourier Anal. Appl. **3**(1), 83–102 (1997)
16. Landauer, R.: The physical nature of information. Phys. Lett. A **217**, 188–193 (1996)

17. Lecerf, Y.: Machine de Turing réversible. Insolubilité récursive en $n \in N$ de l'équation $u = \theta^n u$, où θ est un "isomorphisme de codes". C. R. Acad. Sci. Paris **257**, 2597–2600 (1963)
18. Morita, K.: Computation universality of one-dimensional reversible cellular automata. Inf. Proc. Lett. **42**, 325–329 (1992)
19. Morita, K.: Reversible cellular automata. J. Inf. Proc. Soc. Jpn. **35**, 315–321 (1994)
20. Morita, K.: Reversible cellular automata. Encyclopedia of Complexity and System Science. Springer, Berlin (2009)
21. Morita, K., Harao, M.: Computation universality of 1 dimensional reversible (injective) cellular automata. Trans. Inst. Electron. Inf. Commun. Eng. E **72**, 758–762 (1989)
22. Nekrashevych, V.: Self-similar Groups. Mathematical Surveys and Monographs, vol. 117. AMS (2005)
23. Nekrashevych, V., Sidki, S.: Automorphisms of the Binary Tree: State-Closed Subgroups and Dynamics of 1/2-Endomorphisms. Cambridge University Press, Cambridge (2004)
24. Okano, T.: Invertible binary transducers and automorphisms of the binary tree. MS Thesis, CMU, May 2015
25. Pethö, A.: Connections between power integral bases and radix representations in alebraic number fields, (2009). https://arato.inf.unideb.hu/petho.attila/cikkek/cnsnagoya_paper_110.pdf
26. Raney, G.N.: J. Assoc. Comput. Mach. Sequential functions **5**(2), 177–180 (1958)
27. Sakarovitch, J.: Elements of Automata Theory. Cambridge University Press, Cambridge (2009)
28. Serre, J.-P.: Arbres, Amalgames, SL_2. Number 46 in Astérisque. Société Mathématique de France, Paris (1977)
29. Sidki, S.: Automorphisms of one-rooted trees: growth, circuit structure, and acyclicity. J. Math. Sci. **100**(1), 1925–1943 (2000)
30. Sutner, K.: Invertible transducers and iteration. In: Juergensen, H., Reis, R. (eds.) Descriptional Complexity of Formal Systems. Lecture Notes in Computer Science, vol. 8031, pp. 18–29. Springer, Berlin (2013)
31. Sutner, K.: Invertible transducers, iteration and coordinates. In: Konstantinidis, S. (ed.) CIAA. LNCS, vol. 7982, pp. 306–318. Springer, Berlin (2013)
32. Sutner, K.: Iteration of invertible transductions. Submitted, 2013
33. Sutner, K., Lewi, K.: Iterating invertible binary transducers. JALC **17**(2–4), 293–213 (2012)
34. Sutner, K., Lewi, K.: Iterating invertible binary transducers. In: Kutrib, M., Moreira, N., Reis, R. (eds.) Descriptional Complexity of Formal Systems. Lecture Notes in Computer Science, vol. 7386, pp. 294–306. Springer, Berlin (2012)

Simulation and Intrinsic Universality Among Reversible Cellular Automata, the Partition Cellular Automata Leverage

Jérôme Durand-Lose

Abstract This chapter presents the use of Partitioned Cellular Automata—introduced by Morita and colleagues—as the tool to tackle simulation and intrinsic universality in the context of Reversible Cellular Automata. Cellular automata (CA) are mappings over infinite lattices such that all cells are updated synchronously according to the states around each one and a common local function. A CA is reversible if its global function is invertible and its inverse can also be expressed as a CA. Kari proved in 1989 that invertibility is not decidable (for CA of dimension at least 2) and is thus hard to manipulate. Partitioned Cellular Automata (PCA) were introduced as an easy way to handle reversibility by partitioning the states of cells according to the neighborhood. Another approach by Margolus led to the definition of Block CA (BCA) where blocks of cells are updated independently. Both models allow easy check and design for reversibility. After proving that CA, BCA and PCA can simulate each other, it is proven that the reversible sub-classes can also simulate each other contradicting the intuition based on decidability results. In particular, it is proven that any d-dimensional reversible CA (d-R-CA) can be expressed as a BCA with $d+1$ partitions. This proves a 1990 conjecture by Toffoli and Margolus (*Physica D* 45) improved and partially proved by Kari in 1996 (*Mathematical System Theory* 29). With the use of signals and reversible programming, a 1-R-CA that is intrinsically universal—able to simulate any 1-R-CA—is built. Finally, with a peculiar definition of simulation, it is proven that any CA (reversible or not) can be simulated by a reversible one. All these results extend to any dimension.

J. Durand-Lose (✉)
LIFO EA 4022, Université d'Orléans, FR-45067 Orleans, France
e-mail: jerome.durand-lose@univ-orleans.fr

J. Durand-Lose
LIX, CNRS-INRIA-École Polytechnique, Palaiseau, France

© Springer International Publishing AG 2018
A. Adamatzky (ed.), *Reversibility and Universality*, Emergence, Complexity
and Computation 30, https://doi.org/10.1007/978-3-319-73216-9_4

61

1 Introduction

In this chapter, it is shown how Partitioned Cellular Automata (PCA) have been the key to tackle simulation with Block Cellular Automata (BCA) and intrinsic universality of Reversible Cellular Automata (R-CA). Partitioned Cellular Automata were introduced [24, 25, 28] to prove computation universality of 1-dimensional R-CA (1-R-CA). Before that, for lack of ways to handle 1-R-CA, computation universality of R-CA was only known in dimension 2 and above [34].

Cellular automata (CA) model parallel phenomena and architectures since their introduction by Ulam and von Neumann in the fifties. They form a model for massively parallel computations and physical phenomena. They have been widely studied for decades and there is a lot of results about them [5, 18, 33, 37].

They operate as iterative systems on d-dimensional infinite arrays of *cells* (the underlying space is $\mathbb{Z}^d$). Each cell takes a value from a finite set of *state* (Q). A configuration is a valuation of the whole array. An iteration of a CA is the synchronous replacement of the state of every cell by the image of the states of the cells around it (following a finite local *neighborhood* $\mathcal{N}$). This replacement is done according to a unique *local function*. The update is local, uniform, parallel and synchronous.

Reversibility is the capability of a dynamical system to be invertible and to have its inverse in the same class of dynamical systems. This is interesting for physics and computation [4, 36]. It allows to unambiguous backtrack a phenomenon to its origin. It preserves information and entropy. It may offer a guide to design computers that consume less energy.

A CA is *reversible* when its global function $\mathcal{G}$ is bijective and its inverse ($\mathcal{G}^{-1}$) is the global transition function of some CA. It is known that if $\mathcal{G}$ is one-to-one then it is bijective [23, 29] and the corresponding CA is reversible [13, 32]. The reversibility of a CA is decidable in dimension 1 [2] whereas it is not true anymore for greater dimensions [14, 15].

Lecerf [19] and Bennett [3] proved that reversible Turing machines can simulate any Turing Machine and are thus computationally universal. In 1977, Toffoli [34] proved that any CA can be simulated by a reversible CA (R-CA) one dimension higher. In particular, this proves the existence of 2-dimensional R-CA which are computationally universal. The computing power of R-CA as well as their simulation capability was particularly investigated in Toffoli and Margolus [36] and Morita [27].

Reversible CA are quite tricky to design and handle in their general form so that other forms were introduced. To built computationally universal R-CA, (in dimension 2 and above) Block CA (BCA) and (in dimension 1) Partitioned CA (PCA) were independently introduced as special CA for which reversibility is decidable. Like regular CA, they work on infinite regular lattices where each point has a value in a finite set of states.

Block CA (BCA) were introduced in the 80s as a model for lattice gases and other reversible physical phenomena [20, 21, 35]. A specific one called the *Billiard Ball Model* was defined. It has only 2 states but is yet computationally universal.

Like for CA, the global function of a BCA is locally defined. The underlying lattice is partitioned into identical hypercubic *blocks* regularly displayed. A partition is fully determined by the *size* of the blocks and the position of a block (or *origin*). A *block transition* is the parallel replacement of all the blocks of a given partition by their images by the *block function*. The *global function* is the sequential composition of various block transitions with the same size and block function.

Since the block function operates over a finite set (blocks and states are finite in number), it can be bijective. The global function is reversible if and only if the block function is a permutation, which is decidable.

Originally, BCA were named "Partitioning CA" and are also known as "CA with the Margolus neighborhood". To avoid any confusion with Morita's Partitioned CA, they are referred to as "Block CA" following Kari [16] that named "Block Permutations" bijective block functions.

Morita and colleagues introduced *Partitioned CA* (PCA) to prove that R-CA are computationally universal in dimension 1 [25, 26, 28]. In PCA, the states are partitioned according to the neighborhood. Each cell swaps its sub-states with neighboring cells and then computes its new state. The local function operates over the finite set of states and can be bijective. The global function is reversible if and only if the local function is a permutation, which is decidable.

Another important topic developed in this chapter is the relations between the different kinds of CA, especially in terms of capability to simulate one another. Following the survey on universalities in CA [31], one wants to consider a homogeneous type of simulation: cellular automata simulated by cellular automata in a shift invariant, time invariant way. Trivially, PCA (R-PCA) are CA (R-CA). By considering macro-cells corresponding to the blocks of the first partition, CA (R-CA) simulates BCA (R-BCA).

Block CA can simulate CA by using partitions to progressively add its next state to each cell. PCA can simulate CA by copying the original state in each sub-part. These constructions always generate non reversible BCA and PCA. Nevertheless the following conjecture was made:

Conjecture 1 (Toffoli and Margolus [36, Conjecture 8.1]) *All invertible cellular automata are structurally invertible, i.e., can be (isomorphically) expressed in space-time as a uniform composition of finite logic primitives.*

A "finite logic primitives" is a representation of a local permutation of blocks t. Kari [16] proved Conjecture 1 for dimensions 1 and 2. The construction is complex but does not need extra states. At the end, Kari conjectures that:

Conjecture 2 (Kari [16, Conjecture 5.3]) *For every $d \geq 1$, all reversible d-dimensional cellular automata are compositions of block permutations and partial shifts.*

(Partial shift means that the blocks can be shifted which is included in the present definition of BCA.) Durand-Lose [6, 7] proved that R-BCA can simulate R-CA in any dimension with extra states and then that R-PCA can also simulate R-CA [8].

An important concept that stems from simulation is *intrinsic universality*: the capability of a single CA to simulate all the others in a class. This is different from computation universality because it addresses infinite configurations. There exist intrinsically universal (regular) CA [1, 22, 30]. Durand-Lose [6, 9, 12] proved that the Billiard Ball model is intrinsically universal among the 2-R-CA. Using PCA, Durand-Lose [8] extended the result to 1-R-CA. Both results extend to higher dimensions.

One natural question is whether R-CA can simulate any (non-reversible) CA. As already mentioned, any d-CA can be simulated by a $d+1$-R-CA [34].

In 95, Morita [24, 26] proved with PCA that any CA can be simulated by R-CA of the same dimension over *finite* configurations but the construction does not extend to infinite configurations. A configuration is finite if all but a finite number of cells are in a defined stable state. This is enough for computing since it only treats finite information. But for physical modeling and as mathematical abstractions, there is no reason to restrict to such configurations. Durand-Lose [10] provided a simulation of any CA by a R-CA but the simulation relation is so peculiar (it is not homogeneous at all) that the problem is still open.

This chapter first provides the formal definition of all kind of CA, of their reversible sub-classes, of simulation and of intrinsic universality.

The simulations between the various kinds of CA are presented. They come naturally but only preserve reversibility when the target is a regular CA. Simulating CA with BCA is done by progressively adding its next state to every cells before discarding all the previous states.

Simulating R-CA with R-BCA is more involving and corresponds to solving Conjectures 1 and 2. It uses the local function of the inverse automaton to ensure reversibility. In this construction, a previous state is only erased when it can be regenerated from the next ones in the block. The construction in Durand-Lose [6] uses $2^{d+1}-1$ partitions with blocks of size $4r$ (r is the greater of the coordinates of the elements of the neighborhood of the CA and of its inverse) in dimension d. The construction presented here is taken from Durand-Lose [11]. It needs $d+1$ partitions with blocks of size $3(d+1)r$. One gets from a partition to the next by a shift of $(3r, 3r, \cdots , 3r)$.

By considering blocks as cells, BCA can be simulated by PCA preserving reversibility.

Since simulation is a transitive relation, it is enough to prove intrinsic universality on one kind of CA. The construction works on 1-R-PCA and comes from [8]. The intrinsically universal 1-R-PCA is organized in 10 layers (for delimitation, identification, table, value, signals, and translation of data). The dynamic is totally driven by signals which exchange values, test for equality, update when it should be done and move data around. It uses a posteriori tests to ensure reversibility.

It is still an open problem whether any CA can be simulated by a R-CA of the same dimension. Nevertheless, for a particular notion of simulation, it is possible.

A *space-time diagram* depicts the whole (infinite) computation of a CA on an initial configuration. It corresponds to the sequence of all the configurations, the orbit of the system. *Space-time simulation* defines an embedding relation between the space-time diagrams of different CA. This is a peculiar simulation relation since configurations can be encoded across infinitely many configurations.

Any CA can be space-time simulated by a R-CA of the same dimension. Unbounded delays are used to provide extra storage for the information needed for reversibility. The proof is given in dimension 1 and generalized to higher dimensions. As a corollary, using the existence of intrinsically universal R-CA, there exists a R-CA which is capable of space-time simulating any CA of the same dimension.

This chapter is based on Durand-Lose [6, 8, 10, 11]. All definitions and proofs can be read without any previous knowledge of the subject.

Section 2 formally defines the various models, simulation and intrinsic universality. Section 3 constructs various simulations between the different classes of (reversible) CA. Section 4 details an intrinsically universal 1-R-CA. Section 5 considers space-time simulation and provides CA simulation by R-CA. Section 6 gathers some concluding remarks.

2 Definitions

In this chapter, the following notations are used: $[\![a, b]\!]$ denotes the integers from a to b included; and $<$ and $\leq$, $+$, $-$, mod, div and $.$ also denote respectively the component-wise comparisons, ordering, addition, modulo, Euclidean division and multiplication over $\mathbb{Z}^d$.

Cellular automata (CA) define mappings over d-dimensional infinite arrays over a finite *set of states* Q. The supporting lattice is denoted by $\mathbb{L}$ $(= \mathbb{Z}^d)$. The points of $\mathbb{L}$ are called *cells* and each has a value in Q. The state of cell x in configuration c is denoted by c_x. The set of configurations is denoted by $\mathscr{C}$ $(= Q^{\mathbb{L}})$. Functions on one state/cell are naturally extended into functions over arrays of states/cells and configurations.

For any configuration c and subset E of $\mathbb{L}$, $c_{|E}$ is the restriction of c to E. For any $\mathbf{x} \in \mathbb{L}$, $\sigma_{\mathbf{x}}$ is the shift by $\mathbf{x}$ over configurations ($\forall c \in \mathscr{C}, \forall \mathbf{i} \in \mathbb{L}, (\sigma_{\mathbf{x}}(c))_{\mathbf{i}} = c_{\mathbf{i}-\mathbf{x}}$).

2.1 Cellular Automata

A *Cellular Automaton* of dimension d (d-CA) is defined by $(Q, \mathcal{N}, f)$. The *neighborhood* $\mathcal{N}$ is a finite subset of $\mathbb{L}$. The *local function* $f : Q^{\mathcal{N}} \to Q$ maps the states of a neighborhood into one state. The *global function* $\mathscr{G} : \mathscr{C} \to \mathscr{C}$ maps configurations into themselves as follows:

$$\forall c \in \mathscr{C}, \ \forall \mathbf{x} \in \mathbb{L}, \ \mathscr{G}(c)_{\mathbf{x}} \ = \ f\left((c_{\mathbf{x}+\mu})_{\mu \in \mathcal{N}}\right) \ .$$

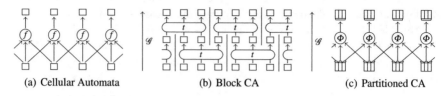

(a) Cellular Automata   (b) Block CA   (c) Partitioned CA

Fig. 1 Schematic CA, BCA and PCA updatings in dimension 1

The new state of a cell depends only on the states of neighboring cells as depicted in Fig. 1a.

The *radius* of a cellular automaton, r, is the maximum absolute value of any coordinate of any element of $\mathcal{N}$. It is the smallest integer r such that: $\mathcal{N} \subseteq [\![-r, r]\!]^d$. By adding dummy entries, the local function can be extended to the domain $[\![-r, r]\!]^d$. Neighborhood and radius can be used equivalently.

2.2 Block Cellular Automata

A *Block CA* of dimension d (d-BCA) is defined by: $(Q, \mathbf{v}, n, (\mathbf{o}^{(j)})_{1 \le j \le n}, t)$. The *size* $\mathbf{v}$ is an element of $\mathbb{L}$ such that $0 < \mathbf{v}$. The *volume* V is the subset $[\![0, \mathbf{v}_1 - 1]\!] \times [\![0, \mathbf{v}_2 - 1]\!] \times \cdots \times [\![0, \mathbf{v}_d - 1]\!]$ of $\mathbb{L}$. A *block* is a mapping from V to Q, or, equivalently, an array of states whose underlying lattice is V. The set of all blocks is Q^V. The *block function* t is a function over blocks. The number of partitions used is n. The origins of the n partitions, $(\mathbf{o}^{(j)})_{1 \le j \le n}$, are elements of V.

The *block transition* T is the following mapping over $\mathscr{C}$: for any $c \in \mathscr{C}$ and $\mathbf{i} \in \mathbb{L}$, let $\mathbf{a} = \mathbf{i} \operatorname{div} \mathbf{v}$ and $\mathbf{b} = \mathbf{i} \operatorname{mod} \mathbf{v}$ ($\mathbf{a} \in \mathbb{L}$ and $0 \le \mathbf{b} < \mathbf{v}$) so that $\mathbf{i} = \mathbf{a}.\mathbf{v} + \mathbf{b}$, then $t(c)_{\mathbf{i}} = t(c_{|\mathbf{a}.\mathbf{v}+V})_{\mathbf{b}}$. In other words, the block containing $\mathbf{i}$ in the regular partition with blocks of size $\mathbf{v}$ is updated according to t. The same happens for all the blocks of this partition. The configuration is partitioned into regularly displayed blocks, then each block is replaced by its image by the block function t as in Figs. 1b and 2.

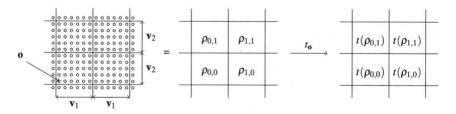

Fig. 2 $t_{\mathbf{o}}$, the block permutation of size $\mathbf{v}$ and origin ($\mathbf{o}$)

The block transition of origin $\mathbf{o}^{(j)}$, T_j is $\sigma_{\mathbf{o}^{(j)}} \circ T \circ \sigma_{-\mathbf{o}^{(j)}}$. It is the original one with the partition shifted by $\mathbf{o}^{(j)}$. The global function is the composition of the block transitions of origins $\mathbf{o}^{(j)}$: $\mathscr{G} = T_n \circ T_{n-1} \circ \cdots \circ T_1$. This is illustrated in Fig. 1b with 2 partitions and $\mathbf{v} = (3)$. The new state of a cell depends only on the states around it.

To see that BCA are indeed CA, consider the blocks of the first partition to be cells. At this scale, the global function commutes with any shift and is continuous for the product topology, according to a theorem of Hedlund [13], Richardson [32], it is a CA. A constructive proof is provided in Sect. 3.1.

2.3 Partitioned Cellular Automata

A *Partitioned Cellular Automaton* of dimension d (d-PCA) is defined by: $(Q, \mathscr{N}, \Phi)$. The set of states is a sub-set product indexed by the neighborhood: $Q = \prod_{\mu \in \mathscr{N}} Q^{(\mu)}$. The μ component of a state q is noted $q^{(\mu)}$. The *state function* Φ operates over Q. The global transition function $\mathscr{G}$ is defined by:

$$\forall c \in \mathscr{C}, \forall \mathbf{x} \in \mathbb{L}, \mathscr{G}(c)_{\mathbf{x}} = \Phi\left(\prod_{\mu \in \mathscr{N}} c_{\mathbf{x}+\mu}^{(\mu)}\right).$$

The local function works only with what remains and what is received. Only partial information is accessible to a cell, even about its own state as depicted in Fig. 1c.

Equivalently, each state is the product of the information to be exchanged. Each component is sent to a single cell. An intermediate state is formed by grouping what is left and what is received. The state function Φ yields the new state from the intermediate state. The cell only keeps a partial knowledge about its own state and only receives a partial knowledge about the states of the neighboring cells, as depicted in Fig. 1(c).

A PCA is indeed a CA: the formalization only prevents it from accessing the full states of its neighbors.

A *space-time diagram* $\mathbb{A} : \mathbb{L} \times \mathbb{N} \rightarrow Q$ is the sequence of the iterated images of a configuration by a CA $\mathscr{A}$ from an initial configuration c_0. It is defined by $\mathbb{A}_{\mathbf{x},t} = \left(\mathscr{G}^t(c_0)\right)_{\mathbf{x}}$ and denoted by $(\mathscr{G}, c_0)$ or $(\mathscr{A}, c_0)$.

2.4 Reversibility

A CA (resp. BCA, PCA) is *reversible* if and only if its global function $\mathscr{G}$ is bijective and $\mathscr{G}^{-1}$ is the global function of some CA (BCA, PCA). Let R-CA (R-PCA, R-BCA) denote the class of reversible CA (BCA, PCA). Myhill [29] and Moore [23] proved that for CA injectivity is equivalent to reversibility. The main decidability result is:

Theorem 3 *(Amoroso & Patt, Kari) The reversibility of CA is decidable in dimension 1 [2] but it is undecidable for higher dimension [14, 15].*

Whereas for BCA and PCA, the following lemmas hold in any dimension.

Lemma 4 (Margolus) *A BCA is reversible iff its block function t is a permutation (which is decidable).*

Proof If the block function t is a permutation, by construction, any block transition is reversible. The global transition as a composition of transitions, is reversible. Otherwise, t is not one-to-one, then neither is any transition, and neither is the global transition.

Decidability comes from the finiteness of the domain of t.   □

Lemma 5 (Morita) *A PCA is reversible iff its state function Φ is a permutation (which is decidable).*

Proof If Φ is a permutation, then the inverse is obtained by reversing Φ and then sending back the pieces to corresponding neighbors. Otherwise, since Φ works on a finite set, it is not one-to-one and it is easy to construct 2 configurations which have the same image.

Decidability comes from the finiteness of the domain of Φ.   □

The inverse PCA is not presented since the proof only assert that, as a CA it is reversible. The inverse is $\left(\prod_{\mu \in -\mathcal{N}} Q^{(-\mu)}, -\mathcal{N}, \Phi^{-1} \right)$ where the state function is computed *before* the sub-states are exchanged. If need, the constructions in the next section can be used to provide the expression of the inverse as a PCA.

As far as reversibility is concerned, BCA and PCA fundamentally differ from CA. It is known that bijectivity for CA is equivalent to reversibility [13, 32] and that there exists CA that are surjective but not reversible. By a local inspection, it is easy to prove that for any surjective BCA or PCA the block or state function must be a permutation.

2.5 Simulation and Intrinsic Universality

The local updating process differs for the various kind of CA. Thus simulation has to be defined at global level. To simplify, the definition is presented in 2 steps. The first one does not allow any shift nor scaling. The second introduces them.

Definition 6 (*Direct simulation*) $\mathcal{A}$ *is directly simulated by $\mathcal{B}$ if there is some onto partial function α from $Q_{\mathcal{B}}$ to $Q_{\mathcal{A}}$ such that:*

$$\forall c \in \mathscr{C}_{\mathscr{A}}, \alpha \circ \mathscr{G}_{\mathscr{B}} \circ \alpha^{-1}(c) = \{\mathscr{G}_{\mathscr{A}}(c)\} \ .$$

This is denoted by $\mathscr{A} \preccurlyeq \mathscr{B}$ or when dealing with functions by $\mathscr{G}_{\mathscr{A}} \preccurlyeq \mathscr{G}_{\mathscr{B}}$.

In this definition, space-time diagrams must match exactly. Any computation that is started on a $\mathscr{B}$-configuration that maps to the initial $\mathscr{A}$-configuration (there exist at least one since α is onto) generates the whole space-time diagram.

The next definition adds scaling and shifting. Let $\mathbf{m}$ be any element of $\mathbb{L}$ with positive coordinates. The $\mathbf{m}$-packing, $p_{\mathbf{m}}$, is the bijective mapping from $Q^{\mathbb{L}}$ to $(Q^{\mathbf{m}})^{\mathbb{L}}$ that correspond to identifying the blocks of the $\mathbf{0}$-partition with cells.

Definition 7 (*Simulation*) $\mathscr{A}$ *is simulated* by $\mathscr{B}$ if there are a positive vector $\mathbf{m}$, an integer τ and a vector $\mathbf{s}$ such that:

$$\mathscr{G}_{\mathscr{A}} \preccurlyeq p_{\mathbf{m}} \circ \mathscr{G}_{\mathscr{B}}^{\tau} \circ p_{\mathbf{m}}^{-1} \circ \sigma_{\mathbf{s}} \ .$$

This is denoted by $\mathscr{A} \lll \mathscr{B}$ or when dealing with functions by $\mathscr{G}_{\mathscr{A}} \lll \mathscr{G}_{\mathscr{B}}$.

Unpacking is used so that the direct simulation works with macro-cells, i.e. blocks. This is used directly in Sect. 3.1 as an example. As the chapter goes, the different elements of the simulation are more and more implicit.

From the definition of simulation comes the following definition:

Definition 8 An XCA is *intrinsically universal* if it can simulate any XCA.

3 Simulations Between Classes of CA

PCA (resp. R-PCA) are CA (resp. R-CA). The remaining simulations have to be expressed or generated by transitivity.

3.1 Simulation of BCA by CA (and R-BCA by R-CA)

Theorem 9 *Any d-BCA can be simulated by a d-CA. This simulation preserves reversibility.*

Proof Let $\mathscr{A} = (Q_{\mathscr{A}}, \mathbf{v}, n, (\mathbf{o}^{(j)})_{1 \le j \le n}, t)$ be any BCA. The set of states of the CA $\mathscr{B}$ is the set of blocks of $\mathscr{A}$ (i.e. $Q_{\mathscr{A}}^{V}$). The cells represent the blocks of the first partition. The traces of the partitions are identical in every cell. The neighborhood is $\mathscr{N} = [\![-n, n]\!]^{d}$ (n is the number of partitions). The cells in $\mathscr{N}$ correspond to the blocks $[\![-n, n]\!]^{d}$ of the first partition centered on the cell.

The local function $f : Q^{(2n+1)^{d}} \to Q$ makes the first block transition on an hypercube of size $(2n)^{d}$. It contains the blocks $[\![-n+1, n-1]\!]^{d}$ of the first partition

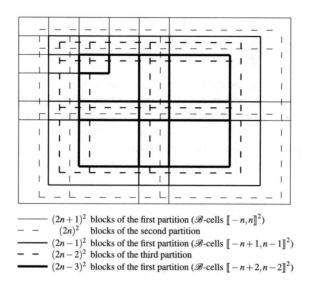

———— $(2n+1)^2$ blocks of the first partition ($\mathscr{B}$-cells $[\![-n,n]\!]^2$)
– – $(2n)^2$ blocks of the second partition
———— $(2n-1)^2$ blocks of the first partition ($\mathscr{B}$-cells $[\![-n+1,n-1]\!]^2$)
– – $(2n-2)^2$ blocks of the third partition
━━━ $(2n-3)^2$ blocks of the first partition ($\mathscr{B}$-cells $[\![-n+2,n-2]\!]^2$)

Fig. 3 First and second cuttings

centered on the cell. Then f makes the second block transition on the $(2(n-1))^d$ blocks containing the blocks $[\![-n+2,n-2]\!]^d[\![-n+2,n-2]\!]^d$ of the first partition centered on the cell. Each subsequent block partition is applied on a smaller part but the central block remains in the middle.

The values corresponding to the updated cells are taken as the image by f of the whole neighborhood. Different partitions in dimension 2 are shown in Fig. 3. All the block partitions are considered in one application of f. In one step of $\mathscr{B}$, the images of the cells in the block are computed.

To be formal with Definition 7 (simulation): $p_{\mathbf{v}}$ is grouping by block, τ is 1, $\mathbf{s}$ is the shift of the first partition and α (for the direct simulation) is the identity. With this simple encoding and this f, there is a natural identification between $\mathscr{A}$ and $\mathscr{B}$: $\mathscr{G}_{\mathscr{A}} = p_{\mathbf{v}} \circ \mathscr{G}_{\mathscr{B}}^{\tau} \circ p_{\mathbf{v}}^{-1} \circ \sigma_{\mathbf{s}}$. Since $p_{\mathbf{v}}$ and σ are reversible, $\mathscr{G}_{\mathscr{B}}$ is reversible if $\mathscr{G}_{\mathscr{A}}$ is. Reversibility is preserved. $\qquad\square$

The size of the blocks defines the number of states of the CA while the radius only depends on the number of partitions.

3.2 Simulation of CA by PCA

Proposition 10 *Any d-CA can be directly simulated by a d-PCA.*

Proof The idea is to duplicate the states in every part. Let $\mathscr{A} = (Q, \mathcal{N}, f)$ be any d-CA. It is simulated by the following d-PCA: $\mathscr{B} = (Q^{\mathcal{N}}, \mathcal{N}, \Phi)$, with:

$$\forall v \in \mathcal{N}, \ \Phi \left(\prod_{\mu \in \mathcal{N}} s_\mu^{(\mu)} \right)^{(v)} = f \left(\left(s_\mu^{(\mu)} \right)_{\mu \in \mathcal{N}} \right) .$$

The onto partial function α is defined by: $\forall s \in Q, \alpha(s^{\mathcal{N}}) = s$. It is undefined for every other value in $Q^{\mathcal{N}}$. ☐

Since Φ only maps onto the diagonal of $Q^{\mathcal{N}}$, the simulating PCA is never reversible.

3.3   Simulation of CA by BCA

Theorem 11 *Any d-CA can be directly simulated by a d-BCA.*

Proof Let $\mathscr{A} = (Q, \mathcal{N}, f)$ be a CA and r be its *radius*: the maximum absolute coordinate of the elements of $\mathcal{N}$. It represents half the size of the "windows" required to gather the information needed to update a cell.

Let $\mathscr{B}$ be the following BCA: $(Q \cup Q^2, (4r, \cdots , 4r), d^2, \{0, 2r\}^d, t)$. The order of the partitions is irrelevant as shows the definition of t below. The state of a $\mathscr{B}$-cell represents either the previous state of a $\mathscr{A}$-cell ($\in Q$) or its previous and next states ($\in Q^2$). The previous configuration is preserved until the next configuration is completely generated.

In each block a part is singled out: the core. It correspond to $[\![r, 3r-1]\!]^d$: the cells that have their full neighborhood in the block. The block function first adds to every cell in the core its next state and then, if all cell in the block have states in Q^2, all previous states are removed and the configuration is ready for next iteration. The choice of block size and partitions ensures that every cell is in the core of exactly one partition.

The onto partial function α is defined by the identity on Q and is undefined on Q^2. ☐

This simulating BCA is not reversible because of the erasement of the previous state.

3.4   Simulation of R-CA by R-BCA

To preserve reversibility, erasing is done progressively. The inverse d-R-CA of $\mathscr{A}$, $\mathscr{A}^{-1}$, is used to impose a condition (using $f_{\mathscr{A}^{-1}}$) to erase injectively.

The inverse CA can be computed: the CA can be effectively enumerated, composition of CA as well as identity test are computable. The algorithm stops in finite time; but in any dimension greater than one this time cannot be bounded by any computable function because of the undecidability of reversibility.

The inverse of a R-BCA is very simple to built as described in the proof of Lemma 4. When simulating a R-CA, the simulation of it inverse is somehow built.

Theorem 12 *Any d-R-CA $\mathscr{A}$ can be simulated by a d-R-BCA $\mathscr{B}$ with states $Q \cup Q^2$, size $3(d+1)\mathbf{r}$ and $d+1$ partitions. The origins of the partitions are:* $\mathbf{0}$, $3\mathbf{r}, 6\mathbf{r}, 9\mathbf{r}, \ldots, 3d\mathbf{r}$ *where* $\mathbf{r}$ *is the vector* $(r, r, \cdots, r)$ *and r is the radius of the CA.*

In the construction, the state of a $\mathscr{B}$-cell represents either the previous or next state of a $\mathscr{A}$-cell ($\in Q$) or both its previous and next states ($\in Q^2$). Before proving the theorem, some lemmas are provided to ensure the distinction between previous and next state for single-state $\mathscr{B}$-cell (i.e.in Q).

The following subsets of $\mathbb{Z}$ and $\mathbb{Z}^d$ are used to locate previous and next states during the iterations. For every θ in $[\![0, d+1]\!]$ and $\kappa \in [\![0, d]\!]$, let

$$F_\kappa = 3\kappa r + [\![r, 3(d+1)r - r - 1]\!] + (3(d+1)r)\,\mathbb{Z} ,$$

$$\overline{F_\kappa} = 3\kappa r + [\![-r, r-1]\!] + (3(d+1)r)\,\mathbb{Z} ,$$

$$F_\kappa^d = 3\kappa\mathbf{r} + [\![r, 3(d+1)r - r - 1]\!]^d + (3(d+1)r)\,\mathbb{Z}^d ,$$

$$E_\theta^{\mathsf{P}} = \bigcup_{\theta \leq \kappa < d+1} F_\kappa^d \qquad \text{and}$$

$$E_\theta^{\mathsf{N}} = \bigcup_{0 \leq \kappa < \theta} F_\kappa^d .$$

These sets are closed under all $\pm 3(d+1)r$ shifts in every direction. This is not always indicated to ease the presentation.

Lemma 13 *For any $0 \leq \theta \leq d+1$, these sets verify the symmetries $E_\theta^{\mathsf{N}} = E_{d+1-\theta}^{\mathsf{P}} - 3(d+1-\theta)\mathbf{r}$ and $E_\theta^{\mathsf{N}} = -3d\mathbf{r} - E_{d+1-\theta}^{\mathsf{P}} - \mathbf{1}$ and the equalities: $E_{d+1}^{\mathsf{P}} = E_0^{\mathsf{N}} = \emptyset$ and $E_0^{\mathsf{P}} = E_{d+1}^{\mathsf{N}} = E_\theta^{\mathsf{P}} \cup E_\theta^{\mathsf{N}} = \mathbb{Z}^d$.*

Proof The symmetries, the equality with $\emptyset$ and $E_0^{\mathsf{P}} = E_{d+1}^{\mathsf{N}} = E_\theta^{\mathsf{P}} \cup E_\theta^{\mathsf{N}}$ are obvious.

It remains to prove that $E_{d+1}^{\mathsf{N}} = \mathbb{Z}^d$. Let $\mathbf{x}$ be any element of $\mathbb{Z}^d$. The $d+1$ sets (of $\mathbb{Z}$) $\overline{F_\theta}$ are non-empty and disjoint. Since $\mathbf{x}$ has d coordinates, there exists θ_0 such that none of the coordinates of $\mathbf{x}$ belongs to $\overline{F_{\theta_0}}$. This means that $\mathbf{x}$ belongs to $F_{\theta_0}^d$, thus to E_{d+1}^{N}. $\qquad\qquad\square$

Let $\bowtie$ be the operator that returns the tuple of defined operand (it returns a pair, a value or undefined). Let following configurations over the alphabet $Q \cup Q^2$ are defined:

$$\forall c \in \mathscr{C}, \ \forall \theta \in [\![0, d+1]\!], \quad \mathscr{E}_\theta(c) = c_{|E_\theta^{\mathsf{P}}} \bowtie \mathscr{G}(c)_{|E_\theta^{\mathsf{N}}} .$$

Lemma 13 implies that: $\mathscr{E}_0(c) = c$, $\mathscr{E}_{d+1}(c) = \mathscr{G}(c)$ and that, for all θ, $\mathscr{E}_\theta(c)$ is everywhere defined. In the following $d+1$ reversible block transitions, $(B_\theta)_{0 \leq \theta < d+1}$,

Fig. 4 Simulation commuting diagram

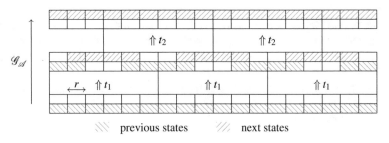

$\diagdown\diagdown$ previous states $\diagup\diagup$ next states

Fig. 5 The 2 steps in dimension 1

are defined so that the diagram in Fig. 4 commutes. Then their block functions are proven compatible to be merge into one reversible block function.

Let B_θ be a block transition of size $3(d+1)\mathbf{r}$ and origin $3\theta\mathbf{r}$. The size of B_θ matches the length of the shift closure of the sets E_θ^P and E_θ^N. The 2 partitions for dimension 1 are given in Fig. 5. (They correspond to the construction of Kari [16].)

The block functions t_θ have to be defined so that B_θ maps reversiblely $\mathscr{E}_\theta(c)$ into $\mathscr{E}_{\theta+1}(c)$. Each one adds next states and erases previous states. The following lemma states that there is enough information to compute and add the next states.

Lemma 14 *For any θ in $[\![0, d]\!]$, there is enough information in $\mathscr{E}_\theta(c)$ to compute $\mathscr{E}_{\theta+1}(c)$ in each block of the partition of B_θ.*

Proof The next states added belong to:

$$\Delta_\theta = E_{\theta+1}^N \setminus E_\theta^N$$
$$= \left(3\theta\mathbf{r} + [\![r, 3(d+1)r-r-1]\!]^d \right) \setminus \bigcup_{0 \le \kappa < \theta} \left(3\kappa\mathbf{r} + [\![r, 3(d+1)r-r-1]\!]^d \right) .$$

For any $\mathbf{x} \in \Delta_\theta, \mathbf{x} \in 3\theta\mathbf{r}+[\![r, 3(d+1)r-r-1]\!]^d$. All the cells of the neighborhood of $\mathbf{x}$ should still hold their previous states in order to compute the next state of $\mathbf{x}$. Cell $\mathbf{x}$ and its neighbors are all in the block $3\theta\mathbf{r}+[\![0, 3(d+1)r-1]\!]^d$. It corresponds to the block of the partition of B_θ since its origin is $3\theta\mathbf{r}$. It remains to verify that the previous states needed to compute the next state of $\mathbf{x}$ are still present.

For any κ in $[\![0, \theta-1]\!]$, since $\mathbf{x} \notin 3\kappa\mathbf{r} + [\![r, 3(d+1)r-r-1]\!]^d$, there is some index j_κ such that $\mathbf{x}_{j_\kappa} \notin 3\kappa r + [\![r, 3(d+1)r-r-1]\!]$. So $\mathbf{x}_{j_\kappa}$ is in $3\kappa r + [\![-r, r-1]\!]$ (all is $3(d+1)r$ periodic following any direction). Since the sets $3\kappa r + [\![-r, r-1]\!]$ are disjoint, all j_κ must be different and there are θ of them.

Let $\mathbf{y}$ be any cell needed to compute the next value in $\mathbf{x}$. It belongs to $\mathbf{x} + [\![-r, r]\!]^d$, then, for all κ in $[\![0, d-1]\!]$, $\mathbf{y}_{j_\kappa}$ must be in $3\kappa r + [\![-2r, 2r-1]\!]$. By contradiction, let us assume that there exists such a $\mathbf{y}$ which does not belong to E_θ^P then for all $\lambda \in [\![\theta, d+1]\!]$, there exists some k_λ such that $\mathbf{y}_{k_\lambda}$ does not belong to $\lambda r + [\![r, 3(d+1)r-r-1]\!]$, or equivalently, $\mathbf{y}_{k_\lambda} \in 3\lambda r + [\![-r, r-1]\!]$. Since the sets $3\lambda r + [\![-r, r-1]\!]$ are disjoint, all the k_λ must be different and there are $d + 1 - \theta$ of them.

Altogether, there are $d+1$ (distinct) j_κ and (distinct) k_λ for d values so there exist κ_0 and λ_0 such that $j_{\kappa_0} = k_{\lambda_0}$. Then the intersection of $3\kappa_0 r + [\![-2r, 2r-1]\!]$ and $3\lambda_0 r + [\![-r, r-1]\!]$ is not empty. This means that $\kappa_0 = \lambda_0$, but by construction, $\kappa_0 < \lambda_0$.

Thus $\mathbf{y}$ belongs to E_θ^P and all the previous states needed to compute the next state of $\mathbf{x}$ are still present in the block. The next state of $\mathbf{x}$ can be computed with the information held inside the block.   □

From the symmetry between E^N and E^P, follows:

Corollary 15 *For any θ in $[\![0, d]\!]$, there is enough information in $\mathcal{E}_{\theta+1}(c)$ to compute $\mathcal{E}_\theta(c)$ in each block of the partition of B_θ.*

This means that the corresponding blocks of $\mathcal{E}_\theta(c)$ and $\mathcal{E}_{\theta+1}(c)$ in the partition of B_θ can be uniquely determined one from the other. The partial function t_θ is one-to-one.

In dimension 2, the partitions are given in Fig. 6a and the positions of previous and next states are detailed in Fig. 6b. (They do not correspond to the construction of Kari any more.)

The following lemma shows that the partial definitions of functions t_θ are compatible so that they can be merged into a unique t to define a reversible BCA.

Lemma 16 *The current block transition B_θ can be identified by the position of the double states inside the blocks of partition.*

Proof If all cells are single, then $\theta = 0$.

For κ in $[\![1, d]\!]$, let ε_κ be the following vector inside the blocks:

$$\varepsilon_1 = 3(d+1)\mathbf{r} + (-3r, \cdots, -3r) ,$$
$$\varepsilon_2 = 3(d+1)\mathbf{r} + (-3r, -6r, \cdots, -6r) ,$$
$$\varepsilon_\kappa = 3(d+1)\mathbf{r} + (-3r, -6r, -9r, \cdots, -3(\kappa-1)r, -3\kappa r, \cdots, -3\kappa r) ,$$
$$\varepsilon_d = 3(d+1)\mathbf{r} + (-3r, -6r, -9r, \cdots, -3dr) .$$

To get the coordinate in $\mathbb{L}$, a translation by $3\theta\mathbf{r}$ have to be applied. The vectors in dimension 2 are indicated in Fig. 6b.

For κ in $[\![1, d]\!]$, no coordinate of ε_κ belongs to $[\![-r, r-1]\!]$, so that $\varepsilon_\kappa + 3\theta\mathbf{r}$ belongs to F_θ and thus to E_θ^P.

For all κ in $[\![1, \theta-1]\!]$, no coordinate of ε_κ belongs in $\overline{F_0}$, so that ε_κ belongs to F_0^d and thus to E_θ^N.

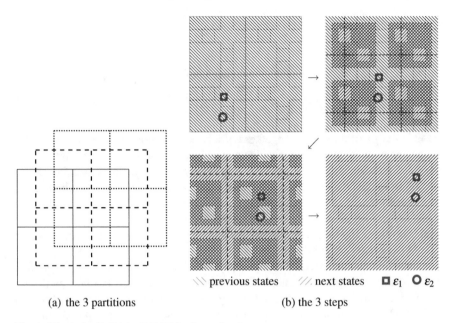

(a) the 3 partitions   (b) the 3 steps

`\\\` previous states   `///` next states   ▫ ε_1 ◯ ε_2

Fig. 6 Simulating R-CA by R-BCA in dimension 2

For all κ in $[\![1, d]\!]$, the coordinate value $-3ir + 3\theta r = 3(\theta - \kappa)r$ prevents ε_κ from being in $F_{\theta-\kappa}$. So that ε_θ does not belong to $F_{\theta-1}^d \cup F_{\theta-2}^d \cup \cdots \cup F_0^d = E_\theta^N$.

Altogether θ is the maximum κ such that ε_κ holds 2 states plus one. If there is no such κ then $\theta = 1$.   □

Corollary 17 *Thanks to the symmetry (Lemma 13), the positions of double states after the block transition indicate which t_θ was used.*

Above Lemma and Corollary show that all the partial definitions of the block permutations of the block transitions are compatible for domains and ranges. They can be grouped and completed in a unique bijective block function.

Altogether, Theorem 12 is proved.

3.5   Simulation of R-BCA (and R-CA) by R-PCA

Lemma 18 *Any d-BCA can be simulated by a d-PCA. This simulation preserves reversibility.*

Proof The idea is to identify cells with blocks. Let $\mathscr{A} = (Q_\mathscr{A}, \mathbf{v}, n, (\mathbf{o}^{(j)})_{1 \le j \le n}, t)$ be any d-BCA. Let $\mathscr{B} = \left(\prod_{\mu \in \mathscr{N}} Q_\mathscr{B}^{(\mu)}, \mathscr{N}, \Phi \right)$ be a PCA where $\mathscr{N} = \{-1, 0, 1\}^d$, i.e., coordinates which differ by at most one in any direction. The block of coordinates $\mathbf{x}$ (at block scale) of the jth partition is $\rho_\mathbf{x}^j$. The block $\rho_\mathbf{0}^j$ holds the cell of coordinates $\mathbf{0}$. The sets of states are defined by:

$$Q_{\mathscr{B}}^{(0)} = \bigcup_{1 \leq j \leq n} \left(\{j\} \times Q_{\mathscr{A}}^{\left(\rho_0^{j-1} \cap \rho_0^j \right)} \right) \text{ , and}$$

$$\forall \mu \in \mathscr{N}, \mu \neq \mathbf{0}, \ Q_{\mathscr{B}}^{(\mu)} = \bigcup_{1 \leq j \leq n} Q_{\mathscr{A}}^{\left(\rho_0^{j-1} \cap \rho_{-\mu}^j \right)} .$$

It holds the partition number together with the intersection of the block that holds the cell of coordinates $\mathbf{0}$ for a partitions and of the one of the next partition holding the cell $\mathbf{0}$ translated by $\mu.\mathbf{v}$. Any intersection may be empty. Blocks are partitioned according to the next partition so that every part is sent to the corresponding cell to form whole blocks of the next partition. Identically, each cell retrieves a full block, uses the local transition and sends the corresponding parts to the neighbors for the next transition. The $\mathbf{0}$-sub-state identifies the partition number which indicates how to split the image block into sub-states.

Configurations are encoded by setting the first components to 1 and by putting states in the corresponding intersections between the last and the first partitions. On the first iteration of $\mathscr{B}$, each cell gets one entire block of the first partition and make the first transition. Then all pieces are sent to the corresponding cells and 2 is recorded in the cell. Each iteration of $\mathscr{B}$ makes a successive transition of $\mathscr{A}$. After n iterations of $\mathscr{B}$, one iteration of $\mathscr{A}$ is made and the first component is 1 again.

This construction preserves reversibility: the partial definition of the PCA state function Φ is one-to-one if the local transition t of the BCA is reversible.   □

From above Lemma and the transitivity of simulation comes:

Theorem 19 *Any d-R-CA can be simulated by a d-R-PCA.*

4 Intrinsic Universality of 1-R-PCA

In this section, a 1-R-PCA $\mathscr{U} = (Q_{\mathscr{U}}, \{-1, 0, 1\}, \Phi_{\mathscr{U}})$ is built such that:

Theorem 20 *The 1-R-PCA $\mathscr{U}$ is intrinsically universal, i.e., able to simulate any 1-R-PCA.*

Let $\mathscr{A} = (Q, \mathscr{N}, \Phi)$ be any 1-R-PCA. With cells grouping, $\mathscr{A}$ can be simulated by a 1-R-PCA with neighborhood $\{-1, 0, 1\}$. From now on, $\mathscr{N} = \{-1, 0, 1\}$.

The construction is first done at macroscopic level (macro-cells at $\mathscr{A}$ scale) to show the reversible process. Then at the microscopic level (at $\mathscr{U}$ scale), the steps of the process are detailed. Macro-cells as well as $\mathscr{U}$-cells are products of different layers. The states and the local function $\Phi_{\mathscr{U}}$ are defined in Tables 1 and 3.

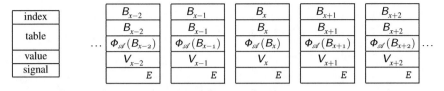

Fig. 7 Initial configuration encoding at $\mathscr{A}$-cell level

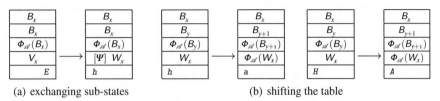

(a) exchanging sub-states   (b) shifting the table

[Ψ] means -1 and 1 parts exchanged with adjacent cells.

Fig. 8 Begining of cycle and table shifting at $\mathscr{A}$-cell level

4.1 Macroscopic Level

Let B_x be the xth element of Q modulo $|Q|$. An $\mathscr{A}$-configuration is encoded by macro-cell as in Fig. 7 in 4 layers: an index to identify the cell in the loop, an entry of the table of $\Phi_{\mathscr{A}}$ (with the same id), the $\mathscr{A}$-state and a signal and mode (lower/upper case) to know the current step of the simulation. The initial configuration presented in Fig. 7 extends infinitely on each sides. The value is denoted by V_x at the beginning and W_x after exchanging parts with neighbors (and $\Phi_{\mathscr{A}}(W_x)$ after updating).

From PCA definition, all $\mathscr{A}$-cells first exchange their -1 and -1 parts and the signal changes from E on right to h on left to denote this (the rule in Fig. 8a). The mode changes from uppercase to lowercase.

The inner loop of the simulation starts then. First the layers holding B_y and $\Phi_{\mathscr{A}}(B_y)$ are shift to the left with the rules in Fig. 8a. The mode is preserved.

If the mode is lowercase (signal a) then if W_x to B_y are equal, W_x is replaced by $\Phi_{\mathscr{A}}(B_y)$ and the signals turn to uppercase (the rules in Fig. 9a). If the mode is an uppercase signal (A) then it ensures that $\Phi_{\mathscr{A}}(B_y)$ to B_y are different (the rules in Fig. 9b). Finally, B_x and B_y are check for equality to know whether the loop is ended (the bottom rules in Fig. 9a, b).

4.2 States, Layers and Configurations at Microscopic Level

The $\mathscr{U}$-cells are organized in 10 layers as detailed in Table 1. Architecture layer (A) holds delimiters for the $\mathscr{A}$-cells ([and]) and for the -1, 0 and 1 parts ($). Layer I holds an index to store where the reading of the table started. Layers B and F hold one

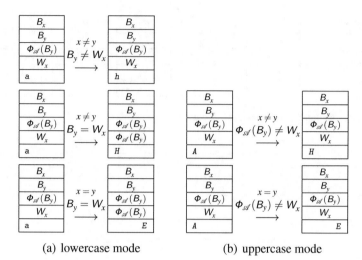

(a) lowercase mode   (b) uppercase mode

Fig. 9 Update inside the loop at $\mathscr{A}$-cell level

Table 1 The 10 layers and corresponding sub-states

Layer	Name	States			Use
		-1	0	1	
1	A			_[$]	Architecture: limits of cells and parts
2	I		01		$\mathscr{A}$-cell identification B_x
3	B		01		Table entry B_y
4	F		01		Image of the table entry B_y, $\Phi_{\mathscr{A}}(B_y)$
5	V		01		Value of the $\mathscr{A}$-cell (V_x or W_x)
6	S	Σ	Σ	Σ	Control signals as detailed in Tab. 2
7–10	L$_1$–L$_4$	01		01	Shift the table of Φ and exchange values (W_x^{-1} & W_x^{1})

entry of the table B_y and its image $\Phi_{\mathscr{A}}(B_y)$. The value of the $\mathscr{A}$-cell (V_x or W_x) is stocked on layer V. Signals are found on layer S. Layers L$_1$ to L$_4$ work like conveyor belts to transfer data. The values in layers A and I never change.

Capital (B, $\Phi_{\mathscr{A}}(B)$, W) are used to address the macroscopic level ($\mathscr{A}$-cells) and small symbols (i, b, f, v) for microscopic level ($\mathscr{U}$-cells). All $\mathscr{A}$-cells are binary encoded. For the exchange, the codes of -1 and 1 parts must have the same length (0s are added if necessary).

The signals are 23 symbols typed in this police as described in Table 2. Uppercase and lowercase signals behave similarly except for the table testing. This mode distinguishes between before and after the replacement. During the simulation, signals are turned from uppercase to lowercase when parts are exchanged and back to uppercase when the value is replaced by its image.

Table 2 Signals of $\mathcal{U}$

Lowercase	Uppercase	Use
a-h		Loop which tests if $W_x = B_y$ and $B_x = B_y$
	A-H	Loop which tests if $W_x = \Phi_{\mathcal{A}}(B_y)$ and $B_x = B_y$
k		Write $\Phi_{\mathcal{A}}(B_y)$ over W_x
m, n	M, N	Shift of the table: B_y and $\Phi_{\mathcal{A}}(B_y)$
	S, T	Exchange of Parts W_x^{-1} and W_x^1

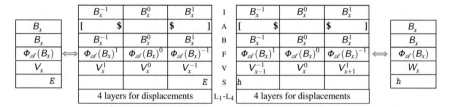

Fig. 10 Encoding of $\mathcal{A}$-cell at coordinate x, before and after the exchange

The encoding of $\mathcal{A}$-cells is given in Fig. 10. It takes care of the particular positions of the -1 and 1 parts from the beginning. Since V_x^{-1} is exchanged with V_{x+1}^1 (V_x^1 with V_{x-1}^{-1}), B_x^1 (B_x^{-1}) should be above it. When $\Phi(W_x)$ replaces W_x, the -1 and 1 parts are directly on the corresponding sides.

4.3 Microscopic Algorithm

It is defined by space-time diagrams driven by signals. Signals in the different $\mathcal{A}$-cells are always exactly synchronized. The duration is the same whether or not a test succeed or fail. The end of loop case is shorter but the test is uniformly satisfied (or not). All the rules for lowercase signals are indicated in Table 3; the rules for the uppercase are similar. The algorithm starts with E signal arriving in the rightmost $\mathcal{U}$-cell of each $\mathcal{A}$-cell.

First, the -1 and 1 parts of the $\mathcal{A}$-cell are exchanged and the signal is switched to h as depicted in Fig. 11. The initial value of the $\mathcal{A}$-Cell is V_x. The bits of V_x^{-1} and V_{x+1}^1 are swapped on the layer L_1 by signals S and T on they way from]. On crossing], the flows are transferred on L_2 (to avoid superposition as explained below). Signals S and T turn back at \$ and on their way back, they retrieve the bits from L_2 and put them in their destination slot with another swap. Synchronization is very important. Signals S and T finally get back together as h at] (switching the mode) and go back to the left end of the $\mathcal{A}$-cell. This is implemented with 11 rules in Table 3.

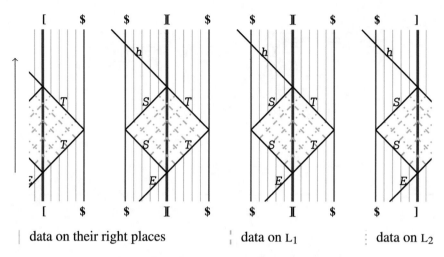

| data on their right places | data on L_1 | data on L_2 |

Fig. 11 Exchanging V_x^{-1} and V_{x+1}^1 to realize the rule in Fig. 8a

Signal h crosses the $\mathscr{A}$-cell and asserts that $B_x = B_y$ (for reversibility). On arriving at [, h splits into m and n. These signals manage the shift of the table by one $\mathscr{A}$-cell rightward using layers L_1 to L_4 as illustrated in Fig. 12. Signal m sets B_{y-1} and $\Phi_{\mathscr{A}}(B_{y-1})$ on movement by swapping then on layers L_1 and L_3. On passing], bits go down a layer so as not to interfere with the moving ones of the next $\mathscr{A}$-cell. On its way back, n sets B_{y-1} and $\Phi_{\mathscr{A}}(B_{y-1})$ on their final places by swapping them from layers L_2 and L_4. Signals m and n gather and form a which starts the test part of the loop. This corresponds to the last 12 rules in Table 3.

The test part of the loop works as follows for the lowercase mode. Value W_x and table entry B_y are in place, bit below bit, to be compared. Signal a crosses the whole $\mathscr{A}$-cell to compare them. If they differ, a marker b is put on the first different bit, and a turns to b. On the way back, b marks d the last bit which differs and collects the marker b back (first column in Fig. 13). If W_x and B_y are equal, the signal reaches [as a, turns to k, writes $\Phi_{\mathscr{A}}(B_y)$ over W_x on the way back and switch mode (second column in Fig. 13). This special behavior takes as much time as the regular one, keeping the synchronization. Equality (with $\Phi_{\mathscr{A}}(B_y)$) is tested on the way back for reversibility: going backward in time, $\mathscr{U}$ must make the correct change at the adequate time, so it needs this as well as inequality for the rest of the iterations (last two columns in Fig. 13).

In uppercase mode, it is exactly the same except that the equality is tested between W_x and $\Phi_{\mathscr{A}}(B_y)$ instead of B_y, and B_y as a requirement. Since $\mathscr{A}$ is reversible, each value of $\Phi_{\mathscr{A}}(B_y)$ appears once and only once in the table. After being copied, $\Phi_{\mathscr{A}}(B_y)$ is never met again.

To know that the table was completely scanned, signal must test whether B_x and B_y are equal. On the second left to right crossing, signal d (or e) gets back the previous marker (if any) and turn to g and marks g the first different bit between

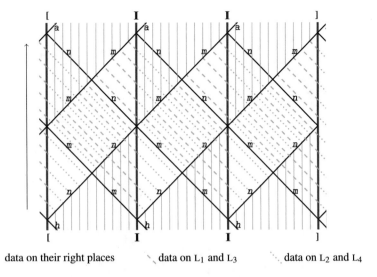

| data on their right places   ╲ data on L_1 and L_3 ⋰ data on L_2 and L_4

Fig. 12 Shifting the table to realize the rule in Fig. 8b

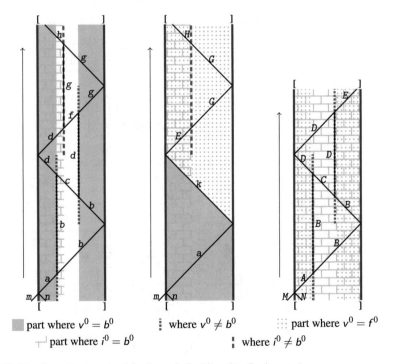

▨ part where $v^0 = b^0$ ┆ where $v^0 \neq b^0$ ⋮ part where $v^0 = f^0$
⌐ part where $i^0 = b^0$ ┇ where $i^0 \neq b^0$

Fig. 13 Test for replacement and for the end of $\mathscr{A}$-iteration (basic cases)

Table 3 Table of $\Phi_{\mathscr{U}}$

		Value				Image		Modification
	Structure	Signals		Condition		Signals		Modification
*	[		a	$v^0 = b^0$			a	
				$v^0 \neq b^0$		b	b	
*	' ',$	a		$v^0 = b^0$			a	
				$v^0 \neq b^0$		b	b	
*	]	a		$v^0 = b^0$			k	$v^0 \leftarrow f^0$
				$v^0 \neq b^0$		d	d	
	' ',$	b					b	
	' ',$		b				b	
*	]	b		$v^0 = b^0$		b		
				$v^0 \neq b^0$		c	d	
*	' ',$		b	$v^0 = b^0$		b		
				$v^0 \neq b^0$		c	d	
*	' ',$	b	b	$v^0 \neq b^0$		d	d	
*	[	b	b	$v^0 \neq b^0$	$i^0 = b^0$		e	
					$i^0 \neq b^0$	g	g	
*	' ',$	b	c	$v^0 \neq b^0$		d		
*	[	b	c	$v^0 \neq b^0$	$i^0 = b^0$		d	
					$i^0 \neq b^0$	g	f	
	' ',$		c			c		
*	' ',$		d	$v^0 = b^0$		d		
*	[		d	$v^0 = b^0$	$i^0 = b^0$		d	
					$i^0 \neq b^0$	g	f	
*	' ',$	d		$v^0 = b^0$	$i^0 = b^0$		d	
					$i^0 \neq b^0$	g	f	
*	' ',$	d	d	$v^0 \neq b^0$	$i^0 = b^0$		e	
					$i^0 \neq b^0$	g	g	
*	]	d	d	$v^0 \neq b^0$	$i^0 = b^0$	S	T	
					$i^0 \neq b^0$	h		
*	' ',$	e		$v^0 = b^0$	$i^0 = b^0$		e	
					$i^0 \neq b^0$	g	g	
*	]	E		$v^0 = b^0$	$i^0 = b^0$	S	T	
					$i^0 \neq b^0$	H		
	' ',$	f					f	
*	' ',$	f	d	$v^0 \neq b^0$			g	
*	]	f	d	$v^0 \neq b^0$		g		
*	' ',$	g		$v^0 = b^0$			g	
*	]	g		$v^0 = b^0$		g		
	[, ' ',$	g					g	
	' ',$		g			g		
	' ',$	g	g		$i^0 \neq b^0$	h		
	[	g	g		$i^0 \neq b^0$	m	n	
	' ',$		h		$i^0 = b^0$	h		
	[		h		$i^0 = b^0$	m	n	
	]	S				S		swap(v^0, l_1^1)
	' '		S			S		swap(v^0, l_1^1)
	$		S				S	swap(v^0, l_1^1)
	$	S					S	swap(v^0, l_2^1)
	' '	S					S	swap(v^0, l_2^1)

(continued)

Table 3 (continued)

	Value				Image		
Structure	Signals		Condition		Signals		Modification
]	s					s	swap(v^0, l_2^1)
[, ' '	T					T	swap(v^0, l_1^{-1})
$	T					T	swap(v^0, l_1^{-1})
$		T			T		swap(v^0, l_2^{-1})
[, ' '			T		T		swap(v^0, l_2^{-1})
]	s	T			h		
* ' ',$			k	$v^0 = b^0$	k		$v^0 \leftarrow f^0$
* [			k	$v^0 = b^0$		E	$v^0 \leftarrow f^0$
], ' '			m		m		swap(b^0, l_1^1), swap(f^0, l_3^1)
' ',$			m		m		swap(b^0, l_1^1), swap(f^0, l_3^1)
' ',$	n		m		m	n	swap(b^0, l_1^1), swap(f^0, l_3^1)
[			m			m	
' ',$]	m					m	
[		n				n	
' ',$	n					n	
]	n				n		
]		n			n		swap(b^0, l_2^{-1}), swap(f^0, l_4^{-1})
' ',$		n			n		swap(b^0, l_2^{-1}), swap(f^0, l_4^{-1})
[		n				n	swap(b^0, l_2^{-1}), swap(f^0, l_4^{-1})
[	m	n				a	

B_x and B_y (last column in Fig. 13). If they differ, g comes back and gets the marker (first two columns in Fig. 13). If there are equal, it turns (or remains) e and the process is restarted. Uppercase signals behave identically.

4.4 Local Function of $\mathcal{U}$

Most of the definition of $\Phi_{\mathcal{U}}$ is given in Table 3. The values of the layers that hold 0 and 1 are not indicated. These values are tested as requirement for rules and are not modified otherwise noted in the last column. These modifications are either swapping or writing on v^0. In the latter case, the previous value is held somewhere else as indicated by a condition. In uppercase mode, the differences are only for the lines with an '*': the test made is $v^0 = f^0$ instead of $v^0 = f^0$, and b^0 (instead of f^0) is copied over v^0. Since the rules are one-to-one, $\Phi_{\mathcal{U}}$ can be completed bijectively.

All rules are combined with the following: for the last 4 layers L_1 to L_4, the -1 and 1 parts are swapped so that -1 (1) parts move at speed 1 to the right (left). For all rules with [: layers L_1 and L_2 (L_3 and L_4) are swapped. This is technical for the flows of the table shift not to collide in the middle in Fig. 12 where 2 flows are traveling together.

The design BCA is reversible: the provided rules are one-to-one. If the simulated BCA is not reversible, then the simulation just does not work because of the backward tests and the duplicate values in images.

4.5 Simulation

For Definition 7, the packing function p is the grouping by macro-cell, $\mathbf{s}$ is the null shift; the onto function α is one-to-one as described in Fig. 10.

Let a be the width of a $\mathscr{A}$-cell and b the width of the exchanged parts ($0 \leq 2b \leq a$ and $\lceil \log |Q| \rceil \leq a \leq 2\lceil \log |Q| \rceil + 2$). The inner loop needs $4(a-1)$ iterations for the tests and $2a$ for the shift of the table. It is done for every $\mathscr{A}$-state, i.e., $|Q|$ times. To make a $\mathscr{A}$-iteration, values are exchanged between neighboring cells, this needs $2b+1$ iterations. All together, τ is a constant bounded by $12|Q|\log(|Q|) + o(|Q|\log(|Q|))$.

The number of states of $\mathscr{U}$ is $2^{13}.24^3$, a little above 113.10^6.

The construction can be extended to greater dimension. The table and test are done in one direction and sub-states exchanged on every directions must be added.

5 Space-Time Simulation of Irreversible CA by Reversible Ones

5.1 Space-Time Approach

A space-time diagram $\mathbb{A}$ is *embedded* into another space-time diagram $\mathbb{B}$ when it is possible to "reconstruct" $\mathbb{A}$ from $\mathbb{B}$ and the way that $\mathbb{A}$ is embedded into $\mathbb{B}$.

The recovering of an embedded $\mathscr{A}$-configuration is done in the following way. A $\mathscr{B}$-configuration is constructed by taking each cell at a given iteration. This $\mathscr{B}$-configuration is decoded to get an iterated configuration for $\mathscr{A}$. More precisely, it is defined as follows:

Definition 21 A space-time diagram $\mathbb{A} = (\mathscr{A}, a)$ is *space-time embedded* into another space-time diagram $\mathbb{B} = (\mathscr{B}, b)$ when there exist two functions $\chi : \mathbb{L} \times \mathbb{N} \to \mathbb{N}$ and $\zeta : \mathscr{C}_{\mathscr{B}} \to \mathscr{C}_{\mathscr{A}}$ such that:

- $\forall (\mathbf{x}, t) \in \mathbb{L} \times \mathbb{N}$, let c^t be the configuration of $\mathscr{B}$ such that $c^t_{\mathbf{x}} = \mathbb{B}_{\mathbf{x}, \chi(\mathbf{x}, t)}$ and
- $\forall t \in \mathbb{N}, \mathscr{G}^t_{\mathscr{A}}(a) = \zeta(c^t)$.

To recover an iterated image of a, the function χ indicates which iteration is to be considered for each cell and ζ decodes the assembled configuration. The generation of c^t and then $\mathscr{G}^t_{\mathscr{A}}(a)$ is illustrated in Fig. 14.

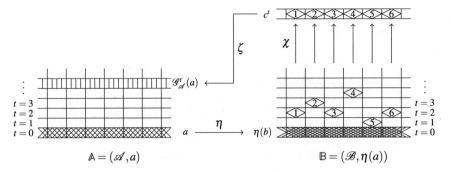

Fig. 14 Space-time diagram $\mathbb{A}$ is embedded into $\mathbb{B}$

The functions χ and ζ could be complex and provide all the computation. To avoid this, in the following definition, they are independent from the initial configuration, thus unable to do much of the $\mathscr{A}$ computation.

Definition 22 A CA $\mathscr{B}$ *space-time simulates* a CA $\mathscr{A}$ when there exists a function $\eta : \mathscr{C}_\mathscr{A} \to \mathscr{C}_\mathscr{B}$ such that any space-time diagram $(\mathscr{A}, a)$ is embedded into the space-time diagram $(\mathscr{B}, \eta (a))$ and all embeddings use the same functions χ and ζ.

This section provides a construction to prove the following lemma:

Lemma 23 *Any d-CA with neighborhood $\{-1, 0, 1\}^d$ can be space-time simulated by a d-R-PCA.*

The proof is only detailed in dimension 1. The generalization to greater dimensions is sketched at the end of this section.

5.2 Macro Dynamics

Let $\mathscr{A} = (Q_\mathscr{A}, \{-1, 0, 1\}, f)$ be any 1-CA. The 1-R-PCA $\mathscr{B} = (Q_\mathscr{B}, \{-1, 0, 1\}, \Phi)$ which space-time simulates $\mathscr{A}$ is progressively constructed. Let a be any configuration in $\mathscr{C}_\mathscr{A}$ and $\mathbb{A}$ the associated space-time diagram. The space-time diagram $\mathbb{B}$ is generated in order to embed $\mathbb{A}$ as follows.

A signal moves forth and back and updates the cells on a finite part of the configuration called the *updating zone*. Outside of this zone, the $\mathscr{B}$-cell are at $\mathscr{A}$-iteration 0. Inside, $\mathscr{A}$-iteration number increases as $\mathscr{B}$-cell are closer to the center of the zone. As $\mathscr{B}$-iterations go, the updating zone is enlarged on both sides (space) and in iteration numbers (time) so that each cell will eventually enter the zone and reach any iteration.

The simulated diagram $\mathbb{A}$ is generated according to diagonal lines, one after the other. The updating lines of $\mathbb{A}$ are depicted in Fig. 15 where the numbers, the arrows and the geometrical symbols on the last column correspond respectively to the

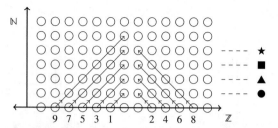

The symbol in the last column identifies the $\mathscr{A}$-iteration.
It corresponds to the embedding in Fig. 16.

Fig. 15 Order of generation inside the simulated diagram $\mathbb{A}$

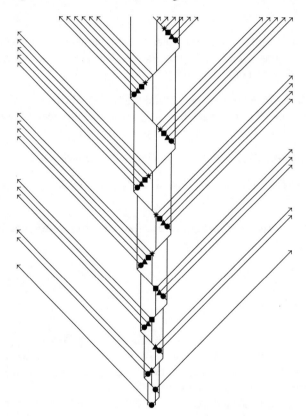

Fig. 16 Scheme of the evacuation of data ($[_x\mathbf{i}]$) on the simulating diagram

order in which updates are made, to their directions and to the identifications of the $\mathscr{A}$-iterations (as on $\mathbb{B}$ in Fig. 16).

The state of a cell x at iteration t in the embedded diagram $\mathbb{A}$ is denoted by $_x\mathbf{i}$ ($_x\mathbf{i} = \mathscr{G}_{\mathscr{A}}^{t}(a)_x$) and the information needed to compute $_x\mathbf{i}$ is denoted by

$[_x\backslash]$ ($[_x\backslash] = (_{x-1}\backslash^{-1}, _x\backslash^{-1}, _{x+1}\backslash^{-1})$). Each time a cell is updated, a $[_x\backslash]$ is generated to keep the data needed for undoing the update. The generated data cannot be disposed off because $\mathscr{G}_{\mathscr{A}}$ is not necessarily one-to-one and the previous configuration might be uncomputable from the actual one. These needed but cumbersome data are evacuated by being sent away outside of the updating zone.

When a signal goes from the left to the right for the nth time on the updating zone, as in Fig. 16, its dynamics work as follows:

Starting from the far left of the updating zone, the first cell encountered by the signal holds $[_x\backslash]$. The signal sets this data moving to the left to save it and evacuate it while generating $_x\backslash$. The next cell holds $[_{x+1}\backslash^2]$ which is also set on movement to the right while $_{x+1}\backslash^2$ is generated. This goes on until the signal reaches the middle of the updating zone (vertical line), then no more updating is done until the signal reaches the right end. On its way back, the signal updates the other half of the updating zone.

The signal makes n updates one way and n updates on its way back. Then it makes $n + 1$ and $n + 1$ updates, then $n + 2$ and so on. The cells corresponding to the iteration 1 (2, 3 and 4 respectively) in $\mathbb{A}$ are generated on a parabola indicated by ● (▲, ■, ★ respectively) on the simulating diagram $\mathbb{B}$ in Fig. 16. This corresponds to the layer-construction of $\mathbb{A}$ depicted in Fig. 15. Figure 16 depicts the evacuation of the $[_x\backslash]$ away from the updating zone for the first 100 iterations. Evacuated data never interact.

5.3 Micro Dynamics

Cells are organized in 3 layers: the upper layer holds the state of the simulated cell, the middle one holds the signal that drives the dynamics and the lower one acts like a conveyor belt to evacuate the $[_x\backslash]$.

The first 26 iterations are depicted in Fig. 17. In the upper layer, the cells alternatively hold 3 times the same state $(_x\backslash)$ or the states of the cell and its 2 closest neighbors at the same iteration $(_{x-1}\backslash^{-1}, _x\backslash^{-1}, _{x+1}\backslash^{-1})$, otherwise some mix over 2 or 3 iterations. A cell can only be updated when it has the information $[_x\backslash]$. By induction from the dynamics in Fig. 17, the possibility to update a cell only depends on the parity of the sum of simulating and simulated iteration numbers.

The signal that rules the dynamics is called the *suit signal*. Depending on its position, it takes the values ♣, ♠, ♥ and ♦ in $\mathbb{B}$. The suit signal only moves forth and back in the updating zone and thus appears as a zigzag in Figs. 16 and 17. It is delayed by one on the left side to keep it synchronized with the presence of $[_x\backslash]$.

The updating zone is delimited by a pair of **I** and its middle is indicated by a ★. The **I** progressively move away from each other while the ★ oscillates in the middle. Starting on the left **I**, the suit signal is ♥. While passing, it makes the updates of the simulated cells until it reaches ★. Afterwards it is ♠ and just moves to the other **I**.

Each time a simulated update is done, 3 values, $_{x-1}\backslash^{-1}, _x\backslash^{-1}$ and $_{x+1}\backslash^{-1}$, are "used up" and become useless. They are gathered in $[_x\backslash]$ and moved to the lower layer to

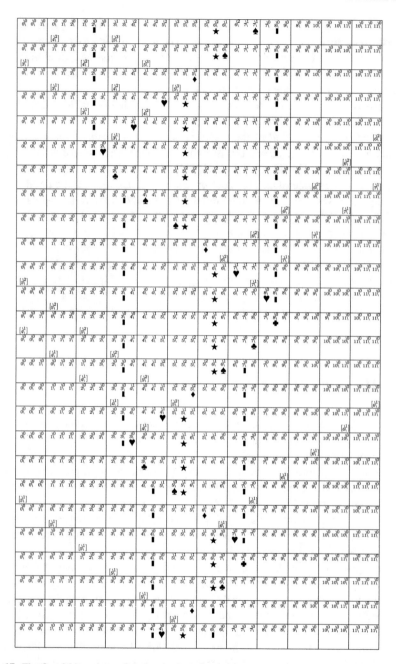

Fig. 17 The first 26 iterations of the space-time simulation

Table 4 States of $\mathscr{B}$

	1	0	-1
	$Q_{\mathscr{A}}$	$Q_{\mathscr{A}}$	$Q_{\mathscr{A}}$
	♣♠♥♦_	✚❙★_	♣♠♥♦_
	$Q_{\mathscr{A}}^3 \cup \{\_\}$	–	$Q_{\mathscr{A}}^3 \cup \{\_\}$

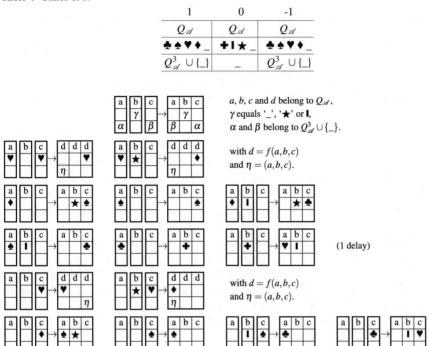

a, b, c and d belong to $Q_{\mathscr{A}}$,
γ equals '_', '★' or ❙,
α and β belong to $Q_{\mathscr{A}}^3 \cup \{\_\}$.

with $d = f(a,b,c)$
and $\eta = (a,b,c)$.

(1 delay)

with $d = f(a,b,c)$
and $\eta = (a,b,c)$.

Fig. 18 Definition of $\Phi_{\mathscr{B}}$

be evacuated. Three copies of the new state $_x\rangle\!\langle$ are made. They will be used for the next update of the simulated cell and of its 2 neighbors.

The endless movement of the suit signal and updates (at correct parities) are deduced by induction. Since the interaction is only local and has radius 1, global properties are not otherwise modified. All the necessary steps for the induction can be found on the two and a half loops of the suit signal in Fig. 17.

5.4 State Function

The R-BCA $\mathscr{B}$ has $100 |Q_{\mathscr{A}}|^3 (|Q_{\mathscr{A}}|^3 + 1)^2$ states detailed in Table 4.

Cells are depicted as 3×3 arrays as in the first line in Fig. 17. The upper layer encodes a configuration of $\mathscr{A}$. The middle layer holds the suit signal. The lower layer is used to store the data away from the updating zone.

The suit signal is alternatively equal to ♥ and ♠. When shifting, ♥ updates cells while ♠ does nothing. The signal becomes ♣ and ♦ to move respectively ❙ and ★.

The transition rules are given in Fig. 18. The first rule corresponds to the lack of any signal. On the lower layer, the 2 values on the side are swapped, this acts like a conveyor belt. As soon as something is put on the lower layer, it is shifted by one cell at each iteration. This is used to evacuate data. The updating rules are on the lines 2 and 5.

The second and third lines in Fig. 18 depicted how ♥ moves to the right and updates cells. When it reaches the middle frontier ★, it moves it one step to the right as ♦ and then turns to ♠.

The signal ♠ turns on the right side, as depicted on the fourth line in Fig. 18. On arriving on I from the left, ♠ grabs it and turns to ♣. On the next iteration, ♣ turns to ✚ and does nothing else. This is the delay of one iteration needed to keep up with parity. Next iteration, ✚ regenerates the I and the signal ♥ which goes back to the left.

The signal turns back one iteration faster on the left side as depicted on the last line in Fig. 18: the state ✚ does not appear.

The rules defined are one-to-one, thus they can be completed so that Φ is a permutation, $\mathscr{A}$ is then reversible (Lemma 5).

The initial configuration is depicted on the first line in Fig. 17. The state of each cell is copied 3 times in the upper layer. Markers I, ★ and I are laid in the center of 3 adjacent cells and the ♥ is together with the left I.

With this construction, the embedded space-time diagram is bent in a parabola shape. This makes it meaningless to access geometrical properties like, e.g., Fisher constructibility or Firing Squad Synchronization.

5.5   Generalization Sketch

This construction can be generalized to any dimension greater than 1. The simulated configurations are still "bent" according to the first direction in the simulating diagram. Along the first direction, the dynamics are exactly as explained above. The signals are duplicated along the other directions. The updatings are still conditioned by parities. There are an infinity of I, ★ and suit signals. They are arranged on hyperplanes orthogonal to the first direction and are exactly synchronized.

Any d-CA can be simulated by a d-CA whose neighborhood is $\{-1, 0, 1\}^d$ (and this simulation is transitively compatible). From Lemma 23 and the fact that d-R-PCA are d-R-CA comes:

Theorem 24 *Any d-CA can be space-time simulated by a d-R-CA.*

Since there are d-R-CA able to simulate all d-R-CA over any configuration and the simulations are compatible enough:

Theorem 25 *There are d-R-CA able to space-time simulate any d-CA.*

6   Conclusion

Conjectures 1 and 2 are true even if states in Q^2 are used in intermediate configurations during the simulation, the input and output are restricted to Q.

For any d, d-CA, d-BCA and d-PCA have the same power over infinite configurations. The same holds for d-R-CA, d-R-BCA and d-R-PCA classes. This is an important result since reversibility is decidable for BCA and PCA while it is not for CA. This is not a contradiction since the inverse CA is needed for the construction.

The proof of Theorem 12 is more involved than the one in Durand-Lose [6]. Nevertheless, the number of block transitions needed is lowered from $2^{d+1}-1$ to $d+1$. Generating and erasing are done concurrently, not one after the other. We conjecture that it is impossible to make a representation with less that $d+1$ block transitions.

The expression with block transitions allows one to use reversible circuitry in order to build R-CA. This was done in Durand-Lose [6] to prove that, for $2 \le d$, there exists d-dimensional R-CA (based on the the Billiard ball model) able to simulate any d-dimensional R-CA on infinite configurations. Kari [17] provides more information on the relation between R-CA and BCA and the inner structure of R-CA in dimensions 1 and 2.

The $\mathcal{U}$ is programmed: loops, tests and conditional executions. Basic programming schemes can be embedded in R-PCA when conceived reversible: a global dynamic of move, test and replace which needs backward tests.

There exist simulations of any Turing machines with R-CA Morita [25] so that all partial recursive functions can be computed by R-PCA, so that $\mathcal{U}$ is computationally universal. The existence of an intrinsically universal R-PCA is proven here with the use of the source code of the R-PCA. So there should be some S-m-n theorem for R-PCA to prove that they form an acceptable programming system as proved for CA by Martin [22].

It is unknown whether the class of d-CA is strictly more powerful than the class of d-R-CA on infinite configurations. Nevertheless, if a 1-R-CA can simulate a non reversible CA, then by transitivity, $\mathcal{U}$ is also able to do it, so that if $\mathcal{U}$ cannot, none can.

With space-time simulation, reversible can simulate irreversible, but this simulation is not homogeneous; it is not shift invariant nor time invariant. An infinite time is required to fully generate the configuration after one iteration. Moreover, it is not possible to go backward before the first configuration if no such configuration were encoded in the initial configuration—anyway, there is no guarantee that any previous configuration does exist. When the significant part of a configuration represents only a finite part of the space, the result of the computation is given in finite time like in Morita [25, 26]. In Toffoli [34], an extra dimension is used to store information for reversibility, here configurations are bent to provide the room.

References

1. Albert, J., Čulik, K. II.: A simple universal cellular automaton and its one-way and totalistic version. Complex Syst. **1**, 1–16 (1987)
2. Amoroso, S., Patt, Y.N.: Decision procedure for surjectivity and injectivity of parallel maps for tessellation structure. J. Comput. Syst. Sci. **6**, 448–464 (1972)
3. Bennett, C.H.: Logical reversibility of computation. IBM J. Res. Dev. **6**, 525–532 (1973)
4. Bennett, C.H.: Notes on the history of reversible computation. IBM J. Res. Dev. **32**(1), 16–23 (1988)
5. Burks, A.W.: Essays on Cellular Automata. University of Illinois Press, Champaign (1970)
6. Durand-Lose, J.: Reversible cellular automaton able to simulate any other reversible one using partitioning automata. In: LATIN 1995. LNCS, vol. 911, pp. 230–244. Springer (1995). https://doi.org/10.1007/3-540-59175-3_92
7. Durand-Lose, J.: Automates Cellulaires, Automates à Partitions et Tas de Sable. Thèse de doctorat, LaBRI (1996). http://www.univ-orleans.fr/lifo/Members/Jerome.Durand-Lose/Recherche/These/index.html. In French
8. Durand-Lose, J.: Intrinsic universality of a 1-dimensional reversible cellular automaton. In: STACS 1997. LNCS, Vol. 1200, pp. 439–450. Springer (1997). https://doi.org/10.1007/BFb0023479
9. Durand-Lose, J.: About the universality of the billiard ball model. In: Margenstern, M. (ed.) Universal Machines and Computations (UMC '98), vol. 2, pp. 118–133. Université de Metz (1998)
10. Durand-Lose, J.: Reversible space-time simulation of cellular automata. Theoret. Comput. Sci. **246**(1–2), 117–129 (2000). https://doi.org/10.1016/S0304-3975(99)00075-4
11. Durand-Lose, J.: Representing reversible cellular automata with reversible block cellular automata. In: Cori, R., Mazoyer, J., Morvan, M., Mosseri, R. (eds.) Discrete Models: Combinatorics, Computation, and Geometry, DM-CCG '01, vol. AA of Discrete Mathematics and Theoretical Computer Science Proceedings, pp. 145–154 (2001a). http://dmtcs.loria.fr/volumes/abstracts/dmAA0110.abs.html
12. Durand-Lose, J.: Back to the universality of the Billiard ball model. Mult. Valued Logic **6**(5–6), 423–437 (2001b)
13. Hedlund, G.A.: Endomorphism and automorphism of the shift dynamical system. Math. Syst. Theory **3**, 320–375 (1969)
14. Kari, J.: Reversibility of 2D cellular automata is undecidable. Phys. D **45**, 379–385 (1990)
15. Kari, J.: Reversibility and surjectivity problems of cellular automata. J. Comput. Syst. Sci. **48**(1), 149–182 (1994)
16. Kari, J.: Representation of reversible cellular automata with block permutations. Math. Syst. Theory **29**, 47–61 (1996)
17. Kari, J.: On the circuit depth of structurally reversible cellular automata. Fund. Inf. **38**(1–2), 93–107 (1999)
18. Kari, J.: Theory of cellular automata: a survey. Theoret. Comput. Sci. **334**, 3–33 (2005)
19. Lecerf, Y.: Machines de Turing réversibles. Récursive insolubilité en $n \in \mathbb{N}$ de l'équation $u = \theta^n u$, où θ est un isomorphisme de codes. Comptes rendus des séances de l'académie des sciences **257**:2597–2600 (1963)
20. Margolus, N.: Physics-like models of computation. Phys. D **10**(1–2), 81–95 (1984)
21. Margolus, N.: Physics and Computation. Ph.D. thesis, MIT (1988)
22. Martin, B.: A universal cellular automaton in quasi-linear time and its S-n-m form. Theoret. Comput. Sci. **123**, 199–237 (1994)
23. Moore, E.F.: Machine models of self-reproduction. Proc. Symp. Appl. Math. **14**, 17–33 (1962)
24. Morita, K.: Any irreversible cellular automaton can be simulated by a reversible one having the same dimension. Tech. Rep. IEICE, Comp. **92–45**(1992–10), 55–64 (1992)
25. Morita, K.: Computation-universality of one-dimensional one-way reversible cellular automata. Inform. Process. Lett. **42**, 325–329 (1992b)

26. Morita, K.: Reversible simulation of one-dimensional irreversible cellular automata. Theoret. Comput. Sci. **148**, 157–163 (1995)
27. Morita, K.: Reversible computing and cellular automata - a survey. Theoret. Comput. Sci. **395**(1), 101–131 (2008). https://doi.org/10.1016/j.tcs.2008.01.041
28. Morita, K., Harao, M.: Computation universality of one-dimensional reversible (injective) cellular automata. Trans. IEICE, E **72**(6), 758–762 (1989)
29. Myhill, J.R.: The converse of Moore's garden-of-eden theorem. Proc. Am. Math. Soc. **14**, 685–686 (1963)
30. Ollinger, N.: Two-states bilinear intrinsically universal cellular automata. In: FCT '01. LNCS, vol. 2138, 369–399. Springer, Berlin
31. Ollinger, N.: Universalities in cellular automata. In: Rozenberg, G., Bäck, T., Kok, J.N. (eds.) Handbook of Natural Computing, pp. 189–229. Springer (2012). https://doi.org/10.1007/978-3-540-92910-9
32. Richardson, D.: Tessellations with local transformations. J. Comput. Syst. Sci. **6**(5), 373–388 (1972)
33. Sarkar, P.: A brief history of cellular automata. ACM Comput. Surv. **32**(1), 80–107 (2000)
34. Toffoli, T.: Computation and construction universality of reversible cellular automata. J. Comput. Syst. Sci. **15**, 213–231 (1977)
35. Toffoli, T., Margolus, N.: Cellular Automata Machine – A New Environment for Modeling. MIT press, Cambridge (1987)
36. Toffoli, T., Margolus, N.: Invertible cellular automata: a review. Phys. D **45**, 229–253 (1990)
37. Wolfram, S.: Theory and Applications of Cellular Automata. World Scientific, Singapore (1986)

A Weakly Universal Cellular Automaton on the Grid {8, 3} with Two States

Maurice Margenstern

Abstract In this chapter we present a result which improves a previous one established by the author. Here we prove that it is possible to construct a weakly universal cellular automaton on the tessellation {8, 3} with two states only. Note that the cellular automaton lives in the hyperbolic plane, that the proof yields an explicit construction and that the constructed automaton is not rotationally invariant.

1 Introduction

The chapter advances papers [9, 10] where the author proved the same result in the tessellation {9, 3}. The reason of this improvement lies in the relatively small number of rules given in those papers and the fact, there noticed, that several rules where uselessly duplicated. Also, as it is usual in this process of reducing the possibilities of the automaton, here its neighbourhood, it is needed to change something in the previous scenario of the simulation. This time, one structure involved in the simulation is replaced by the combination of two other existing structures. The morale of this result compared to that of [6] is that relaxing the rotation invariance allows us to significantly reduce the number of neighbours: from 11 in [6] to 8 in the present paper. As in [9], the new system of coordinates introduced in [8] is used for the tilings $\{p, 3\}$ and $\{p-2, 4\}$.

In the paper, the same model as in [5, 6] and the other quoted papers is used.

In Sect. 2, the model is sketchily described, including the above evoked new structures. In Sect. 3, the rules of the automaton are provided. They allow us to prove the following result:

Theorem 1 *There is a weakly universal cellular automaton on the tessellation* {8, 3} *which is truly planar and which has two states.*

Presently, we turn to the proof of this result.

M. Margenstern (✉)
LITA EA 3097, Campus du Saulcy, Université de Lorraine, 57045 Metz Cédex, France
e-mail: margernstern@gmail.com

© Springer International Publishing AG 2018
A. Adamatzky (ed.), *Reversibility and Universality*, Emergence, Complexity
and Computation 30, https://doi.org/10.1007/978-3-319-73216-9_5

2 The Scenario of the Simulation

The cellular automaton we shall construct simulates a register machine through a railway circuit on which a locomotive is running. Such a circuit assembles infinitely many portions of straight lines, quarters of circles and switches. Switches are a special structure which consist of three tracks meeting at a point P, say a, b and c. In an **active crossing** of the switch, the locomotive arrives at P through a and leaves the point either through b or through c. In a **passive crossing**, the locomotive arrives at P either through b or through c and it leaves the point through a. There are three kinds of switches, see [3, 14]. One switch is called the **fixed switch**: in an active passage, the locomotive is sent to the same track, either always to b or always to c. In the passive crossing, it arrives from whichever track. A second kind of switch is the **flip-flop**. That switch is always crossed actively. Now, the chosen track alternates: if the last passage was through b, c, the next one is through c, b respectively. The third kind of switch is the **memory switch**. The chosen track in an active crossing is defined by the track through which the locomotive arrived at P in the last passive crossing. In [3] we precisely describe how to organize the simulation of a register machine using the above indicated ingredients. In that book, the description is also adapted to the hyperbolic context.

Figure 1 illustrates how to assemble switches belonging to all their kinds in order to mimic an element which contains a one-bit information. We call it the **basic unit**. Combining basic units with appropriate portions of circuits, we can construct a circuit mimicking the simulation of a register machine as proved in the rest of the paper.

As in previous papers, the flip-flop and the memory switch are decomposed into simpler ingredients which we call sensors and control devices. This reinforces the importance of the tracks as their role for conveying key information is more and more decisive. Here too, tracks are blank cells marked by appropriate black cells we call **milestones**. We carefully study this point in Sect. 2.1. Later, in Sect. 2.2, we adapt the configurations described in [9] to the tessellation $\{8, 3\}$.

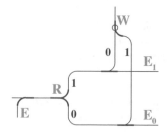

Fig. 1 The basic unit of the circuit. When the locomotive enters the unit through R, it reads either 0 or 1 which corresponds to the exits E_0 or E_1 respectively. When it enters through W, it rewrites the content of the unit, changing 0 to 1 and 1 to 0, exiting through E

2.1 The Tracks

In this implementation, the tracks are represented in a way which is a bit similar to that of [9, 10]. The present implementation is given by Fig. 2.

Here, we explicitly indicate the numbering of the sides in a cell which will be systematically used through the paper. We fix a side which will be, by definition, side 1 in the considered cell. Then, all the other sides are numbered starting from this one and growing one by one while counter-clockwise turning around the cell. Note that, in our setting, the same side, which is shared by two cells, can receive two different numbers in the cells which share it. An example of this situation is given in Fig. 2: in the central cell we denote by 0(0), side 1 is side 7 in the neighbour of the central cell sharing this side, denote it by 1(1). In Sect. 3.1, we go back to the construction of the tracks starting from the elements indicated in Fig. 2. Note that Fig. 2 shows us two rays starting from M, the mid-point of the side 2 of the central cell. Those rays allow us to introduce the numbering of the tiles based on [8]. It will be used in the figures illustrating the paper.

The rays delimit what we call a **sector**. They are defined as follows. The ray u starts from the mid-point M of the side 2 of 0(0) and it passes through the mid-point of its side 1. The ray also passes through the mid-point of the side 8 of 1(1). The ray v, also issued from M, cuts the sides 6 and 5 of 1(1) at their mid-points. Its support also passes through the mid-point of the side 3 of 0(0).

In the figures of the paper, the central cell is the tile whose centre is the centre of the circle in which the figure is inscribed. The central cell is numbered by 0, denoted by 0(0). We number the sides of the tile as indicated in Fig. 2. For $i \in \{1..8\}$,

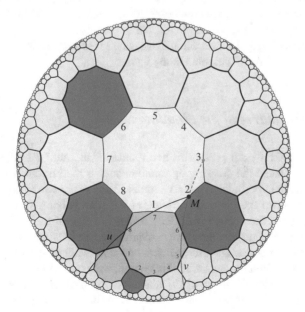

Fig. 2 Element of the tracks in {8, 3}

the cell which shares the side i with the central cell is called **neighbour** i and it is denoted by $1(i)$. The rotation around $0(0)$ allows us to attach a sector to each tile $1(i)$. Number 1 in this notation is the number given to the root of the tree attached to the sector defined from this tile, see [1, 8]. Here, that definition is adapted to the case of the tessellation $\{8, 3\}$. We invite the reader to follow the present explanation on Fig. 1. Consider the sector defined by the rays u and v. The neighbours of the cell $1(1)$ sharing its sides j, $j \in \{1..4\}$ are numbered $j+1$ and are denoted by $j+1(1)$. We say that the cell $1(1)$ is a W-cell and its sons are defined by the rule $W \rightarrow BWWW$, which means that $2(1)$ is a B-cell. This means that the sons of $2(1)$ are defined by the rule $B \rightarrow BWW$, where the B-son has two consecutive sides crossed by u in their midpoints. These sons of $1(1)$ constitute the level 1 of the tree. The sons of $2(1)$, starting from its B-son are numbered 6, 7 and 8 denoted by $6(1)$, $7(1)$ and $8(1)$ respectively. By induction, the level $n+1$ of the tree are the sons of the cells which lie on the level n. The cells are numbered from the level 0, the root, level by level and, on each level from left to right, i.e. starting from the ray u until the ray v. What we have seen on the numbering of the sons of $2(1)$ is enough to see how the process operates on the cells. From now on, we use this numbering of the cells in the figures of the paper.

As can be seen in Figures of Sect. 3, the locomotive is implemented as a single black cell: it has the same colour as the milestones of the tracks. Only the position of the locomotive with respect to the milestones allows us to distinguish it from the milestones. As clear from the next sub-section, we know that besides this **simple locomotive**, the locomotive also occurs as a **double one** in some portions of the circuit. In a double locomotive, we call its first, second cell the **front**, **rear** respectively of the locomotive. In a simple locomotive, the front only is present.

The circuit also makes use of signals which are implemented in the form of a simple locomotive. So that at some point, it may happen that we have three simple locomotives travelling on the circuit: the locomotive and two auxiliary signals involved in the working of some switch. For aesthetic reasons, the black colour which is opposed to the blank is dark blue in the figures.

2.2   The Structures of the Simulation

The crossings of [14] are present in many ones of the author's papers. Starting from [2], he replaced the crossing by round-abouts, a road traffic structure, in his simulations in the hyperbolic plane. At a round-about where two roads are crossing, if you want to keep the direction arriving at the round-about, you need to leave the round-about at the second road. The reader is referred to [3] for references. The structure is a complex one, which requires a fixed switch, a doubler and a selector. Other structures are used to simulate the switches used in [3, 14]: the fork, the controller and the sensor. In this section, we present the implementation of those structures which are those of [9] adapted to the present tessellation with one exception: the doubler, which is here different from that of that paper.

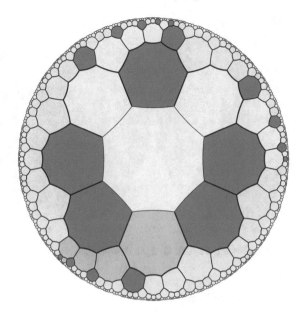

Fig. 3 Idle configuration of the passive fixed switch

2.2.1 The Fixed Switch

As the tracks are one-way and as an active fixed switch always sends the locomotive in the same direction, no track is needed for the other direction: there is no active fixed switch. Now, passive fixed switches are still needed as just seen in the previous paragraph.

Figure 3 illustrates the passive fixed switch when there is no locomotive around: we say that such a configuration is **idle**. We shall again use this term in the similar situation for the other structures. We can see that it consists of elements of the tracks which are simply assembled in the appropriate way in order to drive the locomotive to the bottom direction in the picture, whatever upper side the locomotive arrived at the switch. The path followed by the locomotive to the switch is in yellow until the central cell which is also yellow. The path from the left-hand side consists, in this order of the cells 13(8), 4(8), 5(8), 2(1) and 1(1). From the right-hand side, it consists of the cells 16(6), 4(6), 5(6), 2(7) and 1(7). Of course, 1(1) and 1(7) are neighbours of 0(0). The path followed by the locomotive when it leaves the cell is in pink. It consists of the following cells in this order: 1(4), 2(4), 5(3), 4(3)and 16(3). Note that the cell 0(0) in Fig. 3 has five black neighbours: the cells 1(2), 1(3), 1(5), 1(6) and 1(8). Note that 1(6) and 1(8) are also milestones for the cell 1(7), that 1(3) and 1(5) are milestones for 1(4) and that 1(8) and 1(2) are milestones for 1(1).

From our description of the working of the round-about, a passive fixed switch must be crossed by a double locomotive as well as a simple one. Later, in Sect. 3.2, we shall check that the structure illustrated by Fig. 3 allows those crossings.

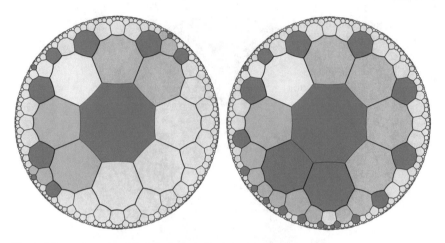

Fig. 4 Idle configurations. To left: the fork. To right: the doubler

2.2.2   The Doubler and the Fork

The fork is the structure illustrated by the left-hand side picture of Fig. 4. Note that its structure is very different from that of the tracks or of the fixed switch. The central cell 0(0) is black and two paths start from 1(1), each one on one side of the central cell with respect to its axis crossing its sides 1 and 5, and traversing a quarter of the cells around 0(0). The cell 1(1) is yellow in the figure. The left-hand side track is green, consisting of the following cells, in this order: 1(2), 1(3), 3(3), 9(3) and 35(3). The right-hand side track is pink. It consists of the cells: 1(8), 1(7), 5(7), 6(8) and 21(8). The locomotive, a simple one, arrives through the yellow path: 21(2), 6(2), 5(1), and 1(1). From 1(1), two simple locomotives appear: one in 1(2), the other in 1(8). The locomotive in 1(2) goes along the light green path while the one in 1(8) goes along the pink path.

The doubler is illustrated by the right-hand side of Fig. 4. This structure is different from that of [9]. Indeed, the even number of sides of a tile does not allow us to divide the path around 0(0) in two equal sub-paths both excluding 1(1). The doubler is a structure which receives a simple locomotive and yields a double one which consists of two consecutive black cells on the track. The idea is to use two already defined structures: the fork and the fixed switch. This combination can make a double locomotive provided that two simple ones arrive at the fixed switch from each side with a delay of one top of the clock. This is realized by the paths illustrated by the figure. The picture uses the same colours as the picture of the fork with the same meaning. Consider the green paths. Its cells are, in this order: 1(2), 2(3), 3(3), 4(3), 5(3), 2(4) and 3(4), which makes 7 cells. The pink path consists of the following cells, in this order: 1(8), 1(7), 1(6), 1(5), 2(5) and 5(4), which makes 6 cells.

We can see that the cells around 4(4) are exactly the neighbours of the central cell of a fixed switch, see Fig. 3. According to this description, the two simple locomotives created at the same time in 1(2) and 1(8) respectively do not arrive at the same time at the cell 4(4). When the locomotive created in 1(8) arrives at the cell 4(4), the locomotive created in 1(2) is at 3(4), so that the two black cells in 3(4) and 4(4) constitute a double locomotive arriving to the fixed switch from the left-hand side. We shall see that the double locomotive does cross the switch. Accordingly the structure works as expected for a doubler. Note the configuration of three black cells around a common vertex.

2.2.3 The Selector

The selector is illustrated by Fig. 5. This structure is not as symmetric as the corresponding structure of [9], which makes another difference with that paper. We have a yellow track through which the locomotive arrives, simple or double, both cases are possible. When a simple locomotive arrives, it leaves the cell through 1(8), via the pink path which consists of the cells 1(8), 2(8), 9(8) and 32(8). When a double locomotive arrives, a simple locomotive leaves the structure through the green path, the cells: 1(4), 2(5), 6(5) and 24(5). Both cells 1(5) and 1(7) can detect whether the locomotive is simple or double. They can do that when the front of the locomotive is in 0(0). Then, if the locomotive is double, its rear is in 1(6). Both cells 0(0) and 1(6) are neighbours of 1(5) and of 1(7) too.

In Sect. 2.2.3, the rules will show that such a working will be observed.

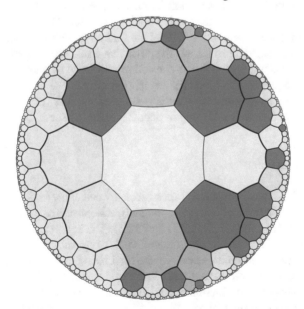

Fig. 5 Idle configuration of the selector. The cells 1(7) and 1(5) detect whether the locomotive is simple or double

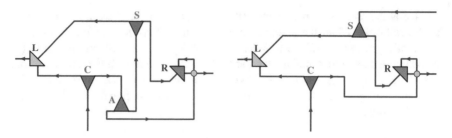

Fig. 6 Idle configuration of the controller of the flip-flop and of the active memory switch

2.3 The Controller and the Controller-Sensor

In this Sub-section, we look at the additional structures used for the flip-flop and for the memory switch, see [3, 14] for the definitions and for the implementation in the hyperbolic plane. As explained in [3], the flip-flop and the active memory switch are implemented by using the fixed switch, the fork and a new structure we shall study in Sect. 2.3.1: the **controller**. Figure 6 illustrates how fixed switch, forks and controllers can be assembled in order to produce a flip-flop, the left-hand side of the figure, and an active memory switch, the right-hand side of the figure. In both cases, the controller has two states, we shall say **black** and **white** for these states. When the controller is white, it stops the locomotive, when it is black, it let it pass. Accordingly, we put a black controller on the selected track and a white one on the non-selected track. Note that once a locomotive entered the fork C, there is a locomotive on both tracks leaving C to the controller which sits on that track, see the pictures of Fig. 6.

In the passive memory switch, the controller is replaced by another device which we call the **sensor**, see Fig. 8, we shall study it in Sect. 2.3.2. The structure is illustrated by Fig. 7. For the passive memory switch, we need the fork, the fixed switch and another new structure we shall study in Sect. 2.3.2: the **sensor** illustrated by Fig. 9.

2.3.1 The Controller

As shown by Fig. 7, the controller sits on an ordinary cell of the track. The locomotive is which runs on that track is always simple. The track consists of the cells 20(6), 6(7), 2(7), 1(6), 0(0), 1(4), 2(5), 3(5), 10(5) and 35(5). The cell 1(3) defines the **colour** of the controller. If it is black, then the cell 0(0) is typically a cell of the track, so that the locomotive goes on its way along the track, leaving the controller. If the cell 1(3) is white, then the cell 0(0) can no more work as a cell of the track. It remains white, which means that the locomotive is stopped at 1(6): after that, it vanishes. This corresponds to the working of a selection in an active passage: the locomotive cannot run along a non-selected track. Here it can for a while, but at some point, it is stopped by the controller. Note that the occurrence of a locomotive in the structure does not change the colour in 1(3). The change of colour in that cell is performed by a signal

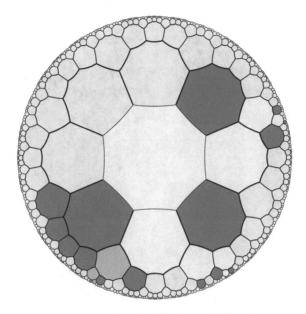

Fig. 7 Idle configuration of the controller of the flip-flop and of the active memory switch

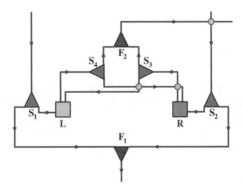

Fig. 8 Idle configuration of the controller of the flip-flop and of the active memory switch

which takes the view of a simple locomotive arriving through another track: 24(4), 6(4) and 2(4), that latter cell being a neighbour of 1(3). When the locomotive-signal arrives at 2(4), it makes the cell 1(3) change its colour: from white to black and from black to white.

2.3.2 The Sensor

This section deals with the passive memory switch. Forks, fixed switch and sensors are assembled as indicated by Fig. 8. The sensor also has two states, **black** and **white**. The sensor works in a different way than the controller. When the sensor is black, it stops the locomotive, when it is white, it let it go. However, we put a black

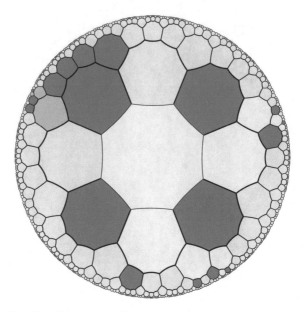

Fig. 9 Idle configuration of the sensor of the passive memory switch

controller and a black sensor on the selected track of the corresponding switches and, symmetrically, a white controller and a white sensor on the non-selected track.

The sensor is illustrated by Fig. 9. As above mentioned, when the locomotive passes on the non-selected track, it does not stop it. It uses it as a messenger for the signal it has to send to the active switch in order to change the selection of the tracks.

This is illustrated by the structure of the figure. The path is the same as in Fig. 7. The cell which plays the role of a sensor is this time the cell 1(1) whose state we call the **colour** of the sensor. Note that the neighbourhood of that cell in Fig. 9 is the same, up to rotation, to the neighbourhood of the cell 1(3) in Fig. 7: the green path here consists of the cells 24(2), 6(2) and 2(2) the latter being a neighbour of 1(1).

Figure 9 shows a very different structure for the cell 0(0) compared with that of Fig. 7. When the sensor is white, its neighbourhood is exactly that of the cell 0(0) when the controller is black: it is an ordinary cell of the track so that the locomotive goes on its way on the track. The difference in both structures lies in the logic of the switches. In the case of the controller, when the locomotive goes on its way, it is the locomotive of the circuit going to another switch or to a round-about. In the case of the sensor, the locomotive which goes on its way on the track becomes a signal sent to the active switch associated to the passive switch.

We can just note that the change of colour is different in the sensor: when the sensor is white, it means that the track is non-selected, if a locomotive passes, it must become black, as the track will now be selected. The locomotive, which is not stopped, is used as a signal sent to the other sensor of the passive switch which, from black must turn to white, and it is also sent to the active switch in order to exchange the colours of both controllers. The change is performed by the arrival of the locomotive through the green path of Fig. 9. As the configuration is the same around the cell 1(1) of that Figure as that around 1(3) in Fig. 7, the change from black to white is performed.

3   Rules

The figures of Sect. 2 help us to establish the rules. Their application is illustrated by figures of this section which were drawn by a computer program which checked the coherence of the rules. The program also wrote the PostScript files of the pictures from the computation of the application of the rules to the configurations of the various type of parts of the circuit. The computer program also established the traces of execution which contribute to the checking of the application of the rules.

Let us explain the format of the rules and what is allowed by the relaxation from rotation invariance. A rule has the form $\underline{X}_o X_1..X_8 \underline{X}_n$, where X_o is the state of the cell c, X_i is the **current** state of the neighbour i of c and X_n is the **new** state of c applied by the rule. As the rules no more observe the rotation invariance, we may freely choose which side has number 1 for each cell. We take this freedom from the format of the rule which only requires to know which neighbour has number 1. In order to restrict the number of rules, it is decided that as a general rule, for a cell which is an element of the track, side 1 is the side shared by the cell and its next neighbour on the track, so that tracks are one way. There can be exceptions when the cell is in a switch or the neighbour of the central cell in a switch. In particular, when a cell belongs to two tracks, side 1 is arbitrarily chosen among the two possible cases. The milestones may have their side 1 shared by an element of the track.

We have to keep in mind that there are two types of rules. Those which keep the structure invariant when it is idle, we call this type of rules **conservative**, and those which control the motion of the locomotive. Those latter rules, which we call **motion rules**, are the rules applied to the cells of the tracks as well as their milestones and, sometimes to the cells of the structures which may be affected by the passage of the locomotive. Next, in each sub-section, we give the rules for the motion of the locomotive in the tracks, then for the fixed switch, then for the doubler and for the fork, then for the selector, then for the controller and, eventually, for the sensor. In each sub-section, we also illustrate the motion of the locomotive in the structure as well as a table giving traces of execution for the cells of the track involved in the crossing.

3.1   The Rules for the Tracks

Figure 2 shows us a single element of the track. Figure 10 shows us how to assemble elements as illustrated in Fig. 2 in order to constitute tracks. In Fig. 10, we can see two tracks a part of which goes around the central cell. We consider the following tracks:

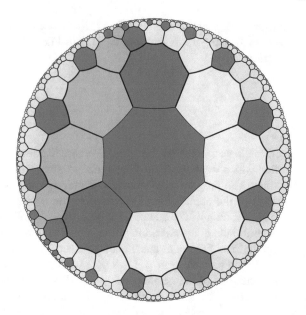

Fig. 10 Assembling elements of the tracks in order to construct tracks, idle configuration

to the right-hand side of 0(0), in yellow: the cells 17(8), 4(8), 3(8), 2(8), 1(7), 1(6), 1(5), 1(4), 2(4), 5(3), 4(3) and 13(3); to the left-hand side of 0(0), in green: 32(3), 9(3), 2(3), 1(2), 1(1), 2(1), 6(1) and 21(1). With this order, both tracks can be seen as a traversal in a clockwise motion. We shall look at both directions as suggested by Table 1 whose rules concern both tracks when a simple locomotive runs over them.

Now, let us turn to the cells of the tracks. There are two kinds of them. There are three-milestoned cells as, for instance, 2(8) and 1(6), and four-milestoned ones as, for instance, 4(8) and 1(4). Below and in the tables too, a red number for a rule means that its new state is opposite to its current one. When this is the case, the current and the new states are also in red. Table 2 shows us that for the cells 2(8) and 1(6), the same rules were applied:

Table 1 Rules managing the motion of a simple locomotive on the tracks

			clockwise motion						
1	WWWWWWWWWW	11	WWBBWBWWBW	21	WBBWWWWWWW	31	BWBBWWBWWB		
2	BWWWWWWWWB	12	WWBWBWWWBW	22	WBBWBWWWBW	32	BWWBWWBWBB		
3	BWWBWBWWWB	13	WBWWWWWWBW	23	BBWWBWBWWB	33	BWWWWWWBWB		
4	WWBWWBWBBW	14	BBWWWWWWWB	24	BWBWWWBWBW	34	BWWBWWBBWB		
5	WWBWWWWWWW	15	WWWWWBWBWW	25	WBBWWBWBBW	35	BBWWBWWWBB		
6	WWWWWWWWBW	16	WWBWWWBWBW	26	BWBWWWWWWB	36	BBWWBWWBWB		
7	WBWWWWWWBW	17	BBWWWBBWWB	27	WWBWWBBBBW	37	BBWWBBWWWB		
8	WWBWWWWWBW	18	WWBWWWBBBB	28	BWWWWWWWBB	38	BBWWWWBWBB		
9	BBWWBWWWWB	19	BWBWWWBBBW	29	BBWBWWBWWB				
10	BBWWWWWBWW	20	BBBWWWBWWB	30	WBBWWWWBWBW				
			counter-clockwise motion						
39	BWBBWBWWBW	41	WBBBWBWWBW	42	BWBWBWWWWBW	43	WWBBBBWWBB		
40	WWBBBWWWBB								

Table 2 Execution of the rules 1 up to 38: motion along the tracks in the clockwise direction, to the right-hand side of 0(0)

	4_8	3_8	2_8	1_7	1_6	1_5	1_4	2_4	5_3	4_3
1	25	24	18	4	16	16	4	16	16	4
2	4	30	24	27	16	16	4	16	16	4
3	4	16	30	19	18	16	4	16	16	4
4	4	16	16	25	24	18	4	16	16	4
5	4	16	16	4	30	24	27	16	16	4
6	4	16	16	4	16	30	19	18	16	4
7	4	16	16	4	16	16	25	24	18	4
8	4	16	16	4	16	16	4	30	24	27
9	4	16	16	4	16	16	4	16	30	19

Table 3 Execution of the rules 1 up to 38 on the left-hand side of 0(0): to left, in the clockwise direction; to right, in the counter-clockwise direction

	9_3	2_3	1_2	1_1	2_1	6_1			6_1	2_1	1_1	1_2	2_3	9_3
1	22	19	27	4	4	4		1	41	39	43	11	11	16
2	12	25	19	27	4	4		2	11	41	39	43	11	16
3	12	4	25	19	27	4		3	11	11	41	39	43	16
4	12	4	4	25	19	27		4	11	11	11	41	39	18

```
16  WWBWWWBWBW    18  WWBWWWBBBB    24  BWBWWWBWBW    30  WBBWWWBWBW
```

Indeed, in those cells, taking into account the numbering of the sides defined in Sect. 2, the milestones are neighbours 2, 6 and 8. Rule 16 is the conservative rule of the cell. Rule 18 can see the locomotive through neighbour 7, rule 24 can see the locomotive in the cell so that its neighbourhood is the same as that of rule 16, at last, rule 30 can see the locomotive in neighbour 1, witnessing that the locomotive is now in the next cell of the track.

Table 2 shows us that for the cells 4(8) and 1(4), the rules are:

```
4  WWBWWBWBBW    27  WWBWWBBBBB    19  BWBWWBWBBW    25  WBBWWBWBBW
```

The milestones of the cell are in neighbours 2, 5, 7 and 8. The conservative rule of the cell is rule 4. Rule 27 can see the arriving locomotive through neighbour 6, rule 19 can see it in the cell, so that its neighbouring is also that of rule 4. At last, rule 25 can see the locomotive in neighbour 1, witnessing that it is now in the next cell of the track. The left-hand side of Table 3 shows us that the same rules for four-milestoned cells are used as in the motion on the left-hand side, still in the clockwise motion.

Now, Table 4 shows us that other rules occur when the motion happens on the same tracks but in the opposite direction: counter-clockwise. For three-milestoned cells we have now the following rules:

```
12  WWBWBWWWBW    40  WWBBBWWWBB    42  BWBWBWWWBW    22  WBBWBWWWBW
```

Table 4 Execution of the rules 1 up to 43: the locomotive on the tracks in the counter-clockwise direction, on the right-hand side of 0(0)

	4_3	5_3	2_4	1_4	1_5	1_6	1_7	2_8	3_8	4_8
1	41	42	40	11	12	12	11	12	12	11
2	11	22	42	43	12	12	11	12	12	11
3	11	12	22	39	40	12	11	12	12	11
4	11	12	12	41	42	40	11	12	12	11
5	11	12	12	11	22	42	43	12	12	11
6	11	12	12	11	12	22	39	40	12	11
7	11	12	12	11	12	12	41	42	40	11
8	11	12	12	11	12	12	11	22	42	43
9	11	12	12	11	12	12	11	12	22	39

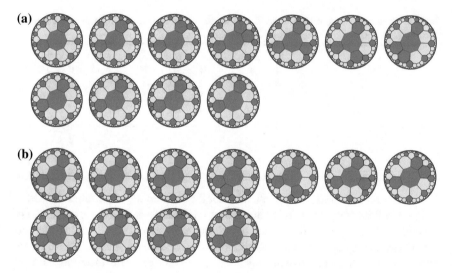

Fig. 11 Illustration of the motion of a simple locomotive in a clockwise motion (**a**) and in a counterclockwise one (**b**)

Note that the milestones are now in neighbours 2, 4 and 8 instead of 2, 6 and 8: the change of side 1 explains the new situation. The rules show that the locomotive is successively seen in neighbours 3, 0 and 1 as expected, where neighbour 0 is the cell itself.

For four-milestoned cells the rules are now:

```
11 W̲WBBWBWWBW̲    43 W̲WBBBBWWBB̲    39 B̲WBBWBWWBW̲    41 W̲BBBWBWWBW̲
```

The milestones are now in neighbours 2, 3, 5 and 8 instead of 2, 5, 7 and 8 for the same reason. The rules indicate that the locomotive is successively seen in neighbours 4, 0 and 1.

Figure 11 illustrates the executions given by Tables 2 and 4. The motion of the simple locomotive is performed on the tracks which lie on the right-hand side of the

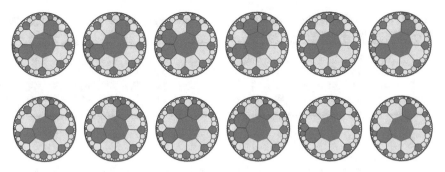

Fig. 12 Illustration of the motion of a simple locomotive on tracks in an counterclockwise motion

central cell in Fig. 10. In the first two rows the motion is clockwise. In the last two rows, the motion is counterclockwise on the same set of cells. Figure 12 illustrates the executions given by Tables 3 where the simple locomotive runs over the tracks illustrated by the left-hand side of cell 0(0) in Fig. 10. In the first row, the motion is clockwise, on the second one, it is counterclockwise.

Let us now look at the motion of a double locomotive on the same tracks, again in both directions. Table 5 gives the rules concerning that situation.

As can be seen on Tables 6 and 7 the rules for a double locomotive on the tracks makes use of the already mentioned rules. To these ones, we have to append rules

Table 5 Rules for the motion of a double locomotive on the tracks

44	BBWWBBBWWB	51	BBWWWWWWBB	58	BBBBWBWWBW	65	BBBWBWWWWB
45	BWBWWWBBBB	52	BBBBWWBWWB	59	BWBBBWWWBB	66	BWWBBWBWWB
46	BBBWBWBBBW	53	BBWBWWBWBB	60	BBBWBWWWBW	67	BBWWWWBBWB
47	BBBWWWWWWB	54	BWWBWWBBBB	61	BWBBBWWWBB	68	BBBBBWWWWB
48	BBWBBWBWWB	55	BWWWWWWBBB	62	WBBWWWWWBW	69	BWWBBBBWWB
49	BBBWWWBWBW	56	BBWBWWWBBB	63	BBWBWWWWWB	70	BBWWWWBBBB
50	BWBWWBBBBB	57	BBWWBWBBWB	64	BWWBWBBWWB		

Table 6 Execution of the rules 1 up to 70: motion in the clockwise direction for a double locomotive

	4_8	3_8	2_8	1_7	1_6	1_5	1_4	2_4	5_3	4_3
1	25	49	45	27	16	16	4	16	16	4
2	4	30	49	50	18	16	4	16	16	4
3	4	16	30	46	45	18	4	16	16	4
4	4	16	16	25	49	45	27	16	16	4
5	4	16	16	4	30	49	50	18	16	4
6	4	16	16	4	16	30	46	45	18	4
7	4	16	16	4	16	16	25	49	45	27
8	4	16	16	4	16	16	4	30	49	50

Table 7 Execution of the rules 1 up to 70, double locomotive, on the left-hand side of 0(0). To left: clockwise; to right: counterclockwise

	9_3	2_3	1_2	1_1	2_1	6_1			6_1	2_1	1_1	1_2	2_3	9_3
1	22	46	50	27	4	4		1	41	58	61	43	11	16
2	12	25	46	50	27	4		2	11	41	58	61	43	16
3	12	4	25	46	50	27		3	11	11	41	58	61	18

which are induced by the structure of the double locomotive. This has an impact on the motion rules and also on several milestones. From Table 6, we can see that for the three-milestoned cells, the sequence of rules is not 16, 18, 24 and 30. Rule 24 is no more used and it is replaced by the occurrence of two rules:

         45   BWBWWWBBBB   49   BBBWWWBWBW

Indeed, rule 18 applies as the cell can only see the front of the arriving locomotive. When the front of the locomotive is in the cell, its rear is in neighbour 7 so that rule 24 cannot be applied. But rule 45 does apply: it can see the front of the locomotive in the cell itself and the rear in neighbour 7. Next, rule 49 applies: the front of the locomotive has been absorbed by the next cell: it is in the neighbour 1 of the cell under examination. The rear is still here, so it must leave the cell, which is performed by rule 49. Now, when the rear of the locomotive is in neighbour 1, the cell cannot see the front of the locomotive. Accordingly, rule 30 again applies, witnessing that the locomotive left the cell.

We can perform a similar analysis for the cells with four milestones:

         50   BWBWWBBBBB   46   BBBWWBWBBW

In that case, where rules 4, 27, 19, and 25 apply for a simple locomotive, here rule 19 is no more used and is replaced by rules 50 and 46 which then allows rule 25 to be again applied.

We noticed for a simple locomotive that the counterclockwise motion involved new rules with respect to those managing a clockwise motion. The same occurs for the double locomotive. In a three-milestoned cell, instead of the sequence of rules 12, 40, 42 and 22 we have the rules 12, 40, 59, 60, and 22. Indeed, rule 42 cannot be applied for the same reason that the locomotive now occupies two consecutive cells of the track. Instead rule 42, two new rules are involved, rule 59 and 60, see below and Table 5. Similarly, for a four-milestoned cell, the sequence of rules 11, 43, 39 and 41 is replace by 11, 43, 61, 58 and 41 where the two new rules 61 and 58, see below and Table 5, replace rule 39 which cannot be applied here (Table 8).

59   BWBBBWWWBB   60   BBBWBWWWBW   61   BWBBBBWWBB   58   BBBBWBWWBW

We conclude the sub-section with a remark on the rules managing the milestones. Some of them have blank neighbours only around them when the locomotive is far from them. We shall see three of them: 2(7), 5(7) and 2(5). We shall see also two of

Table 8 Execution of the rules 1 up to 70 for a double locomotive in a counterclockwise motion

	4_3	5_3	2_4	1_4	1_5	1_6	1_7	2_8	3_8	4_8
1	41	60	59	43	12	12	11	12	12	11
2	11	22	60	61	40	12	11	12	12	11
3	11	12	22	58	59	40	11	12	12	11
4	11	12	12	41	60	59	43	12	12	11
5	11	12	12	11	22	60	61	40	12	11
6	11	12	12	11	12	22	58	59	40	11
7	11	12	12	11	12	12	41	60	59	43
8	11	12	12	11	12	12	11	22	60	61

Table 9 Rules witnessing the passage of the locomotive to the right-hand side of 0(0). Symbols $\curvearrowright$, $\curvearrowleft$ indicate clockwise motion, counterclockwise motions respectively

0(0):

simple, $\curvearrowright$	simple, $\curvearrowleft$	double, $\curvearrowright$	double, $\curvearrowleft$
3 BWWBWWBWWB	3 BWWBWWBWWB	3 BWWBWWBWWB	3 BWWBWWBWWB
31 BWBBWWBWWB	34 BWWBWWBBWB	31 BWBBWWBWWB	34 BWWBWWBBWB
29 BBWBWWBWWB	32 BWWBWWBWBB	52 BBBBWWBWWB	54 BWWBWWBBBB
32 BWWBWWBWBB	29 BBWBWWBWWB	53 BBWBWWBWBB	53 BBWBWWBWBB
34 BWWBWWBBWB	31 BWBBWWBWWB	54 BWWBWWBBBB	52 BBBBWWBWWB
3 BWWBWWBWWB	3 BWWBWWBWWB	31 BWBBWWBWWB	31 BWBBWWBWWB
		3 BWWBWWBWWB	3 BWWBWWBWWB

1(8):

simple, $\curvearrowright$	simple, $\curvearrowleft$	double, $\curvearrowright$	double, $\curvearrowleft$
17 BBWWWBBWWB	10 BBWWWBWWWB	44 BBWWBBBWWB	10 BBWWWBWWWB
23 BBWWBWBWWB	20 BBBWWWBWWB	48 BBWBBWBWWB	20 BBBWWWBWWB
29 BBWBWWBWWB	29 BBWBWWBWWB	52 BBBBWWBWWB	52 BBBBWWBWWB
20 BBBWWWBWWB	23 BBWWBWBWWB	20 BBBWWWBWWB	48 BBWBBWBWWB
10 BBWWWBWWWB	17 BBWWWBBWWB	10 BBWWWBWWWB	44 BBWWBBBWWB

2(7):

simple, $\curvearrowright$	simple, $\curvearrowleft$	double, $\curvearrowright$	double, $\curvearrowleft$
2 BWWWWWWWWB	2 BWWWWWWWWB	2 BWWWWWWWWB	2 BWWWWWWWWB
14 BBWWWWWWWB	26 BWBWWWWWWB	14 BBWWWWWWWB	26 BWBWWWWWWB
26 BWBWWWWWWB	14 BBWWWWWWWB	47 BBBWWWWWWB	47 BBBWWWWWWB
2 BWWWWWWWWB	2 BWWWWWWWWB	26 BWBWWWWWWB	14 BBWWWWWWWB
		2 BWWWWWWWWB	2 BWWWWWWWWB

5(7):

simple, $\curvearrowright$	simple, $\curvearrowleft$	double, $\curvearrowright$	double, $\curvearrowleft$
2 BWWWWWWWWB	2 BWWWWWWWWB	2 BWWWWWWWWB	2 BWWWWWWWWB
28 BWWWWWWWBB	14 BBWWWWWWWB	28 BWWWWWWWBB	14 BBWWWWWWWB
14 BBWWWWWWWB	28 BWWWWWWWBB	51 BBWWWWWWBB	51 BBWWWWWWBB
		14 BBWWWWWWWB	28 BWWWWWWWBB

2(5):

simple, $\curvearrowright$	simple, $\curvearrowleft$	double, $\curvearrowright$	double, $\curvearrowleft$
2 BWWWWWWWWB	2 BWWWWWWWWB	2 BWWWWWWWWB	2 BWWWWWWWWB
33 BWWWWWWBWB	28 BWWWWWWWBB	33 BWWWWWWBWB	28 BWWWWWWWBB
28 BWWWWWWWBB	33 BWWWWWWBWB	55 BWWWWWWBBB	55 BWWWWWWBBB
		28 BWWWWWWWBB	33 BWWWWWWBWB

them, 0(0) and 1(8) which have two milestones among their neighbours, see Fig. 10. Table 9 gives the rules used for those cells during the motion of the locomotive on the right-hand side of 0(0) for the four cases: the simple locomotive clockwise running, then for the same one counterclockwise running, then for the double locomotive when it clockwise traverses the tracks and, at last, for the same locomotive when it counterclockwise traverses the tracks.

Table 10 Rules witnessing the passage of the locomotive to the left-hand side of 0(0). Symbols ↶, ↷ indicate clockwise motion, counterclockwise motions respectively

0(0):	simple, ↷		simple, ↶		double, ↷		double, ↶	
	3	BWWBWWBWWB	3	BWWBWWBWWB	3	BWWBWWBWWB	3	BWWBWWBWWB
	64	BWWBWBBWWB	66	BWWBBWBWWB	64	BWWBWBBWWB	66	BWWBBWBWWB
	66	BWWBBWBWWB	64	BWWBWBBWWB	69	BWWBBBBWWB	69	BWWBBBBWWB
					66	BWWBBWBWWB	64	BWWBWBBWWB

2(2):	simple, ↷		simple, ↶		double, ↷		double, ↶	
	2	BWWWWWWWWB	2	BWWWWWWWWB	2	BWWWWWWWWB	2	BWWWWWWWWB
	14	BBWWWWWWWB	26	BWBWWWWWWB	14	BBWWWWWWWB	26	BWBWWWWWWB
	26	BWBWWWWWWB	14	BBWWWWWWWB	47	BBBWWWWWWB	47	BBBWWWWWWB
					26	BWBWWWWWWB	14	BBWWWWWWWB

(a)

(b)

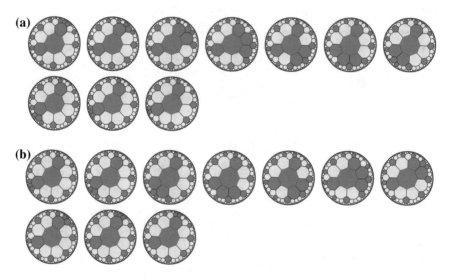

Fig. 13 Illustration of the motion of a double locomotive in a clockwise motion (**a**) and in a counterclockwise one (**b**)

Table 10 gives the rules for the cells 0(0) and 2(2) when the locomotive runs on the left-hand side of the cell 0(0), see Fig. 10, again in the four possible cases. We can see that the rules for 2(2) are those for 2(7) but in the reverse order (Figs. 13 and 14).

It can be noted that all rules given in Tables 9 and 10 do not change the current state of the cell. As they witness the passage of the locomotive, we cannot call them conservative rules. As already noticed, for several milestones, the same rules may be used for the motion of the locomotive in one direction and in the opposite one. The difference appears in the order in which the rules are applied: this is the case, whether a simple or a double locomotive.

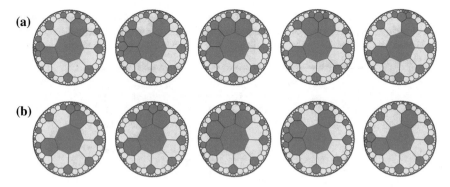

Fig. 14 Illustration of the motion of a double locomotive in a clockwise motion (**a**) and in a counterclockwise one (**b**)

Table 11 Rules for the crossing of a fixed switch

simple locomotive

71	WBBWBWBBWW	76	WBBBWWWWBW	81	BWBWWWWWBB	86	BWBWWWWBWB
72	BWBWBWWWWB	77	BWWWBWBWWB	82	BWWWBWWWBB	87	BWBWWWBWWB
73	WWWWWBWBBW	78	WWBBWWWWBW	83	BBBWBWBBWW	88	BWBWWBWWWB
74	WBWBWWWWWW	79	BWWWBWWBWB	84	WBBWBWBBBW	89	WBBBBWBBWB
75	BWWWBBWWWB	80	WBBWBBBBWB	85	BWWWBWWWWB		

double locomotive

90	BWWWBBBWWB	93	BBBWBBBBWB	96	BWBWWWWBBB	99	BBBBBWBBWB
91	BWWWBWBBWB	94	BBBWWWWWBB	97	BWBWWWBBWB		
92	BWWWBWWBBB	95	BBBWBWBBBW	98	BWBWWBBWWB		

3.2 The Rules for the Fixed Switch

We now turn to the study of the fixed switch, a passive structure as noted in Sect. 2.2. Table 11 gives new rules which are used for the crossing of the structure together with the rules given in Tables 1 and 5.

Table 12 shows us that the rules we have seen for a simple locomotive on the tracks are also used here: both the sequence 16, 18, 24 and 30 and 12, 40, 42 and 22. For 0(0), the central cell of the switch, we have the sequence of rules: 71, 80, 83, 84 when the locomotive arrives to the left-hand side of the switch:

71 WBBWBWBBWW   80 WBBWBBBBWB   83 BBBWBWBBWW   84 WBBWBWBBBW

We can see that rule 71 is the conservative rule of 0(0). The cell has five milestones in its neighbours 1, 2, 4, 6 and 7. A locomotive arriving from the left is seen in neighbour 5 as shown by rule 80. Rule 83 witnesses that the simple locomotive is in the cell: it has the same neighbourhood as rule 71. At last, rule 84, which can see the locomotive in its neighbour 1, witnesses that the locomotive left the cell 0(0).

Table 13 shows us that when the locomotive comes from the right, the sequence 71, 80, 83, 84 is replaced by 71, 89, 83, 84: rule 80 is replaced by rule 89 WBBBBWBBWB,

Table 12 Execution of the rules for the fixed switch when a simple locomotive comes from the left-hand side. Note that the numbers of the cells used by the track on which the locomotive travelled are taken with the standard numbering of the cell 0(0) which is different to the numbering of that cell in the fixed switch

	4_8	5_8	2_1	1_1	0_0	1_4	2_4	5_3	4_3
1	41	42	40	12	71	16	16	16	12
2	11	22	42	40	71	16	16	16	12
3	11	12	22	42	80	16	16	16	12
4	11	12	12	22	83	18	16	16	12
5	11	12	12	12	84	24	18	16	12
6	11	12	12	12	71	30	24	18	12
7	11	12	12	12	71	16	30	24	40

Table 13 Execution of the rules for the fixed switch when a simple locomotive comes from the right-hand side

	4_6	5_6	2_7	1_7	0_0	1_4	2_4	5_3	4_3
1	30	42	40	12	71	16	16	16	12
2	16	22	42	40	71	16	16	16	12
3	16	12	22	42	89	16	16	16	12
4	16	12	12	22	83	18	16	16	12
5	16	12	12	12	84	24	18	16	12
6	16	12	12	12	71	30	24	18	12
7	16	12	12	12	71	16	30	24	40

where we can see that the locomotive is in neighbour 3, which corresponds to an arriving locomotive from the right.

In Table 14 we can see the application of the rules when a double locomotive crosses the switch. The sequences of rules we have noticed for the three-milestoned cells of the track are replaced by the corresponding ones for a double locomotive. Also, the sequence of cells for the cell 0(0) is a bit changed as follows:

93 BBBWBBBBWB   99 BBBBBWBBWB   95 BBBWBWBBBW

When the locomotive comes from the left, rule 80 is replaced by rule 93, when it comes from the right, rule 89 is replaced by rule 99. In both cases, rule 83 is replaced by rule 95 and then, rule 84 applies as in the case of a simple locomotive.

Figure 15 illustrates the four motions we have to consider for the fixed switch for which Tables 12, 13 and 14 give the rules used for such motions.

As in Sect. 3.1, we conclude the sub-section with a study of a specific milestone of the configuration: the cell 1(8), see Table 15. Rule 85 is the conservative rule of the cell. The cell itself is black, as required for a milestone. It has a single black neighbour which is a milestone for the tracks arriving to the switch from the left.

Table 14 Execution of the rules for the fixed switch when a double locomotive crosses the switch

	4_8	5_8	2_1	1_1	0_0	1_4	2_4	5_3	4_3
	\multicolumn{9}{c}{from the left-hand side}
1	41	60	59	40	71	16	16	16	12
2	11	22	60	59	80	16	16	16	12
3	11	12	22	60	93	18	16	16	12
4	11	12	12	22	95	45	18	16	12
5	11	12	12	12	84	49	45	18	12
6	11	12	12	12	71	30	49	45	40

from the left-hand side

	4_8	5_8	2_1	1_1	0_0	1_4	2_4	5_3	4_3
1	41	60	59	40	71	16	16	16	12
2	11	22	60	59	80	16	16	16	12
3	11	12	22	60	93	18	16	16	12
4	11	12	12	22	95	45	18	16	12
5	11	12	12	12	84	49	45	18	12
6	11	12	12	12	71	30	49	45	40

from the right-hand side

	4_6	5_6	2_7	1_7	0_0	1_4	2_4	5_3	4_3
1	30	60	59	40	71	16	16	16	12
2	16	22	60	59	89	16	16	16	12
3	16	12	22	60	99	18	16	16	12
4	16	12	12	22	95	45	18	16	12
5	16	12	12	12	84	49	45	18	12
6	16	12	12	12	71	30	49	45	40

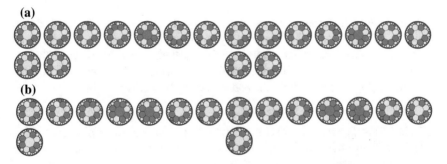

(a)

(b)

Fig. 15 Illustration of the motion of a double locomotive in a clockwise motion (**a**) and in a counterclockwise one (**b**)

Table 15 Rules witnessing the passage of the locomotive. Symbols $\rightarrow$, $\leftarrow$ indicate a motion from the left, the right respectively

1(8):	simple, $\rightarrow$		simple, $\leftarrow$		double, $\rightarrow$		double, $\leftarrow$	
	75	BWWWBBWWWB	85	BWWWBWWWWB	90	BWWWBBBWWB	85	BWWWBWWWWB
	77	BWWWBWBWWB	72	BWBWBWWWWB	91	BWWWBWBBWB	72	BWBWBWWWWB
	79	BWWWBWWBWB	9	BBWWBWWWWB	92	BWWWBWWBBB	65	BBBWBWWWWB
	82	BWWWBWWWBB	85	BWWWBWWWWB	35	BBWWBWWWBB	9	BBWWBWWWWB
	9	BBWWBWWWWB			9	BBWWBWWWWB	85	BWWWBWWWWB
	85	BWWWBWWWWB			85	BWWWBWWWWB		

The cell has two windows on the tracks: its neighbour 2 looks at a cell of the tracks arriving to the switch from the right, see rule 72 in the case of a simple locomotive. Neighbours 5, 6, 7 and 8 offer a view on the tracks arriving to the switch from the left: see rules 75, 77, 79 and 82 in the case of a simple locomotive. At last, neighbour 1 is cell 0(0), the central cell of the switch. Note that rule 9 is common to all motions: the last cell of the locomotive is seen in the cell 0(0) which is neighbour 1 for 1(8).

When the locomotive is double, the rules are different when the locomotive arrives from the left: the window consists of four consecutive cells so that the two cells are visible for three tops of the clock as witnessed by rules 90, 91 and 92. When the last cell of the locomotive is in 1(1), i.e. neighbour 8 for 1(8), the front of the locomotive is in 0(0), which is neighbour 1 for 1(8), see rule 35. Note that rule 35 appears in Table 1. That rule is also used by the cell 1(3) when the locomotive is in the cell 1(4): it occurs in the motions on the track we already studied. It also occurs here, when a simple locomotive crosses the switch whatever the side from which it came.

3.3   The Rules for the Doubler and for the Fork

Presently, we look at the new rules involved by the doubler and by the fork, see Table 16. The rules for the fork are mainly contained in those for the doubler as the doubler contains a fork. It also contains a fixed switch so that the rules of Table 11 should also be involved. The function of the fork is performed by the configuration of the cell 1(1) together with the sequence of rules 100, 104, and 106:

100   WWBWBWBWBW   104   WWBWBBBWBB   106   BWBWBWBWBW   84   WBBWBWBBBW

Rule 100 is the conservative rule for 1(1). Note that two tracks have their origin in 1(1). We chose side 1 to be shared with the cell 1(8). This is why rule 104 can see the arriving locomotive in neighbour 5. Rule 106 witnesses the locomotive in the cell and makes it leave the cell. Rule 84 can see the locomotive in both neighbours 1, i.e. 1(8), and 7 which is 1(2). The left-hand side part of Table 17 shows us that starting from the application of rule 84, each locomotive goes on its way in the tracks: the locomotive created in 1(2) counterclockwise goes on, that created in 1(8) clockwise goes on.

Note that rule 84 is also applied when a locomotive crosses a fixed switch: it witnesses that the locomotive left the central cell of the switch. We also note that

Table 16   Rules for the motion of the locomotive through the doubler and through the fork

100 WWBWBWBWBW	106 BWBWBWBWBW	112 BBBWWWWBWB	118 BWWWWWBBBW	
101 BWWWWBWWWB	107 BWWWBWBBBB	113 BBWBWWWWB		
102 BWWBWWWWWB	108 BWWBWWWBBB	114 BBBWBWWWWB	fork	
103 WWWWWBWBW	109 BBWWBWWBB	115 BWWBWWWWBB		
104 WWBWBBBWBB	110 BBWWWBWWBB	116 BBBWWBWWWB	119 BWWWWBWWWB	
105 BWWWWBWBBB	111 BBWWWBWWWB	117 WWWWWWBBBB	120 BWWBWWWBWB	

Table 17 Execution of the rules for the doubler. To left, around the cell 1(1), the fork; to right, around the cell 4(4), the fixed switch

	5_1	1_1	1_2	2_3	1_8	1_7		2_4	3_4	4_4	5_4	2_5	15_4
1	24	104	11	12	16	16	5	11	12	71	16	16	16
2	30	106	43	12	18	16	6	11	12	71	16	18	16
3	16	84	39	40	24	18	7	43	12	71	18	24	16
4	16	100	41	42	30	24	8	39	40	89	24	30	16
5	16	100	11	22	16	30	9	41	60	93	30	16	18
							10	11	22	95	30	16	45
							11	11	12	84	16	16	49

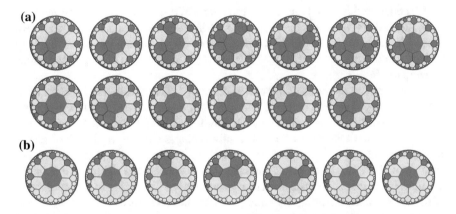

Fig. 16 Illustration of the crossing of the doubler (**a**), and of the fork (**b**) by the locomotive

rule in the right-hand side part of Table 17 which deals with the fixed-switch part of the doubler.

The upper two rows of Fig. 16 illustrate the crossing of the doubler by a locomotive. The last row of the figure illustrates the working of the fork: we can see that it is the same as the one which is installed in the doubler.

The doubler also contains a fixed switch, as already mentioned in Sect. 2. The right-hand side part of Table 17 shows us that the two locomotives arrive at the cell 4(4) with a time difference of one top of the clock as already mentioned. We can see that the locomotive which moved in a clockwise way arrives at the cell 4(4) sooner than the other one, so that at time 8, see Table 17, we have a double locomotive in the fixed switch with its front in the cell 4(4) and its rear in the cell 3(4), the neighbour 5 of the cell 4(4):

```
71  WBBWBWBBWW    89  WBBBBWBBWB    93  BBBWBBBBWB
        95  BBBWBWBBBW    84  WBBWBWBBBW
```

Table 18 Rules used at the milestones 1(3) and 1(4) of the doubler

	1(3)				1(4)		
51	BBWWWWWBB	111	BBWWWBWWBB	47	BBBWWWWWWB	52	BBBBWWBWWB
94	BBBWWWWWBB	38	BBWWWWBWBB	94	BBBWWWWWBB	114	BBBWBBWWWB
109	BBWBWWWWBB	110	BBWWWWWBBB	112	BBBWWWWBWB	116	BBBWBWWWWB
35	BBWWBWWWBB	51	BBWWWWWWBB			47	BBBWWWWWWB

Note that the sequence of rules is a bit different from that of Table 14: the situation is a kind of mix between the two parts of Table 14 as here, the double locomotive arrives as two simple locomotive coming from *both* sides of the switch.

We conclude this subsection by two sequences of rules concerning milestones: the cells 1(3) and 1(4) of the doubler. The former witnesses the passage of the locomotive around it on six steps. The latter is a neighbour of 4(4), the central cell of the fixed switch, which is also a neighbour of 3(4) and 5(4), cells which belong to the tracks arriving to 4(4).

Table 18 shows us the rules applied to the cells 1(3) and 1(4). The rules for each cell are displayed on two columns and they follow each other from top to bottom first in the left-hand side column and then in the right-hand side one. We can see that for 1(3), the window of six consecutive cells is clear as well as the motion of the locomotive in the window. The cell 1(4) also has a window of six cells but their display is a bit different and the rules do not concern the rightmost cell of the window. In the first column devoted to 1(4), we can see the motion of the locomotive which arrives from the right. The motion of both locomotives clearly appears in the second column. We can see the constitution of the double locomotive, see rule 52 followed by rule 114. We also can see that the front corresponds to the right-hand side locomotive which arrives to 4(4) as the first one. Rule 116 shows us that the rear of the double locomotive is in 4(4). The right-hand side part of Table 17 shows us that when this is the case, rule 49 shows us that the front of the double locomotive arrived in 15(4). Table 18 shows us that the cell 4(4) did the job more clearly than Fig. 16.

3.4 The Rules for the Selector

The selector is the last structure we need to implement round-abouts. The new rules needed by the structure are given in Table 19 while the execution of the rules used by the crossing of a locomotive are given in Table 20: the left-, right-hand side sub-table gives the rules used by a simple, double locomotive respectively.

In both sub-tables of Table 20, we can see that the track leading the locomotive to the selector makes use of motion rules examined in Sect. 3.1. We can see that the entrance to the selector, the cell 1(6), makes use of the same neighbourhood as

Table 19 Rules for the locomotive through the selector

simple locomotive

```
121  BWWWWBBBWB  128  BWWWBBWBBB  135  BBWBBBWWWB  142  BWWBBBWWBB
122  WWBWWBBWBW  129  BWBBWBWBBW  136  WWWWBBBWBB  143  BWWWBWBWBW
123  WWBBWBWBBW  130  BBBBWWWWWB  137  WBBWWBWWBW  144  WWBWBWBBBW
124  BWWBBBWWWB  131  BWBBBBWWWB  138  BWWWWWBWBW  145  BWBWWBBBWB
125  WWWWBWBWBW  132  WWBBWWBWBB  139  WWBWWBBBWB  146  BWWBBBWBWB
126  BWWWWBBBBB  133  BBWWWBBBWW  140  WWBWWWWBBW  147  WWWWBWBBBW
127  WWBBWBBBBB  134  WBBBWBWBBW  141  WWWBWBWBBW
```

double locomotive

```
148  BBWBBWWWBB  152  BBBBWBWBBW  156  BBBWWBBBWW  159  BWWWBBWWBW
149  BWBBWBBBBB  153  BBBBBBWWWW  157  WBBBWBWBWW  160  BWWBWBBBWB
150  BBBBWWWWBB  154  BBBWWBBWWW  158  WBWBBBWWBB
151  BBWWWBBBBB  155  BWBBWWBWBW
```

Table 20 Execution of the rules for a locomotive passing through the selector

	6_7	5_6	1_6	0_0	1_8	2_8	9_8		6_7	5_6	1_6	0_0	1_4	2_5	7_5
	\multicolumn{7}{c}{simple locomotive}			\multicolumn{7}{c}{double locomotive}
1	22	24	127	16	125	12	16	1	22	49	149	18	16	16	12
2	12	30	129	18	125	12	16	2	12	30	152	45	132	16	12
3	12	16	134	24	136	12	16	3	12	16	157	154	155	18	12
4	12	16	141	137	143	40	16	4	12	16	123	16	30	24	40
5	12	16	123	16	147	42	18	5	12	16	123	16	16	30	42

the fixed switch. But as side 1 for 1(6) is different, the rules managing the cell are different from those used in the fixed switch for the cell 0(0).

The sequence of rules used by 1(6) when a simple locomotive crosses it is given by: 123, 127, 129, 134, 141:

123 WWBBWBWBBW   127 WWBBWBBBBB   129 BWBBWBWBBW

134 WBBBWBWBBW   141 WWWBWBWBBW

Rule 123 is the conservative rule for 1(6), see Fig. 5. Rule 127 can see the locomotive arrived in neighbour 6 which is the cell 5(6). Rule 129 can see the locomotive in the cell and it makes it leave the cell. Rule 134 witnesses that the locomotive is now in the cell 0(0). Rule 141 can see that the cell 1(5) turned to white and then the next time, rule 123 again applies, witnessing that the cell 1(5) turned back to black. We shall see the effect of this flash issued by the cell 1(5).

The sequence of rules used by 1(6) when the locomotive is double is given by the sequence: 123, 127, 149, 152, 157:

123 WWBBWBWBBW   127 WWBBWBBBBB   149 BWBBWBBBBB

152 BBBBWBWBBW   157 WBBBWBWBWW

The first two rules are the same as previously. Now, after that, rule 149 is needed to make the rear of the double locomotive enter the cell 1(6). Then, rule 152 makes 1(6) turn back to white as the locomotive, seen in neighbour 1, is now in the cell 0(0).

Table 21 The rules used by the sensors 1(5), to left, and 1(7), to right, of the selector

	simple		double		simple		double
121	BWWWWBBBWB	121	BWWWWBBBWB	124	BWWBBBWWWB	124	BWWBBBWWWB
126	BWWWWBBBBB	126	BWWWWBBBBB	131	BWBBBBWWWB	131	BWBBBBWWWB
133	BBWWWBBBWW	151	BBWWWBBBBB	135	BBWBBBWWWB	153	BBBBBBWWWW
139	WWBWWBBBWB	156	BBBWWBBBWB	142	BWWBBBWWWB	158	WBWBBBWWBB
121	BWWWWBBBWB	160	BWWBWBBBWB	146	BWWBBBWBWB	124	BWWBBBWWWB
		121	BWWWWBBBWB	124	BWWBBBWWWB		

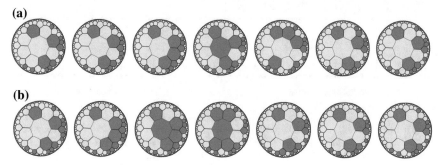

(a)

(b)

Fig. 17 Illustration of the crossing of the selector: (**a**) by a simple locomotive, (**b**) by a double one

Then, rule 157 can see the rear of the locomotive in 0(0) and it also can see that the cell 1(7) turned to white as neighbour 8 in the rule is white. As rule 123 again applies, there is no locomotive in 0(0) and 1(7) turned back to black.

Table 21 gives the rules used by the cells 1(5) and 1(7) which we call the **sensors** of the selector.

For each sensor, we give the rules for a passage of a simple locomotive and those for the passage of a double one. Let us first look at the cell 1(5). The conservative rule is rule 121 and the position of the three consecutive black neighbours shows us that side 1 is shared with 0(0). Rule 126 witnesses that the locomotive is in the cell 1(6). Rule 133 can see the locomotive in 0(0) and as the locomotive is simple, cell 1(6) is now white, so that the sensor 1(5) flashes: it becomes white. Now, rule 139 shows us that there is a locomotive in 1(4) and the rule makes the cell 1(5) return to black. As cell 121 again applies, we can see that the locomotive which was in 1(4) vanished as the cell 2(5) which belongs to the track starting from 1(4) is white. Also see Fig. 17.

Consider the case when a double locomotive crosses the selector: for 1(5), it is the second column of Table 21. We can see that rule 151 detects that the locomotive is double: both neighbours 1, 0(0), and 8, 1(6), are black. Next, rule 156 can see the rear of the locomotive in 0(0) and a new one in 1(4). Then, rule 160 witnesses that there is no locomotive in 0(0), nor in 1(4), but that the locomotive which was previously in 1(4) is now in 2(5). This means that when the locomotive is simple, it does not go through the track starting from 1(4), but when it is double, then a simple locomotive is sent on the track starting from 1(4). This is illustrated by the first row of Fig. 17.

Let us now study what happens at the sensor 1(7). It has the same neighbourhood as the sensor 1(5). Its side 1 is also shared with 0(0), but as the numbering is always defined with the same orientation, rule 121 cannot apply and a new rule is needed: rule 124. The detection of the front of the locomotive is detected by rule 131. Rule 135 detects that a simple locomotive arrived. Accordingly, the sensor remains black. After that, rules 142 and 146 witness that a simple locomotive was in 1(8) and 2(8) respectively: the locomotive goes on the track issued from 1(8).

If a double locomotive arrives, rule 131 again applies and at the next step, when the front of the locomotive is in 0(0), its rear is in 1(6) which is detected by rule 153 which makes the sensor flash: it becomes white. Rule 158 restores the black colour of the sensor and witnesses that a simple locomotive is in 1(8) and that the rear of the double locomotive is in 0(0). Rule 124 shows that, at the next time, there is no locomotive in 0(0), neither in 1(8) nor in 2(8). This is illustrated by the second row of Fig. 17. Accordingly, the rules up to the last one of Table 19 allow the selector to work as indicated in Sect. 2.

3.5   The Rules for the Controller

Let us now consider the rules for the controller of the active switches. The rules are displayed by Table 23. As mentioned in the table itself, the two columns in the left-hand side deal with the passage of the locomotive while the last column deals with the change of colour of the controller. We remind the reader that the colour of the controller is the colour of the cell 1(3) in Fig. 7. Table 24 and Fig. 23 illustrate the crossing of a black controller by the locomotive. All cells of the track obey the rules we have considered for the tracks for three-milestoned rules but the cell 1(4) for which specific rules are used, although it is also a three-milestoned cell. Let us look at those specific rules as well as the rules used for the cell 1(3), also specific rules, all of them repeated in Table 22.

Let us first consider the cell 1(4) when the locomotive crosses a black controller. Rule 164 is the conservative rule. The next cell on the track is 2(5) so that the

Table 22 Rules for 1(4) and for 1(3), when the locomotive passes and when the signal passes

	cell 1(4)			cell 1(3)	
	locomotive black	signal black		locomotive black	signal black
164	WWBWBWBWWW	164 WWBWBWBWWW	68	BBBBBWWWWB	68 BBBBBWWWWB
167	WWBBBWBWWB	180 WWBWBBBWWW	163	BBBBBWWBWB	153 BBBBBWWWWW
168	BWBWBWBWWW	173 WWBWWWBWWW	166	BBBBBWBWWB	172 WBBBBWWWWW
169	WBBWBWBWWW		68	BBBBBWWWWB	
164	WWBWBWBWWW	white			white
	white	173 WWBWWWBWWW		white	172 WBBBBWWWWW
		178 WWBWWBBWWW	172	WBBBBWWWWW	177 WBBBBWWWWB
173	WWBWWWBWWW	164 WWBWBWBWWW			68 BBBBBWWWWB

Table 23 Rules for the control: passage of the locomotive and signal for changing the selected track

	passage of the locomotive			signal
	black		white	$W \to B$
161	WWWWWWWBBW	170	WWBWBWWWWW	175 WWBBBWWWWW
162	WBWWWWWBBW	171	WWWWWWWBWW	176 WWWBBBWWBB
163	BBBBBWWBWB	172	WBBBBWWWWW	177 WBBBBWWWWB
164	WWBWBWBWWW	173	WWBWWWBWWW	178 WWBWWBBWWW
165	WWWWBWWWWW	174	WWWBWBWWBW	179 BWWBWBWWBW
166	BBBBBWBWWB			
167	WWBBBWBWWB			$B \to W$
168	BWBWBWBWWW			180 WWBWBBBWWW
169	WBBWBWBWWW			

Table 24 Execution of the rules used during the traversal of a black controller by the locomotive

	6_7	2_7	1_6	0_0	1_4	2_5	3_5	10_5	1_3
1	22	42	18	12	164	16	16	16	68
2	12	22	24	40	164	16	16	16	68
3	12	12	30	42	167	16	16	16	163
4	12	12	16	22	168	132	16	16	166
5	12	12	16	12	169	24	18	16	68

side shared by those cells is side 1 for 1(4), whence the new rules 164, 167, 168 and 169 although 1(4) is a three-milestoned cell. The reason is that in the usual situation of the tracks, the exit occurs through a cell which stands in between two milestones. This is not the case here: this requires specific rules. To these rules, we could append corresponding rules for the case of a double locomotive: BWBBBWBWWB and BBBWBWBWWW. Note that the following rules:

WWWWBWBWBW   136   WWWWBBBWBB   143   BWWWBWBWBW   WBWWBWBWBW

are compatible with the present rules. To these rules, it would also be possible to use rules for a double locomotive, looking these cells as tracks only: rules BWWWBBBWBB and BBWWBWBWBB. As can be expected, cell 1(4) is not affected when the locomotive arrives to a white controller. Rule 173 is the conservative rule in that case and it is always applied during such a passage: the locomotive does not even arrive to 0(0).

When the signal arrives to the controller in order to change the colour in the cell 1(3), the cell 1(4) remains white and it simply witnesses the arrival of the signal to the cell 2(4) and the corresponding change in 1(3): rule 180 when 1(3) is black, rule 178 when it is white.

Presently, consider the case of the cell 1(3). Rule 68 is the conservative rule when the controller is black, rule 172 is the conservative rule when it is white. When 1(3) is black, rule 163 can see the locomotive in 0(0), its neighbour 7, and rule 166 can see it in 1(4), its neighbour 6. When 1(3) is white, it remains unchanged when the locomotive arrives and remains under rule 172 for the same reason as the cell 1(4) remains under rule 173.

Table 25 Execution of the rules when the locomotive arrives to a white controller and when the signal for changing the colour arrives

	locomotive, 1(3) white				signal for changing the colour							
	6_7	2_7	1_6	0_0		$W \rightarrow B$				$B \rightarrow W$		
						1_3	6_4	2_4		1_3	6_4	2_4
1	22	42	18	170					1	153	30	39
2	12	22	24	175	1	177	30	179	2	172	16	174
3	12	12	16	170	2	68	16	11				

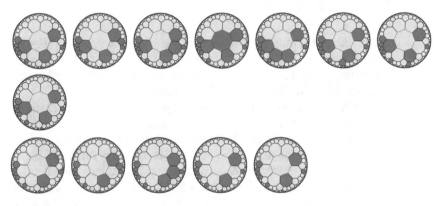

Fig. 18 Illustration of the crossing of the controller by the locomotive: (**a**) and first figure of the second row, when it is black, below after the first figure, when it is white

(a)   **(b)**

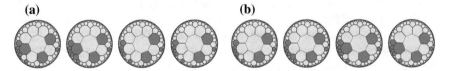

Fig. 19 Illustration of the arrival of the signal to the controller. (**a**) from black to white; (**b**) from white to black

The cell 1(3) can see the signal when it is in the cell 2(4) only: the cell 6(4) is not visible from 1(3). The change from black to white is performed by rule 153 and the change from white to black is performed by rule 177. In both cases, the signal is seen in neighbour 5, see Table 22. See Table 25 too.

Figure 18 illustrates the crossing of controller by the locomotive in both cases, according to the colour of 1(3). Figure 19 illustrates the arrival of the signal for changing the colour of the controller.

Table 26 Rules for the sensor of the passive memory switch

	black		white		B $\rightarrow$ W
181	WWBBWBBWWW	184	WWWBWBBWWW	189	WWBBBBBWWB
182	WWWBWWWWWW	185	WWBWWWWBWW	190	BWBBWBBWWW
183	WWBBBWBWBW	186	WBBBBWWBWB	191	WWBBWWWWWW
		187	WBBWWWWBWW		
		188	WBBWBWBWBW		

3.6   The Rules for the Sensor

In this last subsection of Sect. 3, we examine the rules which manage the working of the sensor, the specific control structure of the passive memory switch. Section 2.3 in Sect. 2 explained the working of the structure, pointing at the differences between the controller and the sensor illustrated by Figs. 7 and 9.

Table 26 illustrates the rules which have to be appended to the already examined ones in order to make the structure working as expected.

A few rules are appended to the 180 previous rules examined in the previous sections. As can be seen in the comparison of Figs. 7 and 9, many rules used for the controller are also used for the sensor. As an example, as long as the sensor is white, the rules executed in the cells of the tracks when the locomotive passes are the same as those used in the same action when the controller is black, see Tables 24 and 27.

Besides the fact that the cell 1(3) is no more considered in Table 27, the difference is in the rules concerning the cell 0(0) starting from time 4. The reason is that the passage of the locomotive through the white sensor transforms it into a black sensor, see the first row of Fig. 20.

Let us look at what happens at the cell 0(0). Instead of the rules 12, 40, 42, 22 for the passage of a simple locomotive, where rule 12 is the conservative rule of 0(0) in the white sensor, we have the sequence 12, 40, 42, 188, 100:

$$188 \quad \text{WBBWBWBWBW} \qquad 100 \quad \text{WWBWBWBWBW}$$

where rule 100 is the conservative rule for 0(0) in a black sensor. The comparison of rule 188 with rule 12, WWBWBWWWBW, shows us that the locomotive is in 1(4), which

Table 27 Execution of the rules when the sensor is white and then a locomotive passes

	6_7	2_7	1_6	0_0	1_4	2_5	3_5	10_5
1	22	42	18	12	164	16	16	16
2	12	22	24	40	164	16	16	16
3	12	12	30	42	167	16	16	16
4	12	12	16	188	168	132	16	16
5	12	12	16	100	169	24	18	16

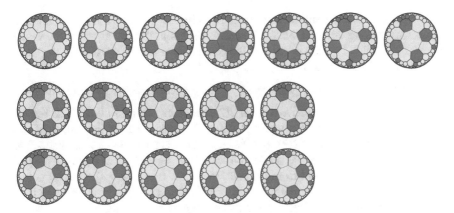

Fig. 20 Illustration of the working of the sensor. First row: passage of the locomotive; second row: the locomotive is stopped; third row: the signal changes a black sensor to a white one

means that it left the cell $0(0)$, and that the cell $1(1)$, neighbour 6 of $0(0)$, is now black. Rule 100 confirms that the cell $1(1)$ remains black: the sensor turned to black.

Table 28 displays the rules in the cells of the tracks when the locomotive arrives at a black sensor as well as the rules used by the cells of the track which conveys the signal for changing the colour, see the second and the last rows of Fig. 20 respectively.

Let us look at the cell $1(1)$ which gives its colour to the sensor. The conservative rule for $1(1)$ is rule 68 when it is black, rule 172 when it is white. The change of colour are induced by rule 186 for W $\rightarrow$ B and by rule 153 for B $\rightarrow$ W. We have already seen rule 153 which operates the same change for the cell $1(3)$ which gives the colour of the controller. In rule 186, WBBBBWWBWB, we can see that the change is triggered by the arrival of the locomotive at the cell $0(0)$ which is neighbour 7 in $1(1)$. In the controller, $0(0)$ is also neighbour 7 for $1(3)$. But rule 186 cannot be applied in the controller: when $1(3)$ is white in the controller, the locomotive never occurs in $0(0)$. In the controller, the change was triggered by the occurrence of the signal in neighbour 5. This is why we again find rule 153 here for turning the colour from black to white, as in the case of the controller. Note that the conservative rules for the cell $1(1)$ are the same as those for the cell $1(3)$ in the controller.

Table 28 Execution of the rules for the black sensor, for the locomotive and for the signal

	locomotive						signal			
	6_7	2_7	1_6	0_0	1_4		10_5	1_1	6_2	2_2
1	22	42	18	100	164	1	16	153	30	190
2	12	22	24	183	164	2	16	172	16	184
3	12	12	16	100	164					

We observe that the sub-table of Table 28 ruling the signal is almost the same as the corresponding one in Table 25. The difference is in the cell at which the signal arrives: the cell 2(4) in the controller, the cell 2(2) here, in the sensor. For the black sensor, the rules for 2(2) are 181, 189, 190, 184:

181 WWBBWBBWWW   189 WWBBBBBWWB   190 BWBBWBBWWW   184 WWWBWBBWWW

Rule 181 is the conservative rule for 2(2) in a black sensor. Rule 189 can see the arriving signal in neighbour 4 which is 6(2) and it makes it enter the cell. Rule 190 restores the white colour of 2(2). But the content of 1(1), neighbour 2, is changed so that the new conservative rule is rule 184, when the sensor is white. Why is the cell 2(2) of the sensor different from the cell 2(4) of the controller? The reason is that according to the natural definition of side 1 which would be the side shared with 1(2), a new rule making the cell enter 2(2) could be accepted but then, the rule BWBBBBWWWW would be in conflict with rule 131, BWBBBBWWWB which is not compatible as a motion rule.

Let us remark that Table 28 shows us that the locomotive is stopped by a black sensor: the conservative rule for 0(0) in the black sensor is rule 100, as already noticed. Also note that we already met the rule in the doubler and in the fork. At some point rule 183, WWBBBWBWBW, is used. The rule shows us that 0(0) can see the arriving locomotive in 1(6), its neighbour 3, so that it remains white as the sensor, neighbour 6, is black. This is enough to cancel that locomotive as the usual motion rules of the tracks erase it from the cell 2(7), see Fig. 9.

We completed the examination of the rules for the sensor. Accordingly, Theorem 1 is proved.   □

4 Conclusion

The result is close to the best: the tessellation $\{7, 3\}$ is the tessellation $\{p, 3\}$ where p has the smallest value as possible for the hyperbolic plane. Is it possible to implement the same model there? The question seems to be very difficult.

Acknowledgements The author is much in debt to Andrew Adamatzky for his interest to the work.

References

1. Margenstern, M.: Cellular Automata in Hyperbolic Spaces, vol. 2. Implementation and Computations, Collection: Advances in Unconventional Computing and Cellular Automata, Adamatzky, A. (ed.), 360 p. Old City Publishing, Philadelphia (2008)
2. Margenstern, M.: A Family of Weakly Universal Cellular Automata in the Hyperbolic Plane with Two States, 83 p, (2012). arXiv:1202.1709

3. Margenstern, M.: Small Universal Cellular Automata in Hyperbolic Spaces: A Collection of Jewels, 331 p. Springer, Berlin (2013)
4. Margenstern, M.: A Weakly Universal Cellular Automaton in the Pentagrid with Three States, 40 p (2015). arXiv:1510.09129
5. Margenstern, M.: A Weakly Universal Cellular Automaton on the Pentagrid with Two States, 38 p (2015). arXiv:1512.07988v1
6. Margenstern, M.: A weakly universal cellular automaton with 2 states in the tiling {11, 3}. J. Cell. Autom. **11**(2–3), 113–144 (2016)
7. Margenstern, M.: Cellular Automata in Hyperbolic Spaces, Chapter in Advances in Unconventional Computing. Springer, Berlin. (to appear)
8. Margenstern, M.: A New System of Coordinates for the Tilings $\{p, 3\}$ and $\{p-2, 4\}$, 33 p (2016). arXiv:1605.03753
9. Margenstern, M.: A Weakly Universal Cellular Automaton on the Tessellation {9, 3}, 37 p (2016). arXiv:1605.09518
10. Margenstern, M.: A weakly universal cellular automaton on the tessellation {9, 3}. J. Cell. Autom. **1605**(09518), 37p (2016)
11. Margenstern, M., Song, Y.: A universal cellular automaton on the ternary heptagrid. Electron. Notes Theor. Comput. Sci. **223**, 167–185 (2008)
12. Margenstern, M., Song, Y.: A new universal cellular automaton on the pentagrid. Parallel Process. Lett. **19**(2), 227–246 (2009)
13. Minsky, M.L.: Computation: Finite and Infinite Machines. Prentice-Hall, Englewood Cliffs, NJ (1967)
14. Stewart, I.: A Subway Named Turing, Mathematical Recreations in Scientific American, pp. 90–92 (1994)

Cellular Automata: Descriptional Complexity and Decidability

Martin Kutrib and Andreas Malcher

Abstract We give a survey on the descriptional complexity of cellular models including one-way and two-way cellular automata, iterative arrays, and models with a fixed number of cells. For the former models so-called *non-recursive trade-offs* can be shown, that is, the savings in size that such automata may provide are not bounded by any recursive function. A consequence is that almost all commonly studied decidability questions are undecidable and not even semidecidable for such automata. On the other hand, for the latter models with a fixed number of cells it is possible to show recursive, in particular, polynomial bounds for mutual transformations which yield the decidability of the standard questions. Finally, we summarize results on the state complexity of the Boolean operations for one-way cellular automata and their variant with a fixed number of cells.

1 Introduction

Cellular automata have been investigated since the late forties as an important area of computer science. In particular, a lot of research has been carried out from a theoretical perspective. Recent surveys highlighting several different points of view may be found in [6, 14, 39]. A specific view on reversible cellular automata is given by Morita in [36], a computational point of view is taken in [17], and formal language aspects of cellular automata are considered in [18].

In this paper, we will complement these surveys with a summary of results on the *descriptional complexity* of cellular automata. Descriptional complexity is a field of theoretical computer science in which one main interest is to investigate how the size of description of a formal language varies when the language is described by different formalisms. A fundamental result is the exponential trade-off between nondeterministic and deterministic finite automata [35]. That is, every nondetermin-

M. Kutrib · A. Malcher (✉)
Institut für Informatik, Universität Giessen, Arndtstr. 2, 35392 Giessen, Germany
e-mail: malcher@informatik.uni-giessen.de

M. Kutrib
e-mail: kutrib@informatik.uni-giessen.de

© Springer International Publishing AG 2018
A. Adamatzky (ed.), *Reversibility and Universality*, Emergence, Complexity and Computation 30, https://doi.org/10.1007/978-3-319-73216-9_6

istic finite automata of size n can be converted into an equivalent deterministic finite automaton of size at most 2^n. Moreover, this bound is known to be tight, that is, there are cases where every equivalent deterministic finite automaton has a size of at least 2^n. Additional exponential or double exponential trade-offs are known, for example, between unambiguous and deterministic finite automata, between alternating and deterministic finite automata, between deterministic pushdown automata and deterministic finite automata, and between the complement of a regular expression and conventional regular expressions. Besides these recursive trade-offs which are bounded by recursive functions, it is also known that non-recursive trade-offs which are not bounded by any recursive function exist. Such trade-offs were first shown in [35] between context-free grammars generating regular languages and finite automata. Up to date, many non-recursive trade-offs concerning many different models have been established. For a summary of known results we refer to [9, 10], where one can also find references for the above-mentioned recursive trade-offs. Another branch of descriptional complexity is the study of the descriptional complexity of operations. For example, given two finite automata M and N and some operation $\circ$ on the languages $L(M)$ and $L(N)$ accepted by M resp. N, how many states are sufficient and necessary in the worst case for a finite automaton to accept the language $L(M) \circ L(N)$? For deterministic and nondeterministic finite automata many results on such questions on the *state complexity* are known and we refer to the surveys [7, 8].

In this survey, we will pose the above-mentioned questions on finite automata also for cellular automata which are basically arrays of interacting finite automata. We will summarize results on non-recursive and recursive trade-offs for different cellular models, we will study the impact of limiting resources such as space or time to the descriptional complexity of the corresponding models, and we will investigate the state complexity of cellular models. In detail, the paper is organized as follows. In Sect. 2 we provide the necessary notations and definitions from the field of both cellular automata and descriptional complexity. Section 3 is devoted to establishing non-recursive trade-offs between several cellular models. In particular, we study cellular automata with one-way and two-way information flow as well as iterative arrays with two-way information flow that process their input sequentially in contrast to cellular automata. Moreover, we differentiate between automata working in minimum time, that is, in realtime, and in lineartime. As by-product of the proofs for the non-recursive trade-offs we obtain that almost all commonly studied decidability questions are undecidable and not even semidecidable for our models. Furthermore, it is shown that neither minimization algorithms nor pumping lemmas exist for these models. In Sect. 4, we consider iterative arrays more intensively by studying restricted models with regard to time, space, dimension, or communication. It turns out that hierarchies concerning the computational power with regard to these parameters are reflected by corresponding non-recursive trade-offs between the models in question. Moreover, it can be shown that non-recursive trade-offs and non-semidecidable questions exist even for iterative arrays with a minimum amount of available time, space, and communication. Hence, in Sect. 5 we introduce different cellular models with a fixed number of cells. With this restriction, the computational power is reduced to the acceptance of regular languages, but the descriptional complexity of such automata with "fixed parallelism" in contrast to "sequential" finite automata becomes more interesting. We describe the costs of mutual transformations between several restricted cellular models and of converting them into finite automata, and it turns out that the costs are at most polynomial. The latter result enables us to show the decidability of several problems and gives that the computational complexity of the problems for restricted cellular models is in the same class as for deterministic finite automata. Finally, in Sect. 6 we investigate in the state complexity of the operations union, intersection, and complementation for one-way cellular automata (with a variable number of cells) working in realtime and for one-way cellular automata with a fixed number of cells. Interestingly, we obtain similar bounds for both models.

2 Definitions and Preliminaries

We denote the set of non-negative integers by $\mathbb{N}$ and the set of positive rational numbers by $\mathbb{Q}^+$. Let A denote a finite set of letters. Then we write A^* for the set of all finite words (strings) built with letters from A. The empty word is denoted by λ, A^+ denotes $A^* \setminus \{\lambda\}$, the reversal of a word w is denoted by w^R, and for the length of w we write $|w|$. For the number of occurrences of a subword x in w we use the notation $|w|_x$. A subset of A^* is called a language over A. We use $\subseteq$ for set inclusion and $\subset$ for strict set inclusion.

In order to avoid technical overloading in writing, two languages L and L' are considered to be equal, if they differ at most by the empty word, that is, $L \setminus \{\lambda\} = L' \setminus \{\lambda\}$. Moreover, we say that two devices are *equivalent* if and only if the languages they accept are equal.

We assume that the reader is familiar with the common notions of formal languages and automata theory as presented in [12]. We will use the abbreviations DFA for deterministic finite automata and PDA for pushdown automata.

2.1 Cellular Automata and Iterative Arrays

Definition 1 A *cellular automaton* (CA) is a system $\langle S, F, A, \#, \delta \rangle$, where S is the finite, nonempty set of *cell states*, $F \subseteq S$ is the set of *accepting states*, $A \subseteq S$ is the finite, nonempty set of *input symbols*, $\# \notin S$ is the *boundary state*, and $\delta : (S \cup \{\#\}) \times S \times (S \cup \{\#\}) \to S$ is the *local transition function*.

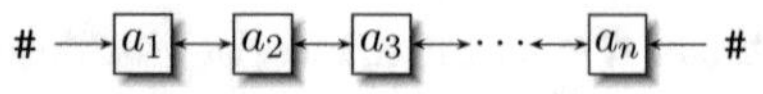

#
a_1
a_2
a_3
a_n
#

Fig. 1 A two-way cellular automaton

a_1
a_2
a_3
a_n
#

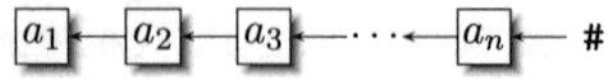

Fig. 2 A one-way cellular automaton

s_0
s_0
s_0
s_0
s_0
$a_1a_2a_3 \cdots a_n\triangledown$

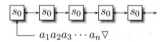

Fig. 3 An iterative array

The CA in this definition are also called two-way CA (see Fig. 1).

A *one-way cellular automaton* (OCA) is a cellular automaton in which each cell receives information from its immediate neighbor to the right only. So, the flow of information is restricted to be from right to left. Formally, δ is a mapping from $S \times (S \cup \{\#\})$ to S (see Fig. 2).

A *configuration* of a cellular automaton $\langle S, F, A, \#, \delta \rangle$ at time $t \geq 0$ is a mapping $c_t : \{1, \dots, n\} \to S$, for $n \geq 1$. For a given input $w = a_1 \cdots a_n \in A^+$ we set the initial configuration $c_0(i) = a_i$, for $1 \leq i \leq n$. Successor configurations are computed according to the global transition function Δ that is induced by δ as follows:

Let c_t, $t \geq 0$, be a configuration. Then its successor configuration $c_{t+1} = \Delta(c_t)$ is as follows.

$$\begin{aligned} c_{t+1}(1) &= \delta(\#, c_t(1), c_t(2)) \\ c_{t+1}(i) &= \delta(c_t(i-1), c_t(i), c_t(i+1)), \text{ for } i \in \{2, \dots, n-1\} \\ c_{t+1}(n) &= \delta(c_t(n-1), c_t(n), \#) \end{aligned}$$

for CA and

$$\begin{aligned} c_{t+1}(i) &= \delta(c_t(i), c_t(i+1)), \text{ for } i \in \{1, \dots, n-1\} \\ c_{t+1}(n) &= \delta(c_t(n), \#) \end{aligned}$$

for OCA.

Definition 2 An *iterative array* (IA) is a system $\langle S, F, A, \triangledown, s_0, \delta, \delta_0 \rangle$, where S is the finite, nonempty set of *cell states*, $F \subseteq S$ is the set of *accepting states*, $A \subseteq S$ is the finite, nonempty set of *input symbols*, $\triangledown \notin A$ is the *end-of-input symbol*, $s_0 \in S$ is the *quiescent state*, $\delta : S^3 \to S$ is the *local transition function for* *non-communication cells* satisfying $\delta(s_0, s_0, s_0) = s_0$, and $\delta_0 : (A \cup \{\nabla\}) \times S^2 \to S$ is the *local transition function for the communication cell* (see Fig. 3).

Let M be an IA. A configuration of M at some time $t \geq 0$ is a description of its global state which is a pair (w_t, c_t), where $w_t \in A^*$ is the remaining input sequence and $c_t : \mathbb{N} \to S$ is a mapping that maps the single cells to their current states. The configuration (w_0, c_0) at time 0 is defined by the input word w_0 and the mapping c_0 that assigns the quiescent state to all cells, while subsequent configurations are chosen according to the global transition function Δ that is induced by δ and δ_0 as follows: Let (w_t, c_t), $t \geq 0$, be a configuration. Then its successor configuration $(w_{t+1}, c_{t+1}) = \Delta(w_t, c_t)$ is as follows.

$$c_{t+1}(i) = \delta(c_t(i-1), c_t(i), c_t(i+1))$$

for all $i \geq 1$, and $c_{t+1}(0) = \delta_0(a, c_t(0), c_t(1))$, where $a = \nabla$ and $w_{t+1} = \lambda$ if $w_t = \lambda$, as well as $a = a_1$ and $w_{t+1} = a_2 \cdots a_n$ if $w_t = a_1 \cdots a_n$.

An input w is accepted by an IA (CA, OCA) M if at some time i during the course of its computation the communication cell (cell 0) enters an accepting state. The *language accepted by* M is denoted by $L(M)$. Let $t : \mathbb{N} \to \mathbb{N}$, $t(n) \geq n+1$ ($t(n) \geq n$ for (O)CA) be a mapping. If all $w \in L(M)$ are accepted with at most $t(|w|)$ time steps, then $L(M)$ is said to be of time complexity t.

The family of all languages which are accepted by some device X with time complexity t is denoted by $\mathscr{L}_t(X)$. If t is the function $n+1$, for IA, or the function n, for (O)CA, acceptance is said to be in *realtime* and we write $\mathscr{L}_{rt}(X)$. Since for nontrivial computations an IA has to read at least one end-of-input symbol, realtime has to be defined as $(n+1)$-time. The *lineartime* languages $\mathscr{L}_{lt}(X)$ are defined according to $\mathscr{L}_{lt}(X) = \bigcup_{k \in \mathbb{Q}^+, k \geq 1} \mathscr{L}_{k \cdot n}(X)$.

2.2 Descriptional Complexity

Let us recall some notions of descriptional complexity. Following [9], we say that a *descriptional system* $\mathcal{S}$ is a set of finite descriptors such that each $D \in \mathcal{S}$ describes a formal language $L(D)$, and the underlying alphabet alph(D) over which D represents a language can be read off from D. The *family of languages represented* (or *described*) by $\mathcal{S}$ is $\mathscr{L}(\mathcal{S}) = \{L(D) \mid D \in \mathcal{S}\}$. For every language L', the set $\mathcal{S}(L') = \{D \in \mathcal{S} \mid L(D) = L'\}$ is the set of its descriptors in $\mathcal{S}$. A *complexity measure* for a descriptional system $\mathcal{S}$ is a total computable mapping $c : \mathcal{S} \to \mathbb{N}$.

Example 1 Cellular automata and iterative arrays can be encoded over some fixed alphabet such that their input alphabets can be extracted from the encodings. The set of these encodings is a descriptional system $\mathcal{S}$, and $\mathscr{L}(\mathcal{S})$ is $\mathscr{L}(\text{CA})$ resp. $\mathscr{L}(\text{IA})$.

Examples for complexity measures for CA and IA are the total number of symbols, that is, the *length of the encoding* (length), or the total *number of transitions* (trans) in δ for CA, and in δ and δ_0 for IA. $\qquad\square$

Here we only use complexity measures that (with respect to the underlying alphabets) are related to **length** by a computable function. If there is a total computable function $g : \mathbb{N} \times \mathbb{N} \to \mathbb{N}$ such that, for all $D \in \mathcal{S}$, $\text{length}(D) \leq g(c(D), |\text{alph}(D)|)$, then c is said to be an *s-measure*. If, in addition, for any alphabet A, the set of descriptors in $\mathcal{S}$ describing languages over A is recursively enumerable in order of increasing size, then c is said to be an *sn-measure*. Clearly, **length** and **trans** are sn-measures for cellular automata and iterative arrays.

Whenever we consider the relative succinctness of two descriptional systems $\mathcal{S}_1$ and $\mathcal{S}_2$, we assume the intersection $\mathcal{L}(\mathcal{S}_1) \cap \mathcal{L}(\mathcal{S}_2)$ to be nonempty. Let $\mathcal{S}_1$ and $\mathcal{S}_2$ be descriptional systems with complexity measures c_1 and c_2, respectively. A total function $f : \mathbb{N} \to \mathbb{N}$ is an *upper bound* for the increase in complexity when changing from a descriptor in $\mathcal{S}_1$ to an equivalent descriptor in $\mathcal{S}_2$, if for all $D_1 \in \mathcal{S}_1$ with $L(D_1) \in \mathcal{L}(\mathcal{S}_2)$, there exists a $D_2 \in \mathcal{S}_2(L(D_1))$ such that $c_2(D_2) \leq f(c_1(D_1))$.

If there is no recursive, that is, computable function serving as upper bound, the *trade-off is said to be non-recursive* and we write $\mathcal{S}_1 \overset{nonrec}{\longrightarrow} \mathcal{S}_2$. That is, whenever the trade-off from one descriptional system to another is non-recursive, one can choose an arbitrarily large recursive function f, but the gain in economy of description eventually exceeds f when changing from the former system to the latter. It should be noted that non-recursive trade-offs are independent of particular sn-measures, since any two sn-measures c_1 and c_2 for some descriptional system $\mathcal{S}$ are related by a recursive function. Hence, a non-recursive trade-off exceeds any difference caused by applying two sn-measures. For proving non-recursive trade-offs the following general result is useful.

Theorem 1 ([9]) *Let $\mathcal{S}_1$ and $\mathcal{S}_2$ be two descriptional systems for recursive languages such that any descriptor D in $\mathcal{S}_1$ and $\mathcal{S}_2$ can effectively be converted into a Turing machine that decides $L(D)$, and let c_1 be a measure for $\mathcal{S}_1$ and c_2 be an sn-measure for $\mathcal{S}_2$. If there exists a descriptional system $\mathcal{S}_3$ and a property P that is not semidecidable for descriptors from $\mathcal{S}_3$, such that, given an arbitrary $D_3 \in \mathcal{S}_3$, (i) there exists an effective procedure to construct a descriptor D_1 in $\mathcal{S}_1$, and (ii) D_1 has an equivalent descriptor in $\mathcal{S}_2$ if and only if D_3 does not have property P, then the trade-off between $\mathcal{S}_1$ and $\mathcal{S}_2$ is non-recursive.*

In the following section, we will show all non-recursive trade-offs by applying Theorem 1. To this end, we will use the non-semidecidability of infiniteness for Turing machines and linear bounded automata as property P. As witness language that is effectively acceptable in the one descriptional system, but acceptable in the other system if and only if property P fails, we will use the set of valid computations of deterministic one-tape one-head Turing machines (DTM) and variations thereof. Formally, let $M = \langle Q, \Sigma, T, \delta, q_0, B, F \rangle$ be a DTM, where T is the set of tape symbols including the set of input symbols Σ and the blank symbol B, Q is the finite set of states and $F \subseteq Q$ is the set of final states. The initial state is q_0 and δ is the transition function. Without loss of generality, we assume that Turing machines can halt only after an even number of moves, halt whenever they enter an accepting state, make at least two moves, and cannot print blanks. At any instant during a computation, M can be completely described by an *instantaneous description* (ID) which is a string $tqt' \in T^*QT^*$ with the following meaning: M is in the state q, the non-blank tape content is the string tt', and the head scans the first symbol of t'. The initial ID of M on input $x \in \Sigma^*$ is $w_0 = q_0 x$. An ID is accepting whenever it belongs to T^*FT^*. The set VALC(M) of valid (accepting) computations of M consists of all finite strings of the form $w_0 \# w_1^R \# w_2 \# w_3^R \# \cdots \# w_{2n-1}^R \# w_{2n}$, where $\# \notin T \cup Q$, w_i, $0 \le i \le 2n$, are instantaneous descriptions of M, w_0 is an initial ID, w_{2n} is an accepting (hence halting) configuration, and w_{i+1} is the successor configuration of w_i, $0 \le i \le 2n - 1$. The set of *invalid computations* INVALC(M) is the complement of VALC(M) with respect to the alphabet $T \cup Q \cup \{\#\}$.

For technical reasons we will also use the set of valid computations of a deterministic linear bounded automaton (DLBA) which is basically a space-bounded DTM that gets its input in between two endmarkers. The benefit of such valid computations is that the length of a configuration is not changing owing to the boundedness of space. Another technical prerequisite is that we want to assure that for any valid computation $w \in$ VALC(M) there are at least $2^{|w|}$ further valid computations of the same length $|w|$ in VALC(M). To this end, let $\#'$ be a new symbol and T' and Q' be primed copies of T and Q. The set of *extended valid computations* VALC$'(M)$ is now defined to be the set of words $\varphi^{-1}(w)$, where $w \in$ VALC(M) is a valid computation and φ is the homomorphism defined by $\varphi(a) = a$ if $a \in T \cup Q \cup \{\#\}$, and $\varphi(a') = a$ if $a' \in T' \cup Q' \cup \{\#'\}$. The set of *extended invalid computations* INVALC$'(M)$ is the complement of VALC$'(M)$ with respect to the *coding alphabet* $\Lambda_M = \{\#, \#'\} \cup T \cup T' \cup Q \cup Q'$. By using the redundant primed symbols we ensure that sufficiently many valid computations of a certain length exist.

It has been shown in [26, 29] that it is possible for a given Turing machine M to effectively construct realtime-OCA and realtime-IA that each accept VALC(M) or INVALC(M). Since both language classes $\mathscr{L}_{rt}$(OCA) and $\mathscr{L}_{rt}$(IA) are effectively closed under inverse homomorphisms [40], it is clear that realtime-OCA and realtime-IA accepting VALC$'(M)$ and INVALC$'(M)$ can effectively be constructed as well.

3 Non-Recursive Trade-Offs

In this section, we are going to establish non-recursive trade-offs between many cellular models. The basic tool to prove such trade-offs is to apply Theorem 1. We start with non-recursive trade-offs of realtime-OCA and show that, for a given Turing machine M, each of the following sets, namely, the set VALC(M) of valid computations, the set INVALC(M) of invalid computations, and the language L_M which is defined as $\{w^{|w|!} \mid w \in \$\text{VALC}(M)\&\}$, where $\$$, $\&$ are two new symbols, can be accepted by some realtime-OCA.

Lemma 1 ([26]) *Let M be a Turing machine. Then realtime-OCA M_1, M_2 and realtime-CA M_3, M_4 can effectively be constructed such that $L(M_1) = VALC(M)$, $L(M_2) = INVALC(M)$, $L(M_3) = L_M$, and $L(M_4) = L_M^R$.*

Proof It is not difficult to construct a realtime-OCA M_2 accepting INVALC(M), since it is shown in [12] (Lemma 8.7) that INVALC(M) is basically a linear context-free language. For such languages it is known [41] that they can be accepted by realtime-OCA. The construction of automaton M_1 accepting VALC(M) follows from the effective closure under complementation for realtime-OCA. Now, we sketch the construction of a realtime-CA M_3 accepting L_M. First, we observe that L_M is the intersection of the following languages $L_1, L_2,$ and L_3 with $\Sigma = T \cup Q \cup \{\#\}$:

$$L_1 = \{\, wx \mid w \in \$VALC(M)\&, x \in (\$\Sigma^+\&)^+ \,\}$$
$$L_2 = \{\, w^n \mid w \in \$\Sigma^+\&, n \geq 1 \,\}$$
$$L_3 = \{\, wx \mid w \in \$\Sigma^+\&, x \in (\$\Sigma^+\&)^+, |wx|_\$ = |w|! \,\}$$

Since $\mathscr{L}_{rt}(CA)$ is closed under intersection, it remains to be shown that each of the languages belongs to $\mathscr{L}_{rt}(CA)$. This is not difficult for the language L_1 using the above-constructed realtime-OCA M_1 and the fact that $\mathscr{L}_{rt}(OCA)$ is closed under concatenation with regular languages. For L_2 we observe that it is basically the complement of a linear context-free language that accepts all words of the form $w_1 w_2 \cdots w_n$ with $w_i \in \$\Sigma^+\&$ and $1 \leq j < k \leq n$ such that $w_j \neq w_k$. Since linear context-free languages can be accepted by realtime-OCA and $\mathscr{L}_{rt}(OCA)$ is closed under complementation, we obtain that L_2 is accepted by some realtime-OCA and, hence, also by some realtime-CA. A realtime-CA for L_3 will use four tracks. In the first track the correct format of the input is checked. Additionally, the input is shifted one cell to the left in every time step. As long as the first cell gets an input up to the first occurrence of $\&$, a binary counter in the second track is increased that eventually carries $|w|$. This binary counter stores its least significant bit in the first cell and the cell storing the most significant bit may move to the right while increasing the counter and may move to the left while decreasing the counter. In the third track, the function $n!$ is calculated following the construction given in [34], that is, the first cell enters a certain state exactly at time steps $1, 2, 6, 24, \ldots$ At every such time step, the binary counter in track 2 is decreased by one, while a binary counter in track 4 is increased as long as the counter in track 2 is decreased to zero. Thus, the counter in track 4 would eventually carry $|w|!$. Simultaneously, the counter in track 4 is decreased by one for every symbol $\$$ seen in the first cell of track 1. If the counter has to be increased and decreased at the same time step, the counter remains unchanged. Finally, the input is accepted if and only if the input is correctly formatted and completely read, and the counter in track 4 has eventually been decreased to zero. To obtain a realtime-CA M_4 accepting L_M^R, we essentially use the same construction as for L_M, observing the VALC(M)^R belongs to $\mathscr{L}_{rt}(OCA)$, since the latter language class is closed under reversal. $\square$

To apply Theorem 1 for the non-recursive trade-offs we need another lemma.

Lemma 2 ([26]) *Let M be a Turing machine. Then*

1. *INVALC(M) is a context-free language and a pushdown automaton accepting it can effectively be constructed.*

2. *INVALC(M) is a regular language if and only if $L(M)$ is finite.*
3. *VALC(M) is a context-free language if and only if $L(M)$ is finite.*
4. *L_M belongs to $\mathscr{L}_{rt}(OCA)$ if and only if $L(M)$ is finite.*
5. *L_M^R belongs to $\mathscr{L}_{rt}(OCA)$ if and only if $L(M)$ is finite.*

Proof Claims 1, 2, and 3 are already shown in [12]. If $L(M)$ is a finite set, then L_M and L_M^R are finite sets as well and can be accepted by realtime-OCA. Thus, it remains to be shown that L_M and L_M^R do not belong to $\mathscr{L}_{rt}(OCA)$, if $L(M)$ is an infinite set. Since the words in L_M and L_M^R are designed to have a cyclic structure, it is possible to prove both claims by using a "pumping lemma for cyclic strings" given in [37]. □

Now, all prerequisites are done to apply Theorem 1 and to obtain $S_1 \overset{nonrec}{\longrightarrow} S_2$. We consider Turing machines as descriptional system S_3 and use "infiniteness" as property P which is known to be not semidecidable for Turing machines. Lemma 1 gives an effective procedure to construct a descriptor in S_1 for a language L depending on some Turing machine, but L belongs to $\mathscr{L}(S_2)$ if and only if the language of the underlying Turing machine is not infinite, that is, if and only if it is finite owing to Lemma 2.

Corollary 1 *The following non-recursive trade-offs hold:*

1. *realtime-OCA $\overset{nonrec}{\longrightarrow}$ DFA by using $L = INVALC(M)$,*
2. *realtime-OCA $\overset{nonrec}{\longrightarrow}$ PDA by using $L = VALC(M)$,*
3. *realtime-CA $\overset{nonrec}{\longrightarrow}$ realtime-OCA by using $L = L_M$, and*
4. *lineartime-OCA $\overset{nonrec}{\longrightarrow}$ realtime-OCA by using $L = L_M^R$.*

Next, we want to improve the non-recursive trade-off between lineartime-OCA and realtime-OCA to a non-recursive trade-off between OCA working in realtime plus log, that is, in time $n + \log(n)$, and realtime-OCA. To this end, we will use, for a given Turing machine M, the language $L_M' = \{ w^{2^{2^{|w|}}} \mid w \in \$VALC(M)\& \}$. With a similar approach as in Lemma 2 (4) it is possible to show that L_M' belongs to $\mathscr{L}_{rt}(OCA)$ if and only if $L(M)$ is finite. Hence, it remains to construct an OCA accepting L_M' in time $n + \log(n)$.

Lemma 3 ([28]) *Let M be a Turing machine. Then an $n + \log(n)$-OCA M_1 can effectively be constructed such that $L(M_1) = L_M'$.*

Proof Similar to the proof of Lemma 1 it can be observed that L_M' is the intersection of the following languages L_1, L_2, and L_3' with $\Sigma = T \cup Q \cup \{\#\}$:

$$L_1 = \{ wx \mid w \in \$VALC(M)\&, x \in (\$\Sigma^+\&)^+ \}$$
$$L_2 = \{ w^n \mid w \in \$\Sigma^+\&, n \geq 1 \}$$
$$L_3' = \left\{ wx \mid w \in \$\Sigma^+\&, x \in (\$\Sigma^+\&)^+, |wx|_\$ = 2^{2^{|w|}} \right\}$$

It is already shown in the proof of Lemma 1 that L_1 and L_2 belong to $\mathscr{L}_{rt}(OCA)$ and, hence, also to $\mathscr{L}_{rt+\log(n)}(OCA)$. Since $\mathscr{L}_{rt+\log(n)}(OCA)$ is closed under intersection [15], it remains for us to construct an $n + \log(n)$-OCA for L_3'. Since only one-way communication is possible, we are not able to apply the approach given in the proof of Lemma 1. We construct an OCA whose state set is divided into four tracks. The first track will be used to check the correct format of the input. In the second track certain signals are generated which ensure in the third track that the number of $-symbols is double exponential. Additionally, it is ensured in the fourth track that the double exponential number calculated in the third track is exactly $2^{2^{|w|}}$.

Let us now explain the tasks of the different tracks in more detail.

In the first track, a signal is started in the rightmost cell which checks cell by cell whether the input read so far is of the form $(\$\Sigma^+\&)^+$ and marks a cell with a permanent state g in the positive case. Thus, the leftmost cell enters after n time steps the state g if and only if the input is correctly formatted.

In the second track, we compute 2^{2^m} for $m \geq 1$, that is, we ensure that cell 2^{2^m} from the right enters a certain state at time step $2^{2^m} + 2^m$. To this end, we modify the construction of an $n + \log(n)$-OCA given in [16] that accepts the unary language $\{a^{2^m} \mid m \geq 1\}$. The basic idea to accept the latter language is to install a vertical binary counter that is initiated in the rightmost cell and moves to the left with maximal speed. The counter is increased by one in every time step and is enlarged by one digit if a carry-over takes place. An accepting state is entered if and only if the counter that moves digit by digit through one cell consists of 1's only and hence represents a number $2^m - 1$ for some $m \geq 1$. The time complexity of that OCA is $n + \log(n)$, since the counter reaches the leftmost cell at time n and $\log(n)$ additional time steps are necessary to inspect the counter. An example can be found in Fig. 4, where the computation on input a^{16} is depicted in the left half of the cells. The modification to identify "double exponential cells" is basically to implement the same construction of a vertical counter in the right half of the cells, but to increase the counter only when a carry-over takes place in the left half and one digit is to be added to the counter. If the counter that moves digit by digit through the right half of one cell consists of 1's only, we know that a number $2^{2^m} - 1$ for some $m \geq 1$ is represented and a certain state can be entered to identify such cells. An example computation may be found in Fig. 4. Since the computation started with one time step delay, we can identify in this way cells 2^{2^m} within time $2^{2^m} + \log(2^{2^m}) + \log(\log(2^{2^m}))$. Thus, the OCA constructed works in time $n + \log(n) + \log(\log(n))$ and can be sped-up by standard methods to work in time $n + \log(n)$. Whenever a cell in the second track identifies a double exponential cell, signals s_2 and s_3 are started in the third resp. fourth track that move with maximal speed to the left. Otherwise, some signal s_1 is started in the third track and moves with maximal speed to the left as well.

In the third track, the original input is kept and signals s_1 and s_2 are processed to count the number of $-symbols and to ensure that their number is double exponential. To this end, both signals move with maximal speed to the left as long as they reach a cell carrying a $. Then, the signals are stopped and the cell is marked with m_1 and m_2, respectively.

In the fourth track, the original input of the form wx with $w \in \$\Sigma^+\&$ and $x \in (\$\Sigma^+\&)^+$ is kept and signal s_3 is processed to ensure that $|w|$ double exponential cells have been identified. Thus, every signal s_3 will wipe off one symbol of w. When eventually the last symbol of w is wiped off, the number of $-symbols is

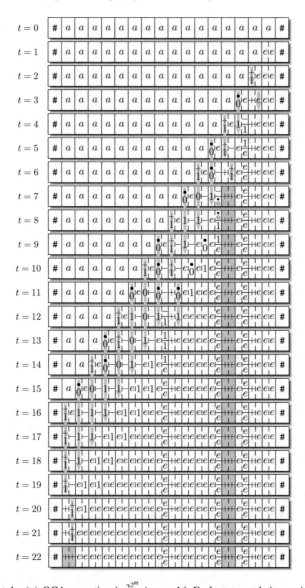

Fig. 4 An $n + \log(n)$-OCA accepting $\{\, a^{2^{2^m}} \mid m \geq 1 \,\}$. Each state not being an input symbol is split into two halves. In the left half, a binary counter is implemented which is increased in every time step. In the right half, a binary counter is implemented which is increased in every time step at which a carry-over takes place in the left half. States $0, \dot{0}, 1,$ and $\overset{+}{1}$ are used to realize the binary counters, e and $-$ denote error and idle states, and $+$ denotes a successful computation

exactly $2^{2^{|w|}}$. Since we have one-way information flow only, we cannot identify w, but apply our construction to all parts of the input of the form $\$\Sigma^+\&$: signal s_3 moves with maximal speed to the left as long as a cell carrying $\&$ is reached. Then, this cell is marked as m and the signal moves to the left with the same behavior. As soon as s_3 reaches a cell already marked with m, this cell is marked with $\overline{m}$, the left neighboring cell is marked with m, and the signal continues to move to the left with the same behavior.

As accepting states we define those states that are marked with g in the first track, with m_2 in the third track and with m in the fourth track. For a computation on input wx it can be observed that $|wx|_\$ = 2^{2^{|w|}}$ if and only if s_2 and s_3 reach the leftmost cell marking it with m_2 resp. m after time $|wx| + 2^{|w|}$. Thus, the OCA constructed accepts L'_3. Concerning the necessary time we know for $wx \in L'_3$ that $|wx| \geq 2^{2^{|w|}}$ which implies that $2^{|w|} \leq \log(|wx|)$. This gives that the OCA constructed is an $n + \log(n)$-OCA, since $|wx| + 2^{|w|} \leq |wx| + \log(|wx|)$. $\qquad\square$

Corollary 2 *Realtime* $+ \log(n)$-*OCA* $\xrightarrow{nonrec}$ *realtime-OCA by using* $L = L'_M$.

Now, we turn to non-recursive trade-offs for iterative arrays. Again, let M be a Turing machine and Σ' be the alphabet over which the extended valid computations VALC$'(M)$ are defined. Then we define a language

$$L''_M = \{ \$x_k\$ \cdots \$x_1\&y_1\$ \cdots \$y_k\$ \mid$$
$$x_i^R = y_i z_i \text{ such that } y_i, z_i \in (\Sigma')^* \text{ and } x_i \in \text{VALC}'(M)^R \}.$$

Lemma 4 ([29]) *Let M be a Turing machine. Then realtime-IA M_1, M_2, and M_3 can effectively be constructed such that $L(M_1) = \text{VALC}(M)$, $L(M_2) = \text{INVALC}(M)$, and $L(M_3) = L_M$. Furthermore, a realtime-OCA M_4 can effectively be constructed such that $L(M_4) = L''_M$.*

Proof It is shown in [12] that the valid computations of a Turing machine can be represented as the intersection of two deterministic context-free languages. A close look on the proof shows that the two languages are realtime languages, that is, their corresponding deterministic PDA accept these languages without using λ-moves. Since it is known due to [18] that every realtime deterministic context-free language effectively belongs to $\mathscr{L}_{rt}(\text{IA})$ and the class is effectively closed under intersection, we obtain that a realtime-IA M_1 accepting VALC(M) can effectively be constructed. The construction of a realtime-IA M_2 accepting INVALC(M) follows from the effective closure under complementation. To construct a realtime-IA M_3 accepting L_M we use an approach similar to the one in the proof of Lemma 1. Since L_M is the intersection of the languages L_1, L_2, and L_3, and $\mathscr{L}_{rt}(\text{IA})$ is effectively closed under intersection it remains for us to construct realtime-IA accepting L_1, L_2, and L_3. For the languages L_1 and L_3 the construction is nearly identical, since $\mathscr{L}_{rt}(\text{IA})$ is closed under right concatenation with regular languages, and binary counters with varying length can analogously be constructed for realtime-IA. To accept L_2, we simulate a queue according to the construction given in [17]. The principal idea is to enqueue a prefix of the form $\$\Sigma^+\&$ in the first track and then to iteratively check the next input portion of the form $\$\Sigma^+\&$ against the stored prefix. To preserve the enqueued prefix for all such checks it is additionally enqueued in the second track while it is dequeued in the first track for the check. For the next check the roles of both tracks are changed.

Finally, the language L''_M can be represented as the intersection of the languages L'_1 and L'_2:

$$L'_1 = \{\, w \mid w \in (\$\mathrm{VALC}'(M)^R)^+ \& ((\Sigma')^*\$)^+ \,\}$$
$$L'_2 = \{\, \$x_k\$ \cdots \$x_1\&y_1\$ \cdots \$y_k\$ \mid x_i^R = y_i z_i \text{ such that } x_i, y_i, z_i \in (\Sigma')^* \,\}$$

Since the set of extended valid computations $\mathrm{VALC}'(M)$ belongs to $\mathscr{L}_{rt}(\mathrm{OCA})$, and the latter class is closed under reversal and marked Kleene plus, it is clear that a realtime-OCA accepting L'_1 can be constructed. The language L'_2 can be accepted by a deterministic pushdown automaton that pushes the complete input up to the $\&$-symbol on its pushdown store. Subsequently, it is checked by popping symbols from the pushdown store, whether y_i is a prefix of x_i^R for $1 \le i \le k$. Since the PDA constructed performs at most one turn, the language accepted is a linear context-free language and an equivalent realtime-OCA can be constructed due to the construction given in [41]. Since $\mathscr{L}_{rt}(\mathrm{OCA})$ is closed under intersection, we obtain that L''_M belongs to $\mathscr{L}_{rt}(\mathrm{OCA})$ as well. □

Lemma 5 ([29]) *Let M be a Turing machine. Then L''_M belongs to $\mathscr{L}_{rt}(\mathrm{IA})$ if and only if $L(M)$ is finite.*

Proof If $L(M)$ is finite, $\mathrm{VALC}'(M)$ is finite as well and each word in $\mathrm{VALC}'(M)$ can be uniquely labeled by a symbol from a set Λ such that $|\Lambda| = |\mathrm{VALC}'(M)|$. A deterministic PDA for L''_M reads the complete input up to the $\&$-symbol and pushes for every word from $\mathrm{VALC}'(M)$ the corresponding label from Λ onto the pushdown store. Subsequently, the PDA checks with the knowledge of the ith popped symbol whether y_i is a prefix of x_i^R, for $1 \le i \le k$, and accepts if all checks are positive. The constructed PDA works deterministically and performs no λ-moves. Thus, L''_M is a realtime deterministic context-free language and can be accepted by some realtime-IA due to the construction given in Theorem 4 of [18].

It remains for us to argue why L''_M does not belong to $\mathscr{L}_{rt}(\mathrm{IA})$ if $L(M)$ is infinite. We consider prefixes of the form $\$x_k\$ \cdots \$x_1\&$ such that all $x_i \in \mathrm{VALC}'(M)^R$ with $1 \le i \le k$ have the same length k for some suitable $k > 0$. Since there are at least 2^k further valid computations of the same length for some $w \in \mathrm{VALC}'(M)$ of length k, we know that there are at least $(2^k)^k = 2^{k^2}$ prefixes of the above form. Next, we consider suffixes of the form $\$^{j-1}x_j^R\$^{k-j+1}$ of length $2k$. Since a realtime-IA with state set S has on a remaining input of length $2k$ at most $|S|^{2k+2}$ different possibilities to differentiate configurations (see, for example, Example 7 in [18]) and $2^{k^2} \ge |S|^{2k+2}$ for k large enough, a realtime-IA can be fooled by adding the same suffix to two different prefixes. □

Now, we apply again Theorem 1 to obtain $S_1 \xrightarrow{nonrec} S_2$. We consider once more Turing machines as descriptional system S_3 and use "infiniteness" as property P. Lemma 4 gives an effective procedure to construct a descriptor in S_1 for a language L depending on some Turing machine, but L belongs to $\mathscr{L}(S_2)$ if and only if the language of the underlying Turing machine is finite owing to Lemmas 2 and 5.

Corollary 3 *The following non-recursive trade-offs hold:*

1. *realtime-IA $\xrightarrow{nonrec}$ DFA by using $L = INVALC(M)$,*
2. *realtime-IA $\xrightarrow{nonrec}$ PDA by using $L = VALC(M)$,*
3. *realtime-CA $\xrightarrow{nonrec}$ realtime-IA by using $L = L''_M$,*
4. *lineartime-IA $\xrightarrow{nonrec}$ realtime-IA by using $L = L''_M$,*
5. *realtime-OCA $\xrightarrow{nonrec}$ realtime-IA by using $L = L''_M$, and*
6. *realtime-IA $\xrightarrow{nonrec}$ realtime-OCA by using $L = L_M$.*

In analogy to realtime-OCA, it is also possible for realtime-IA to improve the non-recursive trade-off between lineartime-IA and realtime-IA to a non-recursive trade-off between realtime$+ \log(n)$-IA and realtime-IA. This will be a consequence of a more general result obtained in Sect. 4.4 that establishes non-recursive trade-offs in between the levels of the time hierarchy for IA shown in [4].

Our next result shows that arbitrary trade-offs do not only exist between the parallel models of realtime-OCA and realtime-IA and the sequential model of finite automata, but can easily be constructed.

Theorem 2 ([26, 35]) *Let f be a recursive function and $n \geq 1$ be an arbitrary, but fixed integer. Then, there is a regular language $L(f, n)$ being accepted by some realtime-OCA resp. realtime-IA with $O(n)$ states, but any finite automaton accepting $L(f, n)$ needs at least $\Omega(f(n))$ states.*

Proof Let f be a recursive function and $n \geq 1$ be a fixed integer. There is a Turing machine M with unary input and output which computes exactly $f(n)$. Hence, $L(f, n) = VALC(M)$ comprises exactly one word w_n and any finite automaton accepting w_n needs at least $|w_n| + 1 \in \Omega(f(n))$ many states.

On the other hand, let M' be a Turing machine which computes the recursive function f on any input of length $n \geq 1$. Due to Lemmas 1 and 4 we can construct a realtime-OCA (resp. realtime-IA) M'' accepting $VALC(M')$, whose size is a constant with respect to length of the input n. By counting the length for some fixed $n \geq 1$ using its state set, we modify M'' to accept $L(f, n)$. Thus, a realtime-OCA (resp. realtime-IA) accepting $L(f, n)$ needs at most $O(n)$ states.   □

The fact that the valid computations and variations thereof can be accepted by realtime-OCA and realtime-IA has enormous consequences for decidability questions on those models. We will show in the following that almost all commonly studied decidability questions are undecidable and not even semidecidable for realtime-OCA and realtime-IA. Here, we say that a decidability problem is *decidable (undecidable)* if the set of all instances for which the answer is "yes" is recursive (not recursive). A decidability problem is said to be *semidecidable* if the set of all instances for which the answer is "yes" is recursively enumerable. It is said to be *co-semidecidable* if its complementary decidability problem is semidecidable. For example, the emptiness problem for Turing machines is known to be not semidecidable [12], whereas the nonemptiness problem is semidecidable. Thus, the emptiness problem for Turing machines is not semidecidable, but co-semidecidable.

Theorem 3 ([26, 29]) *For realtime-OCA or realtime-IA the following questions are not semidecidable: emptiness, universality, finiteness, infiniteness, inclusion, equivalence, regularity, and context-freeness.*

Proof It is a consequence of Rice's Theorem for recursively enumerable index sets [12] that the questions of emptiness, finiteness, and infiniteness are not semidecidable for Turing machines. This enables us to show the claimed non-semidecidability of questions for cellular automata. We prove exemplarily that emptiness and regularity are not semidecidable. The remaining proofs are similar. According to Lemmas 1 and 4 we can construct realtime-OCA resp. realtime-IA M_1 and M_2 such that $L(M_1) = \text{VALC}(M)$ and $L(M_2) = \text{INVALC}(M)$ for some Turing machine M. Obviously, $L(M)$ is empty if and only if $L(M_1) = \text{VALC}(M)$ is empty. Assume that emptiness is semidecidable for realtime-OCA (realtime-IA), then it would be semidecidable for Turing machines as well, which is a contradiction. According to Lemma 2 we know that $\text{INVALC}(M)$ is a regular language if and only if $L(M)$ is a finite language. Thus, the assumption that regularity is semidecidable for realtime-OCA (resp. realtime-IA) implies that finiteness is semidecidable for Turing machines, which is again a contradiction. □

Theorem 4 ([26, 29]) *For realtime-OCA or realtime-IA the questions of emptiness, universality, inclusion, and equivalence are co-semidecidable, whereas the questions of finiteness, infiniteness, regularity, and context-freeness are not co-semidecidable.*

Proof Nonemptiness is semidecidable for realtime-OCA and realtime-IA by successively testing all possible inputs in increasing order of their length. Since both $\mathscr{L}_{rt}(\text{OCA})$ and $\mathscr{L}_{rt}(\text{IA})$ are closed under complementation, the semidecidability of nonuniversality follows from the semidecidability of nonemptiness. Let M_1 and M_2 be two realtime-OCA (realtime-IA). For the non-inclusion of M_1 and M_2 we know $L(M_1) \not\subseteq L(M_2)$ if and only if $L(M_1) \setminus L(M_2) \neq \emptyset$ if and only if $L(M_1) \cap \overline{L(M_2)} \neq \emptyset$. Since both $\mathscr{L}_{rt}(\text{OCA})$ and $\mathscr{L}_{rt}(\text{IA})$ are closed under intersection and complementation, the semidecidability of non-inclusion follows from the semidecidability of nonemptiness. Similarly, non-equivalence can be shown to be semidecidable, because $L(M_1) \neq L(M_2)$ if and only if $L(M_1) \not\subseteq L(M_2)$ or $L(M_2) \not\subseteq L(M_1)$.

Finiteness and infiniteness are obviously not co-semidecidable. Lemma 2 shows that $\text{INVALC}(M)$ is a non-regular language if and only if $L(M)$ is an infinite language. Moreover, $\text{VALC}(M)$ is a non-context-free language if and only if $L(M)$ is an infinite language. Thus, the non-semidecidability of non-regularity and non-context-freeness follows from the non-semidecidability of infiniteness for Turing machines. □

Theorem 5 ([26, 29]) *The following questions are neither semidecidable nor co-semidecidable:*

1. *Is $L(A) \in \mathscr{L}_{rt}(OCA)$ for some realtime-CA A?*
2. *Is $L(A) \in \mathscr{L}_{rt}(OCA)$ for some realtime-IA A?*
3. *Is $L(A) \in \mathscr{L}_{rt}(IA)$ for some realtime-OCA A?*

Proof Lemmas 1 and 4 show that realtime-CA (realtime-IA) accepting L_M and realtime-OCA accepting L_M'' can be effectively constructed for given Turing machines M. On the other hand, it is shown in Lemma 2 that L_M belongs to $\mathscr{L}_{rt}(OCA)$ if and only if $L(M)$ is finite, and it is shown in Lemma 5 that L_M'' belongs to $\mathscr{L}_{rt}(IA)$ if and only if $L(M)$ is finite. Thus, the non-semidecidability of the above questions follows from the non-semidecidability of finiteness for Turing machines, whereas the non-semidecidability of the complementary questions follows from the non-semidecidability of infiniteness for Turing machines. □

The fact that emptiness and infiniteness are not semidecidable for realtime-OCA and realtime-IA has two interesting consequences, namely, the non-existence of a pumping lemma and a minimization algorithm for such automata. We say that a language class $\mathcal{L}$ has a pumping lemma if for every language $L \in \mathcal{L}$ there is an $n \geq 1$ constructible from L such that for every $z \in L$ with $|z| > n$ there is a decomposition $z = uvw$, so that $|v| \geq 1$ and $u'v^iw' \in L$ for infinitely many $i \geq 1$, where u' and w' may depend on u, w and i.

Theorem 6 ([26, 29])

1. $\mathscr{L}_{rt}(OCA)$ *and each language class containing* $\mathscr{L}_{rt}(OCA)$ *do not possess a pumping lemma.*
2. $\mathscr{L}_{rt}(IA)$ *and each language class containing* $\mathscr{L}_{rt}(IA)$ *do not possess a pumping lemma.*
3. *For realtime-OCA, there is no minimization algorithm converting an arbitrary realtime-OCA to an equivalent realtime-OCA with a minimal number of states.*
4. *For realtime-IA, there is no minimization algorithm converting an arbitrary realtime-IA to an equivalent realtime-IA with a minimal number of states.*

Proof We show the claims for realtime-OCA only, since the proofs for realtime-IA are similar. Let M be a realtime-OCA over an alphabet Σ and assume that there is a pumping lemma for $\mathscr{L}_{rt}(OCA)$ with constant n for the language $L(M)$. Then, by applying the pumping lemma we can conclude that $L(M)$ is infinite if and only if there is an $x \in L(M)$ of length $|x| > n$. We construct a Turing machine which successively generates all words from Σ^* with a length of at least n and simulates M on that inputs. As soon as there is an accepted input x with $|x| > n$, the Turing machine halts accepting. Thus, the infiniteness of M is semidecidable which is a contradiction to Lemma 4.

We show that the existence of a minimization algorithm for realtime-OCA would imply the decidability of emptiness for realtime-OCA which is again a contradiction to Lemma 4. Obviously, a realtime-OCA over alphabet A accepting the empty set has exactly $|A|$ states and no accepting states. Let us assume that a minimization algorithm exists, let M be an arbitrary realtime-OCA and $M' = \langle S, F, A, \#, \delta \rangle$ be its minimized equivalent realtime-OCA. Then, we check whether $|S| = |A|$ and $F = \emptyset$. If so, then $L(M') = L(M) = \emptyset$. Otherwise, if $|S| = |A|$ and $F \neq \emptyset$, we know that $L(M') = L(M)$ is not empty since at least one alphabet symbol is an accepting state that may occur in the leftmost cell. If $|S| > |A|$ and $F \neq \emptyset$, we know that M' is not minimal for accepting the empty set. Thus, $L(M') = L(M)$ is not empty in this case. Finally, if $F = \emptyset$, we know that $L(M') = L(M)$ is empty. Altogether, we obtain that the emptiness of M is decidable, which is a contradiction.   □

Next, we study the following reachability questions asking whether during the course of its computation a cellular automaton reaches a certain configuration, or whether a cell ever enters a certain state. Let $M = \langle S, F, A, \#, \delta \rangle$ be a realtime-OCA and $c \in S^+$ be a configuration. Consider the following questions:

(R1) Are there an input $w \in A^+$ and a time step $t \leq |w|$ such that $\Delta^t(w) = c$?
(R2) Are there an input $w \in A^+$ and a time step $t \leq |w|$ such that configuration c is a prefix of $\Delta^t(w)$?
(R3) Are there an input $w \in A^+$ and a time step $t \leq |w|$ such that some cell i with $1 \leq i \leq |w|$ enters some state $s \in S$ at time step t?

Theorem 7 *Question (R1) is decidable, whereas questions (R2) and (R3) are undecidable.*

Proof An algorithm answering (R1) successively generates all inputs of length $|c|$, simulates M on the input generated, and checks whether c occurs during the simulation up to time step $|w|$. If such an input exists, the algorithm stops with answer "yes". Otherwise, having checked all possible inputs, the algorithm stops with answer "no".

Let us assume that (R2) is decidable. Then we can test for every $s \in F$ whether $c = s$ is a prefix of $\Delta^t(w)$. If the test is negative for all $s \in F$, then $L(M)$ is empty. Otherwise, $L(M)$ is not empty. Thus, we can decide the emptiness for a given realtime-OCA, which is a contradiction to Lemma 3.

Let us assume that (R3) is decidable. Then we can test for every $s \in F$ whether the leftmost cell ($i = 1$) enters state s. If the test is negative for all $s \in F$, then $L(M)$ is empty. Otherwise, $L(M)$ is not empty. Thus, we can again decide the emptiness for a given realtime-OCA, which is a contradiction.   □

Finally, we want to discuss some applications of the fact the almost all questions for realtime-OCA, and thus also for OCA without time constraints, are not semidecidable. Interestingly, these results have found applications in showing non-semidecidability results for models other than cellular automata.

In [1] decidability questions of *parallel communicating finite automata* (PCFA) are investigated which are systems of several finite automata that process a common input string in a parallel way and that are able to communicate by sending their states upon request. Such systems have been introduced in [32] and they can work in *returning* or *non-returning* mode. In the former case each automaton which sends its current state is reset to its initial state after this communication step. In the latter case the state of the sending automaton is not changed. It is also distinguished between *centralized* systems where only one designated automaton, called *master*, can request information from other automata, and *non-centralized* systems where every automaton is allowed to request communication with others. Since it is known that non-centralized PCFA are equivalent to one-way multi-head finite automata (see [2, 32]), the undecidability results for the latter class can be translated to the former class, which gives that all classical decidability questions such as emptiness, universality, inclusion, equivalence, finiteness, and infiniteness are not semidecidable for non-centralized PCFA. On the other hand, only little was known for the centralized variants. Some first results for the nondeterministic and non-returning case have been obtained in [33], which are extensively complemented by the results given in [1]. There it is shown that emptiness, universality, inclusion, equivalence, finiteness and infiniteness are not semidecidable for centralized PCFA which are deterministic, nondeterministic, returning, or non-returning. In most cases, the results are obtained for systems with at least two components, and in some cases at least three components are necessary. The basic idea for the non-semidecidability results is first to define suitably encoded valid and invalid computations of one-way cellular automata. Second, deterministic returning and non-returning PCFA with two or three components are constructed accepting these valid and invalid computations. Thus, decidability questions for PCFA are reduced to decidability questions for OCA which are not semidecidable due to Theorem 3. The non-semidecidability results are refined in [3] where communication-bounded PCFA are studied. By once more using the valid and invalid computations of OCA, the main results obtained there are the non-semidecidability of emptiness, universality, inclusion, equivalence, finiteness, and infiniteness for deterministic, returning, and centralized or non-centralized PCFA with at least four components that use in addition at most $O(\log(n))$ communications between components on an input of length n.

A different computational model has been considered in [24], where *one-way multi-head finite automata* are on the one hand enlarged with *pebbles* that can be dropped, sensed, and picked up by the heads. But on the other hand, the state sets of these multi-head automata are restricted to singletons, and these automata are called *stateless*. By defining suitably encoded valid computations of OCA and by constructing stateless one-way two-head finite automata with one pebble for such valid computations, it is again possible to translate non-semidecidability results for OCA to stateless multi-head finite automata with pebbles. In detail, the questions of emptiness, universality, inclusion, equivalence, finiteness, and infiniteness turn out to be not semidecidable for stateless one-way finite automata with at least two heads and one pebble.

4   Resource-Bounded Models

In this section, we will study iterative arrays in more depth with regard to several resources and the consequences for their descriptional complexity. In detail, we will study the parameters dimension, communication, time, and space. It has been shown in several papers [4, 21, 31] that there exist infinite hierarchies concerning the computational capacity with regard to these parameters. We will summarize here some results showing non-recursive trade-offs between automata classes when their corresponding language classes are properly included.

4.1   Multi-Dimensional Iterative Arrays

We start with the parameter dimension and we will show that there is a non-recursive trade-off between $(d+1)$-dimensional and d-dimensional realtime-IA for any $d \geq 1$. To this end, we will adapt the witness language used in [21] to establish the dimension hierarchy. To formally define a multi-dimensional IA (see, for example, [5]), we replace in Definition 2 the local transition functions δ and δ_0 by $\delta : S^{2d+1} \to S$ satisfying $\delta(s_0, s_0, \ldots, s_0) = s_0$, and $\delta_0 : (A \cup \{\triangledown\}) \times S^{d+1} \to S$. A multi-dimensional IA of dimension d is denoted by IA^d and configurations of IA^d and their global transition function are straightforwardly defined, where c_t is now a mapping from $\mathbb{N}^d$ to S.

The witness languages L_d ($d \geq 2$) for the dimension hierarchy obtained in [21] are defined as follows. First, we consider the following series of regular sets:

$$X_1 = \${a, b\}^+, \qquad X_{i+1} = \$X_i^+, \text{ for } i \geq 1.$$

Due to the separator symbol $\$$, every word $u \in X_{i+1}$ can uniquely be decomposed into its subwords from X_i. So, the projection on the jth subword can be defined as usual: Let $u = \$u_1 \cdots u_m$, where $u_j \in X_i$, for $1 \leq j \leq m$. Then $u[j]$ is defined to be u_j, if $1 \leq j \leq m$, otherwise $u[j]$ is undefined. Now, we define the language

$$M_d = \{ u \mathbb{c} e^{x_d} \$ \cdots \$ e^{x_1} \$ e^{2x} \$ v \mid u \in X_d, x_i \geq 1 \text{ for } 1 \leq i \leq d,$$
$$x = x_1 + \cdots + x_d, \text{ and } v = u[x_d][x_{d-1}] \cdots [x_1] \text{ is defined} \}.$$

Example 2 We consider the word $u = \$\$\$aab\$aa\$baa\$\$aaa\$ab \in X_3$. Then, u can uniquely be decomposed into $u = \$u_1u_2$ with $u_1 = \$\$aab\$aa\$baa \in X_2$ and $u_2 = \$\$aaa\$ab \in X_2$. Subsequently, u_1 can uniquely be decomposed into $u_1 = \$u_{1,1}u_{1,2}u_{1,3}$ with $u_{1,1} = \$aab$, $u_{1,2} = \$aa$, $u_{1,3} = \$baa$ and u_2 into $u_2 = \$u_{2,1}u_{2,2}$ with $u_{2,1} = \$aaa$, $u_{2,2} = \$ab$. By using the projections $u[j]$ we can now address every symbol from $\{a, b\}$. We have, for example, $u[1][3][1] = b$ and $u[2][1][3] = a$. Then, the word $u \mathbb{c} e^2 \$ e \$ e^3 \$ e^{12} \$ a$ belongs to M_3.

Finally, the language L_d is defined as the homomorphic image of M_d using a suitable homomorphism h. It is shown in [21], that L_{d+1} can be accepted by a realtime-IA^{d+1}, but not by any realtime-IA^d.

Next, we adapt the set M_d in such a way that the prefix u is interleaved symbol by symbol with some word from the set VALC$'(M)$ for some DLBA M. The remaining symbols $\mathvarphi$, $\$$, e and the suffix letter $v \in \{a, b\}$ are repeated once. Additionally, the prefix u may end with some dummy symbols c.

$$L_{d,M} = \{ u_1 w_1 u_2 w_2 \cdots u_t w_t \mathrm{\mathcal{c}\mathcal{c}} e^{2x_d} \$\$ \cdots \$\$ e^{2x_1} \$\$ e^{4x} \$\$ vv \mid$$

$$u_j \in \{a, b, c, \$\} \text{ and } w_j \in \Lambda_M \text{ for } 1 \le j \le t, u = u_1 u_2 \cdots u_t \in X_d c^*,$$

$$w = w_1 w_2 \cdots w_t \in \text{VALC}'(M), x_i \ge 1 \text{ for } 1 \le i \le d,$$

$$x = x_1 + \cdots + x_d, \text{ and } v = u[x_d][x_{d-1}] \cdots [x_1] \text{ is defined} \}.$$

By using similar ideas as in [21] it is possible to show that a realtime-IA^{d+1} accepting $L_{d+1,M}$ can effectively be constructed, but the language $L_{d+1,M}$ is accepted by some realtime-IA^d if and only if the underlying DLBA accepts a finite language.

Lemma 6 ([22]) *Let M be a DLBA and $d \ge 1$ be a constant number. Then $L_{d+1,M}$ belongs to $\mathscr{L}_{rt}(IA^d)$ if and only if $L(M)$ is finite. Moreover, the language $L_{d+1,M}$ belongs to $\mathscr{L}_{rt}(IA^{d+1})$.*

By using this lemma in connection with Theorem 1, considering DLBA as descriptional system $\mathcal{S}_3$, and "infiniteness" as property P which is known not to be semidecidable for DLBA, we obtain the following non-recursive trade-offs.

Corollary 4 ([22]) *Let $d \ge 1$ be a constant number. Then the trade-off between realtime-IA^{d+1} and realtime-IA^d is non-recursive.*

4.2 Communication-Bounded Iterative Arrays

The dimension hierarchy is refined in [21], where it is shown that there is a double hierarchy concerning dimension and communication. These d-dimensional realtime-IA^d with restricted communication are realtime-IA^d whose bandwidth of the inter-cell communication links is limited in the following way. Whereas in the general case, in every step the states of the cells are transmitted, the limitation is modeled here by a set of messages that can be sent, where the number k of different messages is independent of the number of states. Such devices are denoted by IA$_k^d$ and for the formal definition we add the set of possible messages B and communication functions $b_i : S \to B$, $1 \le i \le 2d$, that determine the messages to be sent to neighbors. Furthermore, we replace δ and δ_0 by $\delta : S \times B^{2d} \to S$ satisfying $\delta(s_0, (b_1(s_0), b_2(s_0), \ldots, b_{2d}(s_0))) = s_0$, and $\delta_0 : (A \cup \{\nabla\}) \times S \times B^d \to S$.

As witness languages for the communication hierarchy the following languages are used. For $d \geq 1$ and any number of messages $k \geq 2$ we define an alphabet $A_{d,k} = \{a_0, \ldots, a_{k^d-1}\}$ and a language $\hat{L}_{d,k}$ as

$$\hat{L}_{d,k} = \{ e^x \$ u_1 u_2 \cdots u_m \mid x \geq 1, m \geq 2x - 1, u_i \in A_{d,k} \text{ for } 1 \leq i \leq m,$$
$$\text{and } u_j = u_{j+2x-1} \text{ for } 1 \leq j \leq m - (2x - 1) \}.$$

It is shown in [21] that for every $d \geq 1$ and $k \geq 2$ a realtime-IA_{k+1}^d accepting $\hat{L}_{d,k+1}$ can be constructed, whereas any realtime-IA_k^d is not able to accept $\hat{L}_{d,k+1}$. Thus, it holds for any $k \geq 1$ that realtime-IA^d that can communicate $k+1$ different messages are more powerful than those that can communicate k different messages only.

Again, we modify these witness languages to our needs for non-recursive trade-offs. Let $c \geq 1$ be a constant, M be a DLBA, and h_c be the homomorphism defined by $h_c(a) = a^c$, for all $a \in \Lambda_M$. By Lemma 4 we know that $\text{VALC}'(M)$ is accepted by some realtime-IA. Let S be the state set of that IA, let $c = \lceil \log_2(|S|) \rceil$, and consider the set $h_c(\text{VALC}'(M))$. It is explained in [20] how a realtime-IA_2 accepting $h_c(\text{VALC}'(M))$ can effectively be constructed. The main idea is to simulate one transition of the IA accepting $\text{VALC}'(M)$ by c transitions. During these transitions the binary encoded states of the cells are communicated. A cell that receives the cth bit with the cth transition eventually simulates the original transition. Finally, the communication cell of the IA can verify whether each input symbol is repeated c times. Next, we define for $d \geq 1$ and $k \geq 2$ the language $L_{d,k,M}$ as follows:

$$L_{d,k,M} = \{ ew_1 ew_2 \cdots ew_x \$^{2x} u_1 u_2 \cdots u_m \mid x \geq 1, m \geq 2x - 1,$$
$$u_i \in A_{d,k} \text{ for } 1 \leq i \leq m, u_j = u_{j+2x-1} \text{ for } 1 \leq j \leq m - (2x - 1),$$
$$w_t \in \Lambda_M \text{ for } 1 \leq t \leq x, \text{ and } w_1 w_2 \cdots w_x \in h_{c+1}(\text{VALC}'(M)) \}.$$

Again, it is possible to use similar ideas as in [21], and we obtain that a realtime-IA_{k+1}^d accepting $L_{d,k+1,M}$ can effectively be constructed, whereas the language $L_{d,k+1,M}$ is accepted by some realtime-IA_k^d if and only if the underlying DLBA accepts a finite language.

Lemma 7 ([22]) *Let M be a DLBA and $d, k \geq 1$ be constants. Then, the language $L_{d,k+1,M}$ belongs to $\mathscr{L}_{rt}(\text{IA}_k^d)$ if and only if $L(M)$ is finite. Moreover, the language $L_{d,k+1,M}$ belongs to $\mathscr{L}_{rt}(\text{IA}_{k+1}^d)$.*

This lemma and Theorem 1 give the following non-recursive trade-offs.

Corollary 5 ([22]) *Let $d, k \geq 1$ be constant numbers. Then the trade-off between realtime-IA_{k+1}^d and realtime-IA_k^d is non-recursive.*

This means that there is a non-recursive trade-off between realtime-IA_2 and realtime-IA_1. Since every cell of an IA_1 sends always the same and only message at every time step, it is clear that usable communication is impossible and the computational capacity of such iterative arrays is reduced to the regular languages. Thus, the minimal number of nontrivial messages is two and the above result shows that this amount is already sufficient to obtain non-recursive trade-offs. The above-described construction of a realtime-IA_2 accepting $h_c(\text{VALC}'(M))$ and the similar construction of a realtime-IA_2 accepting $h_c(\text{INVALC}'(M))$ has the consequence that the results of Theorem 3 can be extended to hold for realtime-IA_2 as well. The proofs are almost identical, considering the fact that the properties of $\text{VALC}'(M)$ and $\text{INVALC}'(M)$ to be empty, finite, infinite, regular, or context-free are preserved under the homomorphism h_c. Thus, we obtain that even the minimum amount of nontrivial messages makes the model very powerful so that non-recursive trade-offs exist and decidability questions are not semidecidable.

Theorem 8 ([20]) *For realtime-IA_2 the following questions are undecidable and not semidecidable: emptiness, universality, finiteness, infiniteness, inclusion, equivalence, regularity, and context-freeness.*

4.3 Space-Bounded Iterative Arrays

Another restriction of realtime-IA is to bound the available space, that is, the number of cells that may be used in realtime computations. In detail, we count the number of different cells entering a state different from the quiescent state during a realtime computation. We say that a realtime-IA is working in $S(n)$ *strong space*, if the maximum number of cells used in any computation of length n is bounded by $S(n)$, where $S : \mathbb{N} \to \mathbb{N}$ is a function. Such iterative arrays have been introduced in [31] and it is known, for example, that the language $\{ a^m b^m c^m \mid m \geq 1 \}$ can be accepted in $\log(n)$ strong space and $\{ a^{m^2} \mid m \geq 1 \}$ can be accepted in $\sqrt{n}$ strong space. Furthermore, an infinite proper space hierarchy has been obtained in [31], which is based on the notion of *IA-ts-constructibility*. Roughly speaking, a strictly increasing function $f : \mathbb{N} \to \mathbb{N}$ is IA-ts-constructible if there is an IA whose communication cell enters on empty input an accepting state exactly at all time steps $f(i)$ for $i \geq 1$, while the IA must not use more than $f^{-1}(n)$ cells for all computations on inputs of length n. Examples of IA-ts-constructible functions are the polynomials $f(n) = n^k$, for $k \geq 2$. The witness languages for the infinite hierarchy are $L_f = \{ c^{f(|w|)} w d w^R \mid w \in \{a, b\}^* \}$, where f is an IA-ts-constructible function such that $f^{-1}(n) \in \Omega(\log(n))$. It can be shown that L_f can be accepted by some realtime-IA in $f^{-1}(n)$ strong space, but not in $g(n)$ strong space, when $g(n) \in o(f^{-1}(n))$ is a non-decreasing function.

For the non-recursive trade-offs, the language L_f is modified, for some Turing machine M, to $L_{M,f} = \{ c^{f(|w|)} w d w^R \mid w \in \text{VALC}'(M) \}$. The following lemma can be shown similar to the result for the infinite hierarchy.

Lemma 8 ([31]) *Let M be a Turing machine, $f(n)$ an IA-ts-constructible function satisfying $f^{-1}(n) \in \Omega(\log(n))$, and $g(n) \in o(f^{-1}(n))$ a non-decreasing function. Then a realtime-IA recognizing $L_{M,f}$ in $f^{-1}(n)$ strong space can effectively be constructed. Moreover, $L_{M,f} \in \mathcal{L}_{rt}(g(n)\text{-}IA)$ if and only if $L(M)$ is finite.*

This result together with Theorem 1 gives the following general statement about non-recursive trade-offs between space-bounded realtime-IA.

Theorem 9 ([31]) *Let $f(n)$ be an IA-ts-constructible function satisfying $f^{-1}(n) \in \Omega(\log(n))$, and let $g(n) \in o(f^{-1}(n))$ be a non-decreasing function. Then there is a non-recursive trade-off between realtime-IA working in $f^{-1}(n)$ strong space and realtime-IA working in $g(n)$ strong space.*

Since 2^n and the polynomials n^k are IA-ts-constructible functions, we immediately obtain the following corollary.

Corollary 6 *For any integer $k \geq 2$, the following non-recursive trade-offs hold:*

1. *realtime-$\log(n)$-IA $\overset{nonrec}{\longrightarrow}$ DFA,*
2. *realtime-$\sqrt[k]{n}$-IA $\overset{nonrec}{\longrightarrow}$ realtime-$\log(n)$-IA,*
3. *realtime-$\sqrt[k]{n}$-IA $\overset{nonrec}{\longrightarrow}$ realtime- $\sqrt[k+1]{n}$-IA,*
4. *realtime-IA $\overset{nonrec}{\longrightarrow}$ realtime-$\sqrt[k]{n}$-IA.*

The question whether the non-recursive trade-off between realtime-$\log(n)$-IA and DFA can be improved to realtime-IA working in sublogarithmic strong space has been answered negatively by showing that at least logarithmic strong space is necessary to accept non-regular languages [31].

Finally, the non-semidecidability results for realtime-IA can directly be translated to realtime-$\log(n)$-IA, since for every realtime-IA M a realtime-$\log(n)$-IA accepting the "padded" language $\{ c^{2^{|w|}-1} w c^{|w|} \mid w \in L(M) \}$ with padding symbol c can effectively be constructed [31]. Thus, we obtain that even the minimum amount of space necessary for non-regular languages makes the model again so powerful that non-recursive trade-offs exist and decidability questions are not semidecidable.

Theorem 10 ([31]) *For realtime-$\log(n)$-IA the following questions are not semidecidable: emptiness, universality, finiteness, infiniteness, inclusion, equivalence, regularity, and context-freeness.*

4.4 Time-Bounded Iterative Arrays

It has been shown in [4] that there exists an infinite time hierarchy in between realtime and lineartime. Thus, strictly stronger classes of iterative arrays are obtained by adding more and more time. It is a result of [22] that there are also non-recursive trade-offs between these classes of iterative arrays. For the hierarchy we need the notion of *IA-constructability* which is a relaxation of the notion of IA-ts-constructability. We say that a strictly increasing function $f : \mathbb{N} \to \mathbb{N}$ is IA-constructible, if there is an IA whose communication cell enters on empty input an accepting state exactly at all time steps $f(i)$ for $i \geq 1$.

Now, let M be a DLBA, r^{-1} be an IA-constructible function. Then we define a function h_r as $h_r(n) = r^{-1}((n + 1)^2)$, and a language

$$L_{r,M} = \{\, \$^{h_r(m)-(m+1)^2+1} w_1 \$ w_2 \$ \cdots \$ w_m \textcent y \mid m \geq 1, w_i \in \Lambda_M^m \text{ for } 1 \leq i \leq m,$$
$$y \in \text{VALC}'(M), \text{ and } \exists\, 1 \leq j \leq m : y = w_j \,\}.$$

Lemma 9 ([4, 22]) *Let M be a DLBA and $r_1, r_2 : \mathbb{N} \to \mathbb{N}$ be two increasing functions such that $r_2 \in o(r_1)$ and r_1^{-1} is IA-constructible. Then an IA working in time $n + r_1(n)$ and recognizing $L_{r_1,M}$ can effectively be constructed. Moreover, the language $L_{r_1,M} \in \mathcal{L}_{n+r_2(n)}(IA)$ if and only if $L(M)$ is finite.*

This lemma together with Theorem 1 gives the following general statement about non-recursive trade-offs between time-bounded IA.

Theorem 11 *Let $r_1, r_2 : \mathbb{N} \to \mathbb{N}$ be two increasing functions with $r_2 \in o(r_1)$ and r_1^{-1} is IA-constructible. Then the trade-off between realtime$+r_1(n)$-time IA and realtime$+r_2(n)$-time IA is non-recursive.*

It is known that the family of IA-constructible functions is very rich. For example, it includes the functions $n!$, k^n, and n^k, where $k \geq 1$ is an integer. Furthermore, it is closed, for example, under the operations addition, multiplication, composition, minimum, and maximum [34]. As an analogon to Corollary 2, we have the following corollary.

Corollary 7 *For any integer $k \geq 2$, the following non-recursive trade-offs hold:*

1. *realtime$+\log(n)$-IA $\overset{nonrec}{\longrightarrow}$ realtime-IA,*
2. *realtime$+\sqrt[k]{n}$-IA $\overset{nonrec}{\longrightarrow}$ realtime$+\log(n)$-IA,*
3. *realtime$+\sqrt[k]{n}$-IA $\overset{nonrec}{\longrightarrow}$ realtime$+\sqrt[k+1]{n}$-IA,*
4. *lineartime-IA $\overset{nonrec}{\longrightarrow}$ realtime$+\sqrt[k]{n}$-IA.*

The fact that 2^n is IA-constructible and the IA-constructible functions are closed under composition implies that also the kth iterated power of 2, for some fixed $k \geq 1$, is IA-constructible. Thus, we obtain as corollary also a non-recursive trade-off between realtime$+\log^{[k]}(n)$-IA and realtime-IA, where $\log^{[k]}(n)$ denotes the kth iterated logarithm. Hence, even a very little amount of additional time leads to a non-recursive trade-off.

Corollary 8 *For any integer $k \geq 1$, the following non-recursive trade-off holds:*
realtime$+\log^{[k]}(n)$-IA $\overset{nonrec}{\longrightarrow}$ realtime-IA.

4.5 Iterative Arrays with Set Storage

A recent generalization of IA is to additionally provide the data structure of a set [23] and to consider so-called set iterative arrays (SIA). In this model, the IA is equipped with a one-way writing tape where strings for the set operations are assembled, and the data storage of a set where words of arbitrary length can be stored. The communication cell additionally controls the set storage by adding, removing, or testing strings written on the tape to or from the set. It is shown in [23] that under real-time conditions such devices are more powerful than conventional IA and classical set automata. We will now sketch how this proper inclusion between the language classes accepted by realtime-IA and realtime-SIA can be extended to a non-recursive trade-off between both automata classes. The witness language used in [23] for the separation of both classes is

$$L_s = \{\, \$x_1\$x_2\$\cdots\$x_k\&y \mid k \geq 1, x_i, y \in \{a, b\}^* \text{ and } y = x_j \text{ for some } 1 \leq j \leq k\,\}.$$

We will modify this language to the language $L_{s,M}$ for a given DLBA M:

$$L_{s,M} = \{\, \$x_1\$x_2\$\cdots\$x_k\&y \mid k \geq 1, x_i, y \in \text{VALC}'(M) \text{ and}$$
$$y = x_j \text{ for some } 1 \leq j \leq k\,\}.$$

Since $\text{VALC}'(M)$ belongs to $\mathscr{L}_{rt}(\text{IA})$, it is straightforward to construct a realtime-SIA accepting $L_{s,M}$ that successively tests all x_i whether they belong to $\text{VALC}'(M)$, adds them to the set, and finally tests whether y belongs to $\text{VALC}'(M)$ and is in the set. If $L(M)$ is finite, $L_{s,M}$ can already be accepted by some DFA and hence by some realtime-IA as well. The basic idea is to test whether all x_i belong to the finite set of elements in $\text{VALC}'(M)$ and additionally to mark which elements have already been tested. When reading y it has to be tested again whether y belongs to the finite set $\text{VALC}'(M)$ and whether y has already been marked in the previous steps. If so, the input is accepted and otherwise rejected. It is clear that all these tests and markings can be realized in the finite state set of a DFA, since $L(M)$ and hence $\text{VALC}'(M)$ are finite sets. Finally, one has to argue that $L_{s,M}$ is not accepted by any realtime-IA, if $L(M)$ is infinite. This can be done in a similar way as in [23] for the language L_s. The basic idea there is to consider subsets of the form $\{x_1, x_2, \ldots, x_n\}$ of $\{a, b\}^n$, to observe that there are at least $\binom{2^n}{n}$ such subsets, and to derive a contradiction using a counting argument. Here, we have to consider subsets of the form $\{x_1, x_2, \ldots, x_n\}$, where each $x_i \in \text{VALC}'(M)$ and $|x_i| = n$, for $1 \leq i \leq n$. Since we consider the extended valid computations $\text{VALC}'(M)$, we know that for each $x \in \text{VALC}'(M)$ of length n, there are at most 2^n further valid computations of the same length. Thus, we can conclude that there are at least $\binom{2^n}{n}$ such subsets and the proof can be complemented along the lines given in [23]. Altogether, we obtain the following lemma.

Lemma 10 *Let M be a DLBA. Then an SIA recognizing $L_{s,M}$ can effectively be constructed. Moreover, the language $L_{s,M} \in \mathscr{L}_{rt}(\text{IA})$ if and only if $L(M)$ is finite.*

This lemma together with Theorem 1 gives the following non-recursive trade-off.

Corollary 9 *The following non-recursive trade-off holds:*
realtime-SIA $\xrightarrow{nonrec}$ realtime-IA.

5   Recursive Trade-Offs

The previous section showed that non-recursive trade-offs exist already between very restricted cellular models such as, for example, space-bounded iterative arrays and communication-bounded iterative arrays. Thus, the question arises which restrictions are necessary to obtain *recursive* trade-offs. One answer to this question is the special case of realtime-OCA accepting unary languages, that is, languages over an alphabet of one symbol only. It is known owing to [40] that such languages are always regular languages and an equivalent DFA can effectively be constructed. Its number of states is bounded by n^2, where n is the number of states of the given realtime-OCA. Another approach is to limit the number of available cells. This has partly been discussed in Sect. 4.3 for iterative arrays, where non-recursive trade-offs are shown to exist if at least a logarithmic number of cells is provided. In this section, we will consider a more severe constraint, namely, the restriction to a fixed number of cells. This restriction may be motivated by the fact that all models studied so far have to provide an arbitrary number of cells since an input may be arbitrarily long. This may not be realistic from a practical point of view. It will turn out that the restriction to a fixed number of cells reduces the computational capacity to the regular languages. Thus, with regard to the computational power a fixed number of cells is as powerful as one DFA, that is, one cell. It is therefore of particular interest to investigate the descriptional complexity of this restricted parallel model for the regular languages.

Basically, this restricted cellular model consists of a finite number of cells, where the rightmost cell acts additionally as communication cell to the input. We will consider here three variants with regard to the effect of allowing more and more communication on the size of description. In the most restricted variant, called kC-OCA, we allow only one-way information flow. That is, every cell gets information from its right neighbors only. In the variant with a minimum amount of two-way communication, called kC-OCA$_r$, every cell but the rightmost cell, gets information from their right neighbors only, whereas the rightmost cell gets in addition information from its left neighbor. Finally, we consider a variant with a maximum amount of two-way communication, called kC-CA, where every cell gets information from its left and right neighbors. We start with the definition of the three restricted variants as given in [27, 30].

Definition 3   A k-cell one-way cellular automaton (kC-OCA) M is a system $M = \langle S, F, A, s_0, \triangledown, k, \delta_r, \delta \rangle$, where S is the finite, nonempty set of cell states, $F \subseteq S$ the set of accepting states, $A \subseteq S$ is the finite, nonempty set of input symbols, $s_0 \in S$ is the quiescent state, $\triangledown \notin A$ is the end-of-input symbol, k is the number of cells, and $\delta_r : S \times (A \cup \{\triangledown\}) \to S$ is the local transition function for the rightmost cell. We require that the pair $(s_0, \triangledown)$ is mapped to s_0. $\delta : S \times S \to S$ is the local transition function for the other cells. We require that the pair (s_0, s_0) is mapped to s_0.

Introducing two-way communication implies that the leftmost cell has to handle the fact that it has no left neighbor. Therefore, we consider an additional boundary state $\# \notin S \cup A$.

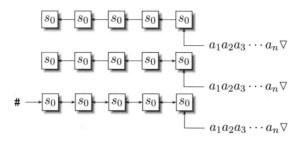

Fig. 5 A 5C-OCA (top), 5C-OCA$_r$ (middle), and 5C-CA (bottom)

Definition 4 A k-cell one-way cellular automaton with two-way communication cell (kC-OCA$_r$) M is identical to a kC-OCA except that δ_r is redefined as $\delta_r : (S \cup \{\#\}) \times S \times (A \cup \{\nabla\}) \rightarrow S$. We require that the tuples (s_0, s_0, ∇) as well as $(\#, s_0, \nabla)$ are mapped to s_0.

A k-cell two-way cellular automaton (kC-CA) M is identical to a kC-OCA except that δ_r and δ are redefined: $\delta_r : (S \cup \{\#\}) \times S \times (A \cup \{\nabla\}) \rightarrow S$ is the local transition function for the rightmost cell. We require that the tuples (s_0, s_0, ∇) and $(\#, s_0, \nabla)$ are mapped to s_0. $\delta : (S \cup \{\#\}) \times S \times S \rightarrow S$ is the local transition function for the other cells. We require that the tuples (s_0, s_0, s_0) and $(\#, s_0, s_0)$ are mapped to s_0.

The restricted models work similar to the unrestricted model (see Fig. 5). The next state of each but the rightmost cell depends on the current state of the cell itself and its right neighbor. The next state of the rightmost cell in kC-OCA$_r$ and of all cells in kC-CA additionally depend on the state of the left neighboring cell. The transition rule is applied synchronously to each cell at the same time. In contrast to unrestricted cellular automata the input is processed as follows. In the beginning all cells are in the quiescent state. The rightmost cell is the communicating cell to the input and thus also called communication cell. At every time step one input symbol is processed by the rightmost cell. All other cells behave as described. The input is accepted, if the leftmost cell enters an accepting state. Since the minimal time to read the input and to send all information from the rightmost cell to the leftmost cell is the length of the input plus k, we feed a special end-of-input symbol ∇ to the rightmost cell after reading the input. The size of an automaton $M = \langle S, F, A, s_0, \nabla, k, \delta_r, \delta \rangle$ is defined as the number of states in S, that is, $|M| = |S|$.

A configuration of a kC-OCA (kC-OCA$_r$, kC-CA) at some time step $t \geq 0$ is a pair (c_t, w_t) where $w_t \in A^*$ denotes the remaining input sequence and c_t is a description of the k cell states, formally a mapping $c_t : \{1, \ldots, k\} \rightarrow S$. The configuration (w_0, c_0) at time 0 is defined by the input word w_0 and the mapping c_0 that assigns the quiescent state to all cells, while subsequent configurations are chosen according to the global transition function Δ that is induced by δ and δ_r as follows: Let (w_t, c_t), $t \geq 0$, be a configuration. Then its successor configuration $(w_{t+1}, c_{t+1}) = \Delta(w_t, c_t)$ is as follows.

$$c_{t+1}(i) = \delta(c_t(i), c_t(i+1)), \text{ for } i \in \{1, \ldots, k-1\}$$
$$c_{t+1}(k) = \delta_r(c_t(k), a)$$

for kC-OCA, $k \geq 1$,

$$c_{t+1}(i) = \delta(c_t(i), c_t(i+1)), \text{ for } i \in \{1, \ldots, k-1\}$$
$$c_{t+1}(k) = \delta_r(c_t(k-1), c_t(k), a)$$

for kC-OCA$_r$, $k \geq 2$, and

$$c_{t+1}(1) = \delta(\#, c_t(1), c_t(2))$$
$$c_{t+1}(i) = \delta(c_t(i-1), c_t(i), c_t(i+1)), \text{ for } i \in \{2, \ldots, k-1\}$$
$$c_{t+1}(k) = \delta_r(c_t(k-1), c_t(k), a)$$

for kC-CA, $k \geq 2$, where $a = \#$ and $w_{t+1} = \lambda$ if $w_t = \lambda$, as well as $a = a_1$ and $w_{t+1} = a_2 \cdots a_n$ if $w_t = a_1 \cdots a_n$.

An input w is accepted by a kC-OCA (kC-OCA$_r$, kC-CA) M if at some time step during the course of its computation cell 1 enters an accepting state. The *language accepted by M* is denoted by $L(M)$. If all $w \in L(M)$ are accepted with at most $|w|+k$ time steps, we say that M accepts in realtime. By $\mathscr{L}_{rt}(k\text{C-OCA})$, $\mathscr{L}_{rt}(k\text{C-OCA}_r)$, and $\mathscr{L}_{rt}(k\text{C-CA})$ we denote the family of all languages accepted by kC-OCA, kC-OCA$_r$, and kC-CA in realtime.

Let us first examine the upper and lower bounds for converting the models kC-OCA, kC-OCA$_r$, and kC-CA to DFA. The construction for the upper bound is basically the construction of the Cartesian product of the k cells, where some states are never entered and may be removed. For the lower bound the language considered is $L_{n,k} = \{a^m \mid m \geq n^k\}$ for $n \geq 1$ and $k \geq 1$. It is not difficult to construct kC-OCA (kC-OCA$_r$, kC-CA) with $n + 1$ states accepting $L_{n,k}$ by implementing an n-ary counter in the k cells. On the other hand, any DFA for $L_{n,k}$ clearly needs at least $n^k + 1$ states.

Theorem 12 ([27, 30] *Every n-state kC-OCA (kC-OCA$_r$, kC-CA) can be converted to an equivalent DFA with at most $n^k - n^{k-1} + 1 \in O(n^k)$ states.*

The language $L_{n,k}$ can be accepted by some kC-OCA (kC-OCA$_r$, kC-CA) with $n + 1$ states, but every DFA accepting $L_{n,k}$ needs at least $n^k + 1 \in \Omega(n^k)$ states.

On the other hand, simulating DFA, that is, one cell by k cells, gives quadratic savings in case of at least minimal two-way communication such as in kC-OCA$_r$ and in kC-CA. It leads to no savings in case of one-way communication as in kC-OCA. In the former case, the quadratic savings can essentially be achieved by encoding

n states of the DFA by two bits of a $\lceil\sqrt{n}\rceil$-ary alphabet and simulating the DFA in the last two cells. In the latter case, the DFA is simulated in the rightmost cell and its states are shifted to the left. Moreover, an additional accepting state is necessary that is entered when the DFA accepts and the input is read completely. Thus, we obtain the following upper bounds for the conversion of DFA to kC-OCA, kC-OCA$_r$, and kC-CA.

Theorem 13 ([27, 30]) *Every n-state DFA can be converted to an equivalent kC-OCA with at most $n + 1$ states.*

Every n-state DFA can be converted to an equivalent kC-OCA$_r$ (kC-CA) with $O(\sqrt{n})$ states.

For the lower bounds we first consider the language $L_p = \{ a^n \mid n \equiv 1 \bmod p \}$, where $p \geq 2$ is a fixed prime number. It is shown in [27] by a thorough case analysis that any kC-OCA accepting L_p needs at least $p + 1$ states. On the other hand, L_p is accepted by some p-state DFA and, moreover, any DFA accepting L_p needs at least p states. Since there are infinitely many prime numbers, we obtain the following lower bound for the simulation of DFA by kC-OCA.

Theorem 14 ([27]) *There is an infinite sequence of languages L_n with $n \in N \subseteq \mathbb{N}$ such that each L_n is accepted by some n-state DFA, but every kC-OCA accepting L_n needs at least $n + 1$ states.*

This result shows that the additionally introduced accepting state is inevitable in some cases. Furthermore, we obtain that in some cases additional efforts in terms of states are necessary to govern an array of DFA in contrast to one single DFA. The witness language L_p shows also that a parallel model does not help to reduce the size of description in comparison with a sequential model. Since the above result is not depending on a specific number k of cells, the language L_p can be seen as an inherently sequential language.

The lower bounds for the simulation of DFA by kC-OCA$_r$ and kC-CA are shown with the help of an incompressibility argument from the field of Kolmogorov complexity. The basic idea is to consider a language that is a singleton $\{x\}$. A DFA accepting that language obviously needs $|x| + 1$ states. The goal is to show that any kC-OCA$_r$ for $\{x\}$ needs $\Omega(\sqrt{|x|/\log(|x|)})$ states. Let $C(\{x\} \mid |x|)$ denote the minimal size of a binary program describing x and knowing its length $|x|$. It is known from the field of Kolmogorov complexity (see, for example, [25]) that incompressible binary strings of arbitrary length exist. That is, for each $n \geq 1$ there is some binary string x of length n such that $C(x \mid n) \geq n$. On the other hand, the language $\{x\}$, with x being incompressible of length n, can be described by some kC-OCA$_r$. It is shown in [30] that under the assumption that $\{x\}$ is accepted by some kC-OCA$_r$ with less than $\sqrt{n/(3\log(n))} - 1$ states it can be concluded that $C(x \mid n)$ is less than n. This contradiction gives that any kC-OCA$_r$ for $\{x\}$ needs $\Omega(\sqrt{n/\log(n)})$ states. Similar considerations show a lower bound of $\Omega(\sqrt[3]{n/\log(n)})$ states for kC-CA.

Theorem 15 ([30]) *There is an infinite sequence of languages L'_n with $n \in N' \subseteq \mathbb{N}$ such that each L'_n is accepted by some n-state DFA, but every kC-OCA_r accepting L'_n needs $\Omega(\sqrt{n/\log(n)})$ states, whereas every kC-CA accepting L'_n needs $\Omega(\sqrt[3]{n/\log(n)})$ states.*

It is interesting that the lower bounds again do not depend on the specific number of cells of the given kC-OCA_r or kC-CA. That is, there are languages where at most quadratic resp. cubic savings can be obtained by kC-OCA_r resp. kC-CA regardless of the number of cells provided. Thus, the constructions leading to the upper bounds are asymptotically optimal for kC-OCA_r and nearly optimal for kC-CA.

Next, we consider the costs for the simulation of kC-OCA, kC-OCA_r, and kC-CA among themselves. First, it is straightforward to embed an n-state kC-OCA into an n-state kC-OCA_r or kC-CA and to embed an n-state kC-OCA_r into an n-state kC-CA. This upper bound is tight by considering the language $L_{n,k} = \{a^m \mid m \geq n^k\}$ for $n \geq 1$ and $k \geq 1$ already used in Theorem 12. We know that the language $L_{n,k}$ is accepted by some kC-OCA or kC-OCA'_r with $n + 1$ states, and it is shown in [30] that any kC-OCA_r or kC-CA for $L_{n,k}$ needs at least $n + 1$ states as well. Second, we can convert an n-state kC-OCA_r into an $(n^2 + n)$-state kC-OCA by simulating the last but one cell of the kC-OCA_r additionally in the last cell of the kC-OCA. The conversion of an n-state kC-CA into a kC-OCA_r is done by dividing the k cells of the kC-CA into two halves and to simulate the left half in the last but one cell and the right half in the last cell of the kC-OCA_r. The equivalent kC-OCA_r needs $O(n^{k/2})$ states. Finally, an n-state kC-CA can be converted into an kC-OCA by first converting the kC-CA into a DFA (see Theorem 12) and secondly simulating the DFA by a kC-OCA that needs at most $O(n^k)$ states.

To establish lower bounds for the conversions we will make use of a general construction given in [30] showing that *unary* regular languages can always be accepted by kC-CA with polynomial savings of degree k. This is no longer true for binary languages as is shown by the incompressibility argument used in Theorem 15. It is sufficient for our purposes to apply the general construction to the unary language $L'_p = \{a^n \mid n \equiv 0 \bmod p\}$, where $p \geq 2$ is a prime number.

Lemma 11 *Each language L'_p can be accepted by a kC-CA with $O(\sqrt[k]{p})$ states provided that $p \geq 2k - 1$.*

Similar to Theorem 14 it can be shown that every kC-OCA accepting L'_p needs at least p states. Since there are infinitely many prime numbers, we immediately obtain a lower bound of $\Omega(n^k)$ states for the conversion of kC-CA into kC-OCA. For the lower bound of $\Omega(n^{k/2})$ for the conversion of kC-CA into kC-OCA_r we use again the language L'_p. We know that L'_p is accepted by a kC-CA with $n \in O(\sqrt[k]{p})$ states. Assuming that L'_p is accepted by a kC-OCA_r with $m \in o(n^{k/2})$ states, we can construct an equivalent kC-OCA with $m^2 + m \in o(n^k) = o(p)$ states which is a contradiction. Finally, we use once more the language L'_p for the lower bound of $\Omega(n^2)$ for the conversion of kC-OCA_r into kC-OCA. By Theorem 13, L'_p is accepted by some kC-OCA_r with $n \in O(\sqrt{p})$ states. On the other hand, every kC-OCA for L'_p needs at least $p \in \Omega(n^2)$ states.

Table 1 Upper and lower bounds for conversions between DFA, kC-OCA, kC-OCA$_r$, and kC-CA

	DFA	kC-OCA	kC-OCA$_r$	kC-CA
DFA	—	$O(n^k)$	$O(n^k)$	$O(n^k)$
		$\Omega(n^k)$	$\Omega(n^k)$	$\Omega(n^k)$
kC-OCA	$\leq n+1$	—	$\leq n^2+n$	$O(n^k)$
	$\geq n+1$		$\Omega(n^2)$	$\Omega(n^k)$
kC-OCA$_r$	$O(\sqrt{n})$	$\leq n$	—	$O(n^{k/2})$
	$\Omega(\sqrt{n/\log(n)})$	$\geq n$		$\Omega(n^{k/2})$
kC-CA	$O(\sqrt{n})$	$\leq n$	$\leq n$	—
	$\Omega(\sqrt[3]{n/\log(n)})$	$\geq n$	$\geq n$	

The upper and lower bounds discussed in this section are summarized in Table 1. An entry in column A and row B describes the upper and lower bounds when converting type-A automata to type-B automata.

It has already been discussed that kC-OCA (kC-OCA$_r$, kC-CA) accept exactly the regular languages. Moreover, this result is not depending on a specific number of cells. Thus, from the perspective of computational capacity kC-OCA (kC-OCA$_r$, kC-CA) are equally powerful for all $k \geq 1$. However, from the perspective of descriptional complexity the natural question arises whether $k+1$ cells are more powerful than k cells, when the number of states is fixed to some natural number n. It is not clear yet whether an n-state kC-OCA can be simulated by an equivalent n-state $(k+1)$C-OCA. Thus, it is an open question whether the former class of automata describes a language class that is properly contained in the language class described by the latter automata. Nevertheless, it is known that there are languages accepted by n-state $(k+1)$C-OCA, but not by any n-state kC-OCA.

Theorem 16 ([27, 28]) *For $n \geq 4$ and $k \leq n$, $L_{n-1,k} = \{ a^m \mid m \geq (n-1)^k \}$ is accepted by some n-state kC-OCA (kC-OCA$_r$, kC-CA), but every equivalent jC-OCA (jC-OCA$_r$, jC-CA) with $1 \leq j < k$ needs more than n states.*

It has been shown in Theorem 6 that there are no minimization algorithms for realtime-OCA and realtime-IA which convert an arbitrary automaton into an equivalent automaton with a minimal number of states. On the other hand, it is known that for DFA such minimization algorithms exist which work in addition in an effective way. For example, in [11] a minimization algorithm for DFA is presented that works in time $O(n \cdot \log(n))$ where n is the number of states. Thus, the question arises whether a, possibly efficient, minimization algorithm exists for kC-OCA, kC-OCA$_r$, or kC-CA. It is straightforward to devise a brute-force algorithm which basically lists the finite number of all kC-OCA (kC-OCA$_r$, kC-CA) of smaller size than the given automaton and subsequently tests their equivalence with the given automaton. The equivalence test is possible, since kC-OCA (kC-OCA$_r$, kC-CA) describe regular languages only. Such algorithms are far from being efficient. However, it is currently not known whether there exists a more efficient one. It is not difficult (see, for example, [27, 28]) to construct non-isomorphic minimal kC-OCA (kC-OCA$_r$, kC-CA). Hence, a minimal kC-OCA (kC-OCA$_r$, kC-CA) needs not to be unique. This may be a hint for the non-existence of efficient minimization algorithms for kC-OCA (kC-OCA$_r$, kC-CA).

It is clear by Theorem 12 that the questions of emptiness, universality, inclusion, or equivalence are decidable for kC-OCA (kC-OCA$_r$, kC-CA), since they are decidable for DFA. Moreover, the problems are efficiently solvable for DFA and with regard to their computational complexity they are known to be **NLOGSPACE**-complete (see, for example, [42]). Now, we also want to investigate the computational complexity of these problems for kC-OCA (kC-OCA$_r$, kC-CA). It turns out that all problems are again **NLOGSPACE**-complete. Hence, the computational complexity of the problems is identical for DFA and the restricted parallel model of kC-OCA, kC-OCA$_r$, and kC-CA.

Theorem 17 ([19]) *Let $k \geq 2$ be an integer. Then for kC-OCA the problems of testing emptiness, universality, inclusion, and equivalence are* **NLOGSPACE**-*complete.*

Proof First, we show that the problem of non-emptiness belongs to **NLOGSPACE**. Since the class **NLOGSPACE** is closed under complementation, emptiness belongs to **NLOGSPACE** as well. We describe a two-way nondeterministic Turing machine M which receives an encoding of some kC-OCA N on its read-only input tape and produces on its write-only output tape an answer *yes* or *no* while the space used on its working tape is bounded by $O(\log(|cod(N)|))$. Then the work space is bounded by $O(\log(n))$ as well where n denotes the maximum of the number of states in N and the size of the input alphabet of N, since both parameters are part of the encoding of N on the input tape of M. It is shown in Theorem 12 that kC-OCA N can be converted to an equivalent DFA N' having at most $n^k - n^{k-1} + 1$ states. It has been shown in [12] by using the pumping lemma for regular languages that $L(N')$ is not empty if and only if $L(N')$ contains a word of length of at most n^k. Thus, the idea for the Turing machine M is to guess a word of length of at most n^k and to check whether it is accepted by N. We implement on M's working tape a binary counter C which counts up to n^k. With the usual construction this needs at most $O(\log(n^k)) = O(k \cdot \log(n)) = O(\log(n))$ tape cells. Additionally, we have to keep track of the current states of the k cells of N. Clearly, the state of each cell can be represented by $O(\log(n))$ tape cells. Altogether, a configuration of N can be represented by $O(\log(n))$ tape cells. Now, M guesses one input symbol a, M increases the counter C, and updates all cells of N according to the transition function of N encoded on the input tape. This behavior is iterated until either the simulated leftmost cell of N enters an accepting state of N or the counter C has been counted up to n^k. In both cases, the Turing machine M halts and outputs *yes* in the former case and outputs *no* in the latter. Altogether, M decides the non-emptiness of N and uses at most a logarithmic number of tape cells with regard to the length of the input. Similar considerations show that the problems of non-universality, inclusion, and equivalence belong to **NLOGSPACE** as well.

The hardness results follow directly from the hardness results for DFA (see, for example, the summary in [42]), since any DFA can be effectively converted to an equivalent kC-OCA by Theorem 13 which simulates the given DFA in the rightmost cell and sends an additional accepting state to the leftmost cell when the end-of-input symbol is read and the input is accepted by the DFA. Obviously, this construction can be done in deterministic logarithmic space.   □

The constructions for kC-OCA used in the proof of Theorem 17 are identical for kC-OCA$_r$ and kC-CA. Thus, we immediately obtain the following corollary.

Corollary 10 *Let $k \geq 2$ be an integer. Then for kC-OCA$_r$ and kC-CA the problems of testing emptiness, universality, inclusion, and equivalence are* **NLOGSPACE**-*complete.*

6 State Complexity

In the previous sections, the descriptional complexity of several cellular models has been investigated in terms of non-recursive and recursive trade-offs. Another important branch of the field of descriptional complexity is the *state complexity* of operations, where language operations under which the models in question are closed, such as union, intersection, or complementation, are investigated with regard to optimal constructions concerning their descriptional size. Hence, the goal is to find constructions leading to upper bounds that give the size needed to represent the result of the operation, and to prove the optimality of the construction by providing suitable lower bounds. It is known by the results given in Theorem 6 and the discussion after Theorem 16 that minimization of cellular models is not possible in general or at least computationally hard. Thus, we will have to confine ourselves with bounds that are tight in order of magnitude. In the following, we will summarize the known results on the state complexity of realtime-OCA and kC-OCA obtained in [19].

It should be noted that state complexity is a vivid area in the field of descriptional complexity. It has been initiated by studying the state complexity of deterministic finite automata in [43, 44]. In the meantime, there exists a vast literature on the deterministic and nondeterministic state complexity of regular languages, and we refer to the recent survey [7]. First results on the state complexity of two-way deterministic finite automata have been obtained in [13]. For results on the state complexity of input-driven pushdown automata we refer to the survey given in [38].

6.1 State Complexity of Realtime-OCA

In this subsection, we will investigate the state complexities of the Boolean operations union, intersection, and complementation for realtime-OCA. The following family of witness languages will be used for the lower bounds. For all integers $k \geq 2$ we consider languages $L_k = \{ 0^i a^{j \cdot k^i} \mid i, j \geq 1 \}$ and their complements $\overline{L_k}$ with regard to the alphabet $\{0, a\}$. It is shown in [19] that L_k and $\overline{L_k}$ can be accepted by realtime-OCA with $k + 4$ states, and every realtime-OCA for L_k or $\overline{L_k}$ needs at least $k + 3$ states.

Theorem 18 *Let $k \geq 2$ be an integer. Then $k + 4$ states are sufficient for a realtime-OCA to accept L_k or $\overline{L_k}$. On the other hand, at least $k + 3$ states are necessary for a realtime-OCA to accept L_k or $\overline{L_k}$.*

Let us now consider the operations intersection and union. The construction uses the well-known two-track technique. That is, a realtime-OCA with two tracks is constructed that simulates in its first track the first given automaton and in its second track the other automaton. Finally, the accepting states are suitably defined. For union it is sufficient to define a state as accepting if an accepting state in one track is entered. For intersection the situation is slightly more complicated. The basic idea is to define those states as accepting, when an accepting state is entered at the same time in both tracks. However, since a realtime-OCA accepts as soon as an accepting state is entered, one has to provide additional states which indicate that the current track has already passed through an accepting state. Both constructions lead to the following upper bounds.

Theorem 19 *Let $m, n \geq 1$ be integers, M_1 be an m-state realtime-OCA with r_1 non-accepting states, and M_2 be an n-state realtime-OCA with r_2 non-accepting states. Then $m \cdot n + r_1 \cdot n + m \cdot r_2 + r_1 \cdot r_2 \in O(m \cdot n)$ states are sufficient for a realtime-OCA to accept $L(M_1) \cap L(M_2)$ and $m \cdot n \in O(m \cdot n)$ states are sufficient for a realtime-OCA to accept $L(M_1) \cup L(M_2)$.*

For the lower bounds we can utilize the languages L_k and $\overline{L_k}$, and we obtain that the upper bounds are tight in order of magnitude.

Theorem 20 *Let $m, n \geq 6$ be integers such that $m - 4$ and $n - 4$ are relatively prime. Then at least $(m - 4)(n - 4) + 3 \in \Omega(m \cdot n)$ states are necessary in the worst case for a realtime-OCA to accept the intersection as well as the union of an m-state realtime-OCA and an n-state realtime-OCA language.*

Proof Let $k = m - 4$ and $\ell = n - 4$ be relatively prime. The witness languages for intersection are L_k accepted by an m-state realtime-OCA and L_ℓ accepted by an n-state realtime-OCA. The intersection $L_k \cap L_\ell$ is $L_{k \cdot \ell} = \{ 0^i a^{j \cdot k^i \cdot \ell^i} \mid i, j \geq 1 \}$. By using Theorem 19 we obtain that every realtime-OCA accepting $L_{k \cdot \ell}$ needs at least $k \cdot \ell + 3 = (m - 4)(n - 4) + 3 \in \Omega(m \cdot n)$ states.

For the union we consider the witness languages $\overline{L_k}$ accepted by an m-state realtime-OCA and $\overline{L_\ell}$ accepted by an n-state realtime-OCA. The union of both languages is $\overline{L_{k \cdot \ell}}$, for which at least $k \cdot \ell + 3 \in \Omega(m \cdot n)$ states are necessary by Theorem 19. $\qquad\square$

For the state complexity of complementation we have to take care again of the way a realtime-OCA accepts. Normally, the closure under complementation is shown by interchanging accepting and non-accepting states. This is not possible for realtime-OCA, since the leftmost cell may enter accepting as well as non-accepting states during the course of its computation. Thus, we add copies of the non-accepting states indicating that the leftmost cell has already passed through an accepting state. Additionally, a signal is sent from the rightmost cell to the leftmost cell along the diagonal of the space-time diagram. The goal of the signal is to set each cell passed through as accepting if that cell has not entered an accepting state before. To this end, the states that appear on the diagonal have to be identified as signal. Clearly, this is trivial for the states which appear at the diagonal only. For those states that can appear on the diagonal as well as at other positions, we use copies. Moreover, on the diagonal may now appear new non-accepting states indicating that the cell has entered an accepting state before. These new copies are now used as accepting states. This construction gives the following upper bound for complementation.

Theorem 21 *Let $n \geq 1$ be an integer and M be an n-state realtime-OCA with r non-accepting states, d states that can appear on the diagonal and also at other positions, from which g are non-accepting. Then $n + r + d + g \in O(n)$ states are sufficient for a realtime-OCA to accept $\overline{L(M)}$.*

The upper bound is witnessed by the family $L_{c,k} = \{ 0^i a^j \mid i \geq 1, j \geq k^i \}$, where $k \geq 2$. The construction of a realtime-OCA for $L_{c,k}$ gives that $n = k + 3$ states are sufficient. On the other hand, it can be shown that $2k + 3 = 2n - 3 \in \Omega(n)$ are necessary for every realtime-OCA to accept $L_{c,k}$. Thus, we obtain the following lower bound for complementation.

Theorem 22 *Let $n \geq 5$ be an integer. Then at least $2n - 3 \in \Omega(n)$ states are necessary in the worst case for a realtime-OCA to accept the complement of an n-state realtime-OCA language.*

6.2 State Complexity of kC-OCA

In this subsection, we will discuss the state complexity of the Boolean operations and reversal for kC-OCA, that is, for one-way cellular automata with a fixed number of cells. It turns out that the results are similar to those for realtime-OCA shown in the previous subsection, whereas the proofs for the lower bounds are different. For the detailed proofs we refer to [19].

The upper bounds for union and intersection are obtained by again applying the two-track technique whereby taking into account the peculiarities for the acceptance in case of intersection. We obtain the same upper bounds.

Theorem 23 *Let $k \geq 2$ and $m, n \geq 1$ be integers, M_1 be an m-state kC-OCA with r_1 non-accepting states, and M_2 be an n-state kC-OCA with r_2 non-accepting* states. Then $m \cdot n + r_1 \cdot n + m \cdot r_2 + r_1 \cdot r_2 \in O(m \cdot n)$ states are sufficient for a kC-OCA to accept $L(M_1) \cap L(M_2)$ and $m \cdot n \in O(m \cdot n)$ states are sufficient for a kC-OCA to accept $L(M_1) \cup L(M_2)$.

For the lower bounds we consider the family $L_{n,k} = \{ a^i \mid i \equiv 0 \bmod n^k \}$, where $k \geq 2$ and $n \geq 2$. It can be shown that $n + 2$ states are sufficient for a kC-OCA to accept $L_{n,k}$, whereas n states are necessary for a kC-OCA to accept $L_{n,k}$. For the lower bound of intersection, one considers for relatively prime numbers $m, n \geq 4$ the language $L_{m,k} \cap L_{n,k} = L_{mn,k}$ which needs at least mn states to be accepted by some kC-OCA. On the other hand, $m - 2$ resp. $n - 2$ states are sufficient for a kC-OCA to accept $L_{m,k}$ and $L_{n,k}$, respectively. The same languages can be used for the lower bound of union. Here, one considers the language $L_{m,k} \cup L_{n,k}$ and a thorough analysis in [19] gives that mn states are necessary for a kC-OCA to accept the union.

Theorem 24 *Let $k \geq 2$ be an integer and let $m, n \geq 4$ be integers such that m and n are relatively prime. Then at least $(m - 2)(n - 2) \in \Omega(m \cdot n)$ states are necessary in the worst case for a kC-OCA to accept the intersection as well as the union of an m-state kC-OCA and an n-state kC-OCA.*

The upper bound for complementation is constructed in a similar way as in the unrestricted case taking again into account that each cell has to remember whether an accepting state has passed through itself and that a signal from the rightmost cell to the leftmost cell has to be realized. We obtain the following upper bound.

Theorem 25 *Let $k \geq 2$ and $n \geq 1$ be integers and M be an n-state kC-OCA with r non-accepting states. Then $2(n + r) \in O(n)$ states are sufficient for a kC-OCA to accept $\overline{L(M)}$.*

For the lower bound we consider the family $L'_{n,k} = \{ a^i \mid i \geq n^k \}$, where $k \geq 2$ and $n \geq 2$. The language $L'_{n,k}$ can be accepted by an $(n + 1)$-state kC-OCA by implementing an n-ary counter. On the other hand, it is easy to show that every kC-OCA for the complement $\overline{L'_{n,k}} = \{ a^i \mid i < n^k \}$ needs at least n states.

Theorem 26 *Let $k \geq 2$ and $n \geq 3$ be integers. Then at least $n - 1 \in \Omega(n)$ states are necessary in the worst case for a kC-OCA to accept the complement of an n-state kC-OCA language.*

Finally, we look at the reversal operation. In realtime-OCA this operation has a quadratic upper bound (see, for example, Theorem 40 in [18]). For kC-OCA the construction is completely different and results in an exponential upper bound. The simple idea is to convert a given kC-OCA M to an equivalent DFA M' using the construction of Theorem 12. This gives a polynomial blow-up of degree k. Subsequently, a nondeterministic finite automaton N' accepting the reversal of $L(M)$ is constructed. Finally, automaton N' is determinized to an equivalent DFA which in turn is embedded into an equivalent kC-OCA N by Theorem 13. Since the determinization of a nondeterministic finite automaton leads to an exponential blow-up, the costs for reversal are bounded by some exponential function whose exponent is a polynomial of degree k.

Theorem 27 *Let $k \geq 2$ and $n \geq 1$ be integers and M be an n-state kC-OCA. Then at most $2^{n^k - n^{k-1} + 1} + 1 \in O(2^{n^k})$ states are sufficient for a kC-OCA to accept $L(M)^R$.*

The conversion described above seems to be very circuitous and does not make much use of the parallelism provided by kC-OCA. Hence, one may ask whether the upper bound presented could be improved. However, the following lower bound of reversal shows that the above-described procedure is in a way the best possible. We consider the family $L''_{n,k} = \{ a^{n^k} \{a, b\}^i \mid i \geq 0 \}$, where $k \geq 2$ and $n \geq 2$. Similar to the family $L'_{n,k}$, it is not difficult to construct an $(n + 1)$-state kC-OCA for $L''_{n,k}$. Finally, one can show that any kC-OCA for $L''_{n,k}$ has to distinguish between 2^{n^k} configurations which leads to a lower bound of $2^{n^{k-1}}$ states.

Theorem 28 *Let $k \geq 2$ and $n \geq 3$ be integers such that $n \geq k$. Then at least $\Omega(2^{(n-1)^{k-1}})$ states are necessary in the worst case for a kC-OCA to accept the reversal of an n-state kC-OCA language.*

6.3 Decomposition Problems

In the previous two subsections, we have obtained tight bounds, at least in order of magnitude, for the Boolean operations in realtime-OCA and kC-OCA. Thus, we know the costs that may arise when we compose two automata to a new automaton. However, the converse question is of large interest as well. For example, how many states can we save when we decompose a realtime-OCA language into the union of two realtime-OCA languages? It is clear from the upper bound that at least $2\sqrt{n}$ states are necessary for two realtime-OCA to describe a language accepted by an n-state realtime-OCA. It is an immediate question whether such decompositions can algorithmically be obtained. Thus, given an m-ary operation under which the families of realtime-OCA and kC-OCA languages are closed, does there exist an algorithm that decomposes any given realtime-OCA (kC-OCA) into m smaller ones if such a decomposition exists? We will call such problems *operation decomposition problems*. It turns out that such algorithms cannot exist for the operations in question for realtime-OCA, but exist for kC-OCA.

Theorem 29 ([19]) *The following decomposition problems for realtime-OCA are algorithmically unsolvable.*

1. *The union decomposition problem.*
2. *The intersection decomposition problem.*
3. *The complementation decomposition problem.*
4. *The reversal decomposition problem.*

Proof We exemplarily present the proof of the first assertion. In contrast to the assertion, we assume that there is an algorithm that solves the union decomposition. We obtain a contradiction by showing that in this case the emptiness for realtime-OCA is decidable which is a contradiction to Theorem 3.

Clearly, any OCA has at least as many states as input symbols. Moreover, there is an OCA accepting the empty language which has exactly as many states, where none of them is accepting.

In order to decide whether a given realtime-OCA accepts no input, we proceed as follows. First we inspect the set of accepting states. If it is empty, the answer is *yes*. If it contains at least one input symbol, the answer is *no*. Otherwise, we apply the union decomposition algorithm. If as a result the algorithm reports that there is no decomposition, the answer is *no*. If the algorithm results in two smaller OCA, we recursively apply the decision process to these devices. Now the answer is *yes* if and only if both smaller OCA accept the empty language.

Why does this procedure give the correct answer? This is evident for the cases where the set of accepting states is empty or contains at least one input symbol. Otherwise, when the union decomposition algorithm is applied, we know that there is at least one accepting non-input state. So, if the OCA accepts the empty language, there is always a possible decomposition into two smaller OCA having only input states which are all non-accepting. □

Finally, we would like to remark that all decomposition problems are algorithmically solvable for kC-OCA, since a brute-force algorithm can find all possible decompositions by checking all kC-OCA of smaller size, constructing the composition, and testing their equivalence with the given kC-OCA.

Theorem 30 *The following decomposition problems for kC-OCA are algorithmically solvable.*

1. *The union decomposition problem.*
2. *The intersection decomposition problem.*
3. *The complementation decomposition problem.*
4. *The reversal decomposition problem.*

References

1. Bordihn, H., Kutrib, M., Malcher, A.: Undecidability and hierarchy results for parallel communicating finite automata. Int. J. Found. Comput. Sci. **22**, 1577–1592 (2011)
2. Bordihn, H., Kutrib, M., Malcher, A.: On the computational capacity of parallel communicating finite automata. Int. J. Found. Comput. Sci. **23**, 713–732 (2012)
3. Bordihn, H., Kutrib, M., Malcher, A.: Returning parallel communicating finite automata with communication bounds: hierarchies, decidabilities, and undecidabilities. Int. J. Found. Comput. Sci. **26**, 1101–1126 (2015)
4. Buchholz, T., Klein, A., Kutrib, M.: Iterative arrays with small time bounds. In: M. Nielsen, B. Rovan (eds.) Mathematical Foundations of Computer Science (MFCS 2000), *LNCS*, vol. 1893, pp. 243–252 (2000)
5. Cole, S.N.: Real-time computation by n-dimensional iterative arrays of finite-state machines. IEEE Trans. Comput. **C–18**(4), 349–365 (1969)
6. Delorme, M., Mazoyer, J. (eds.): Cellular Automata – a Parallel Model. Kluwer Academic Publishers, Dordrechtd (1999)

7. Gao, Y., Moreira, N., Reis, R., Yu, S.: A survey on operational state complexity. J. Autom. Lang. Comb. **21**(4), 251–310 (2017)
8. Holzer, M., Kutrib, M.: Nondeterministic finite automata - Recent results on the descriptional and computational complexity. Int. J. Found. Comput. Sci. **20**, 563–580 (2009)
9. Holzer, M., Kutrib, M.: Descriptional complexity – an introductory survey. In: Martín-Vide, C. (ed.) Scientific Applications of Language Methods, pp. 1–58. Imperial College Press, London (2010)
10. Holzer, M., Kutrib, M.: Descriptional and computational complexity of finite automata - A survey. Inf. Comput. **209**, 456–470 (2011)
11. Hopcroft, J.E., et al.: An $n \log n$ algorithm for minimizing states in a finite automaton. In: Kohavi, Z. (ed.) Theory of Machines and Computations, pp. 189–196. Academic Press, Cambridge (1971)
12. Hopcroft, J.E., Ullman, J.D.: Introduction to Automata Theory, Languages, and Computation. Addison-Wesley, Cambridge (1979)
13. Jirásková, G., Okhotin, A.: On the state complexity of operations on two-way finite automata. In: M. Ito, M. Toyama (eds.) Developments in Language Theory (DLT 2008), *LNCS*, vol. 5257, pp. 443–454. Springer (2008)
14. Kari, J.: Theory of cellular automata: a survey. Theor. Comput. Sci. **334**, 3–33 (2005)
15. Klein, A., Kutrib, M.: Fast one-way cellular automata. Theoret. Comput. Sci. **295**, 233–250 (2003)
16. Klein, A., Kutrib, M.: Cellular devices and unary languages. Fund. Inform. **78**, 343–368 (2007)
17. Kutrib, M.: Cellular automata – a computational point of view. In: Bel-Enguix, G., Jiménez-López, M.D., Martín-Vide, C. (eds.) New Developments in Formal Languages and Applications, vol. 6, pp. 183–227. Springer, Berlin (2008)
18. Kutrib, M.: Cellular automata and language theory. In: Meyers, R. (ed.) Encyclopedia of Complexity and System Science, pp. 800–823. Springer, Berlin (2009)
19. Kutrib, M., Lefèvre, J., Malcher, A.: The size of one-way cellular automata. In: Fatès, N., Kari, J., Worsch, T. (eds.) AUTOMATA 2010, Discrete Mathematics and Theoretical Computer Science Proceedings, pp. 71–90. DMTCS (2010)
20. Kutrib, M., Malcher, A.: Computations and decidability of iterative arrays with restricted communication. Parallel Process. Lett. **19**, 247–264 (2009)
21. Kutrib, M., Malcher, A.: Cellular automata with limited inter-cell bandwidth. Theoret. Comput. Sci. **412**, 3917–3931 (2011)
22. Kutrib, M., Malcher, A.: The size impact of little iterative array resources. J. Cell. Autom. **7**, 489–507 (2012)
23. Kutrib, M., Malcher, A.: Iterative arrays with set storage. J. Cell. Autom. **12**, 7–26 (2017)
24. Kutrib, M., Malcher, A., Wendlandt, M.: Stateless one-way multi-head finite automata with pebbles. Int. J. Found. Comput. Sci. **25**, 1141–1160 (2014)
25. Li, M., Vitányi, P.M.B.: An Introduction to Kolmogorov Complexity and Its Applications. Springer, Berlin (1993)
26. Malcher, A.: Descriptional complexity of cellular automata and decidability questions. J. Autom. Lang. Comb. **7**, 549–560 (2002)
27. Malcher, A.: On one-way cellular automata with a fixed number of cells. Fund. Inform. **58**, 355–368 (2003)
28. Malcher, A.: Beschreibungskomplexität von Zellularautomaten. Ph.D. thesis, Institut für Informatik, Johann Wolfgang Goethe-Universität Frankfurt am Main (2004)
29. Malcher, A.: On the descriptional complexity of iterative arrays. IEICE Trans. Inf. Syst. **E87–D**, 721–725 (2004)
30. Malcher, A.: On two-way communication in cellular automata with a fixed number of cells. Theor. Comput. Sci. **330**, 325–338 (2005)
31. Malcher, A., Mereghetti, C., Palano, B.: Sublinearly space bounded iterative arrays. Int. J. Found. Comput. Sci. **21**, 843–858 (2010)
32. Martín-Vide, C., Mateescu, A., Mitrana, V.: Parallel finite automata systems communicating by states. Int. J. Found. Comput. Sci. **13**, 733–749 (2002)

33. Martín-Vide, C., Mitrana, V.: Some undecidable problems for parallel communicating finite automata systems. Inf. Process. Lett. **77**, 239–245 (2001)
34. Mazoyer, J., Terrier, V.: Signals in one-dimensional cellular automata. Theor. Comput. Sci. **217**, 53–80 (1999)
35. Meyer, A.R., Fischer, M.J.: Economy of description by automata, grammars, and formal systems. In: Symposium on Switching and Automata Theory (SWAT 1971), pp. 188–191. IEEE (1971)
36. Morita, K.: Reversible computing and cellular automata - A survey. Theor. Comput. Sci. **395**, 101–131 (2008)
37. Nakamura, K.: Real-time language recognition by one-way and two-way cellular automata. In: M. Kutyłowski, L. Pacholski, T. Wierzbicki (eds.) Mathematical Foundations of Computer Science (MFCS 1999), *LNCS*, vol. 1672, pp. 220–230. Springer (1999)
38. Okhotin, A., Salomaa, K.: Complexity of input-driven pushdown automata. SIGACT News **45**, 47–67 (2014)
39. Rozenberg, G., Bäck, T., Kok, J.N. (eds.): Handbook of Natural Computing. Springer, Berlin (2012)
40. Seidel, S.R.: Language recognition and the synchronization of cellular automata. Tech. Rep. 79-02, Department of Computer Science, University of Iowa (1979)
41. Smith III, A.R.: Cellular automata and formal languages. In: Symposium on Switching and Automata Theory (SWAT 1970), pp. 216–224. IEEE (1970)
42. Yu, S.: Regular languages. In: Rozenberg, G., Salomaa, A. (eds.) Handbook of Formal Languages, vol. 1, pp. 41–110. Springer, Berlin (1997)
43. Yu, S.: State complexity of regular languages. J. Autom. Lang. Comb. **6**, 221–234 (2001)
44. Yu, S., Zhuang, Q., Salomaa, K.: The state complexities of some basic operations on regular languages. Theoret. Comput. Sci. **125**, 315–328 (1994)

Invertible Construction of Decimal-to-Binary Converter Using Reversible Elements

Tai-Ran He, Jia Lee and Teijiro Isokawa

Abstract We design a logic circuit that is capable of converting from a one-digit decimal number to equivalent four-bit binary number. The circuit is composed by reversible elements each of which takes an equal number of input and output lines, associated with a memory to store a binary state. Reversible elements, as claimed by Morita, (International Conference on Machines, Computations, and Universality, 2001) [8], allow more straightforward and efficient constructions of circuits than conventional reversible logic gates. In particular, the circuit is invertible in the sense that reversing the functionalities of each element in it directly gives rise to another circuit which conducts the inverse conversion.

1 Introduction

A sequential machine (SM) is a finite-state machine with output, of which the output and the next state is determined by its current state and current input [4]. A reversible sequential machine (RSM) is a special type of SMs of which the operation reveals a backward determinism, in the sense that from the present state and the output of an RSM, it is always possible to uniquely determine what the previous state and input were [9]. RSMs can be realized by the reversible logic gates: Fredkin gate [2] or Toffoli gate [12]. Morita [8] introduced a reversible element, called Rotary Element (RE), and showed that the RE is capable of constructing a certain reversible Turing machine (RTM). After that, a pair of mutually inverse elements which take a minimal number of input and output lines was proposed in [6]. These elements are capable of constructing any RTM [7] and RSM [11] based on a general design scheme.

T.-R. He
College of Computer Science, Chongqing University, Chongqing, China

J. Lee (✉)
Chongqing Key Lab. Software Theory & Technology, Chongqing University, Chongqing, China
e-mail: lijia@cqu.edu.cn

T. Isokawa
Graduate School of Engineering, University of Hyogo, Himeji, Japan

© Springer International Publishing AG 2018
A. Adamatzky (ed.), *Reversibility and Universality*, Emergence, Complexity and Computation 30, https://doi.org/10.1007/978-3-319-73216-9_7

Unlike reversible logic gates, a reversible element is associated with a memory inside to record a finite number of states, and hence, it can be formulated in itself as an RSM. Also, instead of processing binary-valued signals as a reversible gate, a reversible element takes a finite number of input and output lines, each of which allows a token to appear on it, albeit not at the same time. Thus, at any time, each element receives at most one token from one of its input lines, changing its state and transferring the token to an appropriate output line, in accordance with a bijective transition function. Since each element can be employed as a primitive RSM, constructions of the RTMs (RSMs) based on reversible elements tend to be much more straightforward and simpler than the constructions based on reversible logic gates [7, 8].

Another remarkable feature of the reversible-element-based constructions that distinguishes them from the reversible logic circuits is their possibility for asynchronous operations [5, 9]. Because at most one token is allowed to run around in the entire construction of an RTM or an RSM at any time, delays involved in the token's transmission on any line or in any element's local operation, are unable to affect the global input and output behavior of the construction. Though the single token may increase latency and diminish concurrency, it actually removes the need of a central clock to synchronize the operations of all reversible elements [3].

This paper presents a scheme for building the logic function: binary/decimal conversions out of reversible elements. To this end, we transform the decimal-to-binary converter into a RSM, which enables the construction using reversible elements. Rather than following the general design in [7], our construction is more efficient in the sense that it requires a much smaller number of elements than the general scheme. Moreover, the construction implementing the decimal-to-binary converter is invertible [5, 9], whereby simply reversing the functionalities of each element in it directly results in the circuit's inverse counterpart, which can convert from binary bits to decimal digit.

2 Preliminaries

2.1 Reversible Sequential Machines

Definition 1 A *sequential machine* is a finite-state machine with output, which can be defined as:

$$M = (Q, \Sigma, \Gamma, q_0, \delta)$$

where Q is a finite set of states with $q_0 \in Q$ being the initial state. Σ and Γ are non-empty finite sets of input symbols and output symbols ($\Gamma \cap \Sigma = \emptyset$), respectively. Also, $\delta : \Sigma \times Q \to \Gamma \times Q$ is a function called *transition function*. Assume $\delta(\sigma, p) = (\gamma, q)$ with $p, q \in Q$, $\sigma \in \Sigma$ and $\gamma \in \Gamma$. This transition describes a situation where the machine is in state p and operates on an input symbol σ, then it will change the state to q and produce an output symbol γ.

A sequential machine $(Q, \Sigma, \Gamma, q_0, \delta)$ is a reversible sequential machine, if its transition function δ is a one-to-one mapping. Because of the bijectivity of the function, it is reasonable to assume that $|\Sigma| = |\Gamma|$, i.e., an RSM takes an equal number of input and output symbols.

2.2 Reversible Elements

A reversible element is a module which consists of a finite number of input and output lines. Communication between an element and other elements is done via exchanging tokens through interconnection lines among them.

Definition 2 A *reversible element* can be defined as:

$$(M, I, O, \rho, \upsilon, \Psi)$$

in which $M = (Q, \Sigma, \Gamma, q_0, \delta)$ is an RSM. I and O are finite sets of input and output lines ($I \cap O = \emptyset$), respectively, and $\rho : I \to \Sigma$ and $\upsilon : O \to \Gamma$ are injective functions. In addition, $\Psi \subseteq I \times Q \times O \times Q$ is a set of *operations*. For any $a \in I$, $b \in O$ and $p, q \in Q$,

$$a, p \to b, q \in \Psi \Rightarrow \delta(\rho(a), p) = (\upsilon(b), q).$$

The operation $a, p \to b, q \in \Psi$ corresponds to a situation where the element, when the RSM M is in state p and a token appears on its input line a, will assimilate the token from line a, then generates a token on output line b and changes the state to q.

For simplicity, a reversible element is said to be in state p if its core RSM is in that state. Assume a reversible element $(M, I, O, \rho, \upsilon, \Psi)$. Let

$$\theta_1 = a_1, p_1 \to b_1, q_1$$
$$\theta_2 = a_2, p_2 \to b_2, q_2$$

be any two operations in Ψ (θ_1 and θ_2 may be the same). Due to the reversibility of the RSM M, it is easy to verify that

$$(a_1 = a_2 \wedge p_1 = p_2) \vee (b_1 = b_2 \wedge q_1 = q_2) \Rightarrow \theta_1 = \theta_2.$$

Thus, each operation of a reversible element does not overlap on the left-hand side or on the right-hand side with any other operations [9]. This paper employs those simple elements associated with no more than two states, examples of which listed in Figs. 1, 2 and 3.

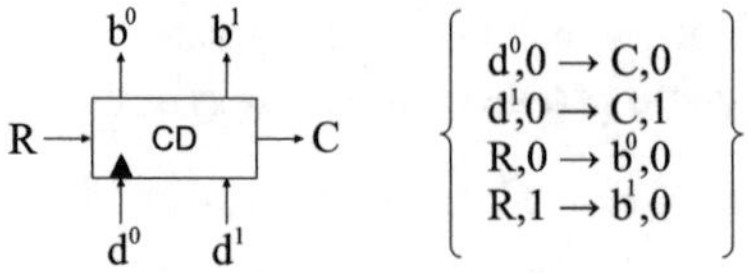

Fig. 1 CD and its respective sets of operations

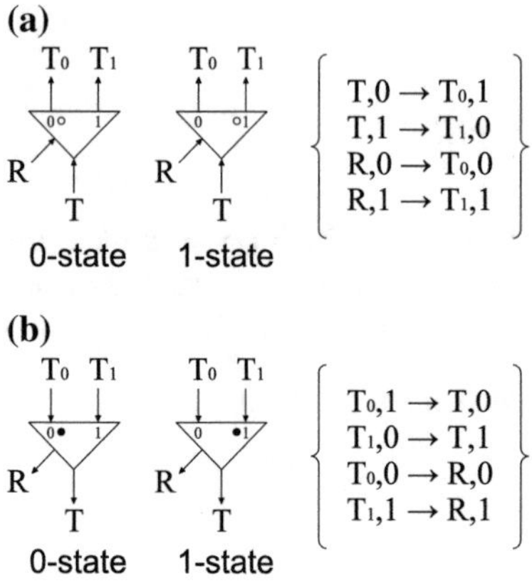

Fig. 2 Reversible elements: **a** RT, **b** IRT and their respective sets of operations

The reversible element given in Fig. 1 is called Coding-Decoding (CD [6]), which has two states: 0 and 1, with 0 being the initial state. Assume $i \in \{0, 1\}$. When a CD is in state 0, a token arriving on input line d^i updates the state to i, and gives rise to a token on output line C. When the CD is in state i, a token arriving on input line R reverts the state to 0, and results in a token on output line b^i. It is easy to verify that the inverse of a CD is still a CD.

The RT and IRT given in Fig. 2a and b, respectively, are mutually inverse to each other [6]. Both elements consist of two input lines and two output lines. Assume $i \in \{0, 1\}$. When an RT is in state i, a token arriving on input line T switches the state to $1 - i$, and results in a token on output line T_i; a token arriving on input line R does not change the state and gives rise to a token on output line T_i. Likewise, when an IRT is in state i, a token arriving on input line T_{1-i} changes the state to $1 - i$, and results in a token on output line T; a token arriving on input line T_i does not change the state, and gives rise to a token on output line R.

Another pair of mutually inverse elements: called Redirector (RD) and Inverse Redirector (IRD) [7], are provided in Fig. 3. An RD (IRD) has input (resp. output) lines: T, R_0, R_1, output (resp. input) lines: T_0, T_1, S_0, S_1, and two states: 0 and 1. Assume $i \in \{0, 1\}$. When an RD (resp. IRD) is in state 0 (the initial state), a token arriving on input line R_i (resp. S_i) does not change the state, and results in an output token on line S_i (resp. R_i); an input token on line T (resp. S_0) changes the state to 1, and gives rise to a token on output line S_0 (resp. T). When the RD (resp. IRD) is in state 1, a token arriving on input line R_i (resp. T_i) reverts the state to 0, and results in an output token on line T_i (resp. R_i).

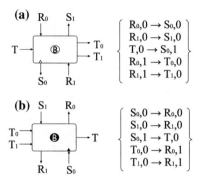

Fig. 3 Reversible elements: **a** RD, **b** IRD, and their respective sets of operations

3 Invertible Conversions Between Decimal and Binary Numbers

It is well known that every one-digit decimal number can be uniquely represented as a four-bit binary number, as illustrated in Table 1. Conversions between decimal and binary numbers can be accomplished using logic circuits composed by conventional logic gates [1, 10]. In this paper, we construct the invertible conversions between decimal digits and binary bits using reversible elements: RT and IRT in Fig. 2. For this purpose, we represent each digit of numbers in based N ($N \geq 2$) in terms of N lines, labelled with $0, 1, \ldots, N - 1$, respectively. Hence, a token arriving on line i with $0 \leq i < N$ denotes the value i of the corresponding digit.

To build the logic circuit converting decimal digit to binary bits out of reversible elements, we design a reversible element, called CT, as given in Fig. 4. This element can be characterized as an RSM with input line set: $\{D_i \mid 0 \leq i \leq 9\}$, output line set: $\{d_j^k \mid 0 \leq k \leq 1 \wedge 1 \leq j \leq 4\} \cup \{C_l \mid 1 \leq l \leq 4\} \cup \{S\}$. CT has states: $0, 1, \ldots, 9$

Table 1 One-to-one correspondence between 1-digit decimal numbers and 4-bit binary numbers

Decimal	Binary
0	0000
1	0001
2	0010
3	0011
4	0100
5	0101
6	0110
7	0111
8	1000
9	1001

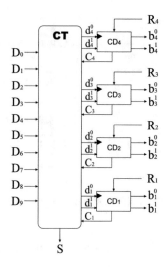

Fig. 4 Invertible decimal-to-binary converter composed by reversible elements CT and CD

with 0 being the initial state. The element include the following operations: For any $m \in \{0, 1, \ldots, 9\}$ and $x, y, z, w \in \{0, 1\}$ that satisfy $m = (xyzw)_2$, we obtain:

$$D_m, 0 \rightarrow d_4^x, (yzw)_2$$
$$C_4, (yzw)_2 \rightarrow d_3^y, (zw)_2$$
$$C_3, (zw)_2 \rightarrow d_2^z, w$$
$$C_2, w \rightarrow d_1^w, 0$$
$$C_1, 0 \rightarrow S, 0.$$

It can be verified that no operation of the CT overlaps with any other operation at the right-hand sides, i.e., the element is reversible. Especially, a token arriving on input line D_m encodes the decimal number m that is equal to the 4-bit binary number $xyzw$. The CT processes the input token, producing a token on output line d^x and changing its state to $(yzw)_2$. The most significant bit of the number $(xyzw)_2$ or m, therefore, is extracted and gives rise to a token on line d_4^x, while the remaining bits $(yzw)_2$ are recorded by the CT's state. After that, the arrivals of tokens on input lines C_4, C_3 and C_2 of the CT will extract each bit y, z and w, respectively and successively, resulting in a token on each of the output lines d_3^y, d_2^z and d_1^w.

As a result, an input token arriving on line D_m of the CT eventually gives rise to a token on each of the output lines d_4^x, d_3^y, d_2^z and d_1^w, which actually converts the 1-digit decimal number m into equivalent 4-bit binary number $(xyzw)_2$. The conversion is accompanied with input tokens to the CT received one by one from the lines C_4, C_3, C_2 and C_1.

Based on the CT and CD in Fig. 1, we can construct a logic circuit, called Invertible decimal-to-binary converter, to conduct the conversion from decimal digit to binary bits, as shown in Fig. 4. In this case, the token appearing on line d_4^x of the CT in response to the input token arriving on line D_m changes the state of the element CD_4 to x, and results in a token on line C_4. Thus, the most significant (the 4th) bit x of the number $(xyzw)_2$ or m is recorded in terms of the internal state of the top CD element, which can be retrieved afterwards via an input token from line R_4 to the CD that will revert the state of CD_4 to 0 and produce a token on the output line b_4^x. Likewise, the k-th bit of the same number with $1 \leq k \leq 4$ is stored via the state of the CD element CD_k that can be retrieved by assigning a token to the input line R_k of the CD_k in a similar way.

Obviously, the Invertible decimal-to-binary converter in itself can be defined as an reversible element (RSM), which takes input line set: $\{D_0, \ldots, D_9, R_1, \ldots, R_4\}$, output lines: $\{b_i^k \mid 1 \leq i \leq 4 \land 0 \leq k \leq 1\} \cup \{S\}$, and states: $0, 1, \ldots, 9$ with initial state 0. It include the following operations: For any $x, y, z, w \in \{0, 1\}$ and $0 \leq (xyzw)_2 \leq 9$,

$$D_{(xyzw)_2}, 0 \rightarrow C, (xyzw)_2$$
$$R_4, (xyzw)_2 \rightarrow b_4^x, (yzw)_2$$
$$R_3, (xyzw)_2 \rightarrow b_3^y, (x0zw)_2$$
$$R_2, (xyzw)_2 \rightarrow b_2^z, (xy0z)_2$$
$$R_1, (xyzw)_2 \rightarrow b_1^w, (xyz0)_2.$$

We proceed to decompose the Invertible decimal-to-binary converter in Fig. 4 into much simpler elements: RT and IRT in Fig. 3. For this purpose, a new reversible element, called D, is given in Fig. 5, which has input lines: T_0 and T_1, output lines: S_0 and S_1, and two states: 0 and 1. When the element is in state k with $k \in \{0, 1\}$, a token arriving on input line T_j with $j \in \{0, 1\}$ changes the state to j and results in a token on output line S_k.

Based on the elements: IRD in Fig. 3b, CD in Fig. 1 and D in Fig. 5, we can construct the Invertible decimal-to-binary converter in Fig. 4, as illustrated in Fig. 6. This scheme is much more efficient than the general design scheme in [7], in the sense that it requires a much smaller number of reversible elements to construct the converter than the latter. For simplicity, a typical process in which the converter in Fig. 6 operates on a token received from one of its input lines is shown in the Appendix. Moreover, the construction in Fig. 6 allows the Invertible decimal-to-binary converter to be realized by the pair of mutually inverse elements: RT and IRT

$$\begin{aligned}T_0, 0 &\rightarrow S_0, 0 \\ T_1, 0 &\rightarrow S_0, 1 \\ T_0, 1 &\rightarrow S_1, 0 \\ T_1, 1 &\rightarrow S_1, 1\end{aligned}$$

Fig. 5 Reversible element D and its respective sets of operations

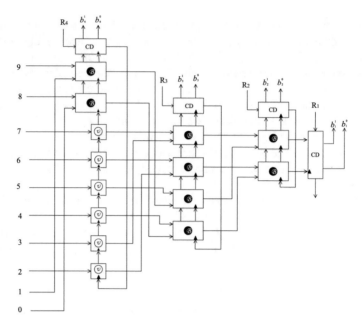

Fig. 6 Construction of the Invertible decimal-to-binary converter using IRD, CD and D elements

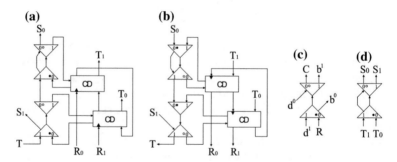

Fig. 7 Constructions of **a** RD and **b** IRD in Fig. 3 using RT, IRT and CD elements. Constructions of **c** CD and **d** D in Fig. 5 using RT and IRT, in each of which both RT and IRT being in state 0 (resp. 1) represent that the construction is in state 0 (resp. 1)

in Fig. 2. This is achieved by replacing each IRT, CD and D elements in Fig. 6 with respective constructions in Fig. 7.

Finally, a logic circuit composed by reversible elements is definitely invertible in the sense that reversing the functionalities of each element in the circuit gives rise to the circuit's inverse counterpart [5, 9]. This remarkable feature of reversible elements allows to obtain the Invertible binary-to-decimal converter, as given in Fig. 8, through switching the directions of each line in the construction in Fig. 6, as well as replacing each IRT, CD and D by their inverse elements. In this case, the IRT's inverse is the RT in Fig. 2b, while the CD (Fig. 1) and D (Fig. 5) are their own inverses.

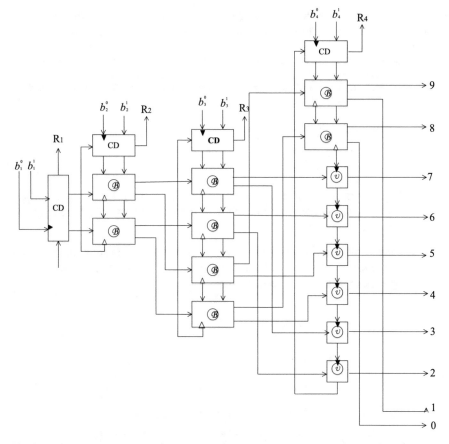

Fig. 8 Construction of Invertible binary-to-decimal converter using RT, CD and D elements. This converter is used to convert every 4-bit binary number $xyzw$ with $x, y, z, w \in \{0, 1\}$ to its equivalent decimal number m where $0 \le m = (xyzw)_2 \le 9$

4 Conclusion

This paper provides an effective scheme for constructing decimal-to-binary converter and binary-to-decimal converter using reversible elements. This is achieved through translating both converters into reversible sequential machines, whereby they can be realized by reversible elements, due to the elements' universality for RSMs [7]. On the other hand, our scheme differs from the general design in [7], which employs a much smaller number of elements to construct the converters than the latter. Furthermore, witness the construction shown in Fig. 6, the scheme may be extended to conduct conversions between multi-digit decimal numbers with equivalent binary numbers.

Acknowledgements We are grateful to Prof. Kenichi Morita at Hiroshima University, Japan, for the valuable guidance and advice. This research work was supported by International Exchange Program of National Institute of Information and Communications (NICT), and Natural Science Foundation of Chongqing (No. cstc2016jcyjA1315).

Appendix

To verify the correctness of the construction in Fig. 6, the following figures illustrate the major steps in converting the decimal number 9 to binary bits 1001 by the Invertible decimal-to-binary converter.

(1) The Invertible decimal-to-binary converter receives a token from input line D_9 that encodes the number 9. Initially, all elements are in state 0, with their states being explicitly denoted in each element.

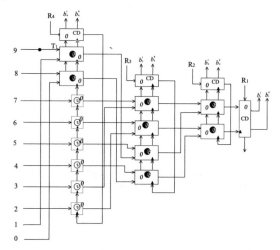

(2) The 4th (most significant) bit of the number 9 (1001) is extracted and encoded by the state of the leftmost CD. The token is transferred on the line representing the binary bits 001.

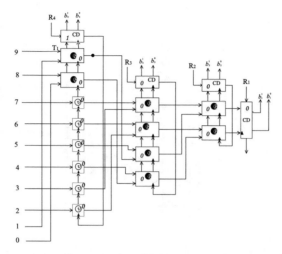

(3) The 3rd bit of the number 9 (1001) is extracted and encoded by the state of the corresponding CD. The token appears on an interconnection line encoding the binary bits 01.

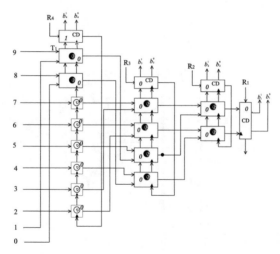

(4) The 2nd bit of the number 9 (1001) is extracted and encoded by the state of the corresponding CD. The token runs on a line representing the binary number 1.

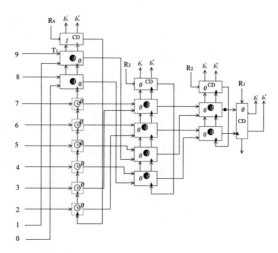

(5) The lowest bit of the number 9 (1001) is extracted and encoded by the state of the rightmost CD. Finally, a token appears on the output line S of the construction, which shows the completion of the conversion process.

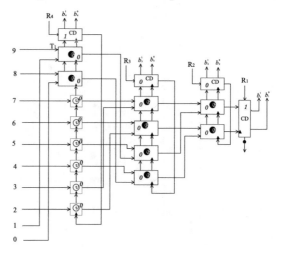

References

1. Bhattacharya, J., Gupta, A., Singh, A.: A high performance binary to BCD converter for decimal multiplication. In: International Symposium on VLSI Design Automation and Test, pp. 315–318 (2010)
2. Fredkin, E., Toffoli, T.: Conserv. log. J. Theor. Phys. **21**(3–4), 219–253 (1982)
3. Hauck, S.: Asynchronous design methodologies: an overview. Proc. IEEE **83**(1), 69–93 (1995)
4. Hopcroft, J.E., Ullman, J.D.: Introduction to Automata Theory, Languages, and Computation. Addison-Wesley, Boston (1979)

5. Lee, J., Adachi, S., Xia, Y.N., Zhu, Q.S.: Emergence of universal global behavior from reversible local transitions in asynchronous systems. Inf. Sci. **282**, 38–56 (2014)
6. Lee, J., Peper, F., Adachi, S., Morita, K.: An asynchronous cellular automaton implementing 2-state 2-input 2-output reversed-twin reversible elements. In: International Conference on Cellular Automata for Research and Industry, pp. 67–76 (2008)
7. Lee, J., Yang, R.L., Morita, K.: Design of 1-tape 2-symbol reversible turing machines based on reversible logic elements. Theor. Comput. Sci. **460**(1), 78–88 (2012)
8. Morita, K.: A simple universal logic element and cellular automata for reversible computing. In: International Conference on Machines, Computations, and Universality (2001)
9. Morita, K.: Reversible computing and cellular automata – a survey. Theor. Comput. Sci. **395**(1), 101–131 (2008)
10. Rhyne, V.T.: Serial binary-to-decimal and decimal-to-binary conversion. IEEE Trans. Comput. **C-19**(8), 808–812 (1970)
11. Tang, M.X., Lee, J., Morita, K.: General design of reversible sequential machines based on reversible logic elements. Theor. Comput. Sci. **568**, 19–27 (2015)
12. Toffoli, T.: Reversible computing. In: de Bakker, J., van Leeuwen J (eds.) Automata, Languages and Programming. LNCS, vol. 85, pp. 632–644 (1980)

Power Consumption in Cellular Automata

Georgios Ch. Sirakoulis and Ioannis Karafyllidis

Abstract Cellular Automata (CAs) have been established as one of the most intriguing and efficient computational tools of our era with unique properties to fit well with the most of the upcoming nanotechnological and parallel computation aspects. Algorithms based on CAs are ideally suited for hardware implementation, due to their discreteness and their simple, regular and modular structure with local interconnections. On the other hand, power dissipation is considered as a rather limiting parameter for the advancement of high performance hardware design. In this chapter the undergoing relationship between CAs and the corresponding power consumption would be exploited as a matter of importance for their hardware design analysis with many promising aspects. First of all in order to establish a clear connection, a power estimation model for combinational logic circuits using CA and focused on glitching estimation will be presented to elucidate the application of CA model to hardware power dissipation measurements. Following that, the power consumption of CA based logic circuits and namely of 1-d CAs rules logic circuits will be analytically investigated. In particular, CMOS power consumption estimation measurements for all the Wolfram 1-d CAs rules as well as entropy variation measurements were conducted for various study cases and different initial conditions and the findings are discussed in detail and in terms of 1-d CAs rules categorization.

1 Introduction

Cellular Automata (CAs) were originally proposed by John von Neumann [1] as formal models of self-reproducing organisms. He was inspired by Stanislaw Ulam's research at Los Alamos laboratory in the 1940s. Following a suggestion byUlam [2],

G. Ch. Sirakoulis (✉) · I. Karafyllidis
Department of Electrical and Computer Engineering, Democritus University
of Thrace, Thrace, Greece
e-mail: gsirak@ee.duth.gr

I. Karafyllidis
e-mail: ykar@ee.duth.gr

© Springer International Publishing AG 2018   183
A. Adamatzky (ed.), *Reversibility and Universality*, Emergence, Complexity
and Computation 30, https://doi.org/10.1007/978-3-319-73216-9_8

von Neumann adopted a fully discrete approach, in which space, time and even the dynamical variables were defined as discrete [1]. The next important milestone in the history of the development of the homogeneous structure of CAs is due to Stephen Wolfram [3]. He suggested simplification of the cell structure with local interconnections and in the early 80s proposed two different approaches to investigate CAs. The following two approaches are quoted from [4]. The mechanisms for information processing in natural system appear to be much closer to those in CAs than in conventional serial-processing computers: CAs may, therefore, provide efficient media for practical simulations of many natural systems [5].

Prior works proved that CAs are very effective in simulating physical systems and solving scientific problems, because they can capture the essential features of systems where global behaviour arises from the collective effect of simple components which interact locally [3, 6–10]. Nontrivial CAs are obtained whenever the dependence on the values at each site is nonlinear. As a result, any physical system satisfying differential equations may be approximated by a CA, by introducing finite differences and discrete variables [11–16]. CAs have been extensively used as a hardware architecture [12, 17–23]. In contrast to the serial computers, the implementation of the model is motivated by parallelism, an inherent feature of CAs that contributes to further acceleration of the model's operation. In CAs, memory (CA cell state) and processing unit (CA local Rule) are inseparably related to a CA cell [17, 24–26].

However, one of the most important parameters of nowadays hardware design is power consumption. Additionally, in terms of low power design, issues such electromigration, reliability, chip overheating and packing selection strongly depend on the power dissipation. Consequently, in recent years, efficient low power design techniques and power estimation methodologies have been developed to solve certain issues at all design levels. In this paper, a first effort of power estimation of VLSI implementation based on CAs circuits is presented. In particular, the proposed model for power estimation of the combinational logic circuits is based on complex-time CAs first introduced by Karafyllidis et al. [27]. In order to bridge the gap between continuous time, in which the combinational logic circuits operate, and the discrete time, in which CA operate, the concept of complex time is introduced. Both the logic combinational circuit and the 2-d CA model that simulates it are assumed to exist in a linear complex space, where the real part of the complex time is the continuous time and the imaginary part is the discrete time [28]. Using the complex time concept, the proposed model can take full advantage of both the powerful local operation capabilities of CAs and the convenient formulation of VLSI circuit description using two-dimensional matrices and netlists.

On the other hand, the power consumption of CA logic circuits and namely of 1-d CAs rules logic circuits was also investigated in detail. More specifically, CMOS power consumption estimation measurements for the entireness of 1-d CAs rules as well as entropy variation measurements were conducted for different study cases [29].The proposed methods used the same logic CA circuit, as ground truth for our experiments. The specific synthesizable Very High Speed Integrated Circuits Hardware Description Language (VHDL) code was produced automatically as output of a CA automation design tool able to implement each one of these rules depending on different inputs [17]. The simulation results on the power consumption of 1-d CA rules circuits were produced with the help of two different models, CMOS power estimation model and entropy variation model, respectively.

2   Cellular Automata Basics

Cellular Automata are dynamical systems in which space and time are discrete, that operate according to local interaction rules [9]. In this section a formal definition of a CA will be presented.

In this paper, we focus on 1-d CA with two possible states per cell, i.e. $S = (0, 1)$. In this case, the local transition rule f is a function $f : (0, 1)^n \rightarrow 0, 1$ and the neighborhood size n is usually taken to be $n = 2r + 1$ such that:

$$s_i (t + 1) = f (s_{i-r}(t), \ldots, s_i(t), \ldots, s_{i+r}(t)) \tag{1}$$

where r (positive integer) is a parameter, known as the radius, representing the standard 1-d cellular neighborhood. We shall furthermore limit ourselves to the $r = 1$ case, i.e. so-called elementary CA, for which the neighborhood size is $n = 3$:

$$f : \{0, 1\}^3 \rightarrow \{0, 1\} \quad s_i (t + 1) = f (s_{i-1} (t), s_i(t), s_{i-r}(t)) \tag{2}$$

The domain of f is the set of all 2^3 3-tuples, which gives rise to $2^8 = 256$ distinct elementary rules. We will use Wolfram's decimal numbering convention for describing these rules [4], e.g. $f(111) = 1, f(110) = 0, f(101) = 1, f(100) = 1, f(011) = 1, f(010) = 0, f(001) = 0, f(000) = 0$, is denoted rule 184. For two-state CA a configuration of a size N grid at time t is a binary sequence $C(t)$.

For a 2-d CA, two neighborhoods are often considered, von Neumann and Moore neighborhood. von Neumann neighborhood is a diamond shaped neighborhood and can be used to define a set of cells surrounding a given cell (x_0, y_0). Equation 3 defines the von Neumann neighborhood of range r.

$$N^v_{(x_0, y_0)} = \{(x, y) : |x - x_0| + |y - y_0| \leq r\} \tag{3}$$

For a given cell (x_0, y_0) and range r, Moore neighborhood can be defined by the following equation:

$$N^M_{(x_0, y_0)} = \{(x, y) : |x - x_0| \leq r, |y - y_0| \leq r\} \tag{4}$$

The local rule, f, in all cases determines the way in which each cell of the 2-d CA is updated. Every cell's state is affected by the cell values in its neighborhood and its value on the previous time step, according to the transition rule or a set of rules.

3   Power Estimation Model

It is well known that the combinational logic circuits that simulate the functionality of 1-d CAs circuits operate in continuous time, whereas CAs operate in discrete time (i.e. the states of all cells are updated simultaneously at discrete time steps). To bridge the gap between the continuous time (logic circuit time) and the discrete time (CA model time), the concept of complex time is introduced. The circuit-model system is assumed to exist in a linear complex space, where the real part of the complex time is the continuous time and the imaginary part is the discrete time.

A combinational logic circuit with n elements (nodes) and m nets is represented by a graph $G(Q, E)$, where $Q = \{q_j | j = 1, 2, 3, ..., n\}$ is the element (node) set and $E = \{e_j | j = 1, 2, 3, ..., m\}$ is the net (edge) set. The connectivity of the circuit is described by the connectivity matrix M, the elements of which are given by the following equation:

$$m_{jk} = \left\{ \begin{array}{l} 1, \text{ if element } j \text{ is connected to element } k \\ 0, \text{ if element } j \text{ is not connected to element } k \end{array} \right\} \tag{5}$$

The netlist of the circuit is given in the same way as in any VLSI circuit.

To estimate the power dissipation the logic circuit is also represented by a CA by considering each circuit logic gate as a CA cell. The resulting CA now is 2-d and has not the regular matrix-like topography. The circuit connectivity matrix M gives the connections between CA cells. Both the logic combinational circuit and the resulted 2-d CA, which simulates it, are assumed to exist in a linear complex space which is spanned by V and t. V is the complex voltage given by:

$$V = V_r + iV_i \tag{6}$$

where V_r and V_i are the real and imaginary parts of the voltage, respectively, and $i^2 = -1$. Since there is no use for the imaginary part of the voltage, it is taken to be equal to zero. The following relations hold:

$$\begin{array}{c} V_i = 0 \\ V_{ss} \leq V_r \leq V_{dd} \end{array} \tag{7}$$

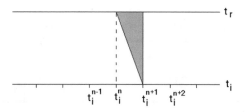

Fig. 1 All the events that, in real time, take place between two successive time steps, t_i^n and t_i^{n+1}, are mapped to t_i^{n+1} in the discrete imaginary time

where V_{ss} and V_{dd} are the ground and power voltages, respectively. Complex time t is given by:

$$t = t_r + it_i \tag{8}$$

t_r and t_i are the real and imaginary parts of the time, respectively. The real part of the time t_r represents the time in which the 1-d CA logic circuit operates, whereas the imaginary part t_i is the discrete time in which the 2-d CA model operates.

An event in this complex linear space is defined as a voltage change in any of the edges of the graph (circuit wires). All the events that take place in real (circuit) time between two successive time steps t_i^n and t_i^{n+1} are assumed to take place at t_i^{n+1} in the discrete imaginary (model) time. Figure 1 shows the relation between t_r and t_i. Each CA cell checks its inputs and executes its local rule at discrete time steps $(t_i^n, t_i^{n+1}, \ldots)$. The local rule for a CA cell with m inputs and one output is given by:

$$O\left(t_i^{n+1}\right) = F_B\left\{I_1\left(t_r\right), I_2\left(t_r\right), \ldots, I_m\left(t_r\right)\right\} \tag{9}$$

where $I_1, I_2, \ldots, I_m$ are the m inputs of the logic circuit and O is the output. F_B is a Boolean function that describes precisely the function of the 1-d CA logic circuit. Different F_B may apply to different CA cells, i.e. different local rules may apply to different CA cells. It is reminded that a 2-d CA cell represents an element of a combinational circuit, i.e. a logic gate. As a result, continuous and discrete times are introduced in Boolean functions, in the solid mathematical framework of linear complex spaces.

Finally, the proposed method is applied to two simple circuits shown in Figs. 2 and 3. The gate delays have been obtained from the technology files of a 0.7 μm process. Figure 2a shows a logic circuit. The input–output waveforms obtained using the proposed CA models in the cases of synchronized and unsynchronized inputs are shown in Fig. 2b, c, respectively. Figure 3a show logic combinational circuits and Fig. 3b show the input–output waveforms obtained using the proposed method. In all cases the glitching in the waveforms has been successfully obtained and the estimation of power dissipation is straightforward.

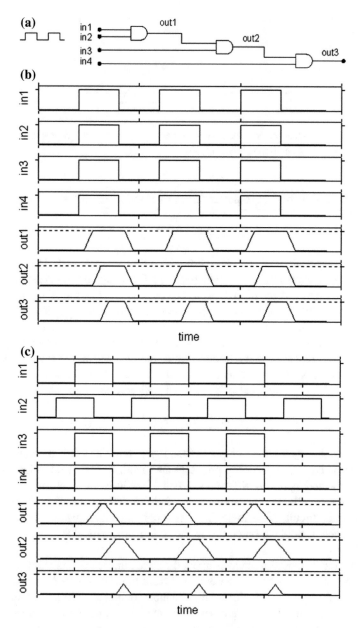

Fig. 2 **a** A logic circuit, and the input–output waveforms obtained using the proposed CA model in the cases of **b** synchronized and **c** unsynchronized inputs

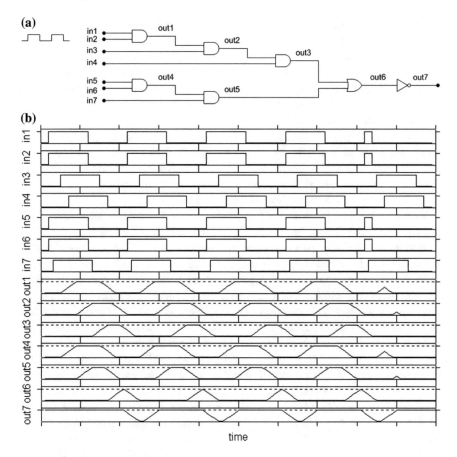

Fig. 3 a Another logic circuit, and **b** the corresponding input–output waveforms obtained using the proposed CA model

4 Power Consumption of 1-d CA Rules

4.1 CMOS Power Consumption Estimation

In order to understand how the proposed architectural strategies can provide high performance for 1-d CA applications used in contemporary combinatorial logic circuits at low power levels, it is necessary to look at the CMOS circuit dynamic power consumption equation:

$$P = ACV^2F \tag{10}$$

where P is the power consumed, A is the switching activity factor, i.e. the fraction of the circuit that is switching, C is the switched capacitance, V is the supply voltage, and F is the clock frequency [30]. If a capacitance of C is charged and discharged by a clock signal of frequency F and peak voltage V, then the charge moved per cycle is CV and the charge moved per second is CVF. Since the charge packet is delivered at voltage V, the energy dissipated per cycle, or the power, is CV^2F. The data power for a clocked flip-flop, which can toggle at most once per cycle, will be V^2F. When capacitances are clock gated or when flip-flops do not toggle every cycle, their power consumption will be lower. Hence, a constant called the activity factor ($0 \leq A \leq 1$) is used to model the average switching activity in the circuit. Equation 10 is derived by incorporating this term into the power consumption. However, it should be also noticed that custom Application Specific Integrated Circuits (ASICs) can drastically reduce the power consumption by using specialized circuit structures and concurrency to lower C and F respectively.

With reference to previous analysis, several simulation tests were carried out to prove the flexibility of the proposed method and provide fruitful information regarding the 1-d CA logic circuit implementations in terms of power estimation. In the following simulations, each one of the 256 1-d CA rules was modeled as described before leading to a 2-d CA power model. Each rule was applied to 100 1-d CA cells with periodic (cyclic) and zero (null) boundary conditions, respectively. Four (4) Study Cases for both boundary conditions with different initial conditions were taken under consideration: (i) the middle CA cell initial stage is set on 1 while all the rest CA cells initial stages remain at 0, (ii) the previous one reversed, meaning that the middle CA cell initial stage is set on 0 while all the rest CA cells initial stages remain at 1, (iii) the middle CA cell initial stage, its left neighbor as well as its right neighbor are set on 1 while all the rest CA cells initial stages remain at 0 and (iv) a random initialization sequence was applied to all CA cells. Regarding the last random sequence, a 1-d CA for pseudorandom number generation (PRNG) based on the real time clock sequence (analytical time description) was used to generate high-quality random numbers [31]. The aforementioned CA- PRNG can pass all of the statistical tests of DIEHARD[1] as well as NIST[2], which seem to be the most powerfully complete general test suites for randomness. In order to accomplish the aforementioned requirements so as to generate high-quality random numbers, a random sequence which originates from the real time computer clock sequence was used. An analytical description of the dependency of CA parameters on the actual clock sequence is shown in Table 1. More specifically, the time sequence is depicted as $yyyy/mm/dd/hh/min/sec$. Two CA rules are executed during the CA evolution. The first or main CA rule originates from the minutes by seconds product. The second one is derived by the division of minutes and seconds, and the multiplication of the final division result by a chosen constant, i.e. $2^{10} = 1024$. The CA execution times of these two CA rules depend on the seconds. If sec is the number of seconds, then the CA execution time of the main CA rule t will be: $t = sec(60 - sec)$. The second CA rule will be effective between sec and $(60 - sec)$. The initial CA configuration

---

[1] https://web.archive.org/web/20160125103112/http://stat.fsu.edu/pub/diehard/.

[2] http://csrc.nist.gov/groups/ST/toolkit/rng/index.html.

Table 1 Dependency of CA parameters on the actual clock sequence

Variables Used	Effect on the CA		
	Parameters	Formulae	
sec	2nd CA rule effective length	$2 \times (sec/8)$	
	1st and 2nd CA rules execution times	$sec \times (60 - sec)$	$(60 - sec)$, if $sec < 60 - sec$, sec, if $sec \geq 60 - sec$
min	1st and 2nd CA rules	$min \times sec$	$min/sec \times 1024$
$yyyy, mm, dd, hh$	Initial state of the CA	$yyyy \times mm \times dd \times hh \times min \times ss$	

and the CA length are given by the product of all the time sequence numbers (year, month, day, hour, minutes and seconds). While the main CA rule is applied to all the cells of the initial CA configuration and for the aforementioned CA execution time, the second CA rule is applied only to the cells which lie within $(sec/2^3)$ cells from the initial configuration centre. Finally, periodic boundary conditions were applied, since their use results in better performance in random number generation compared to the use of null boundary conditions. The result of this operation produced a binary number which indicated the initial CA configuration and simultaneously the length of the CA. More details regarding the usage of the aforementioned CA PRNG can be found on [31]. In order to extend the presented results based on the above proposed basic studies each rule was also applied to 300 1-d CA cells with the more common, periodic (cyclic) as well as zero (null) boundary conditions, respectively.

In the following graphs (Figs. 4, 5, 6 and 7), the horizontal axis is the total range of 1-d CA rules, while the vertical axis provides the final power estimation results of the applied method, for each CA rule, taking also into consideration the switching activity of the examined logic circuit for 100 and 300 simulation steps, respectively.

The already existing symmetries of the 256 CAs rules, i.e. left/right (0/1) also appear on the presented results giving the opportunity of diminishing the power consumption analysis of the examined CAs rules to only 88 of them. Furthermore based on the above simulation results, different grid sizes seem not to affect the power estimation in different study cases, as expected, except the case of application of random initialization sequence to CA cells, where some slight differences can be noticed. It is clear that these simulations based on 100 cells, in case of random initialization sequences, limit considerably some ECA dynamics. Of course, it is known that a number of ECA rules display their dynamics in small lattices comprising only four cells, but others not. Mainly complex ECA seem to be able to represent significant dynamics in lattices of 300 or more cells. Moreover, the effect of the zero and periodic boundary conditions result to the expression of different dynamics of the examined CA rules. In the periodic boundary conditions the CA rules in general obviously and as expected present greater power dissipation in the lower levels (check the base of the simulation figures) compared to zero boundary conditions, while

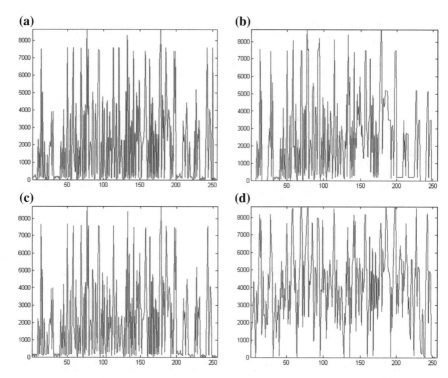

Fig. 4 Simulation results for the applied method in 100 1-d CA cells with zero boundary conditions for each one of the 256 1-d CA rules for **a** Case i, **b** Case ii, **c** Case iii and **d** Case iv. The horizontal axis depicts the total range of 1-d CA rules, while the vertical axis provides the final power estimation results of the applied method, for each CA rule, taking also into consideration the switching activity of the examined logic circuit for 100 simulation steps

on the other hand the upper limit of power dissipation is smaller in all cases of periodic boundary conditions simulations when compared with the zero boundary conditions. It is rather important to notice once again that the presented simulation results of the proposed power analysis reveal in most cases, as described above, the same characteristics and meet the basic concepts and principles of CA rules' analyses as already expressed in literature by various studies. More specifically and regarding the study of CAs rules, a significant amount of work has been presented in the literature by Chris Langton [32], Robert Gilman [33], Howard Gutowitz [34], Andrew Adamatzky [35], David Eppstein [36], Louis D'Alloto [37] and Hector Zenil [38], just to name a few researchers who provided insightful and fruitful studies of CA rules' evolution.

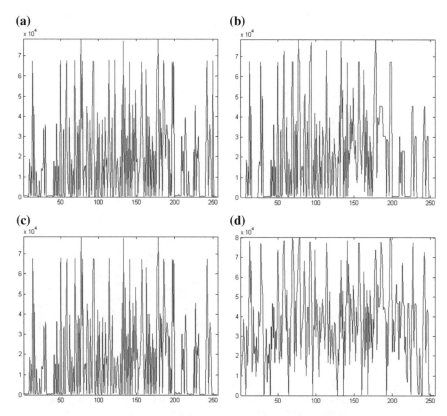

Fig. 5 Simulation results for the applied method in 300 1-d CA cells with zero boundary conditions for each one of the 256 1-d CA rules for **a** Case i, **b** Case ii, **c** Case iii and **d** Case iv. The horizontal axis depicts the total range of 1-d CA rules, while the vertical axis provides the final power estimation results of the applied method, for each CA rule, taking also into consideration the switching activity of the examined logic circuit for 300 simulation steps

4.2 Entropy Variation

It is well known that in computations, energy is dissipated when information is erased, i.e. when the state of the computing element goes from logic 1 to logic 0. In the case of CMOS computing elements (gates), transition from 1 to 0 is obtained by driving the charges stored to the output capacitance to the ground through the n-MOSFET [30] part of the circuit. The energy dissipation during CA evolution can be represented by the entropy variation as a function of the evolution (discrete) time. We calculate this entropy variation according to the following model:

At each evolution time step t, the probability of transition of the state of the ith CA cell from 0 to 1, P_i^t, is given by:

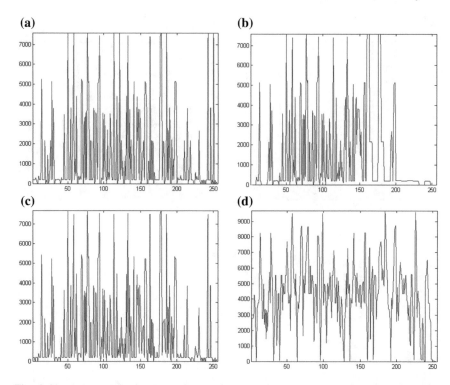

Fig. 6 Simulation results for the applied method in 100 1-d CA cells with periodic boundary conditions for each one of the 256 1-d CA rules for **a** Case i, **b** Case ii, **c** Case iii and **d** Case iv. The horizontal axis depicts the total range of 1-d CA rules, while the vertical axis provides the final power estimation results of the applied method, for each CA rule, taking also into consideration the switching activity of the examined logic circuit for 100 simulation steps

$$P_i^t = \frac{n_{tr}}{n_{stp}} \tag{11}$$

where, n_{tr} is the number of state transitions from 1 to 0 and n_{stp} is the number of time steps.

In Eq. 11 we define as probability of transition, the switch of a CA cell state from 0 to 1. It is well known that energy is dissipated when information is destroyed, i.e. when a state from 1 becomes 0. Consequently, even if the number of transitions in two rules is the same, the entropy variation may differ because only transitions from 0 to 1 count in the calculation of the probability of transition. The entropy of the evolution of a CA with n cells is derived from the number of 1 to 0 transitions in all CA cells according to:

$$H^t = -\sum_{t=1}^{n} P_i^t log_2 \left(P_i^t\right) \tag{12}$$

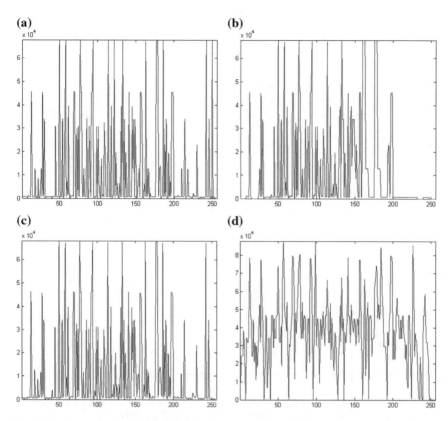

Fig. 7 Simulation results for the applied method in 300 1-d CA cells with periodic boundary conditions for each one of the 256 1-d CA rules for **a** Case i, **b** Case ii, **c** Case iii and **d** Case iv. The horizontal axis depicts the total range of 1-d CA rules, while the vertical axis provides the final power estimation results of the applied method, for each CA rule, taking also into consideration the switching activity of the examined logic circuit for 300 simulation steps

We used this model to compute the entropy variation of the previously described 1-d CA with 100 cells evolving under the Wolfram rules 30, 110, 218 and 250. The results are shown in Figs. 8a–d, respectively. From the presented simulation results, there seems to be some relations to well-known descriptors, such as lambda parameter [32], i.e. a dimensionless measure of complexity and computation potential in CAs, for each evolution of CAs and this could be an interesting starting point for the continuation of our research.

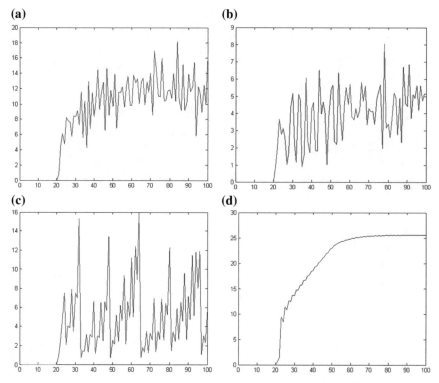

Fig. 8 Entropy variation in a 1-d CA with 100 cells and zero boundary conditions for **a** rule 30, **b** rule 110, **c** rule 218 and **d** rule 250, respectively for Case i after 20 time steps

5   Conclusions

In this paper a different method of power estimation of VLSI combinational logic circuits based on a CA model is presented. In this method the concept of complex time was introduced. Both the logic combinational circuit and the CA model that simulates it are assumed to exist in a linear complex space, where the real part of the complex time is the continuous time and the imaginary part is the discrete time. The presented simulation results prove the robustness of the aforementioned model. Furthermore, the power consumption of CAs based logic circuits and namely of 1-d CAs rules logic circuits was investigated in detail. More specifically, detailed measurements for the entireness of 1-d CAs rules were conducted for different study cases with the help of a CMOS power estimation model and an entropy variation model, respectively. As future work, a more detailed research on the relation between the power consumption of hardware implementation of more complicated 1-d as well as 2-d CAs circuits and the corresponding entropy will be also considered.

References

1. von Neumann, J.: Theory of self-reproducing automata. University of Illinois Press, Champaign (1966)
2. Ulam, S.: Random processes and transformations. Int. Congr. Math. **2**, 264–275 (1952)
3. Wolfram, S.: Theory and applications of cellular automata. World Scientific, Singapore (1986)
4. Wolfram, S.: Cellular automata as models of complexity. Nature **311**, 419–424 (1984)
5. Di Lena, P., Margara, L.: Computational complexity of dynamical systems: the case of cellular automata. Inf. Comput. Inf. Control **206**, 1104–1116 (2008)
6. Hurley, M.: Attractors in cellular automata. Ergod. Theory Dyn. Syst. **10**, 131–140 (1990)
7. Culik II, K., Hurd, L.P., Yu, S.: Computation theoretic aspects of cellular automata. Phys. D. Nonlinear Phenom. **45**, 357–378 (1990)
8. Feynman, R.P.: Simulating physics with computers. Int. J. Theor. Phys. **10**, 467–488 (1982)
9. Sirakoulis, G.Ch., Bandini, S.: Cellular automata. In: 10th International Conference on Cellular Automata for Research and Industry, ACRI 2012, Santorini Island, Greece, September 24–27, 2012. Proceedings. Lecture Notes in Computer Science. Springer, Berlin (2012)
10. Was, J., Sirakoulis, G.Ch., Bandini, S.: Cellular automata. In: 11th International Conference on Cellular Automata for Research and Industry, ACRI 2014, Krakow, Poland, September 22–25, 2014, Proceedings. Lecture Notes in Computer Science. Springer International Publishing, Berlin (2014)
11. Toffoli, T.: Cellular automata as an alternative to (rather than an approximation of) differential equations in modeling physics. Physica D **10**, 117–127 (1984)
12. Chaudhuri, P.P., Chaudhuri, D.R., Nandi, S., Chattopadhyay, S.: Theory and Applications: Additive Cellular Automata. IEEE Press, New York (1997)
13. Sirakoulis, G.Ch., Karafyllidis, I., Soudris, D., Georgoulas, N., Thanailakis, A.: A new simulator for the oxidation process in integrated circuit fabrication based on cellular automata. Model. Simul. Mater. Sci. Eng. **7**(4), 631 (1999)
14. Sirakoulis, G.Ch., Karafyllidis, I., Thanailakis, A.: A cellular automaton methodology for the simulation of integrated circuit fabrication processes. Futur. Gener. Comput. Syst. **18**(5), 639–657 (2002)
15. Tsompanas, M.-A.I., Sirakoulis, G.Ch., Adamatzky, A.I.: Evolving transport networks with cellular automata models inspired by slime mould. IEEE Trans. Cyber. **45**(9), 1887–1899 (2015)
16. Dourvas, N.I., Sirakoulis, G.Ch.: Cellular automaton Belousov–Zhabotinsky model for binary full adder. Int. J. Bifurc. Chaos **27**(06), 1750089 (2017)
17. Sirakoulis, G.Ch., Karafyllidis, I., Thanailakis, A.: A CAD system for the construction and VLSI implementation of cellular automata algorithms using VHDL. Microprocess. Microsyst. **27**(8), 381–396 (2003)
18. Jendrsczok, J., Ediger, P., Hoffmann, R.: A scalable configurable architecture for the massively parallel GCA model. Int. J. Parallel Emerg. Distrib. Syst. **24**, 275–291 (2009)
19. Georgoudas, I.G., Kyriakos, P., Sirakoulis, G.Ch., Andreadis, I.T.: An FPGA implemented cellular automaton crowd evacuation model inspired by the electrostatic-induced potential fields. Microprocess. Microsyst. **34**(7), 285–300 (2010)
20. Tsompanas, M.-A.I., Sirakoulis, G.Ch.: Modeling and hardware implementation of an amoeba-like cellular automaton. Bioinspiration Biomim. **7**(3), 036013 (2012)
21. Vourkas, I., Sirakoulis, G.Ch.: FPGA based cellular automata for environmental modeling. In: 2012 19th IEEE International Conference on Electronics, Circuits, and Systems (ICECS 2012), pp. 93–96 (2012)
22. Kalogeropoulos, G., Sirakoulis, G.Ch., Karafyllidis, I.: Cellular automata on FPGA for real-time urban traffic signals control. J. Supercomput. **65**(2), 664–681 (2013)
23. Progias, P., Sirakoulis, G.Ch.: An FPGA processor for modelling wildfire spreading. Math. Comput. Model. **57**(5), 1436–1452 (2013)

24. Mardiris, V., Sirakoulis, G.Ch., Mizas, C., Karafyllidis, I., Thanailakis, A.: A CAD system for modeling and simulation of computer networks using cellular automata. IEEE Trans. Syst. Man Cyber. Part C Appl. Rev. **38**(2), 253–264 (2008)
25. Sirakoulis, G.Ch.: A TCAD system for VLSI implementation of the cvd process using VHDL. Int. VLSI J. **37**(1), 63–81 (2004)
26. Georgoudas, I.G., Sirakoulis, G.Ch., Scordilis, E.M., Andreadis, I.T.: On-chip earthquake simulation model using potentials. Nat. Hazard. **50**(3), 519–537 (2009)
27. Karafyllidis, I., Mavridis, S., Soudris, D., Thanailakis, A.: Estimation of power dissipation in glitching using complex-time cellular automata. In: 6th IEEE International Conference on Electronics, Circuits and Systems, vol. 3, pp. 1639–1642 (1999)
28. Sirakoulis, G.Ch., Karafyllidis, I.: Power estimation of 1-d cellular automata circuits. In: 2010 International Conference on High Performance Computing Simulation, pp. 691–697 (2010)
29. Sirakoulis, G.Ch., Karafyllidis, I.: Cellular automata and power consumption. J. Cell. Autom. **7**(1), 67–80 (2012)
30. Weste, N., Harris, D.: CMOS VLSI Design: A Circuits And Systems Perspective, 4th edn. Addison-Wesley Publishing Company, USA (2010)
31. Kotoulas, L.G., Tsarouchis, D., Sirakoulis, G.Ch., Andreadis, I.: 1-d cellular automaton for pseudorandom number generation and its reconfigurable hardware implementation. In: IEEE International Symposium on Circuits and Systems, pp. 4627–4630 (2006)
32. Langton, C.G.: Computation at the edge of chaos: phase transitions and emergent computation. Phys. D **42**, 12–37 (1990)
33. Gilman, R.H.: Classes of linear automata. Ergod. Theory Dyn. Syst. **7**(1), 105118 (1987)
34. Gutowitz, H., Langton, C.: Mean field theory of the edge of chaos. In: Proceedings of ECAL3, pp. 52–64. Springer, Berlin (1995)
35. Adamatzky, A.: Identification of cellular automata. In: Encyclopedia of Complexity and Systems Science, pp. 4739–4751 (2009)
36. Eppstein, D.: Growth and decay in life-like cellular automata. In: Adamatzky, A. (ed.) Game of Life Cellular Automata, pp. 71–97. Springer, London (2010)
37. DAlotto, L.: A classification of one-dimensional cellular automata using infinite computations. Appl. Math. Comput. **255**, 15–24 (2015). (Special issue devoted to the international conference Numerical computations: Theory and Algorithms June 1723, 2013. Falerna, Italy)
38. Zenil, H.: Compression-based investigation of the dynamical properties of cellular automata and other systems. CoRR (2009). arXiv:0910.4042

Logical Gates via Gliders Collisions

Genaro J. Martínez, Andrew Adamatzky and Kenichi Morita

Abstract An elementary cellular automaton with memory is a chain of finite state machines (cells) updating their state simultaneously and by the same rule. Each cell updates its current state depending on current states of its immediate neighbours and a certain number of its own past states. Some cell-state transition rules support gliders, compact patterns of non-quiescent states translating along the chain. We present designs of logical gates, including reversible Fredkin gate and controlled NOT gate, implemented via collisions between gliders.

1 Preliminaries

When designing a universal cellular automata (CA) we aim, similarly to designing small universal Turing machines, to minimise the number of cell states, size of neighbourhood and sizes of global configurations involved in a computation. History of 1D universal CA is long yet exciting.[1] In 1971 Smith III proved that a CA for which a number of cell-states multiplied by a neighbourhood size equals 36 simulates a Turing machine [54]. Sixteen years later Albert and Culik II designed a universal 1D CA with just 14 states and totalistic cell-state transition function [6]. In 1990 Lindgren and Nordahl reduced the number of cell-states to 7 [21]. These proofs were obtained using signals interaction in CA. Another 1D universal CA employing

---

[1]Complex Cellular Automata Repository http://uncomp.uwe.ac.uk/genaro/Complex_CA_repository.html.

G. J. Martínez (✉)
Escuela Superior de Cómputo, Instituto Politécnico Nacional, México and Unconventional
Computing Centre, University of the West of England, Bristol, UK
e-mail: genaro.martinez@uwe.ac.uk

A. Adamatzky
Unconventional Computing Centre, University of the West of England, Bristol, UK
e-mail: andrew.adamatzky@uwe.ac.uk

K. Morita
Hiroshima University, Higashi Hiroshima, Japan
e-mail: km@hiroshima-u.ac.jp

© Springer International Publishing AG 2018
A. Adamatzky (ed.), *Reversibility and Universality*, Emergence, Complexity
and Computation 30, https://doi.org/10.1007/978-3-319-73216-9_9

signals was designed by Kenichi Morita in 2007 [45]: reversible CAs which evolves in partitioned spaces.

In 1998, Cook demonstrated that elementary CA, i.e. with two cell-states and three-cell neighbourhood, governed by rule 110 is universal [13, 66]. He did this by simulating a cyclic tag system in a CA.[2] Operations were implemented with 11 gliders and a glider gun.

We will show that an elementary CA with memory, where every cells updates its state depending not only on its two immediate neighbours but also on its own past states, exhibits gliders which collision dynamics allows for implementation of logical gates just with one glider. We demonstrate implementation of NOT and AND gates, DELAY, NAND gate, MAJORITY gate, and CNOT and Fredkin gates.

2 Elementary Cellular Automata

One-dimensional elementary CA (ECA) [65] can be represented as a 4-tuple $\langle \Sigma, \varphi, \mu, c_0 \rangle$, where $\Sigma = \{0, 1\}$ is a binary alphabet (cell states), φ is a local transitions function, μ is a cell neighbourhood, c_0 is a start configuration. The system evolves on an array of *cells* x_i, where $i \in Z$ (integer set) and each cell takes a state from the $\Sigma = \{0, 1\}$. Each cell x_i has three neighbours including itself: $\mu(x_i) = (x_{i-1}, x_i, x_{i+1})$. The array of cells $\{x_i\}$ represents a *global configuration c*, such that $c \in \Sigma^*$. The set of finite configurations of length n is represented as Σ^n. Cell states in a configuration $c(t)$ are updated to next configuration $c(t + 1)$ simultaneously by a the local transition function φ: $x_i^{t+1} = \varphi(\mu(x_i))$. Evolution of ECA is represented by a sequence of finite configurations $\{c_i\}$ given by the global mapping, $\Phi : \Sigma^n \to \Sigma^n$.

3 Elementary Cellular Automaton Rule 22

Rule 22 is an ECA with the following local function:

$$\varphi_{R22} = \begin{cases} 1 \text{ if } 100, 010, 001 \\ 0 \text{ if } 111, 110, 101, 011, 000 \end{cases} . \tag{1}$$

The local function φ_{R22} has a probability of 37.5% to get states 1 in the next generation and much higher probability to get state 0 in the next generation. Examples of evolution of ECA Rule 22 from a single cell in state '1' and from a random configurations are shown in Fig. 1.

---

[2]A reproduction of this machine working in rule 110 can be found in http://uncomp.uwe.ac.uk/genaro/rule110/ctsRule110.html.

(a)

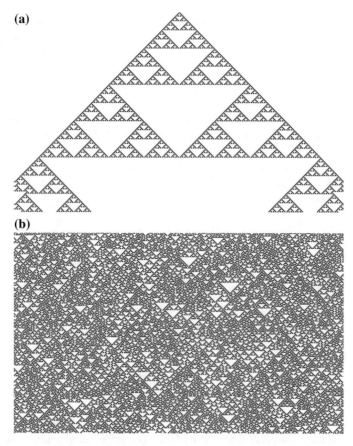

(b)

Fig. 1 Typical evolution of ECA rule 22 **a** from a single cell in state '1' and **b** from a random initial configuration where half of the cells, chosen at random, are assigned state '1'. The ECA consists of 598 cells and evolves for 352 generations. White colour represents state '0' and dark colour represents state '1'

4 Elementary Cellular Automata with Memory

Conventional CA are ahistoric (memoryless): the new state of a cell depends on the neighbourhood configuration solely at the preceding time step of φ. CA with *memory* (CAM) can be considered as an extension of the standard framework of CA where every cell x_i is allowed to remember some period of its previous evolution. CAM was introduced by Ramon Alonso-Sanz, see overview in [53]. To implement a memory function we need to specify the kind of memory ϕ, as follows:

$$\phi(x_i^{t-\tau}, \ldots, x_i^{t-1}, x_i^t) \to s_i. \tag{2}$$

The parameter $\tau < t$ determines the depth, or a degree, of the memory and each cell $s_i \in \Sigma$ being a state function of the series of states of the cell x_i with memory up to time-step. To execute the evolution we apply the original rule as:

$$\varphi(\ldots, s_{i-1}^t, s_i^t, s_{i+1}^t, \ldots) \rightarrow x_i^{t+1}. \tag{3}$$

The main feature in CAM is that the mapping φ remains unaltered, while historic memory of all past iterations is retained by featuring each cell in the context of history of its past states in ϕ. This way, cells *canalise* memory to the map φ. For example, we can consider memory function ϕ as a *majority memory* $\phi_{maj} \rightarrow s_i$, where in case of a tie given by $\Sigma_1 = \Sigma_0$ in ϕ we will take the last value x_i. In this case, function ϕ_{maj} represents the classic majority function for three values [39] as follows:

$$(a \wedge b) \vee (b \wedge c) \vee (c \wedge a) \tag{4}$$

that represents the cells $(x_i^{t-\tau}, \ldots, x_i^{t-1}, x_i^t)$ and define a temporal ring of s cells, before reaching the next global configuration c.

5 Elementary Cellular Automaton with Memory Rule $\phi_{R22maj:4}$

ECA with memory (ECAM) rule $\phi_{R22maj:4}$ employ the majority memory (*maj*) and degree of memory $\tau = 4$.

Figure 2 shows a typical evolution of ECAM rule $\phi_{R22maj:4}$ from a random initial condition. There we observe emergence of gliders travelling and colliding with each other.

The set of gliders $\mathcal{G}_{\phi_{R22maj:4}} = \{g_L, g_R\}$ are defined in a square of 11×11, their properties are shown in Table 1.

We undertook a systematic analysis of binary collisions between gliders and found 44 different types of the collision outcomes, we selected some types of the collisions to realise logical gates presented in this paper (Fig. 3).

6 Logic Gates via Gliders Collisions

We assume presence of a glider at a given site of space and time is a logical '1' (TRUE) and absence is '0' (FALSE). Logic gates $f(a, b) \rightarrow \neg a \mid \neg b \mid a \wedge b$ are realised via binary collisions of gliders in rule $\phi_{R22maj:4}$. Figure 4b, c shows how AND- NOT gate is realised. Figure 4d demonstrates implementations of AND gate. Also a DELAY operator can be implemented by delaying glider a and conserving momentum of glider b as shown in Fig. 4e

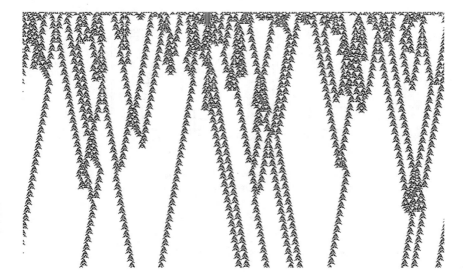

Fig. 2 Typical evolution of $\phi_{R22maj:4}$ from a random initial configuration at 50% on a ring of 598 cells for 352 generations

Table 1 Properties of gliders in ECAM rule $\phi_{R22maj:4}$

Localisation	Period	Shift	Velocity	Mass
g_R	11	2	2/11	38
g_L	11	−2	−2/11	38

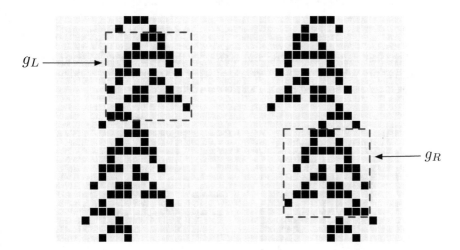

Fig. 3 g_L and g_R gliders in ECAM rule $\phi_{R22maj:4}$

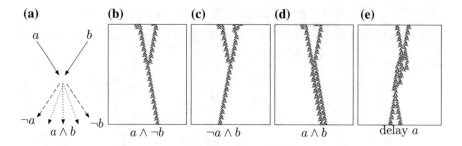

Fig. 4 Examples of logical gates implemented via glider collisions. Presence of a glider is logical TRUTH '1', absence logical FALSE '0'. **a** Scheme of the gate. **b, c** AND- NOT gate, **d** AND, **e** DELAY

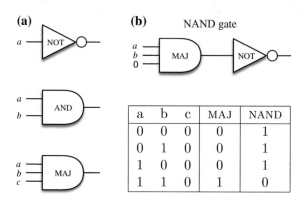

Fig. 5 Gate NAND made of MAJORITY and NOT gates. **a** Schematics of gates. **b** Design of NAND gate

A NAND gate can be cascaded from MAJORITY and NOT gates. To design NAND gate, we fix third value in MAJORITY gate to produce AND gate as illustrated in Fig. 5b. Later a NOT gate is cascaded to get a NAND gate (Fig. 5a).

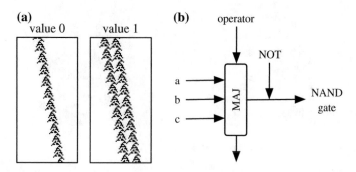

Fig. 6 Gate MAJORITY. **a** Binary values represented by gliders, **b** scheme of an NAND gate using MAJORITY and NOT gates

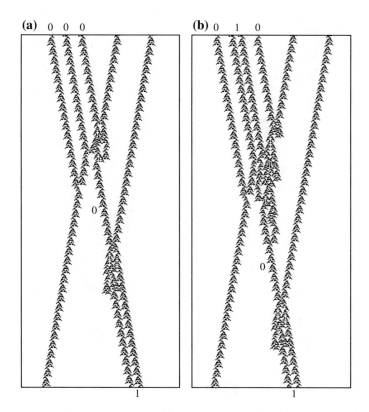

Fig. 7 Gate NAND implemented with MAJORITY and NOT gates in ECAM rule $\phi_{R22maj:4}$ following the scheme proposed in Fig. 6b. **a** Input values $f(0, 0, 0)$ and **b** input values $f(0, 1, 0)$. Intervals between gliders in the initial condition are the same in all operations

Figure 6a represents a value '0' as one glider and a value '1' as a couple of gliders. Figure 6b presents a scheme of NAND gate in $\phi_{R22maj:4}$. We have three-input values/gliders that will be evaluated by one control glider travelling perpendicularly to input gliders. The control glider transforms $f(a, b, c)$ into a majority function. Further, other glider acts as an active signal in NOT gate.

Figures 7 and 8 show the implementation of NAND gate in $\phi_{R22maj:4}$, encoded in its initial condition set of gliders in six positions. As was shown in the scheme (Fig. 6c), gliders coming from the left side represent binary values and a glider coming from the right acts as a control glider for the values of the MAJORITY gate. The control glider continues its travel undisturbed, after interaction with input gliders, and therefore it can be recycled in further MAJORITY gate.

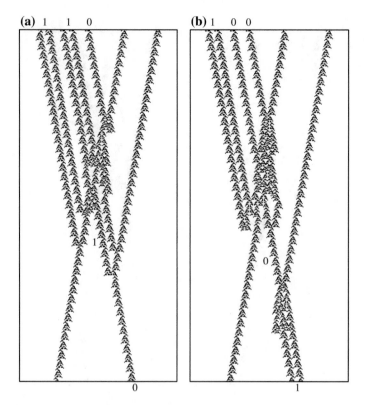

Fig. 8 Gate NAND implemented with MAJORITY and NOT gates in ECAM rule $\phi_{R22maj:4}$ following the scheme proposed in Fig. 6b. **a** Input values $f(1, 1, 0)$ and **b** input values $f(1, 0, 0)$

6.1 Ballistic Collisions

A concept of ballistic logical gates represent Boolean values by *balls* which preserve their identity during collision but change their velocity vectors; the balls are routed in the computing space using mirrors [60]. The *billiard ball model* was advanced by Margolus in his designs of partitioned CA to demonstrate a logical universality of a billiard-ball model and to implement Fredkin gate with soft spheres in billiard-ball model [22, 23].

Ballistic collisions of gliders, or particles, permit to represent functions of two arguments (general signal-interaction scheme is conceptualized in Fig. 4a), as follows [60]:

1. $f(u, v)$ is a product of one collision (Fig. 4b);
2. $f_i(u, v) \mapsto (u, v)$ identity (Fig. 9c);
3. $f_r(u, v) \mapsto (v, u)$ reflection (Fig. 9d);

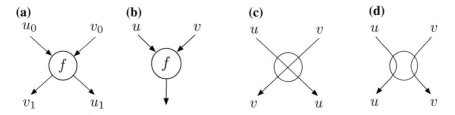

Fig. 9 Toffoli's notation for basic ballistic collision with particles [60]

where Fig. 9a represents a collision not preserving identity of gliders u and v; Fig. 9b shows a collision where identities of input gliders are preserved.

Below we show that ECAM rule $\phi_{R22maj:4}$ reproduces ballistic collisions. Figure 10 demonstrates identity collisions $f_i(u, v) \mapsto (u, v)$. The collisions between gliders are illustrated in (Fig. 10). The collisions are of soliton nature [20, 56]. A carry-ripple adder [57] can be implemented in ECAM Rule $\phi_{R22maj:4}$ by solitonic reactions with pairs of gliders and identity function (Fig. 10a).

Figure 11 illustrates elastic collisions $f_r(u, v) \mapsto (v, u)$ with a pair of gliders. The pairs of gliders reflect in the same manner in all collisions (Fig. 11a). These elastic collisions are robust to phase changes and initial positions of gliders (Fig. 11b).

The ballistic interaction gate (Fig. 12a) is invertible (Fig. 12b) [9]. Therefore inverse and elastic collisions in $\phi_{R22maj:4}$ may model the interaction gate: gliders g_R and g_L are equivalent to balls. To allow gliders to return to the original input locations we may constrain them with boundary conditions (Figs. 10 and 11) or route them with mirrors.

6.2 Fredkin Gate

A *conservative logic gate* is a Boolean function that is invertible and preserve signals [15]. Fredkin gate is a classical conservative logic gate. The gate realises the transformation $(c, p, q) \mapsto (c, cp + \bar{c}q, cq + \bar{c}p)$, where $(c, p, q) \in \{0, 1\}^3$. Schematic functioning of Fredkin gate is shown in Fig. 13 and the truth table is in Table 2.

Fredkin gate is logically universal because one can implement a functionally complete set of logical functions with this gate (Fig. 14). Other gates implemented with Fredkin gate are shown below:

- $c = u, p = v, q = 1$ (or $c = u, p = 1, q = v$) yields the IMPLIES gate $y = u \rightarrow v$ ($z = u \rightarrow v$).
- $c = u, \ p = 0, \ q = v$ (or $c = u, \ p = v, \ q = 0$) yields the NOT IMPLIES gate $y = \overline{(v \rightarrow u)}$ $(z = \overline{(v \rightarrow u)})$ [15].

(a) **(b)**

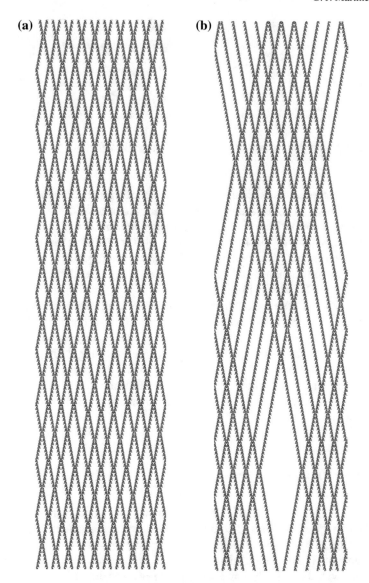

Fig. 10 Implementation of ballistic collisions in ECAM rule $\phi_{R22maj:4}$. **a** 20 gliders are synchronised to simulate the function identity $f_i(u, v) \mapsto (u, v)$, **b** 18 gliders preserve their identity during collisions. ECAM is a ring of 418 cells. It evolved in 1697 time steps

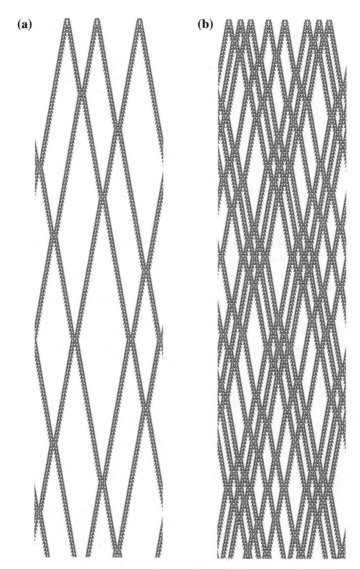

Fig. 11 Implementation of ballistic collisions in ECAM rule $\phi_{R22maj:4}$. **a** Pairs of six gliders are synchronised to simulate the reflections or elastic collisions $f_r(u, v) \mapsto (v, u)$, **b** 16 pairs of gliders preserve their identity during collisions. ECAM is a ring of 418 cells. It evolved in 1697 time steps

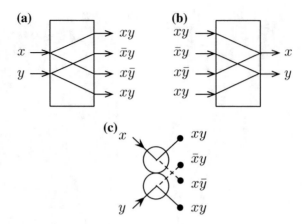

Fig. 12 In ballistic collisions we can implement an (**a**) interaction gate, (**b**) its inverse, and (**c**) its billiard ball model realisation

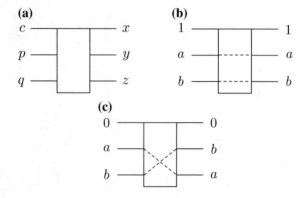

Fig. 13 Fredkin gate. **a** Scheme of the gate, **b** operation for control value 1, **c** operation for control value 0

Table 2 Truth table of Fredkin gate

c	p	q		x	y	z
0	0	0		0	0	0
0	0	1		0	1	0
0	1	0		0	0	1
0	1	1	$\rightarrow$	0	1	1
1	0	0		1	0	0
1	0	1		1	0	1
1	1	0		1	1	0
1	1	1		1	1	1

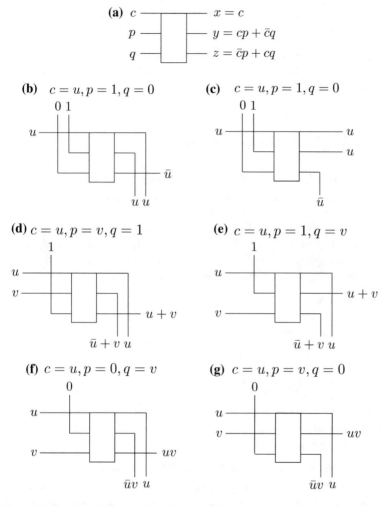

Fig. 14 Realisation of Boolean functions using Fredkin gate. **a** Fredkin gate, **b** NOT gate, **c** FANOUT gate, **d, e** OR gate, and **f, g** AND gate [15]

6.3 Basic Collisions

To simulate a Fredkin gate in a non-invertible ECAM rule $\phi_{R22maj:4}$, we utilise a set of collisions to preserve the reactions:

- *Mirror* reflects a glider, the mirror should be deleted, not to become an obstacle for other signals.
- *Doubler* splits a signal into two signals.
- *Soliton* crosses two gliders preserving its identity but might change its phase.
- *Splitter* separates two gliders into gliders travelling in opposite directions.

- *Flag* is a glider that is generated depending of an input value given.
- *Displacer* moves a glider forward.

Fredkin gates implemented in non-invertible systems were proposed and simulated by Adamatzky in a non-linear medium — Oregonator model of a Belousov–Zhabotinsky medium [3].

6.4   Fredkin Gates in One Dimension

Figure 15 displays the schematic diagram proposed to simulate Fredkin gates in ECAM rule $\phi_{R22maj:4}$. There are three inputs (c, p, q) and three outputs (x, y, z). During the computation auxiliary gliders are generated. They are deleted before reaching outputs. Gliders travel in two directions in a 1D chain of cells, they can be reused only when crossing periodic boundaries or via combined collisions. We use two gliders as flags, travelling from the left 'L_f' and from the right 'R_f'. Flags are activated depending on initial values for c or q as follows

If $c = 0$ then $R_f = 1$,
If $q = 0$ then $L_f = 1$,
In any other case L and $R = 0$.

Mirrors M are defined as follows:

If $c = p = q = 1$ then $M = 2$,
If $c = p = 1$ then $M = 2$,
If $c = q = 1$ then $M = 1$,
If $p = 1$ then $M = 1$,
In any other case $M = 0$.

Distances between gliders are fixed as positive integers determined for a number of cells in the state '0', as $\overset{0^n}{\frown} \forall n \geq 0$. During the computation we split gliders when two gliders travel together, the split gliders travel in opposite directions. We use two gliders as mirrors to change the direction of an argument movement. We use displacer to move an output glider and adjust its distance with respect to other.

Table 3 shows values of inputs, outputs, flags, and mirrors. First column represents INPUTS, the second column OUTPUTS, third column are values of a flag activated in the left side, fourth column are values of the flag activated in the right side, and the last column shows the number of necessary mirrors.

Space-time configurations of ECAM rule $\phi_{R22maj:4}$ implementing Fredkin gate for all non-zero combinations of inputs are shown in Figs. 16, 17, 18 and 19.

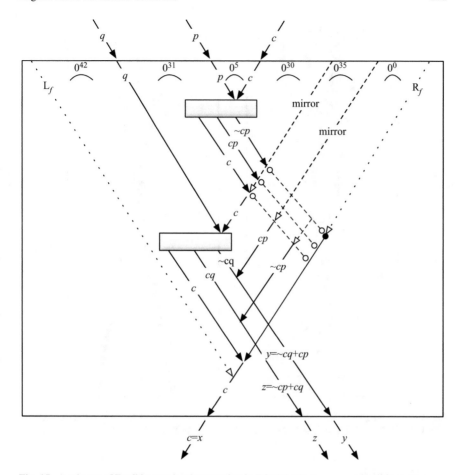

Fig. 15 A scheme of Fredkin gate implementation in ECAM rule $\phi_{R22maj:4}$ via glider collisions

Table 3 Following the scheme in Fig. 15 we specify a sequence of collisions that are controlled with flag gliders (L_f, R_f) and mirrors (M) glider

cpq	xyz	L_f	R_f	M
111	111	0	0	2
110	110	1	0	2
101	101	0	0	1
100	100	1	0	0
011	011	0	1	0
010	001	1	1	1
001	010	0	1	0

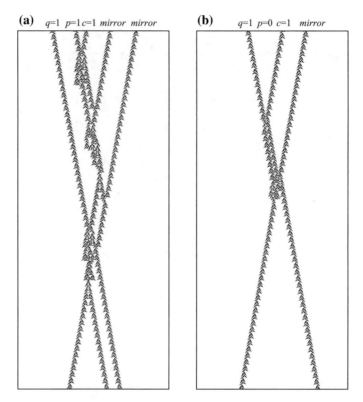

Fig. 16 Fredkin gate in ECAM rule $\phi_{R22maj:4}$. **a** INPUT $c = 1$, $p = 1$, $q = 1$, OUTPUT $x = 1$, $y = 1$, $z = 1$, **b** INPUT $c = 1$, $p = 0$, $q = 1$, OUTPUT $x = 1$, $y = 0$, $z = 1$

7 Discussion

We demonstrated how to implement a functionally complete set of Boolean functions in a one-dimensional cellular automaton with binary cell states, three-cell neighbourhood and memory depth four. We also shown that Fredkin gate can be realised the this automaton. Let us compare complexity of our designs with previously published models of universal one-dimensional cellular automata (Fig. 20).

CA simulating Turing machine, proposed by Smith III in 1971 [54], satisfied the condition: a number of cell-states multiplied by a neighbourhood size equals $\eta = 36$. A universal 1D CA designed by Albert and Culik II [6] has 14 states and totalistic cell-state transition rule, that is for their automaton $\eta = 42$. The Lindgren–Nordahl CA [21] has 7 states, assuming three-cell neighbourhood we have $\eta = 21$. ECAM implementation of a Fredkin gate, proposed in present paper, has two cell states, three cell neighbourhood, and memory depth 4, this implies $\eta = 24$. Thus, in terms of a neighbourhood size and a number of states our automaton is less efficient than the CA proposed by Lindgren–Nordahl [21], however we use gliders

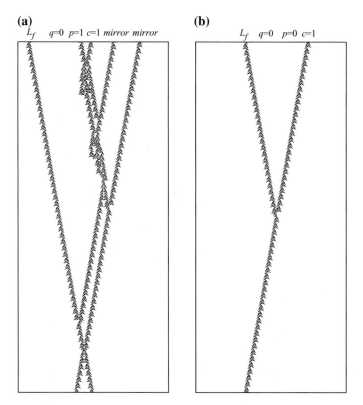

Fig. 17 Fredkin gate in ECAM rule $\phi_{R22maj:4}$. **a** INPUT $c = 1$, $p = 1$, $q = 0$, OUTPUT $x = 1$, $y = 1$, $z = 0$, **b** INPUT $c = 1$, $p = 0$, $q = 0$, OUTPUT $x = 1$, $y = 0$, $z = 0$

as signals. Gliders are discrete analogs of solitons, therefore our design, in principle, can be considered as a blueprint for physical implementation, as we will discuss below. Cook's design [13, 66] of a universal CA via cyclic tag system is the most optimal in terms of neighbourhood size and number of states, $\eta = 6$, however the implementation involves 11 gliders and a glider gun. Our design of Fredkin gate utilises just two types of gliders.

Polymer chains that support propagation of travelling localisations (solitons, kinks, defects) are potential substrates for implementation of glider-based Fredkin gate. There are many potential candidates, here we discuss actin filaments. Actin is a protein presented in all eukaryotic cells in forms of globular actin (G-actin) and filamentous actin (F-actin), see history overview in [58]. Filamentous actin is a double helix of F-actin unit chains. In [4] we proposed a model of actin polymer filaments as two chains of one-dimensional binary-state semi-totalistic automaton arrays and uncovered rules supporting gliders, discrete analogs of ionic waves propagating in actin filaments [61].

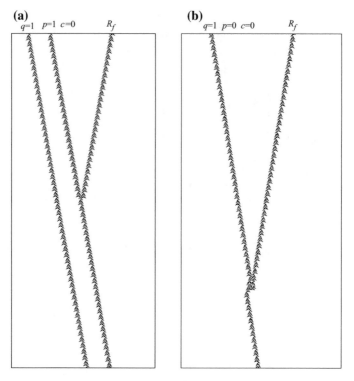

Fig. 18 Fredkin gate in ECAM rule $\phi_{R22maj:4}$. **a** INPUT $c = 0$, $p = 1$, $q = 1$, OUTPUT $x = 0$, $y = 1$, $z = 1$, **b** INPUT $c = 0$, $p = 0$, $q = 1$, OUTPUT $x = 0$, $y = 1$, $z = 0$

Let us evaluate a physical space-time requirement for implementation of Fredkin gate on actin filaments. Assume a unit of F-actin corresponds to a cell of 1D CA. Maximum diameter of an actin filament is 8 nm [42, 55]. An actin filament is composed of overlapping units of F-actin. Thus, diameter of a single unit is c. 4 nm. A glider in our ECAM model occupies 10 cells, that makes a size of a signal in actin filaments 40 nm. Maximum distance between inputs in ECAM Fredkin gate is 200 cells. This makes gate size 880 nm.

With regards to speed of Fredkin gate realisation, being unaware of exact mechanisms of travelling localisations on actin filaments we can propose speculative estimates. Assume the underpinning mechanism of generating a localisation is an excitation of F-actin molecule (single unit of actin polymer). An excitation in a molecule takes place when an electron in a ground state absorbs a photon and moves up to a high yet unstable energy level. Later the electron returns to its ground state. When returning to the ground state the electron releases photon which travels with speed $\times 10^{18}$ Å per second. F-actin molecule (one unit of actin filament) is a polymer chain of at most 3 K nodes. Thus the F-actin unit can be spanned by an excitation in at most 10^{-15} sec. That can be adopted as a physical time step equivalent to one

L_f q=0 p=1 c=0 *mirror* R_f

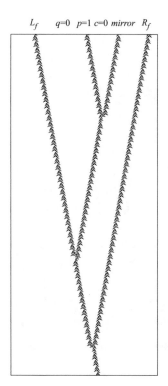

Fig. 19 Fredkin gate in ECAM rule $\phi_{R22maj:4}$. INPUT $c = 0$, $p = 1$, $q = 0$, OUTPUT $x = 0$, $y = 0, z = 1$

step of ECAM evolution. The ECAM Fredkin gate completes its operation on actin filament in 600 time steps, that is at most 10^{-13} sec, i.e. 0.1 picosecond of real time.

Experimental implementation of the Fredkin gate, including cascading of the gates, on actin filaments, or any other soliton-supporting polymer chain makes a very challenging topic of future studies. There methodological approaches to control and monitor attosecond scale molecular dynamics [7, 11, 17, 49] however it is not clear if they can be applied to actin polymers. The problems to overcome include input of data in acting filament with a single F-actin unit precision, keeping the polymer chain stable and insulated from thermal noise, reading outputs from the polymer chain. However exact experimental implementation remains uncertain.

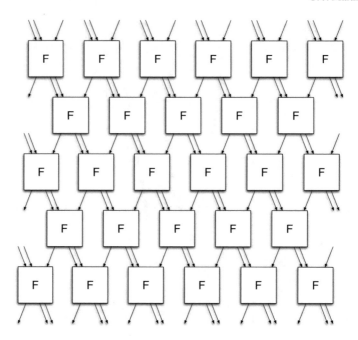

Fig. 20 Cascaded Fredkin gates (It is a modification of Fredkin array proposed in [24])

References

1. Adamatzky, A. (ed.): Collision-Based Computing. Springer, Berlin (2002)
2. Adamatzky, A.: Physarum Machines: Computers from Slime Mould. World Scientific Series on Nonlinear Science Series A, vol. 74. World Scientific, Singapore (2010)
3. Adamatzky, A.: Fredkin and Toffoli gates implemented in oregonator model of Belousov–Zhabotinsky medium. Int. J. Bifurc. Chaos **27**(3), 1750041 (2017)
4. Adamatzky, A., Mayne, R.: Actin automata: phenomenology and localizations. Int. J. Bifurc. Chaos **25**(02), 1550030 (2015)
5. Adamatzky, A., Costello, B.L., Asai, T.: Reaction-Diffusion Computers. Elsevier, Amsterdam (2005)
6. Albert, J., Culik II, K.: A simple universal cellular automaton and its one-way and totalistic versions. Complex Syst. **1**(1), 1–16 (1987)
7. Baltuska, A., Udem, T., Uiberacker, M., Hentschel, M., Goulielmakis, E., Gohle, C., Holzwarth, R., Yakovlev, V.S., Scrinzi, A., Hansch, T.W., Krausz, F.: Attosecond control of electronic processes by intense light fields. Nature **421**(6923), 611–615 (2003)
8. Banks, E.R.: Information and transmission in cellular automata. Ph.D. Dissertion, Cambridge, MA, MIT (1971)
9. Bennett, C.H.: Logical reversibility of computation. IBM J. Res. Dev. **17**(6), 525–532 (1973)
10. Berlekamp, E.R., Conway, J.H., Guy, R.K.: Winning Ways for your Mathematical Plays, vol. 2. Academic Press, London (1982). (chapter 25)
11. Ciappina, M.F., Perez-Hernandez, J., Landsman, A., Okell, W., Zherebtsov, S., Frg, B., Schtz, J., Seiffert, L., Fennel, T., Shaaran, T. and Zimmermann, T.: Attosecond physics at the nanoscale. Reports on Progress in Physics (2017)
12. Codd, E.F.: Cellular Automata. Academic Press Inc, New York (1968)
13. Cook, M.: Universality in elementary cellular automata. Complex Syst. **15**(1), 1–40 (2004)

14. Davis, M.D., Signal, R., Weyuker, E.J.: Computability, Complexity, and Languages. Computer Science and Scientific Computing, 2nd edn. Academic Press, London (1994)
15. Fredkin, E., Toffoli, T.: Conservative logic. Int. J. Theor. Phys. **21**, 219–253 (1982)
16. Fredkin, E., Toffoli, T.: Design Principles for Achieving High-Performance Submicron Digital Technologies, pages 27–46, (2001) (in [2])
17. Goulielmakis, E., Schultze, M., Hofstetter, M., Yakovlev, V.S., Gagnon, J., Uiberacker, M., Aquila, A.L., Gullikson, E.M., Attwood, D.T., Kienberger, R., Krausz, F.: Single-cycle nonlinear optics. Science **320**(5883), 1614–1617 (2008)
18. Hey, A.J.G.: Feynman and Computation: Exploring the Limits of Computers. Perseus Books (1998)
19. Hutton, T.J.: Codd's self-replicating computer. Artif. Life **16**(2), 99–117 (2010)
20. Jakubowski, M.H., Steiglitz, K., Squier, R.: Computing with solitons: a review and prospectus. Mult. Valued Log. **6**(5–6) (2001) (also republished in [2])
21. Lindgren, K., Nordahl, M.: Universal computation in simple one-dimensional cellular automata. Complex Syst. **4**, 229–318 (1990)
22. Margolus, N.H.: Physics-like models of computation. Phys. D **10**(1–2), 81–95 (1984)
23. Margolus, N.H.: Crystalline computation. In: Hey, A.J.G. (ed.) Feynman and Computation: Exploring the Limits of Computers, pp. 267–305. Perseus Books (1998)
24. Margolus, N.H.: Universal cellular automata based on the collisions of soft spheres. In: Adamatzky, A. (ed.) Collision-Based Computing, pp. 107–134. Springer, Berlin (2002)
25. Margolus, N., Toffoli, T., Vichniac, G.: Cellular-automata supercomputers for fluid dynamics modeling. Phys. Rev. Lett. **56**(16), 1694–1696 (1986)
26. Martínez, G.J., Adamatzky, A., McIntosh, H.V.: Phenomenology of glider collisions in cellular automaton rule 54 and associated logical gates. Chaos Solitons Fractals **28**, 100–111 (2006)
27. Martínez, G.J., Adamatzky, A., Sanz, R.A.: Complex dynamics of elementary cellular automata emerging in chaotic rule. Int. J. of Bifurcation and Chaos **22**(2), 1250023–13 (2012)
28. Martínez, G.J., Adamatzky, A., Sanz, R.A.: Designing Complex Dynamics with Memory. Int. J. Bifurcation and Chaos **23**(10), 1330035–131 (2013)
29. Martínez, G.J., Adamatzky, A., McIntosh, H.V.: A computation in a cellular automaton collider rule 110. In: Adamatzky, A. (ed.) Advances in Unconventional Computing Volume 1: Theory, pp. 391–428. Springer, Berlin (2016)
30. Martínez, G.J., Adamatzky, A., Sanz, R.A., Mora, J.C.S.T.: Complex dynamic emerging in rule 30 with majority memory. Complex Syst. **18**(3), 345–365 (2010)
31. Martínez, G.J., Adamatzky, A., Mora, J.C.S.T., Alonso-Sanz, R.: How to make dull cellular automata complex by adding memory: rule 126 case study. Complexity **15**(6), 34–49 (2010)
32. Martínez, G.J., Adamatzky, A., Stephens, C.R., Hoeflich, A.F.: Cellular automaton supercolliders. Int. J. Mod. Phys. C **22**(4), 419–439 (2011)
33. Martínez, G.J., McIntosh, H.V., Mora, J.C.S.T., Vergara, S.V.C.: Reproducing the cyclic tag system developed by Matthew Cook with Rule 110 using the phases f_1_1. J. Cell. Autom. **6**(2–3), 121–161 (2011)
34. Martínez, G.J., Adamatzky, A., Chen, F., Chua, L.: On soliton collisions between localizations in complex elementary cellular automata: rules 54 and 110 and beyond. Complex Syst. **21**(2), 117–142 (2012)
35. Martínez, G.J., Mora, J.C.S.T., Zenil, H.: Computation and universality: class IV versus class III cellular automata. J. Cell. Autom. **7**(5–6), 393–430 (2013)
36. McIntosh, H.V.: Wolfram's class IV and a good life. Phys. D **45**, 105–121 (1990)
37. McIntosh, H.V.: One Dimensional Cellular Automata. Luniver Press, United Kingdom (2009)
38. Mills, J.W.: The nature of the extended analog computer. Phys. D **237**(9), 1235–1256 (2008)
39. Minsky, M.: Computation: Finite and Infinite Machines. Prentice Hall, Englewood Cliffs (1967)
40. Mitchell, M.: Life and evolution in computers. Hist. Philos. Life Sci. **23**, 361–383 (2001)
41. Moore, C., Mertens, S.: The Nature of Computation. Oxford University Press, Oxford (2011)
42. Moore, P.B., Huxley, H.E., DeRosier, D.J.: Three-dimensional reconstruction of F-actin, thin filaments and decorated thin filaments. J. Mol. Biol. **50**(2), 279IN17289–288IN28292 (1970)

43. Morita, K.: A simple construction method of a reversible finite automaton out of Fredkin gates, and its related problem. Trans. IEICE Jpn. **E–73**, 978–984 (1990)
44. Morita, K.: A new universal logic element for reversible computing. IEICE technical report. Theor. Found. Comput. **99**(724), 119–126 (2000)
45. Morita, K.: Simple universal one-dimensional reversible cellular automata. J. Cell. Autom. **2**, 159–165 (2007)
46. Morita, K.: Reversible computing and cellular automata-a survey. Theor. Comput. Sci. **395**, 101–131 (2008)
47. Morita, K.: Universality of 8-State reversible and conservative triangular partitioned cellular automata. Lecture Notes in Computer Science, vol. 9863, pp. 45–54. Springer, Cham (2016)
48. Morita, K., Harao, M.: Computation universality of one-dimensional reversible (injective) cellular automata. Trans. IEICE Jpn. **E–72**, 758–762 (1989)
49. Nabekawa, Y., Okino, T., Midorikawa, K.: Probing attosecond dynamics of molecules by an intense a-few-pulse attosecond pulse train. In: 31st International Congress on High-Speed Imaging and Photonics, p. 103280B. International Society for Optics and Photonics (2017)
50. Park, J.K., Steiglitz, K., Thurston, W.P.: Soliton-like behavior in automata. Phys. D **19**, 423–432 (1986)
51. Post, E.L.: The Two-Valued Iterative Systems of Mathematical Logic. Princeton University Press, Princeton (1941)
52. Rendell, P.: Turing Machine Universality of the Game of Life. Springer, Berlin (2016)
53. Sanz, R.A.: Cellular Automata with Memory. Old City Publishing, Philadelphia (2009)
54. Smith III, A.R.: Simple computation-universal cellular spaces. J. Assoc. Comput. Mach. **18**, 339–353 (1971)
55. Spudich, J.A., Huxley, H.E., Finch, J.T.: Regulation of skeletal muscle contraction: II. Structural studies of the interaction of the tropomyosin-troponin complex with actin. J. Mol. Biol. **72**(3), 619IN5IN18621–620IN16IN19632 (1972)
56. Steiglitz, K.: Soliton-guided quantum information processing. In: Adamatzky, A. (ed.) Advances in Unconventional Computing Volume 2: Prototypes, Models and Algorithms, pp. 297–307. Springer, Berlin (2016)
57. Steiglitz, K., Kamal, I., Watson, A.: Embedding computation in one-dimensional automata by phase coding solitons. IEEE Trans. Comput. **37**(2), 138–145 (1988)
58. Szent-Györgyi, A.G.: The early history of the biochemistry of muscle contraction. J. Gen. Physiol. **123**(6), 631–641 (2004)
59. Toffoli, T.: Non-conventional computers. In: Webster, J. (ed.) Encyclopedia of Electrical and Electronics Engineering, pp. 455–471. Wiley, London (1998)
60. Toffoli, T.: Symbol super colliders. In: Adamatzky, A. (ed.) Collision-Based Computing, pp. 1–22. Springer, Berlin (2002)
61. Tuszyński, J.A., Portet, S., Dixon, J.M., Luxford, C., Cantiello, H.F.: Ionic wave propagation along actin filaments. Biophys. J. **86**(4), 1890–1903 (2004)
62. von Neumann, J.: Theory of Self-reproducing Automata (edited and completed by A.W. Burks). University of Illinois Press, Urbana (1966)
63. Wolfram, S.: Universality and complexity in cellular automata. Phys. D **10**, 1–35 (1984)
64. Wolfram, S.: Computation theory of cellular automata. Commun. Math. Phys. **96**, 15–57 (1984)
65. Wolfram, S.: Cellular Automata and Complexity. Addison-Wesley Publishing Company, Reading (1994)
66. Wolfram, S.: A New Kind of Science. Wolfram Media Inc., Champaign (2002)

Computation and Pattern Formation by Swarm Networks with Brownian Motion

Teijiro Isokawa and Ferdinand Peper

Abstract Swarm Networks are networks of cells of which the functionalities are determined by the presence or absence of connections with cells around them. They have a more flexible neighborhood than Cellular Automata, and can thus be regarded as a generalization of the latter. This chapter describes Swarm Networks in which connections can be changed dynamically, and in which the cells are subject to Brownian motion. According to these characteristics, the model mimics behavior typically encountered in biological organisms. Two types of Swarm Networks are described in this chapter: one has the ability of universal computation through the implementation of a universal Brownian circuit, and the other has the ability of pattern formation in which a circle is formed through the serial connection of agents.

1 Introduction

With the increasing interest in molecular robotics [10], wireless sensor networks [1], and artificial cells [3], the field of distributed computing has recently experienced a reawakening. Unlike in conventional computers, computation in such models is distributed over a large number of cells. The computation power of these cells tends to be limited, like in Finite Automata, but when the cells are combined with each other, like in Cellular Automata (CAs) [2, 4, 11], they have the potential to simulate universal Turing Machines [15]. Key to the operation of such models is the interaction

T. Isokawa (✉)
Graduate School of Engineering, University of Hyogo, 2167 Shosha, Himeji,
Hyogo 671-2280, Japan
e-mail: isokawa@eng.u-hyogo.ac.jp

F. Peper
Center for Information and Neural Networks, National Institute of Information and
Communications Technology, 1-4 Yamadaoka, Suita, Osaka 565-0871, Japan
e-mail: peper@nict.go.jp

© Springer International Publishing AG 2018
A. Adamatzky (ed.), *Reversibility and Universality*, Emergence, Complexity
and Computation 30, https://doi.org/10.1007/978-3-319-73216-9_10

between cells. In Cellular Automata these interactions are well-defined, both in terms of the neighborhood of a cell, and the state transition function. Nature, however, tends to be more flexible, with neighborhoods often changing over time.

The *Swarm Network* model [5] has been proposed with this in mind, and its flexibility derives from the ability of agents to have a connection pattern that is less regular than that of CAs. In fact, the variety in neighborhood according to which an agent can be connected to other agents is used as a way to define the functionalities of agents: depending on which of an agent's connection terminals are used, it will behave differently. This has plausibility in nature, where molecules react differently depending on their chemical bonds.

Brownian motion has been incorporated into the original Swarm Network model to provide a driving force to agents in the network [9]. As a result, agents tend to change their states and to form connections occasionally, i.e. when colliding with other agents in the space. Cooperative tasks can thus be performed according to appropriately designed rules that determine how cell states change depending on the neighbors connected to them, in a way resembling the operation of CAs. Unlike with CAs, in Swarm Networks it is also necessary to design transition rules that define how and when connections between agents are formed, and when connections are deleted.

In this chapter, we demonstrate a model for Swarm Networks and two applications based on this model, i.e., computation and pattern formation. One example shows the construction of circuits that are able to conduct universal computation [8, 9]. Computation universality is achieved by designing circuit elements and connecting them in appropriate ways. We opt for a type of circuit that exploits fluctuations of signals; these circuits are called *Brownian*. The other example refers to the formation of patterns by a network of agents. A linear pattern and a circular pattern can be constructed through the serial connection of agents.

This chapter is organized as follows. Section 2 introduces the model of Swarm Networks, with agents being the basic elements of the network and transition rules employed by agents to determine state transitions as well as connectivity with other agents. Computation by the Swarm Network is described in Sect. 3, with signal transmission by cooperative behaviors of agents outlined in detail. Pattern formation by the agents is demonstrated in Sect. 4, whereby protocols for the communication between agents are explained. Section 5 concludes this chapter with perspectives on the applications of Swarm Networks.

2   Models

2.1   Swarm Networks

A Swarm Network is a collection of agents that can freely move around in space, only restricted by connections with other agents. The space containing the agents is two-dimensional, and the agents themselves have a circular shape with six terminals attached at their outsides at identical distances from each other (Fig. 1). The model can be extended to three-dimensions in principle, but we will not consider this case in this chapter. Agents can be connected to each other via their terminals, which are used (when connected) to exchange information between agents. The black color of the terminal in Fig. 1 indicates that it is connected with the terminal of another agent. Each terminal has a state which can be read by an agent connected to it. Each agent is assumed to be a Mealy-type finite automaton, with an internal state denoted by a symbol in it (Fig. 1). The internal state of an agent is only known to the agent itself, not to other agents, whether it is connected to them or not.

The functionality of an agent is determined by the connection pattern of the agent to other agents. For example, an agent being connected to two agents at opposite terminals has a different functionality than an agent connected to three agents via adjacent terminals. The functionality of an agent is expressed in terms of transitions. An agent has a set of transition rules, each of which applies to a certain connectivity pattern of the agent.

The space containing the agents satisfies simple mechanical laws, governing forces between the terminals that are interconnected with each other [9]. The connections between terminals are modeled as springs, and they exert a repulsive force between terminals very near to each other, and an attractive force when terminals are a bit further away from each other. In other words, connections are elastic. The mechanical properties of connections do not directly influence their logical properties, but are mostly used to make the model more realistic. However, due to the mechanical properties of connections, some agents may be biased to move into the direction of other agents, so indirectly there is an influence on the logical properties.

Communication between agents takes place via the terminals through which they are connected to each other in terms of the terminals' states, which are expressed as integers [9]. An agent can read and write the states of its own terminals, and read the states of terminals of agents to which it is connected. A transition of an agent has as its domain the agent's state, the states of its own terminals, and the states of other agent's terminals to which it is connected, and has as codomain the new state of the agent as well as the new states of its own terminals. As stated above, the transition rules depend on the connection pattern of an agent. Formally, a transition rule for an agent is defined as:

$$(q, t, s) \rightarrow (q', t') \qquad (1)$$

where q and q' are the internal states of an agent before and after the transition, respectively, $t = (t_0, \ldots, t_5)$ is the tuple of states of the agent's terminals before the transition, $s = (s_0, \ldots, s_5)$ is the tuple of states input to the terminals of the agent before the transition, whereby these inputs are supplied from the other agents' terminals that connect to the respective terminals of the agent. The state of a terminal in s is indicated by an $\varnothing$ symbol if there is no agent connected to the corresponding terminal. Furthermore, $t' = (t'_0, \ldots, t'_5)$ is the tuple of integers on the terminals of the

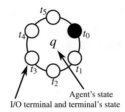

Agent's state
I/O terminal and terminal's state

Fig. 1 Individual agent. The terminals are labeled by the numbers between brackets

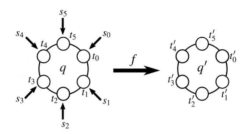

Fig. 2 Transition rule of an agent

agent after the transition. We assume that the transition rules are rotation-symmetric, i.e., one transition rule exists in six varieties, which are rotated analogues of each other. An example of a transition rule is shown in Fig. 2.

Connections between agents are made and broken according to a set of connectivity conditions that describe the influence of the environment on connections [9]. Connectivity conditions come in two types: the first type specifies the condition under which a connection is established, and the second type specifies the condition under which a connection is broken.

A connectivity condition C for establishing a connection is defined as:

$$C(t_1, t_2, r) \tag{2}$$

where t_1 is the state of a terminal in the space, t_2 is the state of another terminal, and r is a real number denoting the distance threshold between the two terminals. If the distance between the terminals is less than the threshold r, then a connection will be created if the respective states of the terminals equal t_1 and t_2. The value of r is chosen such that it is less than the distance between two adjacent terminals of the same agent. This is to prevent an agent's own terminals to become connected to each other.

A connectivity condition D (which stands for *Disconnect*) for breaking a connection is defined as:

$$D(t_1, t_2) \tag{3}$$

where t_1 is the state of a terminal in the space and t_2 is the state of another terminal connected to the first terminal. The connection between the two terminals will be broken if their respective states equal t_1 and t_2.

Whenever there is a match of a connectivity condition anywhere in the space of swarm agents, it will be applied, thus making or breaking a connection. Making or breaking a connection is a stochastic process, so even if condition C or D is valid, no immediate action may be taken with probability 1.

3   Computation by Swarm Networks

This section presents a swarm network with the ability of computational universality. This is achieved by implementing elements of a so-called Brownian circuit, which is introduced in the next subsection. It is important to organize the protocols for communication between agents so that desired collective behavior emerges, thus this section describes details of handshake protocols to be used in signal transmission between agents. Descriptions for implementation of other functionalities can be found in [9].

3.1   Brownian circuits

A Brownian circuit is a circuit in which signals, represented by tokens, fluctuate forward and backward [6, 7, 12]. Through their random fluctuations, tokens search their way from input to output in the circuit. The random search property allows signals to backtrack their way out of deadlocks, which may occur in token-based circuits due to waiting conditions that can not be satisfied. Since the signals in Brownian circuits are represented by tokens, they cannot split or fuse on their own. It has been shown that computational universality in Brownian circuits can be achieved by very limited sets of primitives [6, 7, 12]. In this paper we will use the set in [6, 7].

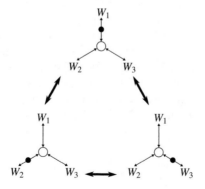

Fig. 3 Hub and its possible transitions. A token is denoted by a black fat dot. Fluctuations cause a token to move between any of the Hub's three wires W_1, W_2, and W_3 in any order [6]

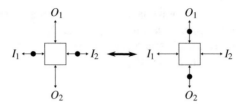

Fig. 4 CJoin and its possible transitions. If there is a token on only one input wire (I_1 or I_2), this token remains pending until a signal arrives on the other wire. These two tokens will then result in one token on each of the two output wires O_1 and O_2 [6]

Fig. 5 Ratchet and its possible transition. The token on the wire W may fluctuate before the ratchet as well as after the ratchet, but once it moves over the ratchet it cannot return. The ratchet thus imposes a direction on a (originally) bi-directional wire [6]

The first primitive is the *Hub*, which contains three wires that are bidirectional (Fig. 3) [6]. There will be at most one signal at a time on any of the Hub's wires, and this signal can move to any of the wires due to its fluctuations.

The second primitive is the *Conservative Join (CJoin)*, which has two input wires and two output wires, all bi-directional (Fig. 4) [6]. The CJoin can be interpreted as a synchronizer of two signals passing through it. Signals may fluctuate on the input wires, and when processed by the CJoin, they will be placed on the output wires where they may also fluctuate. When only one input wire contains a signal, the CJoin will not do anything and the signal remains pending on the wire, fluctuating, but it may fluctuate backwards through another circuit element, thus removing itself as input of the CJoin. The operation of the CJoin may also be reversed, and the forward / backward movement of the two signals through it may be repeated an unlimited number of times. Due to this bidirectionality, there is strictly speaking no distinction between the input and output wires to the CJoin, though we still use the terminology of input and output, since the direction of the process is eventually forward.

The third primitive is the *Ratchet*, which restricts the movement of tokens through itself to one direction [6]. It effectively transforms a bidirectional wire into a uni-directional wire (see Fig. 5). The Ratchet is used to speed up searching in circuits. Searching cannot backtrack over a Ratchet, so a signal will take less time as a result to move along a wire. However, a Ratchet cannot be placed at positions at which backtracking of signals is an essential part of the search process in a circuit, so its placement is restricted.

3.2 Building Signals and Signal Paths by Swarm Agents

To establish the computational universality of a model based on Swarm Networks, we construct the Brownian circuit elements from the agents [9]. Since each agent in itself lacks sufficient complexity to behave like a circuit element, multiple agents are combined to construct the elements. An agent in the model has one of two states $\{q_1, q_2\}$ and each of its terminals has as its state a number from the set with 17 members $\{0, 1, 2, 3, 4, 5, 6, 7, 10, 11, 12, 13, 14, 15, 16, 17, 20\}$. These states have roughly the following meanings; state 0 denotes a quiescent state (no connection), states 1 and 2 are used to prepare for a connection, states 3, 4, 5, 6, 7 are used to establish a connection, state 10 denotes that any agent is connected to a wall agent, state 11 is a permanent state to distinguish the CJoin, the Crossing, and Ratchet from each other according to connectivity pattern, state 12 is used to prepare for a connection in the CJoin, the Crossing, and the Ratchet, states 13, 15, 15, and 16 are used to establish a connection in the CJoin, the Crossing, and the Ratchet, state 17 is used to break and block a connection in the CJoin, the Crossing, and the Ratchet, and state 20 is used by a wall agent connected to an agent. The above-mentioned wall agent is another type of agent, which is passive, unable to change its state or the states of its terminals, and it is as well unable to read states of terminals of other agents. Rather its terminal states are constant but they can be read out by normal (active) agents connected to it. Wall agents are used to form boundaries of the wires along which normal agents may move around. Wall agents are represented as black circles in the figures, and they constitute the walls of wires in the circuits to be constructed. Signals travel in the circuit along these wires.

Figure 6a shows a configuration of a signal with a wire, where four wall agents lined up horizontally correspond to a wire, whereas the normal agent in state q_2, connected to one of the wall agents, is a signal in the circuit [9]. The numbers beside the terminals represent the states of the terminals. Fig. 6a shows a normal agent with state 1 at four of its terminals. This agent also outputs a state 10 to its one terminal that is connected to a wall agent. State 20 of this wall agent can be recognized by the normal agent.

This configuration represents a signal moving eastward or westward (with direction randomly determined) along a wire [9]. There exist other agents in state q_1 around the signal, but they are not shown if they are not involved in the process of signal transmission. There is thus only one agent in state q_1 next to the normal agent, as shown in Fig. 6b.

Connections between the agent and another agent are established with a positive probability when the connectivity condition C is satisfied [9]. To break a connection, on the other hand, condition D is employed. The C and D conditions are important in the interactions between agents, and their actions are implicitly assumed in this chapter. The two normal agents in Fig. 6b, for example, become connected due to the condition $C(1, 2, r)$.

Once two agents are connected, the movement of a signal is achieved by exchanging information between the agents [9]. As a result, the state q_2 moves from one agent

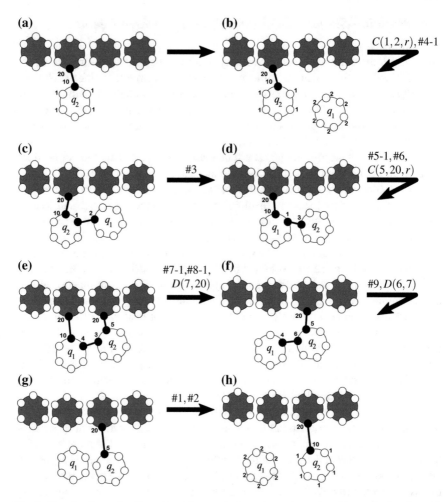

Fig. 6 Signal and signal path configurations. **a** Agent representing a signal is attached to the wall. **b** Agent in the inactive state q_1 floats near the signal. There are many of such inactive agents floating around in the space. **c** Connection between signal and inactive agent is established. **d** Signal and inactive agent reverse roles. **e** The newly active agent establishes connection to the wall. **f** and **g** The old signal that has become inactive is released into the environment. **h** The new signal can accept another inactive agent

to another agent in state q_1 connected to it. Figure 6c, d, e and f show typical steps involving such signal movement. After the agent in state q_1 is connected to the agent in state q_2 (Fig. 6c), they exchange their states (Fig. 6d). Following this, the new agent in state q_2 connects to the wall agent at its nearest neighbor (Fig. 6e). Finally the agent in state q_1 detaches from the normal agent and the wall agent (Fig. 6f, g), and the agent in state q_1 can accept another agent (Fig. 6h). The direction of a signal is randomly determined by the location of the agent in state q_1 that will be connected to the signal agent in state q_2 [9]. Thus, state q_2, acting as a signal, can move in forward and backward directions along the wall agents, which constitute the signal path.

All agents involved in signal transmission are programmed to recognize their stage in the above procedure, such as whether another agent has already been connected to, the other agent has already changed its state to q_2, and so on [9]. The stage is encoded by the states of the terminals of the agents, as well as its connection pattern with the other agents. Agents work independently from each other, so they need to conduct communications and synchronizations with other agents on a local level. We adopt a handshake protocol to transfer data between agents, like is used in many asynchronous (and self-timed) systems [13, 14].

Figure 7 illustrates the protocol for the signal transmission in Fig. 6 [9]. The initial configuration shown in Fig. 7a corresponds to the configuration in Fig. 6a, in which the signal agent in state q_2 connects to one of the wall agents. Unconnected agents in state q_1 have state 2 in all their terminals. So, if an unconnected agent is sufficiently close and the terminal states satisfy a connectivity condition, like in Fig. 7b, a connection will be established. After connecting with the signal agent, this agent changes its state to q_2 and informs this change to the signal agent by setting the state of its corresponding terminal to '3' (Fig. 7c). The signal agent changes its state to q_1 by accepting state '3' via the terminals and emits this acknowledgment by setting its corresponding terminal state to '4' (Fig. 7d). To connect to a wall agent, the new signal agent sets its terminal state to '5', and if successfully connected, the

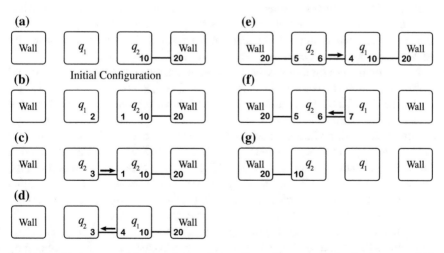

Fig. 7 Handshake protocol for signal transmission. Generally, one agent sends a request, which invokes an action in the other agent, which then replies with an acknowledgment. This acknowledgment is used by the original sender to start an action on its own. **a** Initial configuration. **b** Preparation to make connection between signal and inactive agent. **c** Connection established, after which request for agent state changes is sent. **d** Acknowledgment is returned, and agents have exchanged their states. **e** The new signal establishes connection to the wall. **f** Preparation to disconnect from old signal. **g** Final state, in which the two agents have reversed roles. [9]

agent sends a request to disconnect by setting state '6' on its terminal connected to the previous signal agent (Fig. 7e). This agent then cuts its connection with the wall agent, and acknowledges this to the other agent by setting state '7' on its terminal (Fig. 7f). Finally, the previous signal agent disconnects from the other agent, and the newly connected agent becomes the signal agent (Fig. 7g). As stated previously, making or breaking connections is achieved by conditions C and D. In order to satisfy these conditions, agents set their terminals to appropriate states. Implementation of the above protocol requires quite a few states of the terminals involved to synchronize the agents that are connected to each other. Other types of elements such as crossover and CJoin communicate with each other in a similar way [9].

The implementation of a signal requires 48 transition rules, of which 21 transition rules (divided in nine groups of rules) are used for interactions between agents [9]. The remaining 27 transition rules for a signal are used for correcting inappropriate connections to other agents during signal propagation. These transition rules are listed in Appendix A of this chapter, and the complete list of transition rules and their descriptions can be found in [9].

4 Pattern Formation on Swarm Network

This section introduces a scheme for forming circular patterns by swarm agents, as an example of pattern formation on Swarm Networks. This pattern formation is conducted by first making a linear chain of several swarm agents that forms when the agents collide with each other, then forming a circle by changing the connections between their terminals. Circular pattern formation is finally completed by establishing connections between both ends of the chain of agents.

In the circular configuration scheme, an agent has one of two states $\{q_1, q_2\}$ and each of its terminals has one of 17 states in $\{1, 2, 5, 6, 7, 9, 10, 11, 12, 13, 14, 15, 16, 17, 18, 19, 20\}$. These states have roughly the following meanings. State 1 is used to prepare for a connection, state 2 is used to refuse a connection, states 5 and 6 are used in the formation of the connection that is the backbone in a double connection, and this changes to states 5 and 7 in the final state of the backbone. State 9 is used in the double connection chain, states 11 to 19 are used as counters to connect agents, and states 10 and 20 indicate the respective ends of the chain of agents. Agents with the state q_2 are responsible for initiating the formation of a pattern, and other agents in the state q_1 are to be connected to these agents via two adjacent terminals. The states are fixed in the agents throughout the process of pattern formation. The connections between agents are established through occasional collisions caused by Brownian motion, which acts as a driving force of the process. The terminal states

$\{1, 2, 5, 6, 7, 9, 10, 20\}$ are used for controlling the process of pattern formation and the remaining states $\{11, 12, 13, 14, 15, 16, 17, 18, 19\}$ are the states for counting the number of agents in the pattern.

This pattern formation consists of three stages, described as follows:

1. A linear chain is first formed consisting of 10 agents. When an agent in state q_2 collides with an agent in state q_1, the result is two consecutive agents that are connected by their two terminals, and so on.
2. The linear chain curves to the right. This is achieved by making each of the agents in the chain change the states of its terminals and then reconfigure the connections between the terminals of the agents.
3. Establishing a connection between both ends in the linear chain.

Figures 8, 9, and 10 show networks of swarm agents in stage 1, 2, and 3, respectively. All transition rules for the pattern formation with the connectivity conditions in this section are listed in Appendix B of this chapter.

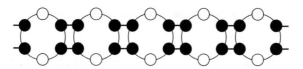

Fig. 8 Linear chain consisting of a series of agents

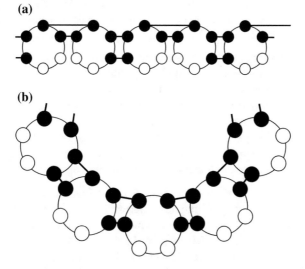

Fig. 9 Curving a linear chain of agents by reconfiguring the connections between the agents. **a** Alternating connectivity required for correct curvature of a circular pattern consisting of 10 agents. **b** Same connectivity as in (**a**), but now bent in a more natural shape so that the connections are less stretched

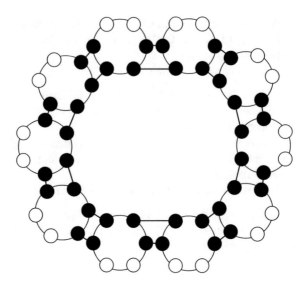

Fig. 10 Circular pattern is completed by connecting both ends of the chain

The first stage of the pattern formation process is as follows. An initiating agent for the pattern is in state q_2; the state of one of its terminals is set to 11 and the other five terminals are set to state 2. An agent is connected to this initiating agent if it is in state q_1 with one terminal being in state 1 and five terminals being in state 2. The following procedures are adopted for completing a linear chain of agents, where n counts up as $n = 11, 12, \ldots, 18, 19$:

1. Two agents of which one is in state q_2 and the other in state q_1 make a connection via the terminals in the states n and 1, respectively.
2. Each pair of agents that are connected by a single connection makes a further connection via their adjacent terminals, resulting in two adjacent terminals being used for a double connection between agents.
3. The agent in state q_1 sets the state of the terminal clockwise next to the terminal in state n to state $(n + 1)$ in order to accept the next agent to be connected.
4. This connection process is finished when the number of agents in the group is 10.

A handshake protocol for this process is depicted in Fig. 11. An initiating agent in state q_2 and its neighboring agent in state q_1 first establish a connection between one of their terminals (Fig. 11a, b and c), then one more connection is established between other terminals (Fig. 11d, e). After making the double connections, the connected agent in state q_1 changes the state of its terminal so that it can accept a connection from an agent in state q_1 (Fig. 11e, f). Counting the number of agents in the network is accomplished by encoding the terminal states for accepting another agent. Finally 10 agents are linearly connected in the network (Fig. 11i).

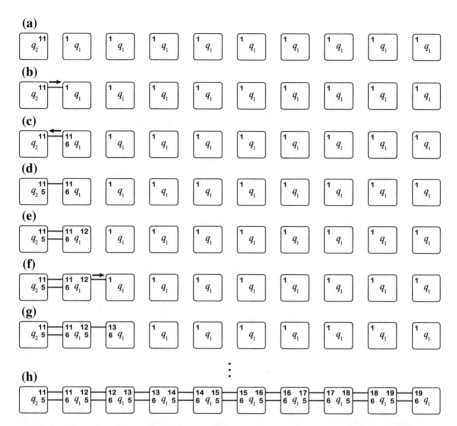

Fig. 11 Handshake protocol for constructing a linear chain of agents. **a** One initiating agent with state q_2 has a terminal in state 1 and tries to attract another agent with state q_1 to connect to. **b, c, d** Request and Acknowledgment to make one connection. **e** Double the connection. **f, g** Making a single connection to the next agent in the chain. **h** State after the linear chain of 10 agents has been constructed

An example of forming a linear patten with three agents is shown in Fig. 12 where the applied transition rules and connectivity conditions are indicated, and the number besides a terminal denotes the state of the terminal. The agents become connected through occasional collisions driven by Brownian motion; the transition rules and connectivity conditions are designed such that they can be applied in an appropriate order.

The second stage for forming a circular pattern by the agents is to curve the linear chain in a certain direction. This is accomplished by reconfiguring the connections between the agents. This step is triggered by the agent in the state q_1 at the end of the group, i.e., the agent lastly connected to a group of agents, and a signal for reconfigurations flows toward the initiating agent in the state q_2. Reconfigurations of the connections between agents are conducted every two neighboring agents in order to ensure that a circular pattern of 10 agents is made.

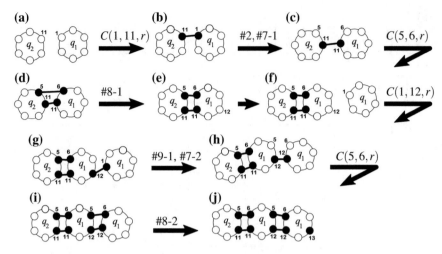

Fig. 12 Three agents forming a linear chain by transmitting terminal states to each other. State 2 on the terminals is not shown. **a** Initially, one agent in state q_2 and one in state q_1 are not connected, but when close enough, **b** become connected. **c** Preparation to make double connection, **d** which is established when terminals in state 5 and 6 are close enough. **e, f** Preparation to connect the third agent. **g** Connection with third agent established. **h** Preparation to establish double connection with third agent. **i** Double connection to third agent established. **j** Preparation to extend the chain further

Figure 13 shows a communication protocol for this stage. The rightmost agent first emits a reconfiguration signal to its left neighboring agent (Fig. 13a), and reconfiguration of connections between these agents is then conducted (Fig. 13b, c and d). The reconfiguration signal then proceeds to the left (Fig. 13e, f) but the reconnection is not conduced in these agents (Fig. 13g). Repeating this signal transmission and reconfiguration, a configuration of the terminal states is obtained as in Fig. 13h. An example of the early steps of this stage is shown in Fig. 14 with transition rules and connectivity conditions indicated explicitly.

The final stage in the pattern formation concerns the connection of both ends of the linear chain of agents. This is conducted in a similar way as the first stage in which two adjacent agents are connected. Figures 15 and 16 show a protocol used in this stage and the steps followed in this stage, respectively. A circular pattern consisting of 10 agents is completed after undergoing the three steps described above.

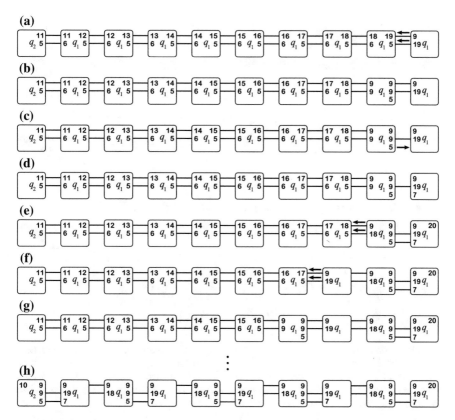

Fig. 13 Handshake protocol for curving a linear chain of agents. **a** Curving process starts at end of chain. **b** Preparation to cut one connection and establish new connection. **c, d, e** Establish new connection that facilitates curve. **f, g** Reconfiguration process skips one agent. **h** Process works backward, until the head of the chain is reached

An example of the circular pattern formation in this section is shown in Fig. 17 where two initiating agents in the state q_2 and many other agents in the state q_1 exist in the space. Initially the agents are distributed in the space as in Fig. 17a. Each of the agents fluctuates and changes its location by Brownian motion, sometimes colliding with other agents. The connections between the agents gradually grow by applying the transition rules when agents collide (Figs. 17b, c), and then they start to form circles (Fig. 17d). Finally two circles of agents are completed as in Figs. 17e, f.

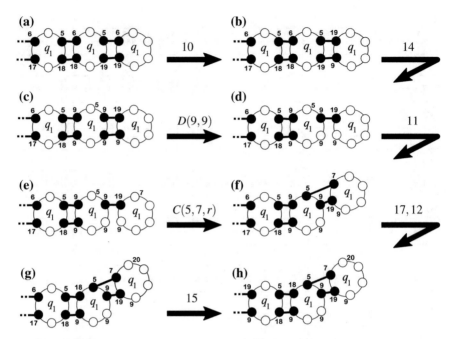

Fig. 14 Reconfiguration of connections between the agents to make a curved chain of agents. State 2 on a terminal is not shown. **a** Initial chain of three agents. **b, c** Prepare to reconfigure connection. **d** One connection is cut, **e** and preparation to establish new connection starts. **f** Double connection has been reestablished but at a shifted location. **g** Prepare to move to next agent, **h** which will not reconfigure its connection, because reconfiguration is done alternately

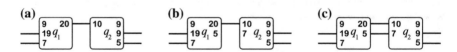

Fig. 15 Handshake protocol for connecting the two end agents in the curved linear chain. **a** Establish one connection. **b** Prepare to establish second connection. **c** Double connection between ends of agent chain established

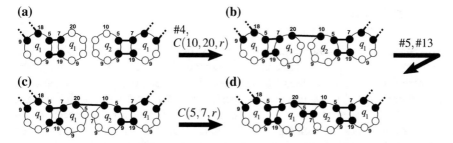

Fig. 16 Connecting both ends of the linear chain to complete the formation of the circular pattern. **a** Initially, both ends of the chain are not connected. **b** One connection is established between the ends of the chain. **c** Preparation to establish second connection. **d** Double connection between chain ends established

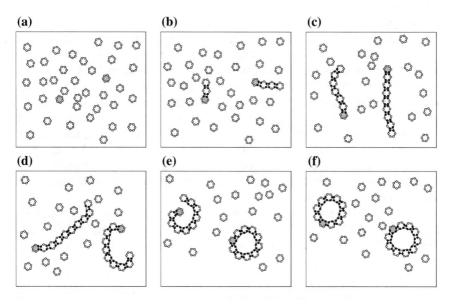

Fig. 17 Example of pattern formation whereby **a** there are two initiating agents in the space. **b** Two chains start to form. **c** Formation of chains continues. **d** Curving of chains start. **e** Both chains have been curved, and one chain has its ends connected. **f** Finally, two circular patterns have formed

5   Conclusion

This chapter presents a swarm network model, which, like cellular automata, consists of a large number of cells that contain Mealy automata, but which, unlike cellular automata, have a very flexible neighborhood. The swarm agents in a swarm network have finite internal states that can be updated in accordance with transition rules. Furthermore, swarm agents can establish new connections with other agents or cut existing connections depending on their internal states and their distance to each other. The behaviors of the agents are partially governed by Brownian motion, i.e., the agents fluctuate and accordingly change their positions. The transition rules are designed such that both the states of the agents as well as the connection patterns with other agents can be changed. Additionally, depending on the connections an agent has with other agents, the applicable transition rules change. This results in a flexible model in which the functionality of an agent depends on the terminals at which it is connected with other agents.

Two types of swarm networks have been presented in this chapter. One model can conduct universal computation by implementing so-called Brownian circuits, and the other model is able to form circular patterns by connecting with other agents. Though the implementation of universal computation in terms of Brownian circuits works in theory, in practical simulations it is quite slow since a lot of stochastic search takes place based on the Brownian motion. Nevertheless the number of states of agents has a fixed upper bound that is independent of the size of circuits. The implementation of circular pattern formation is less general, unfortunately. The size of a circle (i.e., 10 agents) is hard-coded in terms of the number of the states in an agent and its terminals. The challenge is to find a more general enumeration scheme by which patterns can be encoded such that agents and their terminals require only a constant number of states.

Finally we believe that the Swarm Network model outlined in this chapter is a potential candidate for micro/nano-meter scaled distributed computing mechanisms, like molecular robots, wherein power requirements are extremely low.

A. Transition Rules For Signal Transmission

This section provides a list of transition rules for agents, which are responsible for signal transmission in Brownian circuits. A transition rule is written in the following form:

$$(q, (t_0, \cdots, t_5), (s_0, \cdots, s_5)) \rightarrow \left(q', \left(t_0', \cdots, t_5'\right)\right),$$

where q and q' are the states of an agent before and after a transition, respectively, t_i and t_i' are the states of terminal i before and after a transition, respectively, and s_i are the states of other agents' terminals connected to the respective terminals i of the agent, or $\varnothing$ if there is no connection. The symbol $*$ for t_i and s_i represents any

Table 1 List of 21 transition rules for signal transmission in Brownian circuits

Rule #1 $(q_2, (*, *, *, *, *, *),$ $(20, \varnothing, \varnothing, \varnothing, \varnothing, \varnothing)) \rightarrow (q_2, (10, 1, 1, 0, 1, 1))$

Rule #2 $(q_1, (*, *, *, *, *, *),$ $(\varnothing, \varnothing, \varnothing, \varnothing, \varnothing, \varnothing)) \rightarrow (q_1, (2, 2, 2, 2, 2, 2))$

Rule #3 $(q_1, (2, *, *, *, *, *),$ $(1, \varnothing, \varnothing, \varnothing, \varnothing, \varnothing)) \rightarrow (q_2, (3, 0, 0, 0, 0, 0))$

Rule #4-1 $(q_2, (10, *, 1, *, 1, 1),$ $(*, *, \varnothing, \varnothing, \varnothing, \varnothing)) \rightarrow (q_2, (10, \circledast, 0, \circledast, 0, 0))$

Rule #4-2 $(q_2, (10, 1, *, *, 1, 1),$ $(*, \varnothing, *, \varnothing, \varnothing, \varnothing)) \rightarrow (q_2, (10, 0, \circledast, \circledast, 0, 0))$

Rule #4-3 $(q_2, (10, 1, 1, *, *, 1),$ $(*, \varnothing, \varnothing, \varnothing, *, *)) \rightarrow (q_2, (10, 0, 0, \circledast, \circledast, 0))$

Rule #4-4 $(q_2, (10, 1, 1, *, 1, *),$ $(*, \varnothing, \varnothing, \varnothing, \varnothing, *)) \rightarrow (q_2, (10, 0, 0, \circledast, 0, \circledast))$

Rule #5-1 $(q_2, (10, 1, 0, *, 0, 0),$ $(20, 3, \varnothing, \varnothing, \varnothing, \varnothing)) \rightarrow (q_1, (10, 4, 0, \circledast, 0, 0))$

Rule #5-2 $(q_2, (10, 0, 1, *, 0, 0),$ $(20, \varnothing, 3, \varnothing, \varnothing, \varnothing)) \rightarrow (q_1, (10, 0, 4, \circledast, 0, 0))$

Rule #5-3 $(q_2, (10, 0, 0, *, 1, 0),$ $(20, \varnothing, \varnothing, \varnothing, 3, \varnothing)) \rightarrow (q_1, (10, 0, 0, \circledast, 4, 0))$

Rule #5-4 $(q_2, (10, 0, 0, *, 0, 1),$ $(20, \varnothing, \varnothing, \varnothing, \varnothing, 3)) \rightarrow (q_1, (10, 0, 0, \circledast, 0, 4))$

Rule #6 $(q_2, (3, *, *, *, *, *),$ $(4, \varnothing, \varnothing, \varnothing, \varnothing, \varnothing)) \rightarrow (q_2, (3, 5, 5, 0, 5, 5))$

Rule #7-1 $(q_2, (3, 5, 5, 0, 5, 5),$ $(4, 20, \varnothing, \varnothing, \varnothing, \varnothing)) \rightarrow (q_2, (6, 5, 0, 0, 0, 0))$

Rule #7-2 $(q_2, (3, 5, 5, 0, 5, 5),$ $(4, \varnothing, 20, \varnothing, \varnothing, \varnothing)) \rightarrow (q_2, (6, 0, 5, 0, 0, 0))$

Rule #7-3 $(q_2, (3, 5, 5, 0, 5, 5),$ $(4, \varnothing, \varnothing, \varnothing, 20, \varnothing)) \rightarrow (q_2, (6, 0, 0, 0, 5, 0))$

Rule #7-4 $(q_2, (3, 5, 5, 0, 5, 5),$ $(4, \varnothing, \varnothing, \varnothing, \varnothing, 20)) \rightarrow (q_2, (6, 0, 0, 0, 0, 5))$

Rule #8-1 $(q_1, (10, 4, 0, 0, 0, 0),$ $(20, 6, \varnothing, \varnothing, \varnothing, \varnothing)) \rightarrow (q_1, (7, 4, 0, 0, 0, 0))$

Rule #8-2 $(q_1, (10, 0, 4, 0, 0, 0),$ $(20, \varnothing, 6, \varnothing, \varnothing, \varnothing)) \rightarrow (q_1, (7, 0, 4, 0, 0, 0))$

Rule #8-3 $(q_1, (10, 0, 0, 0, 4, 0),$ $(20, \varnothing, \varnothing, \varnothing, 6, \varnothing)) \rightarrow (q_1, (7, 0, 0, 0, 4, 0))$

Rule #8-4 $(q_1, (10, 0, 0, 0, 0, 4),$ $(20, \varnothing, \varnothing, \varnothing, \varnothing, 6)) \rightarrow (q_1, (7, 0, 0, 0, 0, 7))$

Rule #9 $(q_1, (4, *, *, *, *, *),$ $(6, \varnothing, \varnothing, \varnothing, \varnothing, \varnothing)) \rightarrow (q_1, (7, 0, 0, 0, 0, 0))$

value being accepted (so-called *don't care*), and the symbol $\circledast$ used for t_i' denotes that the state of a terminal remains unchanged under a transition. Table 1 shows the transition rules.

The connection condition $C(t_1, t_2, r)$ and disconnection condition $D(t_1, t_2)$ used in this paper are:

$$C(1, 2, r), \ C(5, 20, r),$$

and

$$D(7, 20), \ D(6, 7),$$

B. Transition Rules for Pattern Formation

This section provides a list of transition rules for agents, which are responsible for pattern formation. The notations for transition rules used in this section are the same as the ones in Appendix A. Transition rules for the formation of circles consisting of 10 agents are listed in Table 2. Besides the transition rules, it is necessary to define the connectivity conditions. The connectivity conditions for pattern formation are:

Table 2 List of transition rules for formation of circular patterns of 10 agents

Rule #1 $(q_2, (*, *, *, *, *, *),$ $(\varnothing, \varnothing, \varnothing, \varnothing, \varnothing, \varnothing)) \rightarrow (q_2, (11, 2, 2, 2, 2, 2))$

Rule #2 $(q_2, (*, *, *, *, *, *),$ $(11, \varnothing, \varnothing, \varnothing, \varnothing, \varnothing)) \rightarrow (q_2, (11, 2, 2, 2, 2, 5))$

Rule #3 $(q_2, (*, *, *, *, *, *),$ $(9, \varnothing, \varnothing, \varnothing, \varnothing, 19)) \rightarrow (q_2, (9, 2, 2, 2, 5, 9))$

Rule #4 $(q_2, (*, *, *, *, *, *),$ $(\varnothing, \varnothing, \varnothing, \varnothing, \varnothing, 19)) \rightarrow (q_2, (9, 2, 2, 10, 5, 9))$

Rule #5 $(q_2, (*, *, *, *, *, *),$ $(\varnothing, \varnothing, \varnothing, \varnothing, \varnothing, \varnothing)) \rightarrow (q_2, (9, 2, 7, 10, 5, 9))$

Rule #6 $(q_1, (*, *, *, *, *, *),$ $(\varnothing, \varnothing, \varnothing, \varnothing, \varnothing, \varnothing)) \rightarrow (q_1, (2, 2, 2, 2, 2, 2))$

Rule #7-1 $(q_1, (*, *, *, *, *, *),$ $(11, \varnothing, \varnothing, \varnothing, \varnothing, \varnothing)) \rightarrow (q_1, (11, 6, 2, 2, 2, 2))$

Rule #7-2 $(q_1, (*, *, *, *, *, *),$ $(12, \varnothing, \varnothing, \varnothing, \varnothing, \varnothing)) \rightarrow (q_1, (12, 6, 2, 2, 2, 2))$

Rule #7-3 $(q_1, (*, *, *, *, *, *),$ $(13, \varnothing, \varnothing, \varnothing, \varnothing, \varnothing)) \rightarrow (q_1, (13, 6, 2, 2, 2, 2))$

Rule #7-4 $(q_1, (*, *, *, *, *, *),$ $(14, \varnothing, \varnothing, \varnothing, \varnothing, \varnothing)) \rightarrow (q_1, (14, 6, 2, 2, 2, 2))$

Rule #7-5 $(q_1, (*, *, *, *, *, *),$ $(15, \varnothing, \varnothing, \varnothing, \varnothing, \varnothing)) \rightarrow (q_1, (15, 6, 2, 2, 2, 2))$

Rule #7-6 $(q_1, (*, *, *, *, *, *),$ $(16, \varnothing, \varnothing, \varnothing, \varnothing, \varnothing)) \rightarrow (q_1, (16, 6, 2, 2, 2, 2))$

Rule #7-7 $(q_1, (*, *, *, *, *, *),$ $(17, \varnothing, \varnothing, \varnothing, \varnothing, \varnothing)) \rightarrow (q_1, (17, 6, 2, 2, 2, 2))$

Rule #7-8 $(q_1, (*, *, *, *, *, *),$ $(18, \varnothing, \varnothing, \varnothing, \varnothing, \varnothing)) \rightarrow (q_1, (18, 6, 2, 2, 2, 2))$

Rule #7-9 $(q_1, (*, *, *, *, *, *),$ $(19, \varnothing, \varnothing, \varnothing, \varnothing, \varnothing)) \rightarrow (q_1, (19, 6, 2, 2, 2, 2))$

Rule #8-1 $(q_1, (11, 6, 2, 2, 2, 2),$ $(11, \varnothing, \varnothing, \varnothing, \varnothing, \varnothing)) \rightarrow (q_1, (\circledast, \circledast, \circledast, \circledast, 12, \circledast))$

Rule #8-2 $(q_1, (12, 6, 2, 2, 2, 2),$ $(12, \varnothing, \varnothing, \varnothing, \varnothing, \varnothing)) \rightarrow (q_1, (\circledast, \circledast, \circledast, \circledast, 13, \circledast))$

Rule #8-3 $(q_1, (13, 6, 2, 2, 2, 2),$ $(13, \varnothing, \varnothing, \varnothing, \varnothing, \varnothing)) \rightarrow (q_1, (\circledast, \circledast, \circledast, \circledast, 14, \circledast))$

Rule #8-4 $(q_1, (14, 6, 2, 2, 2, 2),$ $(14, \varnothing, \varnothing, \varnothing, \varnothing, \varnothing)) \rightarrow (q_1, (\circledast, \circledast, \circledast, \circledast, 15, \circledast))$

Rule #8-5 $(q_1, (15, 6, 2, 2, 2, 2),$ $(15, \varnothing, \varnothing, \varnothing, \varnothing, \varnothing)) \rightarrow (q_1, (\circledast, \circledast, \circledast, \circledast, 16, \circledast))$

Rule #8-6 $(q_1, (16, 6, 2, 2, 2, 2),$ $(16, \varnothing, \varnothing, \varnothing, \varnothing, \varnothing)) \rightarrow (q_1, (\circledast, \circledast, \circledast, \circledast, 17, \circledast))$

Rule #8-7 $(q_1, (17, 6, 2, 2, 2, 2),$ $(17, \varnothing, \varnothing, \varnothing, \varnothing, \varnothing)) \rightarrow (q_1, (\circledast, \circledast, \circledast, \circledast, 18, \circledast))$

Rule #8-8 $(q_1, (18, 6, 2, 2, 2, 2),$ $(18, \varnothing, \varnothing, \varnothing, \varnothing, \varnothing)) \rightarrow (q_1, (\circledast, \circledast, \circledast, \circledast, 19, \circledast))$

Rule #9-1 $(q_1, (11, 6, 2, 2, 12, 2),$ $(\varnothing, \varnothing, \varnothing, \varnothing, \varnothing, \varnothing)) \rightarrow (q_1, (\circledast, \circledast, \circledast, 5, \circledast, \circledast))$

Rule #9-2 $(q_1, (12, 6, 2, 2, 13, 2),$ $(\varnothing, \varnothing, \varnothing, \varnothing, \varnothing, \varnothing)) \rightarrow (q_1, (\circledast, \circledast, \circledast, 5, \circledast, \circledast))$

Rule #9-3 $(q_1, (13, 6, 2, 2, 14, 2),$ $(\varnothing, \varnothing, \varnothing, \varnothing, \varnothing, \varnothing)) \rightarrow (q_1, (\circledast, \circledast, \circledast, 5, \circledast, \circledast))$

Rule #9-4 $(q_1, (14, 6, 2, 2, 15, 2),$ $(\varnothing, \varnothing, \varnothing, \varnothing, \varnothing, \varnothing)) \rightarrow (q_1, (\circledast, \circledast, \circledast, 5, \circledast, \circledast))$

Rule #9-5 $(q_1, (15, 6, 2, 2, 16, 2),$ $(\varnothing, \varnothing, \varnothing, \varnothing, \varnothing, \varnothing)) \rightarrow (q_1, (\circledast, \circledast, \circledast, 5, \circledast, \circledast))$

Rule #9-6 $(q_1, (16, 6, 2, 2, 17, 2),$ $(\varnothing, \varnothing, \varnothing, \varnothing, \varnothing, \varnothing)) \rightarrow (q_1, (\circledast, \circledast, \circledast, 5, \circledast, \circledast))$

Rule #9-7 $(q_1, (17, 6, 2, 2, 18, 2),$ $(\varnothing, \varnothing, \varnothing, \varnothing, \varnothing, \varnothing)) \rightarrow (q_1, (\circledast, \circledast, \circledast, 5, \circledast, \circledast))$

Rule #9-8 $(q_1, (18, 6, 2, 2, 19, 2),$ $(\varnothing, \varnothing, \varnothing, \varnothing, \varnothing, \varnothing)) \rightarrow (q_1, (\circledast, \circledast, \circledast, 5, \circledast, \circledast))$

Rule #10 $(q_1, (19, 6, 2, 2, 2, 2),$ $(\varnothing, \varnothing, \varnothing, \varnothing, \varnothing, \varnothing)) \rightarrow (q_1, (9, 19, \circledast, \circledast, \circledast, \circledast))$

Rule #11 $(q_1, (*, *, *, *, *, *),$ $(\varnothing, 9, \varnothing, \varnothing, \varnothing, \varnothing)) \rightarrow (q_1, (9, 19, 7, 2, 2, 2))$

Rule #12 $(q_1, (*, *, *, *, *, *),$ $(\varnothing, 9, 5, \varnothing, \varnothing, \varnothing)) \rightarrow (q_1, (9, 19, 7, 20, 2, 2))$

Rule #13 $(q_1, (*, *, *, *, *, *),$ $(\varnothing, \varnothing, \varnothing, 10, \varnothing, \varnothing)) \rightarrow (q_1, (9, 19, 7, 20, 5, 2))$

Rule #14 $(q_1, (*, *, *, *, *, *),$ $(\varnothing, \varnothing, \varnothing, 19, 9, \varnothing)) \rightarrow (q_1, (9, 9, 5, 9, 9, 2))$

Rule #15 $(q_1, (*, *, *, *, *, *),$ $(\varnothing, \varnothing, \varnothing, 18, 9, \varnothing)) \rightarrow (q_1, (9, 19, 2, 2, 2, 2))$

Rule #16 $(q_1, (*, *, *, *, *, *),$ $(\varnothing, 9, \varnothing, 18, 9, \varnothing)) \rightarrow (q_1, (9, 19, 7, 2, 2, 2))$

Rule #17 $(q_1, (9, 19, *, *, *, *),$ $(\varnothing, \varnothing, \varnothing, 19, \varnothing, \varnothing)) \rightarrow (q_1, (\circledast, 18, 5, 9, 9, 2))$

$$C(5, 6, r), \quad C(5, 7, r), \quad C(1, 11, r), \ C(1, 12, r), \ C(1, 13, r), \ C(1, 14, r),$$
$$C(1, 15, r), \ C(1, 16, r), \ C(1, 17, r), \ C(1, 18, r), \ C(1, 19, r), \ C(10, 20, r),$$

and

$$D(9, 9).$$

References

1. Akyildiz, I.F., Su, W., Sankarasubramaniam, Y., Cayirci, E.: A survey on sensor networks. IEEE Commun. Mag. **40**(8), 102–114 (2002)
2. Banks, E.R.: Universality in cellular automata. In: IEEE 11th Annual Symposium on Switching and Automata Theory, pp. 194–215 (1970)
3. Chang, T.M.S.: Artificial Cells: Biotechnology, Nanomedicine, Regenerative Medicine, Blood Substitutes, Bioencapsulation. World Scientific Publishing, Cell/Stem Cell Therapy (2007)
4. Codd, E.F.: Cellular Automata. Academic Press, Orlando, FL, USA (1968)
5. Isokawa, T., Peper, F., Mitsui, M., Liu, J.Q., Morita, K., Umeo, H., Kamiura, N., N.Matsui: Computing by swarm networks. In: Proceedings of 8th International Conference on Cellular Automata for Research and Industry (ACRI2008), LNCS 5191, pp. 50–59 (2008)
6. Lee, J., Peper, F.: On Brownian cellular automata. Proc. Automata **2008**, 278–291 (2008)
7. Lee, J., Peper, F., Cotofana, S., Naruse, M., Ohtsu, M., Kawazoe, T., Takahashi, Y., Shimokawa, T., Kish, L., Kubota, T.: Brownian circuits: Designs. Int. J. Unconv. Comput. **12**(5–6), 341–362 (2016)
8. Mori, M., Isokawa, T., Peper, F., N.Matsui: On swarm networks in brownian environments. In: Proceedings of the 2nd International Symposium on Computing and Networking (AFCA'14-CANDAR'14), pp. 495–498 (2014)
9. Mori, M., Isokawa, T., Peper, F., Matsui, N.: Swarm networks in brownian environments. New Gener. Comput. **33**(3), 297–318 (2015)
10. Murata, S., Konagaya, A., Kobayashi, S., Hagiya, M.: Molecular robotics: A new paradigm for artifacts. New Gener. Comput. **31**(1), 27–45 (2013)
11. Neumann, J.V.: Theory of Self-Reproducing Automata. University of Illinois Press, USA (1966)
12. Peper, F., Lee, J., Carmona, J., Cortadella, J., Morita, K.: Brownian circuits: Fundamentals. ACM J. Emerg. Technol. Comput. Syst. **9**(1), 3:1–24 (2013)
13. Sparsø, J., Furber, S. (eds.): Handshake Protocols, chap. 2.1. Kluwer Academic Publishers, Netherlands (2001)
14. Sutherland, I.E.: Micropipelines. Commun. ACM **32**(6), 720–738 (1989)
15. Turing, A.M.: On computable numbers, with an application to the entscheidungsproblem. Proc. Lond. Math. Soc. **2**(42), 230–265 (1936)

Clean Reversible Simulations of Ranking Binary Trees

Yuhi Ohkubo, Tetsuo Yokoyama and Chishun Kanayama

Abstract We propose clean reversible simulations of ranking binary trees and unranking as reversible algorithms for reversible computing systems, which are useful for enumerating and randomly generating binary trees. Algorithms for ranking binary trees and their inverses have been studied since the 1970s. Each of these algorithms can be converted into a reversible simulation by saving all data and control information otherwise lost, and each pair of ranking and unranking reversible programs can be combined to realize a clean reversible simulation by using the Bennett method. However, such a clean reversible simulation requires multiple traversal of the given data and/or intermediate data as well as additional storage proportional to the length of the computation. We show that for Knott's ranking and unranking algorithms, additional storage usage can be reduced by using the proper assertions of reversible loops in embedded reversible simulations. We also show a clean reversible simulation that involves only one traversal. The running time and memory usage of the proposed clean reversible simulations are asymptotically equivalent to those of the original programs by Knott with intermediate garbage of constant size. In general, the derivation strategy of efficient reversible programs from irreversible ones has not yet been established, and this study can be seen as one of the case studies. All the reversible programs presented in this paper can be run on an interpreter of the reversible programming language Janus.

Y. Ohkubo · T. Yokoyama (✉) · C. Kanayama
Department of Software Engineering, Nanzan University, Nagoya, Japan
e-mail: tyokoyama@acm.org

© Springer International Publishing AG 2018
A. Adamatzky (ed.), *Reversibility and Universality*, Emergence, Complexity and Computation 30, https://doi.org/10.1007/978-3-319-73216-9_11

1 Introduction

The design of reversible algorithms on reversible computing systems is particularly facilitated when such algorithms are strictly based on reversible operations. Owing to the reversibility constraints, methods for designing and analyzing reversible algorithms should be sufficiently different from ordinary methods. There are general solutions for simulating (irreversible) algorithms on reversible computing systems [5, 10, 30, 31] that share trade-off relations in terms of the time and space complexity. However, the problem is that the size of the garbage output generated by the resulting reversible algorithms (which does not exist in the case of the original algorithms) is proportional to their execution time or the resulting reversible algorithms become space hungry, and the additional costs are typically asymptotically greater than the input size. Indeed, for each generated reversible algorithm, there is usually scope for further optimization. The reversible simulation of a (possibly irreversible) program is a reversible program, implying that all atomic operations are injective, and for any input, the reversible simulation returns the same output as the original program, if any, as well as some additional (garbage) output. It is known that for any program that computes an injective function, it is possible to perform clean (zero output garbage) reversible simulation of the program [2, 6]. However, to the best of the authors' knowledge, thus far, it is not known whether for any injective program p there is a clean reversible simulation of p that is linear-time and linear-space to the original program p. Therefore, at present, the optimization of each clean reversible simulation depends on heuristics, designer's experience, and human invention, and the reduction of the intermediate additional memory usage and the garbage output is facilitated by domain knowledge.

A tree, which is a data structure for arranging information by means of branching nodes, is one of the most widely used data structures in computer science. Its applications include representation of arithmetic expressions and analysis of algorithms [22]. In this paper, we design and analyze clean reversible simulations for ranking binary trees and unranking, which serve as the foundation of algorithms for randomly generating and enumerating binary trees [21, 22, 33]. In spite of the wide range of applications of ranking and unranking binary trees, to the best of the authors' knowledge, reversible counterparts of these basic algorithms have not been proposed yet.

According to the literature [33], the first algorithm for ranking binary trees and unranking was proposed by Knott [21]. In this paper, we consider reversible simulations of the Algol programs by Knott, and we show two clean reversible simulations. We first construct reversible simulations by embedding irreversible algorithms; then, we combine them into a clean reversible simulation by means of the (input-erasing) Bennett method [5]. While embedding saves only the data and control information otherwise lost, the worst-case garbage size of the resulting reversible simulation is proportional to its running time. To remedy the problem, we develop and use proper assertions for embedded reversible simulations. Next, we propose a one-pass clean reversible algorithm for ranking binary trees and unranking. We separate the proposed ranking and unranking algorithms into three processes, and we explain each process. We combine these processes into an efficient clean reversible simulation. The proposed reversible simulation is efficient in the sense that both its asymptotic execution time and memory usage are equal to those in the case of the original irreversible algorithm.

The complete versions of the programs of the reversible imperative programming language Janus presented in this paper can be accessed from our website: http://tetsuo.jp/ref/ranking2017/. All the programs can be run on an online Janus interpreter provided on the website.

The rest of this paper is organized as follows. Section 2 provides the background and related work. Section 3 introduces the reversible programming language Janus, which is used for describing reversible programs in this study. Section 4 presents the definition of binary trees and reviews (irreversible) ranking binary trees and unranking. Section 5 describes the proposed clean reversible simulations of reversible ranking and unranking. Section 6 presents our conclusions.

2 Related Work

One of the initial motivations for research on reversible computing was to identify and reduce the essential energy and heat dissipation required to realize computation [7, 13, 23]. By Landauer's principle, any information lost in an irreversible computation process is accompanied by (a positive amount of) heat dissipation. Whereas much more energy is consumed for transmitting and preserving information in current integrated circuits than in logical operations, the physical reversibility in such circuits at the microscopic level is an important characteristic that we have to consider in order to realize computation with further miniaturization.

The use of the benefit of physical reversibility at the logical level requires all the computation layers to be designed from reversible principles, including reversible gates [16], reversible circuits [32], reversible instruction set architecture [34], reversible machine code and programming languages (e.g., [4, 15, 19, 20, 24, 40]), reversible computation models [27], and reversible algorithms [3, 11]. Each lower level should constitute a clean reversible simulation for the higher levels. Many clean reversible simulations have been realized using various reversible computation models. Examples include reversible Turing machines (RTMs) [5] simulated by reversible logical elements [26], reversible cellular automata [28], or themselves [1], and reversible programming languages simulated by themselves [18, 37]. However, research on reversible algorithms at the highest level has not been conducted extensively thus far, with a few exceptions.

Any irreversible algorithm becomes reversible if it saves all data and control information that are lost during computation [5]. We call the translation saving operational garbage *naive embedding* from irreversible to reversible computation. There are several methods for realizing reversible simulations by embedding in terms of the various amount of time, memory usage, and lost information (e.g., [15, 31, 35] and the references therein).

In most cases, for each embedded reversible simulation, there is a more efficient reversible simulation. A two-pass clean reversible simulation can be constructed if the garbage outputs of the reversible simulations of an injective function and its inverse are the same [39]; two-pass reversible range coding has been realized by the construction. In the context of the design of reversible logic circuits, various reversible simulations have been proposed for specific families of functions (e.g., [36]). For instance, a method has been proposed for generating a reversible in-place multiplier by a constant using Mealy machines [25].

Domain knowledge enables us to construct more efficient reversible algorithms than those constructed using the above-mentioned general methods. While comparison sorts such as bubble sort and quicksort are not injective because many unsorted inputs lead to the same sorted output and inherently require garbage output, reversible comparison sorts with minimum garbage have been targeted in the literatures [3, 11, 24, 30]. Burignat et al. developed the prototype chip of the H.264[1] encoder [8]. The integer transformation in the chip is an efficient reversible transformation.

Besides the original problems in reversible computing, reversibility is a common and convenient concept in computer science. Each reversible computation model corresponds to its irreversible counterpart (e.g., [27] and the references therein). The relevant concepts appear in program inversion [17, 19, 29], and bidirectional transformations [14], including view-updating in databases.

3   Reversible Programming Language Janus

In this paper, the programs are written in the extended version of the imperative reversible programming language Janus [38] (hereafter, Janus). Janus has C-like syntax that guarantees reversibility. It is impossible to write irreversible programs in Janus; Janus programs can only denote injective functions.

We consider an example of a reversible program. The nth Catalan number $B_n = \frac{1}{n+1}\binom{2n}{n}$ is efficiently calculated by the initial value $B_0 = 1$ and a recurrence relation as follows:

$$B_n = 4B_{n-1} - \frac{6B_{n-1}}{n+1} \quad (n \geq 1). \tag{1}$$

The following reversible procedure takes zero-cleared[2] array B[0:n-1] and a positive integer n, and assigns Catalan sequences $B_0, B_1, \ldots, B_{n-1}$ to the array B[0:n-1]:

---

[1] A video compression standard.

[2] A variable is *zero-cleared* if it is set to zero, and an array is *zero-cleared* if all its elements set to zero.

```
1 procedure mk_catalan_tbl(int B[],int n)
2    B[0] ^= 1
3    local int i = 0
4       from  i = 0 do
5          i += 1
6       loop
7          B[i] ^= 4*B[i-1] - 6*B[i-1]/(i+1)
8       until i = n
9    delocal int i = n
```

The syntax and semantics of Janus are as follows. The base types are integers as well as arrays and stacks of integers. The Algol-like simple assignment x := 1 is prohibited. Instead, we must use C-like complex assignment operators +=, -=, or ^= (e.g., at lines 2, 5, and 7), under the condition that for the compound assignment expression $x \oplus = e$, the object pointed by the left operand x must not appear in the right operand e, and when the left operand is the array expression $x[e_i] \oplus = e$, the object pointed by $x[e_i]$ must appear neither in the right operand e nor in the index e_i. This condition syntactically excludes non-injective statements, such as x -= x, which would zero-clear x regardless of its value. The stack operations push and pop are inverse to each other. Push push(x,g) places the value of the variable x on the top of the stack g and makes the variable x zero-cleared. When the element to be pushed is constant, i.e., push(c,g), the value of the constant c is placed on the stack g. All the pop operations in this paper are performed in the inverse invocations of procedures (by **uncall**) with the push operations. The top element of the stack g is obtained by evaluating the expression top(g). At each join point of control flow statements, the associated assertion guarantees reversibility. The reversible selection statement **if** e_1 **then** s_1 **else** s_2 **fi** e_2 behaves like the selection statement of C, except that the value of the assertion e_2 must be true after the execution of the clause **then** and false after the execution of the clause **else**. The reversible loop statement **from** e_1 **do** s_1 **loop** s_2 **until** e_2 checks the assertion e_1 to be true on entry (otherwise, it halts abnormally) and executes the statement s_1. Until e_2 becomes true, it keeps executing s_2, asserting e_1 to be false, and executing s_1. The clauses **then, else, do**, and **loop** are abbreviated only if they have empty statements. The clause **local** allocates the specified local variables and initializes them with specified values. Each **local** clause is paired with the corresponding clause **delocal**, which asserts that the allocated variable has the value of the specified expression and removes the variable from the scope. The **call** statement invokes the procedure with arguments passed by reference. The **uncall** statement invokes the procedure of the **call** statement inversely, i.e., the inverse statements of the body of the procedure are executed in the inverse order. Alternatively, we can always translate any Janus procedure to its inverse procedure of the same time and space complexity by a recursive descent translator [37].

The procedure mk_catalan_tbl repeats the **loop** branch n times; thus, its execution time is $O(n)$. If $B[0:n-1]$ has the sequence $B_0, \ldots, B_{n-1}$, the inverse procedure call **uncall** mk_catalan_tbl(B,n) makes $B[0:n-1]$ zero-cleared.

For any function $f : X \to Y$, there exists an injective function $f' : X \to Y \times G$ such that for any $x \in X$, $fst(f'(x)) = f(x)$ for some G. Here, the function fst is the projection $fst(x, y) = x$. Given a (possibly non-injective) function f, making an injective function f' satisfy the condition is called *injectivization*. The elements of G are called the *garbage output*. If the type of G is stack, the stacks in G are called *garbage stacks*.

Given a programming language L, L's semantic function is denoted by $[\![\cdot]\!]^{L}$, which associates with every L-program p a corresponding partial function $[\![p]\!]^{L}$. A program q of the reversible language R is called the *reversible simulation* of the program p of the (possibly irreversible) language IR, iff for any x, we have $fst([\![q]\!]^{R}(x)) = [\![p]\!]^{IR}(x)$, i.e., the function $[\![q]\!]^{R}$ is an injectivization of $[\![p]\!]^{IR}$. The reversible simulation q is said to be *clean* if the type of the garbage output of $[\![q]\!]^{R}$ is a singleton set. For example, the reversible simulation q of the program p by the input-saving Bennett construction [6, Lemma 1] is not clean, because it returns the original input as garbage: $[\![q]\!]^{R}(x) = ([\![p]\!]^{IR}(x), x)$.

It is possible to construct a reversible simulation for any irreversible program by embedding the computation history of deleted data and lost control information into a garbage stack. For example, the procedure catalanE(b,g), which calculates the bth Catalan number and assigns it to the variable b, saving unnecessary data to the garbage stack g, is realized by embedding[3]:

```
 1  procedure catalanE(int b, stack g)
 2    if b=0 then
 3      b ^= 1                              // Set b to B₀
 4      push(0,g)
 5    else
 6      local int n = b
 7        b -= 1
 8        call catalanE(b,g)
 9        push(b,g)
10        b ^= 4*top(g) - 6*top(g)/(n+1)   // Compute Eq. 1
11        push(n,g)
12      delocal int n = 0
13      push(1,g)
14    fi top(g)=0
```

The garbage stack g is used to remember the control information. The assertion top(g)=0 in the selection statement at line 14 always holds, because of the control information pushed on the garbage stack g at the last lines of the **then** and **else** clauses (lines 4 and 13).

The garbage stack g is also used to remember the otherwise lost data. In Janus, simple assignments, such as b := 4*b - 6*b/(n+1), which directly correspond to the recurrence relation (Eq. 1), are not allowed because of the syntactic restriction on the reversible assignments. The previous Catalan number is pushed on the garbage stack g, the new Catalan number is calculated by the recurrence relation (Eq. 1) at line 10, and stored in the variable b. Furthermore, the value of the local variable n, which remembers what Catalan number the procedure calculates, is moved to the garbage stack g immediately before the deallocation at line 12.

---

[3]In this paper, reversible procedures generated by embedding have the suffix E, e.g., catalanE.

As such, it is always possible to perform reversible simulation by embedding both deleted data and lost control information into garbage stacks. The problem is that, given the value b, the additional space complexity to the original irreversible program is proportional to the time complexity $O(b)$.

In this paper, we use the following notation. Code is denoted by `typewriter` `font`. Reserved words in code are denoted in **`bold typewriter font`**. Mathematical expressions, values of the semantic domain, and metavariables are denoted by *italic font*. For example, the value of the variable `i` is denoted by i. The array `p` of length n is denoted by `p[0:n − 1]`. Given an array `p[0:n − 1]` (in typewriter font), $p[i : j]$ (in math font) is a sequence of the values of `p[i]`, `p[i+1]`, ..., `p[j]` if $0 \leq i \leq j < n$, and an empty sequence if $i > j$. The value of the array that has s in all its elements is denoted by $(s, \ldots, s)$. The value of the subarray that has $p[k] - s$ in the kth element for $i \leq k \leq j$ is denoted by $p[i : j] - (s, \ldots, s)$. Similarly, the value of the subarray that has $p[k] + s$ in the kth element for $i \leq k \leq j$ is denoted by $p[i : j] + (s, \ldots, s)$.

For simplicity, we assume that the elements of the array `B[0:n − 1]` that contain the Catalan numbers $B_0, B_1, \ldots, B_{n-1}$ can be accessed anywhere in the Janus programs in this paper.

4 Binary Trees and Their Ranks

In this section, first, we present the definition of (unlabeled) binary trees and their orders. Next, we introduce a method for computing the ranks of binary trees that correspond to the indices of the sequence of binary trees in the natural order, and conversely, a method for computing the binary trees from their ranks. Furthermore, we explain a representation of binary trees that we use in this paper, i.e., tree permutation. The content of this section is based on [21]. The purpose of the explanation is to construct efficient reversible simulations of ranking and unranking binary trees in the next section.[4]

4.1 Binary Trees

An *(unlabeled) binary tree* T is a structure defined on a finite set of nodes that either is an empty set, or is composed of three disjoint sets of nodes: a *root* node and two binary trees called *a left subtree* and *a right subtree*. We denote the *left and right subtrees* of T by $l(T)$ and $r(T)$, respectively. The number of nodes of the binary tree T is denoted by $|T|$. In particular, T is called an *empty tree* if $|T| = 0$. The number of nodes of the binary tree T is called the *size* of the binary tree T. If the

---

[4]The indices of the tree permutations (see Sect. 4.4) in [21] lie in the range one or above, but in this paper, they lie in the range zero or above for simplicity.

left subtree $l(T)$ is not empty, its root is called the *left child* of the root of the tree T. Similarly, if the right subtree $l(T)$ is not empty, its root is called the *right child* of the root of the tree T. In contrast to an ordered tree, even when a node has only one child, we distinguish whether the child is left or right.

We define the equivalence relation and orders on binary trees. Two trees are called *identical* if either both are empty trees or their left subtrees and right subtrees are, respectively, identical. We denote $T_1 \equiv T_2$ if T_1 and T_2 are identical. The binary relation $T_1 \prec T_2$ of binary trees is recursively defined as follows:

(i) $|T_1| < |T_2|$,
(ii) $|T_1| = |T_2|$ and $l(T_1) \prec l(T_2)$, or
(iii) $|T_1| = |T_2|$, $l(T_1) \equiv l(T_2)$, and $r(T_1) \prec r(T_2)$.

The order $\prec$ is called the *natural order* on binary trees. In this paper, we use the natural order to compare the sizes of binary trees that may have different sizes.

We denote by $\mathcal{T}(n)$ a set of n-node binary trees:

$$\mathcal{T}(n) = \left\{ T \,\middle|\, |T| = n \right\}. \tag{2}$$

The number of elements of $\mathcal{T}(n)$, i.e., the number of n-node binary trees, is the nth Catalan number B_n.

We denote by G_{jn} the number of binary trees that have n nodes ($n \geq 1$) and whose left subtrees have j nodes ($j \geq 0$). The size of right subtrees of such binary trees is $n - j - 1$. G_{jn} is the product of the numbers of the left subtrees of size j and the right subtrees of size $n - j - 1$:

$$G_{jn} = B_j \times B_{n-j-1}. \tag{3}$$

4.2 Ranking Binary Trees

The rank $rank'(T)$ of the binary tree T is the number of binary trees of size $|T|$ that are smaller than T in the natural order[5]:

$$rank'(T) = \#\{T' \mid T' \prec T, T' \in \mathcal{T}(|T|)\}. \tag{4}$$

Specifically, given a sequence of binary trees of the same size in the natural order, the rank of each binary tree is equal to its index in the sequence starting from zero. The value of $rank'(T)$ is an integer lying in the range of zero to $B_{|T|} - 1$. For example, in Fig. 1, the binary trees of size four are exhaustively enumerated in the natural order. We can check that the binary trees of rank three and rank four are aligned in the natural order by recursively checking conditions (iii), (ii), (ii), and (i) in that order. Here, checking condition (ii) for the second time requires the empty tree to be smaller

---

[5]$\#X$ is the number of elements in the set X.

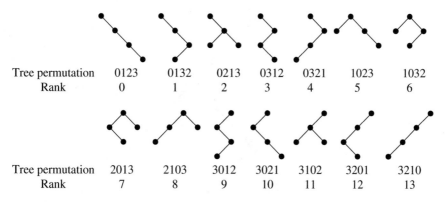

Tree permutation	0123	0132	0213	0312	0321	1023	1032
Rank	0	1	2	3	4	5	6

Tree permutation	2013	2103	3012	3021	3102	3201	3210
Rank	7	8	9	10	11	12	13

Fig. 1 Binary trees of size four in the natural order

than a one-node tree, which is checked by condition (i). The tree permutations in Fig. 1 will be explained in Sect. 4.4.

The function $rank'(T)$ can be calculated recursively:

$$rank'(T) = \textbf{if } |T| = 0 \textbf{ then } 0 \textbf{ else}$$
$$\left(\sum_{0 \le j < |l(T)|} G_{j,|T|} \right.$$
$$+ rank'(l(T)) \times B_{|r(T)|}$$
$$\left. + rank'(r(T)) \right). \tag{5}$$

The term of the **then** clause is zero, which implies that there is no binary tree that is smaller than the empty tree. The **else** clause returns the sum of the numbers of the binary trees T' that satisfy

(a) $|l(T')| < |l(T)|$,
(b) $|l(T')| = |l(T)|$ and $l(T') \prec l(T)$, and
(c) $l(T') \equiv l(T)$ and $r(T') \prec r(T)$,

respectively. The size of the left subtrees $|l(T')|$ that satisfy condition (a) lies in the range of $j = 0, 1, \ldots, l(T) - 1$. The first term is the sum of the sizes, i.e., $G_{j,|T|}$ for the range. The second term is the product of the number of left subtrees $l(T')$ that satisfy $l(T') \prec l(T)$ and the number of possible right subtrees $B_{|r(T)|}$ corresponding to the left subtrees. The third term is the number of right subtrees $r(T')$ that satisfy $r(T') \prec r(T)$. The conditions (a), (b), and (c) are mutually exclusive, and the sum of the binary trees that satisfy the conditions is equal to the number $rank'(T)$ of binary trees of the same size as T, which are smaller than T in the natural order.

For example, the binary tree in Fig. 2 has a left subtree of rank zero and a right subtree of rank seven from Fig. 1. Therefore, the rank of the binary tree in Fig. 2 is 49 ($= G_{0,6} + 0 \times B_4 + 7$).

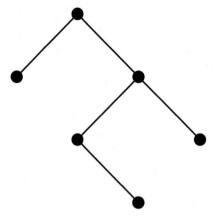

Fig. 2 Binary tree of rank 49 (corresponding to the tree permutation 104235)

Binary trees of different sizes may have the same rank, but non-identical binary trees of the same size do not have the same rank; for any n, $rank' : \mathcal{T}(n) \to \{0, 1, \ldots, B_n - 1\}$ is injective. Hence, the function returning a pair of the size and rank of the given binary trees,

$$rank(T) = (|T|, rank'(T)),\tag{6}$$

is injective. In the next section, we consider a clean reversible program that computes *rank*.

4.3 Constructing Binary Trees from Ranks

Because *rank* is injective, there is an inverse function of *rank*. Given the size $n = |T|$ and the rank r of T, the inverse function *unrank*(n, r) of *rank*(T) can be computed by using the recursive definition of *rank'*(T) in Eq. 5 as follows.

If $n = 0$ and $r = 0$, they are returned by the **then** clause, and T is an empty tree.

We consider the case when the rank is returned by the **else** clause. For any $n \ (\geq 1)$, we have $G_{jn} > 0$ if $0 \leq j \leq n - 1$, and we have $0 \leq rank'(l(T))$ and $0 \leq rank'(r(T)) < B_{|r(T)|}$. Therefore, dividing the sum of the second and third terms uniquely returns the quotient $rank'(l(T))$ and the remainder $rank'(r(T))$. The maximum of the sums of the second and third terms $(B_{|l(T)|} - 1) \times B_{|r(T)|} + (B_{|r(T)|} - 1)$ is $G_{|l(T)|,|T|} = B_{|l(T)|} \times B_{|r(T)|}$ minus one. Hence, j becomes $|l(T)|$ when the sum of the first, second, and third terms becomes less than $G_{j,|T|}$ after repeating the subtraction of $G_{j,|T|}$ from the sum for $j = 0, 1, \ldots$. We obtain $|r(T)|$ by $|r(T)| = |T| - |l(T)| - 1$. Then, we can simply construct T from $l(T)$ and $r(T)$, which are recursively obtained by $rank'(l(T))$ and $rank'(r(T))$, respectively.

It should be noted that the values returned by the **then** clause and the **else** clause of $rank'(T)$ are not exclusive. Even if the rank is zero, when it is not an empty tree, the binary tree must be consumed by the **else** clause.

4.4 Representing Binary Trees in Tree Permutation

In the programs in this paper, like [21], (unlabeled) binary trees are represented by tree permutation. Tree permutations are the permutations of a set $\{0, 1, \ldots, n-1\}$ that satisfy certain conditions (not all permutations are tree permutations). The *tree permutation* $p[0 : n-1]$ either is an empty sequence, which represents an empty tree with zero nodes ($n = 0$), or consists of three parts $s p_l p_g$, i.e., a natural number s, a tree permutation p_l, which contains elements less than s, and a sequence p_g, which becomes a tree permutation if we subtract $s + 1$ from each of the elements. In the latter case, s, p_l, and p_g correspond to the root, left subtree, and right subtree, respectively, and s corresponds to the size of the left subtree. Tree permutations may be stored in a subarray $p[m : n]$. In particular, we denote by an empty tree by $p[a : a-1]$.

For example, the binary tree of rank 49 in Fig. 2 corresponds to the tree permutation 104235, in which the head 1 corresponds to the root and means the size of the left subtree, and thus the tree permutation 0 of length one corresponds to the left subtree, and 2013 (in each element, the head 1 plus one, i.e., two, is subtracted from 4235) corresponds to the right subtree. Next, in the right subtree 2013, the head 2 corresponds to the root and means the size of the left subtree 01, and the tree permutation 0 obtained by subtracting the head two and one from each element of the sequence 3 corresponds to the right subtree. Further, in the left subtree 01, the head 0 corresponds to the root meaning that it does not have the left subtree, and the tree permutation 0 obtained by subtracting one from each element of the sequence 1 corresponds to the right subtree. The tree permutation 0 consists of a node and two empty trees.

In this study, we identify any binary tree with its tree permutation. In the following, we represent an (unlabeled) binary tree T by the tree permutation $p[n : m]$, and we write $T \simeq p[n : m]$ if the binary tree T corresponds to the tree permutation $p[n : m]$.

If we adjust the input and output variables of ranking and unranking programs in [21] and decrease the ranks of the inputs and outputs by one, we can obtain the procedures rankA and unrankA.[6] We assume that $T \simeq p[0 : n-1]$. Then, we have

$$[\![\texttt{rankA}]\!]^{\texttt{Algol}} : (p[0 : n-1], n) \mapsto (n, r), \tag{7}$$

where $r = rank'(T)$. Moreover, we have

$$[\![\texttt{unrankA}]\!]^{\texttt{Algol}} : (n, r) \mapsto (p[0 : n-1], n), \tag{8}$$

---

[6]The suffix A indicates that the procedures are implemented in A̲lgol.

where $unrank(n, r) \simeq p[0 : n - 1]$. In the next section, we develop reversible simulations of the ranking and unranking binary trees, rankA and unrankA, respectively.

5 Ranking Binary Trees Reversibly

In this section, by following the general reversibilization method by Bennett [5], we first obtain reversible simulations of ranking and unranking binary trees by embedding, and merge them into a multi-pass clean reversible simulation. Then, we propose a more efficient reversible simulation, and analyze its complexity.

5.1 Ranking Binary Trees by Embedding

The Janus reversible procedure rankE at lines 38–40 in Fig. 3 is realized by embedding the computation history of rankA. The computation history consists of the unnecessary values in local variables and the original tree permutation. The arguments of the procedure call rankE(p, n, r, g) are an array of a tree permutation p, the length of the tree permutation n, a variable to store the resulting rank r, and a garbage stack to store garbage information during computation g.

The subprocedure rE(p, n, a, s, r, g) at lines 2–35 in Fig. 3 computes the rank of the tree permutation, which is obtained by subtracting the offset s from all the elements of $p[a : a + n - 1]$. After the procedure call rE stores the rank to the variable r, all the elements of p[0 : n − 1] that are not necessary for the original output are moved to the garbage stack g by the procedure call clear_array(p, n, g). Here, the reversible procedure clear_array, defined in Fig. 4, moves the values of the array p[0 : n − 1] to the garbage stack g.

The rank is computed in the subprocedure rE, which consists of the memory allocation of the local variable j, the code corresponding to the above conditions (a), (b), and (c), and the deallocation of the local variable j. Here, we assume that $T \simeq p[a : a + n - 1] - (s, \ldots, s)$. The local variable r_rT stores an intermediate value for computing the rank, and the local variable j stores the number of nodes $|l(T)|$ in the left subtree T. Further, the value of the expression p[a]-s at line 3 is the head of the tree permutation $p[a : a + n - 1] - (s, \ldots, s)$, i.e., the size of the left subtree T.

The program in Fig. 3 is obtained not only by embedding but also by optimization using local control and data flow analysis. For example, at line 8, the compound assignment operator += reversibly overwrites the value that is irreversibly assigned in the original program. Moreover, only one local variable r_rT in addition to the garbage stack g is used to remember an intermediate rank. Furthermore, the assertions of the selections and the loop are composed of variables and arrays that are available

```
 1  // Set the rank of tree permutation p[a:a+n-1]-(s,...,s) to r
 2  procedure rE(int p[],int n,int a,int s,int r,stack g)
 3    local int j = p[a] - s           // j = |l(T)|
 4      // (a) Place encoded |l(T)| to r
 5      local int v = 0
 6        local int i = 0
 7          from i = 0 loop
 8            v += B[i] * B[n-i-1]  // Add G_in to r
 9            i += 1
10          until i = j
11        delocal int i = j
12        r += v
13        push(v,g)
14      delocal int v = 0
15
16      // (b) Add rank'(l(T)) to r
17      if j > 1 then                   // if |l(T)| > 1
18        local int r_lT = 0
19          call rE(p,j,a+1,s,r_lT,g)
20          r += B[n-j-1] * r_lT
21          push(r_lT,g)
22        delocal int r_lT = 0
23      fi j > 1
24
25      // (c) Add rank'(r(T)) to r
26      if n-j-1 > 1 then               // if |r(T)| > 1
27        local int r_rT = 0
28          call rE(p,n-j-1,a+j+1,s+j+1,r_rT,g)
29          r += r_rT
30          push(r_rT,g)
31        delocal int r_rT = 0
32      fi n-j-1 > 1
33
34      push(j,g)
35    delocal int j = 0
36
37  // Set the rank of tree permutation p[0:n-1] to r
38  procedure rankE(int p[],int n,int r,stack g)
39    call rE(p,n,0,0,r,g)
40    call clear_array(p,n,g)          // Clear array p
```

Fig. 3 Ranking tree permutations $p[0 : n - 1]$ using a garbage stack g

```
 1  procedure clear_array(int p[],int n,stack g)
 2    local int i = 0
 3      from i = 0 do
 4        local int t = p[i]
 5          p[i] ^= t // Clear p[i]
 6          push(t,g)
 7        delocal int t = 0
 8        i += 1
 9      until i = n
10    delocal int i = n
```

Fig. 4 Clearing array $p[0 : n - 1]$ using a garbage stack g

locally. In the following subsections, those types of heuristic optimization are applied without any explanation.

These reversible procedures have the same asymptotic number of the executed statements as the original Algol procedures rankA for any input, owing to the same control flow. However, at lines 13, 21, 30, and 34 the originally deleted data is pushed onto the garbage stack g.

It is possible to optimize the memory usage at the expense of additional execution time. We can replace push(r_rT,g) at line 30 with the inverse procedure call **uncall** rE(p,j,a+1,s,r_rT,g), which negates the effect of the procedure call at the previous-but-one line. This realization conforms to the call-copy-uncall format of the local Bennett method [37]. However, each time the procedure is called or uncalled, one procedure call and one procedure uncall are invoked; thus, the number of procedure invocations increases exponentially.

Even if we apply the above optimization for the memory usage, in order to obtain a clean reversible program of *rank*, we have to zero-clear the array p without using the garbage stack g. Such optimization in the procedure rankE requires at least another traverse on the array p. It should be noted that, in general, it might be much more difficult to reversibly delete intermediate data than to compute the original output (e.g., the factorization of two prime numbers looks more difficult than their multiplication). Therefore, it is not possible to achieve an efficient reversible simulation only by replacing the procedure call clear_array.

We ensure that the Janus procedure rankE is the reversible simulation of the Algol procedure rankA on the basis of the following fact. We consider the case in which the tree permutation $p[0 : n − 1]$ is the binary tree T of size n. For this case, we have

$$[\![\mathtt{rankE}]\!]^{\mathrm{Janus}} : (p[0 : n − 1], n) \mapsto ((n, r), g), \qquad (9)$$

where $r = rank'(T)$. Here, we abbreviate the constant inputs on r and g and outputs on $p[0 : n − 1]$, and the original output is represented in the nested pair (n, r). Because for any n and $p[0 : n − 1]$, we have

$$fst([\![\mathtt{rankE}]\!]^{\mathrm{Janus}}(p[0 : n − 1], n)) = [\![\mathtt{rankA}]\!]^{\mathrm{Algol}}(p[0 : n − 1], n), \quad (10)$$

the reversible Janus program rankE is the reversible simulation of the Algol program rankA.

5.2   Unranking Binary Trees by Embedding

The Janus reversible procedure unrankE at lines 55–57 in Fig. 5 is realized by embedding the computation history of unrankA. The procedure unrankE(p,n,r,g) computes the tree permutation $p[0 : n − 1]$ from the rank r and the size of binary trees n. By using the garbage stack g, the procedure stores the tree permutation corresponding to the rank r and the size n to the initially

```
 1 // Store unrank(r,n)+(s,...,s) to p[a:a+n-1]
 2 procedure riE(int p[],int n,int a,int s,int r,stack g)
 3     // {j = 0,n = |T|,r = rank'(T)}
 4     local int j = 0
 5     local int k = 0
 6     local int v = 0
 7     local int h = 0
 8         // (a) Extract |l(T)| from encoded r, and set it to j
 9         from j = 0 loop
10             push(k,g) // Clear k
11             k ^= n-j-1
12             push(h,g) // Clear h
13             h ^= B[k]
14             push(v,g) // Clear v
15             v ^= B[j] * h
16             r -= v    // Subtract G_jn from r
17             j += 1
18         until r < 0
19         r += v
20         j -= 1
21         // {j = |l(T)|,n = |T|,r = rank'(l(T)) × B_{|r(T)|} + rank'(r(T))}
22
23         p[a] ^= j + s // set a root node
24
25         // (b) Construct l(T)
26         local int r_lT = r / h      // r_lT = rank'(l(T))
27             if j > 1 then
28                 call riE(p,j,a+1,s,r_lT,g)
29             else
30                 if j = 1 then          // if |l(T)| = 1
31                     p[a+1] ^= s
32                 fi j = 1
33             fi j > 1
34             push(r_lT,g)
35         delocal int r_lT = 0
36
37         // (c) Construct r(T)
38         local int r_rT = r-1 - ((i-1) / h) * h  // r_rT = rank'(r(T))
39             if k > 1 then
40                 call riE(p,k,a+j+1,s+j+1,r_rT,g)
41             else
42                 if k = 1 then          // if |r(T)| = 1
43                     p[a+j+1] ^= s+j+1
44                 fi k = 1
45             fi k > 1
46             push(r_rT,g)
47         delocal int r_rT = 0
48         push(h,g); push(v,g); push(k,g); push(j,g) // Clear variables
49     delocal int h = 0
50     delocal int v = 0
51     delocal int k = 0
52     delocal int j = 0
53
54 // Compute tree permutation p[0:n-1] from rank r
55 procedure unrankE(int p[],int n,int r,stack g)
56     call riE(p,n,0,0,r,g)
57     push(r,g)
```

Fig. 5 Computing a tree permutation $p[0 : n - 1]$ from a rank r using a garbage stack g

zero-cleared array p[0 : n − 1]. The procedure riE at lines 2–52 stores to the sub-array $p[a : a + n − 1]$ the sequence consisting of the tree permutation of length n incremented by s.

The code corresponding the condition (a) computes the size of the left subtree from the rank r, and stores it to j. If the rank of T is passed to r, $G_{j,|T|}$ is repeatedly subtracted from r for $j = 0, 1, \ldots, |l(T)| − 1$, and eventually r becomes negative. Then, after the loop terminates, j becomes $|l(T)|$ and r becomes the number of subtrees satisfying the conditions (b) and (c).

All variables to be deallocated in the body need to be cleared by pushing their contents on the garbage stack g. Since the code corresponding to condition (a) contains a loop in which unnecessary data is pushed on to the garbage stack, the required additional memory size to maintain the garbage is $O(n^2)$. After the procedure call riE, the variable r takes a value that is neither the original rank nor 0. By moving the value to the garbage stack g at line 57, the variable r is zero-cleared.

To obtain the reversible simulation without using the garbage stack, it is not sufficient to rewrite the code via local analysis. For example, immediately before push (line 57) in the body of unrankE, the value of r is temporary and depends on the content of the tree permutation p. Analysis of the content is necessary to zero-clear r.

As in the previous subsection, where we had seen that rankE is the reversible simulation of the Algol procedure rankA, here, we can see that unrankE is the reversible simulation of the Algol procedure unrankA as follows: For any n-node binary trees of rank r, we have

$$[\![\text{unrankE}]\!]^{\text{Janus}} : (n, r) \mapsto ((p[0 : n − 1], n), g), \tag{11}$$

and $T \simeq p[0 : n − 1]$ where $T = unrank(n, r)$. Here, we again abbreviate the constant inputs on $p[0 : n − 1]$ and g and outputs on r, such as 0 and *nil*, and the original outputs are represented in the nested pair $(p[0 : n − 1], n)$. For any n and r, we have

$$fst([\![\text{unrankE}]\!]^{\text{Janus}}(n, r)) = [\![\text{unrankA}]\!]^{\text{Algol}}(n, r). \tag{12}$$

Therefore, the reversible procedure unrankE is the reversible simulation of the Algol procedure unrankA.

5.3   Clean Reversible Simulation by the Bennett Method

It is known that the general reversible simulation, called the (input-erasing) Bennett method, realizes the clean reversible simulation by combining the reversible simulations of an injective function and its inverse function [5]. By following this method, it is possible to construct a clean reversible simulation by combining two naively embedded reversible simulations, which are inverse to each other. Similarly, we can construct a clean reversible simulation rankB, as shown in Fig. 6, by combining the

```
 1 procedure rankB(int p[],int n,int r)
 2   local stack g = nil
 3     call rankE(p,n,r,g)
 4   local int t = r
 5     uncall rankE(p,n,t,g)   // Clear t and g
 6   delocal int t = 0
 7   local int q[n] = {0}      // Clear all the elements
 8     call unrankE(q,n,r,g)
 9     uncall copy(p,q,n)      // Clear q
10   delocal int q[n] = {0}
11     uncall unrankE(p,n,r,g) // Clear p and g
12   delocal stack g = nil
```

Fig. 6 Input-erasing Bennett reversible simulation of ranking using our embedded reversible simulations

```
 1 procedure copy(int x[],int y[],int n)
 2   local int i = 0
 3     from i = 0 do
 4       y[i] ^= x[i]
 5       i += 1
 6     until i = n
 7   delocal int i = n
```

Fig. 7 Reversibly copying $x[0 : n - 1]$ to zero-cleared $y[0 : n - 1]$

embedded reversible simulations rankE and unrankE, which are reversible programs that compute the injective function *rank* and its inverse *unrank*, respectively. Specifically, given the tree permutation $p[0 : n - 1]$, its rank is stored in r; then, the array $p[0 : n - 1]$ is zero-cleared.

The rank corresponding to the tree permutation $p[0 : n - 1]$ is stored to r by the procedure call rankE at line 3, while the garbage information generated during the computation is stored to g. The rank copied to t and the garbage stored on the garbage stack g are zero-cleared by the inverse procedure invocation at line 5. The procedure call unrankE at line 8 stores the garbage information to g and sets to the variable q the permutation corresponding to the rank stored in r.

The procedure copy, defined in Fig. 7, reversibly copies $x[0 : n - 1]$ to $y[0 : n - 1]$ using the compound assignment operator of exclusive or. Therefore, if all the elements of $x[0 : n - 1]$ and $y[0 : n - 1]$ are the same, we zero-clear $y[0 : n - 1]$. The behavior of the inverse procedure call copy is exactly the same as that of its procedure call. At line 9 of Fig. 6, the uncalled procedure copy(p, q, n) zero-clears the array q. Similarly, at line 11, the inverse invocation of unrankE makes p zero-cleared and g nil-cleared.

As a result, the local variables t, g, and q, and the array for the input data $p[0 : n - 1]$, are all zero- or nil-cleared after the execution.

The (input-erasing) Bennett method requires four-time procedure invocation, i.e., forward and backward invocation of the reversible simulations of a program and its inverse, as well as traversal of the output data when copying. It also requires memory for storing the garbage data and intermediate data. Therefore, the Bennett method is inefficient in terms of both execution time and memory usage.

5.4   Efficient Clean Reversible Simulation of Ranking Binary Trees

We simplify the programs in Figs. 3 and 5 by arranging the code blocks of rE in the inverse order of riE. This is because of the insight that the execution of unranking uses similar conditions as the inverse execution of ranking.

In the procedure rE in Fig. 3, the modified variables are only r and g among the ones passed by the formal parameters. The code blocks corresponding to the conditions (a), (b), and (c) do not use the variable r in the right hand sides of assignments, the tests and assertions, and the initializers and finalizers, and each computed value in each block is added to r. Because of the commutativity of plus, changing the order of the code blocks only affects the order of the garbage values pushed on g, and it does not change the resulting value of r. Therefore, we change the order to the inverse order (c), (b), and (a). Then, after the code blocks (c) and (b) are computed in order, push(r_lT,g) at line 21 is no longer needed. Instead, we can deallocate r_lT by the information that $r = rank'(l(T)) \times B_{|r(T)|} + rank'(r(T))$, $r\_lT = rank'(l(T))$, and $B[n - j - 1] = B_{|r(T)|}$, and hence, $r\_lT = r/B[n - j - 1]$.

In code block (a), the values G_{jn} are first added to v and then r. However, the allocation/deallocation of the variable v is not necessary if we directly add the values G_{jn} to r. Moreover, the final value of r does not change by the commutativity of plus, if instead of adding $G_{0n}, G_{1n}, \ldots, G_{p[a]-s-1,n}$ to r in order we add them in the inverse order.

The rank r does not change if the procedure rE is called for the tree of size zero ($n = 0$). Therefore, we change the procedure body such that the body of the procedure is computed only if $n \geq 1$.

Then, computing the then branches in code blocks (b) and (c) even when the sizes of the left and right children are zero or one does not change the final ranks. Thus, we replace the conditions with their bodies.

Now, the modified procedure rE is supposed to be initially called with r being zero. Therefore, the final rank obtained does not change if we do not use the local variable r_rT and directly pass r as an actual argument.

In addition to the above modification, by changing the procedure name to rE', we obtain the program in Fig. 8.

Next, we consider simplifying riE in Fig. 5.

In code block (a), we use the test of the loop checking that r becomes negative if we next subtract B[j] * B[n-j-1] from r. This change causes the deletion of two lines after the loop. In the loop body, we expand the assignments to the local variables k, h, and v, and remove those local variables. As the result, we do not have to push the values of the variables to the garbage stack g in the body of the loop.

Similar to the above modification to rE, the rank r does not change if the procedure riE is called for the tree of size zero ($n = 0$). Therefore, we change the procedure body such that the body of the procedure is computed only if $n \geq 1$. Moreover, because the tree permutation of size one can be computed by riE, replacing the

```
 1  // Set the rank of the tree permutation p[a:a+n-1]-(s,s,...,s) to
       r
 2  procedure rE' (int p[],int n,int a,int s,int r,stack g)
 3     // {T ≃ p[a : a+n-1] - (s,...,s), r = 0}
 4     if n>=1 then // if |T| ≥ 1
 5        local int j = p[a]-s
 6           // (c) Add rank'(r(T)) to r
 7           // {r = 0, r(T) ≃ p[a+j+1 : a+n-1] - (s+j+1,...,s+j+1)}
 8           call rE' (p,n-j-1,a+j+1,s+j+1,r,g)
 9           // {r = rank'(r(T))}
10
11           // (b) Add rank'(l(T)) to r
12           local int r_1T = 0
13              // {l(T) ≃ p[a+1 : a+j] - (s,...,s)}
14              call rE' (p,j,a+1,s,r_1T,g)
15              // {r_1T = rank'(l(T))}
16              r += B[n-j-1] * r_1T
17              // {r = rank'(l(T)) × B_{|r(T)|} + rank'(r(T))}
18           delocal int r_1T = r/B[n-j-1]
19
20           // Clear j, and (a) add encoded |l(T)| to r
21           // {j = |l(T)|, n = |T|, r = rank'(l(T)) × B_{|r(T)|} + rank'(r(T))}
22           from j = p[a]-s loop
23              j -= 1
24              r += B[j] * B[n-j-1] // subtract G_{jn} from r
25           until j = 0
26           // {j = 0, n = |T|, r = rank'(T)}
27           delocal int j = 0
28     fi n>=1
29     // {r = rank'(T)}
```

Fig. 8 Simplified embedded reversible simulation of ranking

conditionals in code blocks (b) and (c) with the bodies of their then clauses does not change the meaning of the procedure.

At the first line of block (c), the initializer of the variable r_rT can be reduced to r%B[n-j-1].

In addition to the above modification, by changing the procedure name to riE', we obtain the program in Fig. 9. The procedure clT counts the number of the nodes of the left tree from the encoded information in r. We also use clT later.

We compare the inverse of the procedure rE' in Fig. 8 and the procedure riE' in Fig. 9 and infer invariants to make the reversible unranking program clean by removing the push operations. There are three push operations in riE'. First, the push operation at line 26 is unnecessary because the corresponding line 12 in rE' asserts r_1T is zero where the paired delocal clause in rE' has the same finalizer as the initializer in riE'. Next, the push operation at line 36 can be removed by using the relation at line 5 of rE'. Finally, by the invariant at line 9 in rE', the push operation at line 34 can be removed if we directly use r instead of setting r_rT to r%B[n-j-1].

While admittedly it is not a systematic method, these observations suggest efficient clean reversible unranking. For the case of a non-empty binary tree T, now the

```
 1 // Extract |l(T)| from encoded r, and set it to j
 2 procedure clT(int j,int n,int r)
 3     // {j=0,n=|T|,r=rank'(T)}
 4     from j = 0 loop
 5         r -= B[j]*B[n-j-1]     // Subtract G_jn from r
 6         j += 1
 7     until r < B[j]*B[n-j-1]
 8     // {j=|l(T)|,n=|T|,r=rank'(l(T))×B_|r(T)|+rank'(r(T))}
 9
10 // Store unrank(r,n)+(s,...,s) to p[a:a+n-1]
11 procedure riE'(int p[],int n,int a,int s,int r,stack g)
12     // {p[a:a+n-1]=(0,...,0),r=rank'(T)}
13     if n>=1 then // if |T|≥1
14         local int j = 0
15             // {j=0,n=|T|,r=rank'(T)}
16             call clT(j,n,r)    // (a) Set |l(T)| to j
17             // {j=|l(T)|,n=|T|,r=rank'(l(T))×B_|r(T)|+rank'(r(T))}
18
19             p[a] ^= j+s         // Set the root p[a]=|l(T)|+s
20
21             // (b) Construct l(T)
22             local int r_lT = r/B[n-j-1]
23                 // {p[a+1:a+j]=(0,...,0),r_lT=rank'(l(T))}
24                 call riE'(p,j,a+1,s,r_lT,g)
25                 // {l(T)≃p[a+1:a+j]-(s,...,s)}
26                 push(r_lT,g)
27             delocal int r_lT = 0
28
29             // (c) Construct r(T)
30             local int r_rT = r%B[n-j-1]
31                 // {p[a+j+1:a+n-1]=(0,...,0),r_rT=rank'(r(T))}
32                 call riE'(p,n-j-1,a+j+1,s+j+1,r_rT,g)
33                 // {r(T)≃p[a+j+1:a+n-1]-(s+j+1,...,s+j+1)}
34                 push(r_rT,g)
35             delocal int r_rT = 0
36             push(j,g)
37         delocal int j = 0
38     fi n>=1
39     // {T≃p[a:a+n-1]-(s,...,s)}
```

Fig. 9 Simplified embedded reversible simulation of unranking

unranking computation corresponding to the conditions (a), (b), and (c) changes the data as shown in Fig. 10. Here, we let $ra = rank'(l(T)) \times B_{|r(T)|} + rank'(r(T))$.

The procedure unrank in Fig. 11 is based on this concept. It is the reversible simulation computing from the rank and the size of the binary trees to the corresponding tree permutation. The reversible simulation does not generate garbage output; thus, it is clean. In the remainder of this section, we assume that T is a binary tree corresponding to the tree permutation $p[0 : n - 1]$, i.e., $T \simeq p[0 : n - 1]$.

When n is greater than or equal to one ($|T| \geq 1$), we recursively construct the tree permutation. By calling the procedure clT(j,n,r) defined in Fig. 9, we extract the value $|l(T)|$ from the encoded variable r and store it to j, and subtract the number of binary trees satisfying the condition (a) from r. We initialize the allocated

$$rank(T)$$
$$\updownarrow (a)\ (|T|, \Sigma_{0 \le j < |l(T)|} G_{j,|T|} + ra) = rank(T)$$
$$|l(T)|, ra, |T|$$
$$\updownarrow rank'(l(T)) = ra/B_{|T|-|l(T)|-1}$$
$$|l(T)|, rank'(l(T)), ra, |T|$$
$$\updownarrow (b)$$
$$|l(T)|, l(T), rank'(r(T)), |T|$$
$$\updownarrow (c)$$
$$|l(T)|, l(T), |r(T)|, r(T), |T|$$
$$\updownarrow \text{arithmetic operations}$$
$$T, |T|$$

Fig. 10 Reversible calculation from the rank of the binary tree T to its tree permutation

```
1  // Store tree permutation incremented by s to p[a:a+n-1]
2  procedure ri(int p[],int n,int a,int s,int r)
3    if n>=1 then                      // if |T| >= 1
4      local int j = 0
5        call clT(j,n,r)               //(a)
6        local int r_lT = r/B[n-j-1]   //(b)
7          r -= r_lT*B[n-j-1]          // r becomes rank'(r(T))
8          call ri(p,j,a+1,s,r_lT)
9        delocal int r_lT = 0
10       call ri(p,n-j-1,a+j+1,s+j+1,r) //(c)
11       p[a] ^= j+s                    // Set the root
12     delocal int j = p[a]-s          // j is |l(T)|
13   fi n>=1
14
15 // Set to p[0:n-1] the tree permutation of the binary tree of
      rank r and size n
16 procedure unrank(int p[],int n,int r)
17   call ri(p,n,0,0,r)
```

Fig. 11 Clean reversible simulation for unranking binary trees

variable `r_lT` to $rank'(l(T))$ and subtract the number of binary trees satisfying the condition (b) from `r` so that `r` becomes $rank'(r(T))$, which is the number of binary trees satisfying the condition (c). The recursive procedure call `ri` sets the subarray `p[a+1:a+n-j-2]` to the sequence of the tree permutation corresponding to the left subtree $l(T)$ incremented by the offset s, and makes the variable `r_lT` zero-cleared. Similarly, the recursive procedure call `ri` at line 10 sets the subarray `p[a+j+1:a+n-1]` to the sequence of the binary tree corresponding to the right subtree $r(T)$ incremented by the offset $s + j + 1$, and makes the variable `r` zero-cleared. The element `p[a]` corresponding to the root of the considered tree permutation stores the size of the left subtree $|l(T)|$ plus the offset s. Therefore, we have $j = p[a] - s$. This information is used to deallocate the variable `j` at line 12.

The procedure `unrank` is more efficient than the reversible simulation `unrankE` constructed by the (input-erasing) Bennett reversible simulation `rankB`. This is because the procedure `unrank` traverses the array p only once, and the intermediate variables and garbage stack, i.e., `t` and `q`, are not required.

Owing to the reversibility of Janus, the ranking computation is simply realized by the inverse invocation, using an **uncall** statement, of the procedure `unrank` or the invocation of the inverse program of `unrank`. It should be noted that the program inversion of well-formed Janus programs is always possible [38].

The procedure `unrank` is more efficient and concise than the unranking procedure in [21] in the sense that it has a smaller number of local variables and fewer lines of code.

5.5 Complexity of the Proposed Reversible Simulation

The irreversible ranking and unranking programs of Knott [21], on which our reversible simulations are based, have asymptotic time complexity $O(n^2)$ for length n of the given tree permutation [12]. Since embeddings keep their time complexity [6], the reversible simulations `rankE` in Fig. 3 and `unrankE` in Fig. 5, which are the embeddings of the programs, and their simplifications in Figs. 8 and 9 have time complexity $O(n^2)$.

Owing to the following facts, the time complexity of `unrank` in Fig. 11 is also $O(n^2)$. The second argument n of the procedure `ri` is equal to the length of the permutation to be constructed. The procedure `ri` constructs tree permutations of length j and $n - j - 1$ at lines 8 and 10, respectively. In the body of the procedure `clT`, which is called at line 5, a loop is repeated $|l(T)|$ times, where $0 \le |l(T)| \le n - 1$. Therefore, for the worst-case execution time $W(n)$ of the procedure `ri`, we have the recurrence relation

$$W(n) = \max_{0 \le j \le n-1} (W(j) + W(n - j - 1) + \Theta(j)). \tag{13}$$

Hence, we have

$$W(n) \le \max_{0 \le j \le n-1} (W(j) + W(n - j - 1)) + \Theta(n). \tag{14}$$

The inequality also holds when $W(n)$ is the execution time for quicksort of length n (see [9, Sect. 7.4.1]). Therefore, the worst-case execution time of the reversible procedure `rank` for ranking binary trees is the same as that of quicksort, i.e., $O(n^2)$. If we fix j to $n - 1$ and solve Eq. 13, we have $W(n) \in \Theta(n^2)$. As a result, the worst-case execution time of `unrank` is $\Theta(n^2)$.

When we use a naive embedding for ranking binary trees and its inverse by remembering all information otherwise lost, the resulting reversible simulation in Figs. 3 and 5 needs space complexity proportional to its time complexity $\Theta(n^2)$. Then, the (input-erasing) Bennett reversible simulation in Fig. 6 requires the spacecomplexity $\Theta(n^2)$. However, our simplified embedded ranking and unranking programs in Figs. 8 and 9 use $\Theta(n)$ extra space to the corresponding irreversible programs. Therefore, their asymptotic space complexity would be the same as the original irreversible programs $\Theta(n)$, and combining them by the Bennett method leads to a clean reversible simulation with the space complexity $\Theta(n)$ and the intermediate garbage size $\Theta(n)$.

The clean reversible simulation `unrank` of `unrankA` is optimal in the following sense. A reversible simulation q of p is said to be *faithful* with garbage bound g iff the garbage output is asymptotically bounded by $g(|x|)$ for all the inputs x, there is no asymptotic time overhead incurred by p for all the inputs x, and p asymptotically uses at most $g(|x|)$ additional space compared to q for all the inputs x [3]. Here, $|x|$ is the size of data x in the binary representation. A faithful reversible simulation q is said to be *hygienic* if q is asymptotically optimal in its garbage usage [3]. The procedure `unrank` does not have garbage output, the time complexity of `unrank` is the same as that of `unrankA`, and `unrank` does not use the asymptotically larger memory. Namely, g becomes a constant function, and the size of intermediate garbage becomes $\Theta(n)$. Therefore, the reversible simulation `unrank` of `unrankA` is hygienic. In contrast, the (input-erasing) Bennett reversible simulation of embedded reversible simulations are not hygienic because it uses asymptotically extra space $(g(x) = \Theta(|x|))$.

6 Conclusion

We aimed at deriving efficient clean reversible simulations from irreversible programs. We obtained a clean hygienic reversible simulation of ranking binary trees and its inverse of the Knott method [21], which is more efficient in time and space than the solutions obtained by the (input-erasing) Bennett method of naively embedded reversible simulations [5]. In particular, the space complexity of our embedded programs is asymptotically lower than naive embedding, which saves all data and control information otherwise lost. While it is unknown whether the linear-time and linear-space reversible simulation exists for any problem, in this paper, we show that there is a linear-time and linear-space reversible simulation for ranking binary trees and its inverse by the Knott method [21]. To the best of the authors' knowledge, thus far, there is no report on a clean reversible version of the algorithms.

Because ranking and unranking binary trees are fundamental algorithms, we expect the proposed reversible algorithms to be useful for future research on reversible algorithms. We do not presently realize the systematic derivation of clean reversible simulation `rank` from two embedded programs `rankE` and `unrankE`. Another possible future research would be to mechanize the process, and investigate strategies for systematically deriving efficient reversible programs and methods for constructing efficient reversible programs in reversible programming languages.

Acknowledgements The authors would like to thank Kota Kimura for insightful comments on an early draft of this paper, and the anonymous reviewers for their helpful and constructive comments.

Preliminary versions of this paper were presented at the 33rd Workshop of the Japan Society for Software Science and Technology and the 19th JSSST Workshop on Programming and Programming Languages. This work was supported by JSPS KAKENHI Grant Number JP25730049 and Nanzan University Pache Research Subsidy I-A-2 for the 2017 academic year.

References

1. Axelsen, H.B., Glück, R.: A simple and efficient universal reversible Turing machine. In: Dediu, A.H., Inenaga, S., Martín-Vide, C. (eds.) Language and Automata Theory and Applications. Lecture Notes in Computer Science, vol. 6638, pp. 117–128. Springer, Berlin (2011). https://doi.org/10.1007/978-3-642-21254-3_8
2. Axelsen, H.B., Glück, R.: What do reversible programs compute? In: Hofmann, M. (ed.) Foundations of Software Science and Computation Structures. Proceedings. Lecture Notes in Computer Science, vol. 6604, pp. 42–56. Springer, Berlin (2011). https://doi.org/10.1007/978-3-642-19805-2_4
3. Axelsen, H.B., Yokoyama, T.: Programming techniques for reversible comparison sorts. In: Feng, X., Park, S. (eds.) Programming Languages and Systems. Lecture Notes in Computer Science, vol. 9458, pp. 407–426. Springer, Berlin (2015). https://doi.org/10.1007/978-3-319-26529-2_22
4. Baker, H.G.: NREVERSAL of fortune — the thermodynamics of garbage collection. In: Bekkers, Y., Cohen, J. (eds.) International Workshop on Memory Management. Proceedings. Lecture Notes in Computer Science, vol. 637, pp. 507–524. Springer, Berlin (1992). https://doi.org/10.1007/BFb0017210
5. Bennett, C.H.: Logical reversibility of computation. IBM J. Res. Dev. **17**(6), 525–532 (1973). https://doi.org/10.1147/rd.176.0525
6. Bennett, C.H.: Time/space trade-offs for reversible computation. SIAM J. Comput. **18**(4), 766–776 (1989). https://doi.org/10.1137/0218053
7. Bennett, C.H., Landauer, R.: The fundamental physical limits of computation. Sci. Am. **253**(1), 48–56 (1985)
8. Burignat, S., Vermeirsch, K., De Vos, A., Thomsen, M.K.: Garbageless reversible implementation of integer linear transformations. In: Glück, R., Yokoyama, T. (eds.) Reversible Computation. Proceedings. Lecture Notes in Computer Science, vol. 7581. Springer, Berlin (2013)
9. Cormen, T.H., Leiserson, C.E., Rivest, R.L., Stein, C.: Introduction to Algorithms, 3rd edn. The MIT Press, Cambridge (2009)
10. De Vos, A.: Reversible Computing: Fundamentals, Quantum Computing, and Applications. Wiley-VCH, New York (2010)
11. Early, D., Gao, A., Schellekens, M.: Frugal encoding in reversible $\mathcal{MOQA}$: A case study for quicksort. In: Glück, R., Yokoyama, T. (eds.) Reversible Computation. Proceedings. Lecture Notes in Computer Science, vol. 7581, pp. 85–96. Springer, Berlin (2013). https://doi.org/10.1007/978-3-642-36315-3_7
12. Er, M.C.: Enumerating ordered trees lexicographically. The Comput. J. **28**(5), 538–542 (1985)
13. Feynman, R.P.: Reversible computation and the thermodynamics of computing (Chapter 5). In: Hey, A.J.G., Allen, R.W. (eds.) Feynman Lectures on Computation, pp. 137–184. Addison-Wesley, Reading (1996)
14. Foster, J.N., Greenwald, M.B., Moore, J.T., Pierce, B.C., Schmitt, A.: Combinators for bi-directional tree transformations: a linguistic approach to the view update problem. ACM Trans. Program. Lang. Syst. **29**(3), 1–65 (2007). https://doi.org/10.1145/1232420.1232424
15. Frank, M.P.: Reversibility for efficient computing. Ph.D. thesis, Massachusetts Institute of Technology (1999)
16. Fredkin, E., Toffoli, T.: Conservative logic. Int. J. Theor. Phys. **21**, 219–253 (1982). https://doi.org/10.1007/BF01857727

17. Glück, R., Kawabe, M.: Derivation of deterministic inverse programs based on LR parsing. In: Kameyama, Y., Stuckey, P.J. (eds.) Functional and Logic Programming. Proceedings. Lecture Notes in Computer Science, vol. 2998, pp. 291–306. Springer, Berlin (2004). https://doi.org/10.1007/978-3-540-24754-8_21
18. Glück, R., Yokoyama, T.: A linear-time self-interpreter of a reversible imperative language. Comput. Softw. **33**(3), 108–128 (2016). https://doi.org/10.11309/jssst.33.3_108
19. Gries, D.: Inverting programs. The Science of Programming. Texts and Monographs in Computer Science, pp. 265–274. Springer, Berlin (1981)
20. James, R.P., Sabry, A.: Information effects. Principles of Programming Languages. Proceedings, pp. 73–84. ACM Press (2012). https://doi.org/10.1145/2103656.2103667
21. Knott, G.D.: A numbering system for binary trees. Commun. ACM **20**(2), 113–115 (1977). https://doi.org/10.1145/359423.359434
22. Knuth, D.E.: The Art of Computer Programming, Volume 1: Fundamental Algorithms, vol. 1, 3rd edn. Addison-Wesley Professional, Reading (1997)
23. Landauer, R.: Information is physical. Phys. Today **44**(5), 23–29 (1991)
24. Lutz, C.: Janus: A time-reversible language. Letter to R, Landauer (1986)
25. Mogensen, T.Æ.: Garbage-free reversible multipliers for arbitrary constants. ACM J. Emerg. Technol. Comput. Syst. **11**(2), 1–18 (2014)
26. Morita, K.: A simple universal logic element and cellular automata for reversible computing. In: Margenstern, M., Rogozhin, Y. (eds.) Machines, Computations, and Universality. Lecture Notes in Computer Science, vol. 2055, pp. 102–113. Springer, Berlin (2001). https://doi.org/10.1007/3-540-45132-3_6
27. Morita, K.: Reversible computing and cellular automata – a survey. Theor. Comput. Sci. **395**(1), 101–131 (2008). https://doi.org/10.1016/j.tcs.2008.01.041
28. Morita, K.: Simulating reversible Turing machines and cyclic tag systems by one-dimensional reversible cellular automata. Theor. Comput. Sci. **412**(30), 3856–3865 (2011). https://doi.org/10.1016/j.tcs.2011.02.022
29. Nishida, N., Sakai, M., Sakabe, T.: Partial inversion of constructor term rewriting systems. In: Giesl, J. (ed.) Term Rewriting and Applications. Proceedings. Lecture Notes in Computer Science, vol. 3467, pp. 264–278. Springer, Berlin (2005)
30. Perumalla, K.S.: Introduction to Reversible Computing. CRC Press, Boca Raton (2013)
31. Pesu, T., Phillips, I.: Real-Time Methods in Reversible Computation, pp. 45–59. Springer International Publishing, Berlin (2015)
32. Saeedi, M., Markov, I.L.: Synthesis and optimization of reversible circuits — a survey. ACM Comput. Surv. **45**(2), 21:1–21:34 (2013). https://doi.org/10.1145/2431211.2431220
33. Sprugnoli, R.: The generation of binary trees as a numerical problem. J. ACM **39**(2), 317–327 (1992). https://doi.org/10.1145/128749.128753
34. Thomsen, M.K., Axelsen, H.B., Glück, R.: A reversible processor architecture and its reversible logic design. In: De Vos, A., Wille, R. (eds.) Reversible Computation. Proceedings. Lecture Notes in Computer Science, vol. 7165, pp. 30–42. Springer, Berlin (2012). https://doi.org/10.1007/978-3-642-29517-1_3
35. Vitányi, P.: Time, space, and energy in reversible computing. Computing Frontiers. Proceedings, pp. 435–444. ACM Press (2005). https://doi.org/10.1145/1062261.1062335
36. Wille, R., Drechsler, R.: Towards a Design Flow for Reversible Logic. Springer, Berlin (2010)
37. Yokoyama, T., Glück, R.: A reversible programming language and its invertible self-interpreter. Partial Evaluation and Semantics-Based Program Manipulation. Proceedings, pp. 144–153. ACM Press (2007). https://doi.org/10.1145/1244381.1244404
38. Yokoyama, T., Axelsen, H.B., Glück, R.: Principles of a reversible programming language. Computing Frontiers. Proceedings, pp. 43–54. ACM Press (2008). https://doi.org/10.1145/1366230.1366239
39. Yokoyama, T., Axelsen, H.B., Glück, R.: Optimizing clean reversible simulation of injective functions. J. Multi. Valued Log. Soft Comput. **18**(1), 5–24 (2012)
40. Yokoyama, T., Axelsen, H.B., Glück, R.: Towards a reversible functional language. In: De Vos, A., Wille, R. (eds.) Reversible Computation. Proceedings. Lecture Notes in Computer Science, vol. 7165, pp. 14–29. Springer, Berlin (2012). https://doi.org/10.1007/978-3-642-29517-1_2

On Radius 1 Nontrivial Reversible and Number-Conserving Cellular Automata

Katsunobu Imai, Bruno Martin and Ryohei Saito

Abstract Reversibiliity and number-conservation are widely studied physics-like constraints for cellular automata (CA). Although both seem to be 'natural' constraints for a CA, it was conjectured that one-dimensional reversible and number-conserving CA (RNCCA) only has a limited computing ability. Particularly in the case of radius 1/2 (2-neighbor), it was shown that the class of RNCCA is equal to a trivial class of CA, so called shift-identity product cellular automata (SIPCA). But recently it was also shown that a RNCCA of neighborhood size four is computation-universal. In this paper, we list radius 1 (3-neighbor) RNCCAs up to 4-state by exhaustive search. In contrast to the radius 1/2 case, there are three new types of nontrivial RNCCA rules in the case of 4-state. We also show that it is possible to compose new nontrivial RNCCAs by modifying a SIPCA even when the state number is larger than four.

1 Introduction

A reversible cellular automaton (RCA) is a cellular automaton (CA) whose global function is injective. A number-conserving cellular automaton (NCCA) is a cellular automaton whose states are integers and whose transition function keeps the sum of all cells constant throughout its evolution. Both can be seen as some kind of modeling of the physical conservation laws of mass or energy. Although both constraints seem to be 'natural' constraints for a cellular automaton [6], it was conjectured that one-dimensional reversible and number-conserving cellular automata (RNCCA) only have a limited computing ability. Particularly in the case of radius 1/2 (2-neighbor), it was shown that RNCCAs are characterized by shift-identity product cellular automata (SIPCA) [5]. But recently it was also shown that a 4-neighbor

K. Imai (✉) · R. Saito
Graduate School of Engineering, Hiroshima University, Higashihiroshima, Japan
e-mail: imai@hiroshima-u.ac.jp

B. Martin
I3S-CNRS, Université Côte d'Azur, Nice, France
e-mail: Bruno.Martin@unice.fr

© Springer International Publishing AG 2018
A. Adamatzky (ed.), *Reversibility and Universality*, Emergence, Complexity and Computation 30, https://doi.org/10.1007/978-3-319-73216-9_12

RNCCA is computation-universal [7]. In this paper we explore the whole list of radius 1 (3-neighbor) RNCCAs up to 4-state by exhaustive search. In contrast to the radius 1/2 case, there are three new types of nontrivial RNCCAs in the case of 4-state. We also show that one of the types is useful because it is possible to compose new associated nontrivial RNCCAs of this type by modifying a trivial RNCCA when the number of states is larger than four.

2 Preliminaries

Definition 1 A n-state *deterministic one-dimensional cellular automaton* is defined by $A = (\mathbb{Z}, N, Q, f)$, where $\mathbb{Z}$ is the set of all integers, the neighborhood $N = \{n_1, \ldots, n_k\}$ is a finite subset of $\mathbb{Z}$, $Q = [\![0, \ldots, n-1]\!]$ is the set of states of each cell, $f : Q^k \to Q$ is a mapping called the *local transition function*.

A *configuration* over Q is a mapping $\alpha : \mathbb{Z} \to Q$. Then $\mathrm{Conf}(Q) = \{\alpha \mid \alpha : \mathbb{Z} \to Q\}$ is the set of all configurations over Q. The *global function* of A is $F : \mathrm{Conf}(Q) \to \mathrm{Conf}(Q)$ defined as $F(\alpha)(i) = f(\alpha(i + n_1), \ldots, \alpha(i + n_k))$ for all $\alpha \in \mathrm{Conf}(Q)$, $i \in \mathbb{Z}$.

We call the value $r = (\max_{1 \le i \le k} n_i - \min_{1 \le i \le k} n_i)/2$ the *radius* of A. A neighborhood $\{-1, 0, 1\}$ corresponds to radius 1, while a neighborhood $\{0, 1\}$ corresponds to radius 1/2. In these cases we replace N by r in the notation of cellular automata. We employ the Wolfram numbering [9] $W(f)$ to represent the local function f: $W(f) = \sum f(x_1, \ldots, x_{2r+1}) n^{n^{2r} x_1 + n^{2r-1} x_2 + \ldots + n^0 x_{2r+1}}$ where the sum is applied on $(x_1, \ldots, x_{2r+1}) \in Q^{2r+1}$. Note that the Wolfram numbering of a n-state CA, expressed in radix n form is the concatenation of the values of its local transition function.

In the sequel, we denote any finite configuration with the finite sequence of its states: $[x_0, x_1, x_2, \ldots, x_m]$ where $x_i \in Q$.

Definition 2 A CA A is *reversible* iff its global function F is injective. A is said to be number-conserving iff $\forall c \in \mathrm{Conf}_F(Q)$, $\sum_{i \in \mathbb{Z}} \{F(c)(i) - c(i)\} = 0$ where $\mathrm{Conf}_F(Q)$ denotes the set of all *finite configurations* (i.e. configurations in which the number of nonzero cells is finite.)

Definition 2 is equivalent to the general definition in the infinite case [3].

Definition 3 For a state $q \in Q \setminus \{0\}$ of $A = (\mathbb{Z}, 1, Q, f)$, we call q a *stable* state if $f(0, q, 0) = q$, $f(q, 0, 0) = 0$, $f(0, 0, q) = 0$, q is a *right* (resp. *left*) propagating state, if $f(0, q, 0) = 0$, $f(q, 0, 0) = q$, $f(0, 0, q) = 0$ (resp. $f(0, q, 0) = 0$, $f(q, 0, 0) = 0$, $f(0, 0, q) = q$). For a state $q \in Q \setminus \{0\}$ of $A = (\mathbb{Z}, 1/2, Q, g)$, we call q a stable state if $g(q, 0) = q$, $g(0, q) = 0$, q is a left propagating state, if $g(q, 0) = 0$, $g(0, q) = q$. We denote the set of stable, right propagating, left propagating states by $Q_{\downarrow}$, $Q_{\searrow}$, and $Q_{\nearrow}$ ($\subset Q$) respectively.

We recall a necessary and sufficient condition for radius 1 CAs to be number-conserving, as a special case of results from [2].

Proposition 1 *A deterministic one-dimensional CA* $A = (\mathbb{Z}, 1, Q, f)$ *is number-conserving iff there exists a function* $\varphi : Q^2 \to \mathbb{Z}$ *such that*

$$f(\ell, c, r) = c - \varphi(\ell, c) + \varphi(c, r) \text{ for all } c, \ell, r \in Q.$$

The function φ represents the movement of numbers between two neighboring cells. We call this function φ the *flow function*.

3 Radius 1 RNCCAs and Their Composition Methods

In this section we denote each RNCCA with an expression of the form $q-n-i$ where q is the size of its neighborhood, n the number of states, and i $(= 1, 2, \ldots)$ is an index which indicates the local rules arranged in the ascending order of the Wolfram numbering ($W(f_i) < W(f_{i+1})$). See Table 1 and Appendix.

As far as the radius $1/2$ case, RNCCAs are characterized by shift-identity product cellular automata (SIPCA) [5]. The possible function by SIPCA signals is only a crossing of a stable state and a left propagating state. Thus a typical behavior is trivial and its time evolution is similar to the one depicted in Fig. 1 3-4-4 (Type I). Its state set Q is $Q_\downarrow \cup Q_\nearrow \cup Q_{\nearrow\downarrow}$ where $Q_{\nearrow\downarrow} = \{a + b \mid a \in Q_\downarrow, b \in Q_\nearrow\}$. An RNCCA whose state number is a prime, is only a left shift or the identity [5], thus $2-p-1$ is a left shift and $2-p-2$ is the identity, when p is a prime number. Appendix 1 proposes the cases up to 10-state.

Schranko and Oliveira [8] introduced a way of constructing RNCCA rules with a large neighborhood by the composition of radius $1/2$ RNCCA rules. Their scheme in the cases ranging from radius $1/2$ to 1 is as follows:

Proposition 2 [8] *Given two radius* $1/2$ *RNCCAs:* $A_1 = (\mathbb{Z}, 1/2, Q, g_1)$ *and* $A_2 = (\mathbb{Z}, 1/2, Q, g_2)$, $A = (\mathbb{Z}, 1, Q, f)$ *is a radius 1 RNCCA, where* $f(\ell, c, r) \equiv g_1 \circ g_2(\ell, c, r) = g_1(g_2(\ell, c), g_2(c, r))$ *for any* $\ell, c, r \in Q$.

This procedure does not produce any complex behavior. If $a \in Q$ is a left propagating (resp. a stable) state of both A_1 and A_2 then a is a left propagating (resp. a right propagating) state of A. Otherwise a is a stable state of A. So they also conjectured that every q-state n-neighbor RNCCA rule is a composition of q-state radius $1/2$ RNCCA rules. This is not true because there is a counter-example in [5] and we show it even in the case of radius 1 and 4-state. By exhaustive search, there are 21 RNCCAs with radius 1 and 4-state (among 89588 NCCAs). Appendix 2 shows the result. Although enumeration of RCAs is hard even in the 4-state case, enumeration of NCCAs is easier because it is possible to reduce the search domain by Proposition 1 (see also [2]). So we first enumerated NCCAs and checked their reversibility by [1].

Table 1 Flow function values of radius 1 (3-neighbor) 4-state RNCCAs

No. 3-4-	1	2	3	4	5	6	7	8	9	10	11	12	13	14	15	16	17	18	19	20	21
(3,3)	3	1	1	1	-1	2	2	2	1	0	-1	0	-2	0	0	0	-2	-2	-1	-1	-3
(3,2)	2	0	0	0	-2	2	2	2	1	0	-1	0	-2	0	0	0	-2	-2	-1	-1	-3
(3,1)	1	1	1	1	-1	0	0	0	-1	0	-1	0	-2	0	0	0	-2	-2	-1	-1	-3
(3,0)	0	-2	0	0	-2	-1	0	0	-1	-2	-3	-1	-3	0	0	0	-2	-2	-1	-1	-3
(2,3)	3	1	1	1	-1	2	2	2	2	0	0	0	-2	0	0	0	-2	-2	0	0	-2
(2,2)	2	0	0	0	-2	2	2	2	2	0	0	0	-2	0	0	0	-2	-2	0	0	-2
(2,1)	1	1	-1	1	-1	1	0	0	0	0	0	1	-1	-2	0	0	-2	-2	-2	0	-2
(2,0)	0	0	0	0	-2	0	0	0	0	0	0	0	-2	0	0	0	-2	-2	0	0	-2
(1,3)	3	1	1	1	-1	2	2	2	1	0	-1	0	0	0	-1	0	0	0	-1	-1	-1
(1,2)	2	2	0	0	0	2	1	2	1	2	1	0	0	0	0	0	-1	0	-1	-1	-1
(1,1)	1	1	1	1	1	0	0	0	-1	0	-1	0	0	0	0	0	0	0	-1	-1	-1
(1,0)	0	0	0	0	0	0	0	0	-1	0	-1	0	0	0	0	0	0	0	-1	-1	-1
(0,3)	3	1	3	1	1	2	3	2	2	0	0	0	0	2	1	0	1	0	2	0	0
(0,2)	2	0	0	0	0	2	2	2	2	0	0	0	0	0	0	0	0	0	0	0	0
(0,1)	1	1	1	1	1	0	0	0	0	0	0	0	0	0	0	0	0	0	0	0	0
(0,0)	0	0	0	0	0	0	0	0	0	0	0	0	0	0	0	0	0	0	0	0	0

0	0	2	0	1	2	0	0	1
0	0	2	1	0	2	0	1	0
0	0	3	0	0	2	1	0	0
0	1	2	0	0	3	0	0	0
1	0	2	0	1	2	0	0	0
0	0	2	1	0	2	0	0	0
0	0	3	0	0	2	0	0	0
0	1	2	0	0	2	0	0	0

4 − 4

0	0	2	0	1	2	0	0	1
0	0	2	1	2	0	0	1	0
0	0	3	2	0	0	1	0	0
0	1	2	2	0	1	0	0	0
1	2	0	2	1	0	0	0	0
2	0	0	3	0	0	0	0	0
0	0	1	0	2	0	0	0	0
0	1	0	0	2	0	0	0	0

4 − 2

0	0	2	0	1	2	0	0	1
0	0	2	1	0	2	0	1	0
0	0	1	2	0	2	1	0	0
0	1	0	2	0	1	2	0	0
1	0	0	2	1	0	2	0	0
0	0	0	1	2	0	2	0	0
0	0	1	0	2	0	2	0	0
0	1	0	0	2	0	2	0	0

4 − 3

0	0	2	0	1	2	0	0	1
0	0	2	0	3	0	0	0	1
0	0	2	0	1	2	0	0	1
0	0	2	0	3	0	0	0	1
0	0	2	0	1	2	0	0	1
0	0	2	0	3	0	0	0	1
0	0	3	0	1	2	0	0	1
0	0	2	0	3	0	0	0	1

4 − 10

Fig. 1 Time space diagrams of 4-state RNCCAs of each type

Their flow function values are shown in Table 1. They are classified into four types (see Fig. 1). 3-1, 4, 5, 8, 9, 16, 18, 20, 21 (type I) are composed by the above method. Type II rules: 3-2,6,17,19 have one left (or right) propagating state and one stable state and their collision shifts the stable state to the right (or to the left). Each type II rule can be induced by a related type I rule: $3-4-4 \rightarrow 3-4-2, 3-4-8 \rightarrow 3-4-6, 3-4-18 \rightarrow 3-4-17, 3-4-20 \rightarrow 3-4-19$ respectively by replacing two flow functions values. Type III rules: $3-4-3, 7, 11, 13$ have two left (or right) propagating states and one stable state. Their collision is performed by swapping two signals (i.e. it does not make the sum of two signal values). Type IV rules: 3-4-10, 12, 14, 15 have no propagating state.

Next, we show that a type II RNCCA can be induced by a type I RNCCA even when the state number is larger than four.

Proposition 3 *Let $A = (\mathbb{Z}, 1, Q, f)$. Suppose A is a RNCCA composed by Proposition 2 and $s \in Q$ is a stable state and $r \in Q$ is a right (resp. left) propagating state, then there is an injective NCCA which has an associated type II rule.*

Proof We only show the result when r is a right propagating state. The proof in the case of a left propagating state is similar.

Let $r, r', r'' \in Q_\searrow \cup \{0\}$ and $s, s' \in Q_\downarrow \cup \{0\}$. Choose two states r_0 and s_0 such that $r_0 \in Q_\searrow$ and $s_0 \in Q_\downarrow$. The CA A must attain the transition: $[r + s, 0, r_0 + s_0, r' + s'] \xrightarrow{f} [r'' + s, r, s_0, r_0 + s']$ since r and r_0 are right-propagating states and the flow function φ of f should have the value: $\varphi(0, r_0 + s_0) = 0$. If we modify the value of the flow function $\varphi(0, r_0 + s_0)$ to s_0, the transition becomes $[r + s, 0, r_0 + s_0, r' + s'] \xrightarrow{f} [r'' + s, r + s_0, 0, r_0 + s']$. But the transition to the configurations already exists, i.e., $[r + s, s_0, r_0, r' + s'] \xrightarrow{f} [r'' + s, r + s_0, 0, r_0 + s']$. These transitions have to be changed to the transition $[r + s, s_0, r_0, r' + s'] \xrightarrow{f} [r'' + s, r, s_0, r_0 + s']$. This can be done by changing the flow function value of $\varphi(s_0, r_0) = 0$ to $-s_0$. This modification swaps the configurations $[r'' + s, r + s_0, 0, r_0 + s']$ and $[r'' + s, r, s_0, r_0 + s']$. Even if a configuration contains the sequence more than once (maybe infinitely often), the modification only affects every finite length sequence, keeping thus injectivity because the number of configurations for which preimages are changed is equal. The modification also keeps the number-conservation since it is applied to the values of the flow function. □

Example 1 A type I 6-state radius 1 RNCCA A_6 can be combined from two radius 1/2 RNCCAs 2-6-4 and 2-6-6 (in Appendix 1). 2-6-6 is the identity and 2-6-4 has a stable state 1 and two left propagating state 2 and 4. Thus A_6 has $Q_\searrow = \{1\}$, $Q_\downarrow = \{2, 4\}$, and $Q_\swarrow = \varnothing$. Its flow function $\varphi(x, y)$ has the following values:

$(5, 5)$ -1, $(5, 4)$ -1, $(5, 3)$ -1, $(5, 2)$ -1, $(5, 1)$ -1, $(5, 0)$ -1, $(4, 5)$ 0, $(4, 4)$ 0,
$(4, 3)$ 0, $(4, 2)$ 0, $(4, 1)$ -4, $(4, 0)$ 0, $(3, 5)$ -1, $(3, 4)$ -1, $(3, 3)$ -1, $(3, 2)$ -1,
$(3, 1)$ -1, $(3, 0)$ -1, $(2, 5)$ 0, $(2, 4)$ 0, $(2, 3)$ 0, $(2, 2)$ 0, $(2, 1)$ 0, $(2, 0)$ 0,
$(1, 5)$ -1, $(1, 4)$ -1, $(1, 3)$ -1, $(1, 2)$ -1, $(1, 1)$ -1, $(1, 0)$ -1, $(0, 5)$ 4, $(0, 4)$ 0,
$(0, 3)$ 0, $(0, 2)$ 0, $(0, 1)$ 0, $(0, 0)$ 0.

Replacing two values, $\varphi(0, 3) = 2$, $\varphi(2, 1) = -2$ gives a 6-state type II RNCCA. Moreover replacing two values, $\varphi(0, 5) = 4$, $\varphi(4, 1) = -4$ gives another 6-state type II RNCCA.

4 Conclusion

In this paper, we present the exhaustive search result of radius 1 RNCCAs up to four states. There are three types of nontrivial RNCCA rules in the case of four states. Although our result deny Schranko and Oliveira's conjecture [8], it remains still open whether the complexity of radius 1 RNCCA is sufficient to provide a universal CA or not. Collision of two propagating (or stable) signals cannot change their states as far as this method is employed. To obtain a more complex behavior, another way of collisions seems to be required.

Acknowledgements Katsunobu Imai thanks Artiom Alhazov from the Academy of Science of Moldova for the helpful discussions and gratefully acknowledges the support of the Japan Society for the Promotion of Science and the Grant-in-Aid for Scientific Research (C) 22500015. Bruno Martin was partially supported by the French ANR, project EMC (ANR-09-BLAN-0164).

Appendix: The List of RNCCAs

Each rule is described in Wolfram numbering and its suffix is the radix of the number. The sets followed by $\swarrow$, $\downarrow$ and $\searrow$ denote left propagating, stable and right propagating state sets, respectively.

1. Radius 1/2
$n = 4$:
2-4-1 : 32103210321032104 $\swarrow$ $\{1, 2, 3\}$
2-4-2 : 32323232101010104 $\swarrow$ $\{1\}$ $\downarrow$ (2)
2-4-3 : 33112200331122004 $\swarrow$ $\{2\}$ $\downarrow$ $\{1\}$
2-4-4 : 33332222111100004 $\downarrow$ $\{1, 2, 3\}$

$n = 6$:

2-6-1 : $5432105432105432105432105432105432105432105432105432105432105432105432105_6$ ⤢ $\{1, 2, 3, 4, 5\}$

2-6-2 : $543543543543543543210210210210210210_6$ ⤢ $\{1, 2\}$ ↓ $\{3\}$

2-6-3 : $545454545454323232323232321010101010101010_6$ ⤢ $\{1\}$ ↓ $\{2, 4\}$

2-6-4 : $553311442200553311442200553311442200_6$ ⤢ $\{2, 4\}$ ↓ $\{1\}$

2-6-5 : $555222444111333000555222444111333000_6$ ⤢ $\{3\}$ ↓ $\{1, 2\}$

2-6-6 : $555555444444333333222222111111000000_6$ ↓ $\{1, 2, 3, 4, 5\}$

$n = 8$:

2-8-1 : $765432107654321076543210765432107654321076543210765432107654321076543210_8$ ⤢ $\{1, 2, 3, 4, 5, 6, 7\}$

2-8-2 : $765476547654765476547654765476543210321032103210321032103210321032103210_8$ ⤢ $\{1, 2, 3\}$ ↓ $\{4\}$

2-8-3 : $767632327676323254541010545410107676323276763232545410105454101010_8$ ⤢ $\{1, 4, 5\}$ ↓ $\{2\}$

2-8-4 : $767676767676767654545454545454543232323232323232321010101010101010_8$ ⤢ $\{1\}$ ↓ $\{2, 4, 6\}$

2-8-5 : $775533116644220077553311664422007755331166442200775533116644220_8$ ⤢ $\{2, 4, 6\}$ ↓ $\{1\}$

2-8-6 : $775577556644664477557755664466443311331122002200331133112200220_8$ ⤢ $\{2\}$ ↓ $\{1, 4, 5\}$

2-8-7 : $777733336666222255551111444400007777333366662222555511114444000_8$ ⤢ $\{4\}$ ↓ $\{1, 2, 3\}$

2-8-8 : $777777776666666655555555444444443333333322222222111111110000000_8$ ↓ $\{1, 2, 3, 4, 5, 6, 7\}$

$n = 9$:

2-9-1 : $876543210876543210876543210876543210876543210876543210876543210876543210_9$
⤢ $\{1, 2, 3, 4, 5, 6, 7, 8\}$

2-9-2 : $876876876876876876876876765435435435435435435435435432102102102102102102102102_9$
⤢ $\{1, 2\}$ ↓ $\{3, 6\}$

2-9-3 : $888555222777444111666333000888555222777444111666333000888555222777444111666333000_9$
⤢ $\{3, 6\}$ ↓ $\{1, 2\}$

2-9-4 : $888888887777777766666666655555555544444444433333333222222222111111111000000000_9$
↓ $\{1, 2, 3, 4, 5, 6, 7, 8\}$

$n = 10$:

2-10-1 : $9876543210987654321098765432109876543210987654321098765432109876543210987654321098765$
432109876543210_{10}, ⤢ $\{1, 2, 3, 4, 5, 6, 7, 8, 9\}$

2-10-2 : $9876598765987659876598765987659876598765987654321043210432104321043210432104321043210$
432104321043210_{10}, ⤢ $\{1, 2, 3, 4\}$ ↓ $\{5\}$

2-10-3 : $9898989898989898989876767676767676767676765454545454545454545454543232323232323232323210101$
010101010101010_{10}, ⤢ $\{1\}$ ↓ $\{2, 4, 6, 8\}$

2-10-4 : $9977553311886644220099775533118866442200997755331188664422009977553311886644220099775$
533118866442200_{10}, ⤢ $\{2, 4, 6, 8\}$ ↓ $\{1\}$

2-10-5 : $9999944444888883333377777222226666611111555550000099999444448888833333777772222266666$
111115555500000_{10}, ⤢ $\{5\}$ ↓ $\{1, 2, 3, 4\}$

2-10-6 : $9999999999888888888877777777776666666666555555555544444444443333333333222222222211111$
111110000000000_{10}, ↓ $\{1, 2, 3, 4, 5, 6, 7, 8, 9\}$

2. Radius 1

$n = 2$: $n = 3$:

3-2-1 : 10101010_2 ⤢ $\{1\}$ 3-3-1 : $210210210210210210210210_3$ ⤢ $\{1, 2\}$

3-2-2 : 11001100_2 ↓ $\{1\}$ 3-3-2 : $222111000222111000222111000_3$ ↓ $\{1, 2\}$

3-2-3 : 11110000_2 ↘ $\{1\}$ 3-3-3 : $222222221111111110000000000_3$ ↘ $\{1, 2\}$

$n = 4$:

3-4-1 : $3210321032103210321032103210321032103210321032103210321032103210_4$ ⤢ $\{1, 2, 3\}$

3-4-2 : $3230323212103232303232121010103230101210101032303232121010104_4$ ⤢ $\{1\}$ ↓ $\{2\}$

3-4-3 : $3232321210103010323232123232301032323212101030101010321210103010_4$ ⤢ $\{1, 3\}$ ↓ $\{2\}$

3-4-4 : $3232323210101010323232321010101032323232101010103232323210101010_4$ ⤢ $\{1\}$ ↓ $\{2\}$

3-4-5 : $3232323232323232323232323232323232321010101010101010101010101010_4$ ⤢ $\{1\}$ ↘ $\{2\}$

3-4-6 : $33102210331133113310221022002200331022103311220033102210331 12200_4$ ↗ {2} ↓ {1}

3-4-7 : $331122003211320033112200321132003311331132113200220022003 2113200_4$ ↗ {2, 3} ↓ {1}

3-4-8 : $331122003311220033112200331122003311220033112200331122003311 2200_4$ ↗ {2} ↓ {1}

3-4-9 : $33113311331133112200220022002200331133113311331122002200220022 00_4$ ↗ {2} ↘ {1}

3-4-10 : $333122221311222233312222131100003331000013110000333122221311 0000_4$ ↓ {1, 2}

3-4-11 : $3331333313113333222022220200000003331111113111111222022220200 0000_4$ ↓ {2} ↘ {1, 3}

3-4-12 : $3332223211111111333222320000000033322232111100003332223211110 000_4$ ↓ {1, 2}

3-4-13 : $333222323333333333222322222222111000101111000011100010111100 00_4$ ↓ {1} ↘ {2, 3}

3-4-14 : $333322021111200033332202333320000333220201111200011112202111 12000_4$ ↓ {1, 2}

3-4-15 : $333322221011100033332222101110003333333310111000222222210111 0000_4$ ↓ {1, 2}

3-4-16 : $333322221111000033332222111100003333222211110000333322221111 0000_4$ ↓ {1, 2, 3}

3-4-17 : $33332222323332223333222232333322211111111011100000000000010111 1000_4$ ↓ {1} ↘ {2}

3-4-18 : $3333222233332222333322223333222211110000111100001111000011110 000_4$ ↓ {1} ↘ {2}

3-4-19 : $333333131111311122222200222222000333333131111311100002020000020 000_4$ ↓ {2} ↘ {1}

3-4-20 : $3333333311111111122222220000000003333333311111111222222200000000 00_4$ ↓ {2} ↘ {1}

3-4-21 : $3333333333333333322222222222222222211111111111111110000000000000000 00_4$ ↘ {1, 2, 3}

$n = 5$:

3-5-1 : $4321043210432104321043210432104321043210432104321043210432104321043 2104321043210432104321043210432104321043210432104321$
$0432104321043210432104321043210432104321043210_5$

3-5-2 : $4442433332420233333020004442433332420211111020004442411111242023333302000044424333 3$
$3242021111102000444241111124202111111020000_5$

3-5-3 : $44444333332220021111102000044444333332220211111020000444443333322200233333020000444443333$
$32220021111102000444441111122202111111020000_5$

3-5-4 : $44444333332222100111100004444433333222221001110000444443333322222100111100004444444444$
$433333100111100003333322222222222221001111000_5$

3-5-5 : $44444333332222211011010000444443333322222110110100004444433333222221101101000044444 3333$
$3333331101101000044444222222222211011010000_5$

3-5-6 : $4444433303222221111300004444333032222224444430000044444333032222211111300004444443330$
$322222111113000011111330032222211111300000_5$

3-5-7 : $44444333332122220111100004444433333212222011110000444444444421222201111000033333444$
$4212222011110000333333333212222011110000_5$

3-5-8 : $44444333332222210111100004444433333222221011110000444443333322222101111000044444444$
$42222210111100003333333333222221011110000_5$

3-5-9 : $44444333312222211311000004444433333122222113112222444443333122222113110000044444333 3$
$100000131110000044444333312222211311000005_5$

3-5-10 : $44444331313122222313100000444443131444443131122222444443131222231311000002222233 13$
$1000003131100000444443313122222313110000005_5$

3-5-11 : $444413333322222141113333344441333332222214111000004444133333222221411100000444410 000$
$022222141110000044441333332222214111000005_5$

3-5-12 : $44444331332222231111000004444433133444443111100000444443313322222311110000022222331 3$
$32222231111000004444433133222223111100000_5$

3-5-13 : $44444333332122221111000004444433333212222111100000444444444421222211111000003333333 3$
$3212222111110000044444333332122221111100000_5$

3-5-14 : $4444333334222232111111111144443333342222320000011111444433334222232000000000044443333 4$
$22223211111000004444333334222232111110000_5$

3-5-15 : $44444333332222321111100000444443333322222321111111111444443333322222320000000000444443333$
$222232111110000044444333332222321111100000_5$

3-5-16 : $44424333332422233333000004442433333242221111100000444241111124222111110000044424333 3$
$3242221111100000444243333324222111110000_5$

3-5-17 : $444433333432222211111111111444433333432222000000000004444333334322222111110000044443333 4$
$3222221111100000444433333432222211111100000_5$

3-5-18 : $4443433433222222222000004443433433311111111110000044434334332222211111100000444343343$
$32222211111000004443433433222222220000000_5$

3-5-19 : $444443333322222111110000044444333332222211111100000444443333322222111110000044444 3333$
$32222211111000004444433333222221111100000_5$

3-5-20 : $4444444444444444444444443333333333333333333333333332222222222222222222222222221111111111$
$1111111111111111000000000000000000000000000_5$

References

1. Amoroso, S., Patt, Y.N.: Decision procedures for surjectivity and injectivity of parallel maps for tessellation structures. J. Comput. Syst. Sci. **6**(5), 448–464 (1972)
2. Boccara, N., Fukś, H.: Number-conserving cellular automaton rules. Fundamenta Informaticae **52**, 1–13 (2003)
3. Durand, B., Formenti, E., Róka, Z.: Number conserving cellular automata I: decidability. Theor. Comput. Sci. **299**(1–3), 523–535 (2003)
4. Fukś, H., Sullivan, K.: Enumeration of number-conserving cellular automata rules with two inputs. J. Cell. Autom. **2**(2), 141–148 (2007)
5. García-Ramos, F.: Product decomposition for surjective 2-block NCCA. In: Proceedings of 17th International Workshop on Cellular Automata and Discrete Complex Systems (AUTOMATA 2011), pp. 221–232 (2011)
6. Kari, J., Taati, S.: A particle displacement representation for conservation laws in two-dimensional cellular automata. In: Proceedings of Journées Automates Cellulaires (JAC 2008) (Uzès), pp. 65–73 (2008)
7. Morita, K.: Universality of 1-D reversible number-conserving cellular automata. Resume for International meeting on cellular automata, pp. 1–8. Osaka (2012)
8. Schranko A., Oliveira P.: Derivation of one-dimensional, reversible, number-conserving cellular automata rules. In: Proceedings of 15th International Workshop On Cellular Automata and Discrete Complex Systems (Automata 2009). pp. 335–345 (2009)
9. Wolfram S.: A New Kind of Science, Wolfram Media (2002)

The Computing Power of Determinism and Reversibility in Chemical Reaction Automata

Fumiya Okubo and Takashi Yokomori

Abstract Chemical reaction automata (CRAs) are computing models with multiset storage based on multiset rewriting introduced in Okubo, Yokomori, (DNA20, LNCS, vol. 8727, pp. 53–66, (2014), [25]). A CRA consists of a finite set of reactions (or pairs of multisets called reactants and products, respectively) and an initial multiset as well as a set of final multisets. Taking an input symbol in the current configuration (multiset) a CRA changes it into a new configuration. Thus, a CRA offers an automaton-like computing model to investigate the computational analysis of chemical reactions. On the other hand, since any (irreversible) Turing machine was proven to be effectively simulated by a reversible Turing machine in Bennett, (IBM J Res Dev, 17(6), 525–532, (1973), [4]), reversible computing has become a research field that has been receiving increased attention. In this paper we introduce the notions of determinism and reversibility into CRAs, and investigate the computational powers of those classes of CRAs in comparison with the language classes of Chomsky hierarchy. The computing power of reversible CRAs involves the physical realization of molecular programming of chemical reaction networks (Thachuk, Condon, DNA 18, LNCS, vol. 7433, pp. 135–149, (2012), [32]) with DNA strand displacement system implementation (Qian, Winfree, Science, 332, 1196–1201, (2011), [29]), and therefore, it is of great significance to elucidate the computing capabilities of both deterministic and reversible CRAs from the theoretical viewpoint of molecular computing.

F. Okubo
Faculty of Arts and Science, Kyushu University, 744 Motooka, Nishi-ku,
Fukuoka 819-0395, Japan
e-mail: fokubo@artsci.kyushu-u.ac.jp

T. Yokomori (✉)
Department of Mathematics, Faculty of Education and Integrated Arts and Sciences,
Waseda University, 1-6-1 Nishiwaseda, Shinjuku-ku, Tokyo 169-8050, Japan
e-mail: yokomori@waseda.jp

© Springer International Publishing AG 2018
A. Adamatzky (ed.), *Reversibility and Universality*, Emergence, Complexity
and Computation 30, https://doi.org/10.1007/978-3-319-73216-9_13

1 Introduction

During the past two decades, biochemical computing theory has attracted much attention in the area of "natural computation", in particular, molecular-based computing (e.g., [31]). There are two major approaches to the theoretical research on biochemical computing. One is to utilize an analytical framework based on ordinary differential equations (ODEs) in which macroscopic behavior of molecules is formulated as ODEs by approximating a massive number of molecules (or molecular concentration) as a continuous quantity. The other is to construct a discrete framework based on multiset rewriting in which a volume of various molecular species in relatively small quantities is represented by a multiset and a biochemical reaction is simulated by replacing the multiset with another one, under prescribed conditions [5, 7, 26].

Among many models studied from the viewpoint of the latter approach, a most seminal paper [9] introduced formal models, called reaction systems, for investigating the functioning of living cells, where two basic components (reactants and inhibitors) play a key role as a regulation mechanism in controlling interactions. Inspired by the paper, we first introduced reaction automata (RAs) in [23] as computing devices for accepting string languages and it was shown that RAs are computationally Turing universal. Then, in subsequent articles [22, 24], we continued to investigate the computing powers of subclasses of RAs as well as the closure properties of their language families. Further, in our recent paper [25] a simplified version of RAs, called *chemical reaction automata* (CRAs), has been studied, which has clarified the difference of computational capability between sequential and parallel manners of multiset rewriting within the framework of CRAs.

In the theory of CRAs, a chemical reaction is formulated as a pair $\mathbf{a} = (R_{\mathbf{a}}, P_{\mathbf{a}})$, denoted by $R_{\mathbf{a}} \rightarrow P_{\mathbf{a}}$, where $R_{\mathbf{a}}$ is the multiset of molecules called *reactants*, and $P_{\mathbf{a}}$ is the multiset of molecules called *products*. Let T be a multiset of molecules, then the result of applying a reaction $\mathbf{a}$ to T, denoted by $Res_{\mathbf{a}}(T)$, is given by $T'(= T - R_{\mathbf{a}} + P_{\mathbf{a}})$ if $\mathbf{a}$ is enabled by T (i.e., if T completely includes $R_{\mathbf{a}}$). A computation step in CRAs is performed in such a way that each time receiving an input molecule a_{i+1} $(i = 0, 1, 2, \ldots)$ provided from the environment, the system incorporates it to the current multiset T_i and transforms into $T_{i+1}(= Res_{\mathbf{a}}(T_i \cup \{a_{i+1}\}))$. When starting with the initial state T_0 (of a multiset), an external input string $a_1 \ldots a_n$ is accepted if it induces a sequence of configurations T_i that ends in a final configuration predesignated. As is mentioned below, this computing model CRA is closely related to an important model of molecular programming languages called chemical reaction networks (CRNs) in [32].

CRNs are formal models for molecular programming and defined as a finite set of chemical reactions consisting of pairs of multisets (molecules) which prescribes how given sets of molecules can be transformed into new sets of molecules caused by activating reactions. In this sense, CRAs can be regarded as *online* computational models of CRNs to whom an input signal is fed one by one from the environment. An advantage of CRNs is that they can be in principle implemented by a molecular reaction primitive called DNA strand displacement (DSD) systems in [29]. For example, a bimolecular reaction $X + Y \to A + B$ is successfully materialized by a DSD system implementation, and in fact, an additionally modified implementation enables its reverse reaction, i.e., $X + Y \rightleftarrows A + B$ is realizable with DNA molecules [30]. CRNs with DSD system implementation also have a great potential for realizing energy efficient computation due to the physical reversibility that is one of the essential principles in microscopic nature.

Since any (irreversible) Turing machine was proven to be effectively simulated by a reversible Turing machine in [4], reversible computing has become a research field that has been receiving increased attention. The physical significance of reversible computation is theoretically supported by the results that non-injective operations entail an irreducible thermodynamic cost, while injective ones do not [4, 15]. One may find an informatively complete survey [19] that details results on reversible Turing machines, reversible cellular automata, and reversible computing models of logic gates and circuits.

With these academic backgrounds, in this paper we shall first introduce the notions of determinism and reversibility into CRAs, and then investigate the computational powers of those classes of CRAs. The computing power of reversible CRAs involves the physical realization of molecular programming of CRNs with DSD system implementation, and therefore, it is of great significance to elucidate the computing capabilities of deterministic and reversible CRAs from the theoretical viewpoint of molecular computing.

2 Preliminaries

2.1 Basic Definitions

We assume that the reader is familiar with the basic notions of formal language theory. For unexplained details, refer to [11]. Let V be a finite alphabet. For a set $U \subseteq V$, the cardinality of U is denoted by $|U|$. The set of all finite-length strings over V is denoted by V^*. The empty string is denoted by λ.

We use the basic notations and definitions regarding multisets that follow [5, 13]. A multiset over an alphabet V is a mapping $\mu : V \to \mathbf{N}$, where $\mathbf{N}$ is the set of non-negative integers and for each $a \in V$, $\mu(a)$ represents the number of occurrences of a in the multiset μ. The set of all multisets over V is denoted by $V^\#$, including the empty multiset denoted by μ_λ, where $\mu_\lambda(a) = 0$ for all $a \in V$. We can represent the multiset μ by any permutation of the string $w = a_1^{\mu(a_1)} \ldots a_n^{\mu(a_n)}$, where $V = \{a_1, a_2, \ldots, a_n\}$. Conversely, with any string $x \in V^*$ one can associate the multiset $\mu_x : V \to \mathbf{N}$ defined by $\mu_x(a) = |x|_a$ for each $a \in V$. In this sense, we often identify a multiset μ with its string representation w_μ or any permutation of w_μ. Note that the string representation of μ_λ is λ, i.e., $w_{\mu_\lambda} = \lambda$. A set $U(\subseteq V)$ is regarded as a multiset μ_U such that $\mu_U(a) = 1$ if a is in U and $\mu_U(a) = 0$ otherwise. In particular, for each symbol $a \in V$, a multiset $\mu_{\{a\}}$ is often denoted by a itself.

For two multisets μ_1, μ_2 over V, we define one relation and two operations as follows:

- Inclusion: $\mu_1 \subseteq \mu_2$ iff $\mu_1(a) \leq \mu_2(a)$, for each $a \in V$,
- Sum: $(\mu_1 + \mu_2)(a) = \mu_1(a) + \mu_2(a)$, for each $a \in V$,
- Difference: $(\mu_1 - \mu_2)(a) = \mu_1(a) - \mu_2(a)$, for each $a \in V$
  (for the case $\mu_2 \subseteq \mu_1$).

2.2 Chemical Reaction Automata

Inspired by the work of reaction systems [9], we have introduced the notion of reaction automata in [23] by extending sets in each reaction to multisets. In addition, as a special version of a reaction automaton we defined a *chemical reaction automaton* (CRA) in [25]. Here, we start by recalling basic notions concerning CRAs.

Definition 1 For a set S, a *reaction rule* (or *reaction*) in S is an ordered pair $\mathbf{a} = (R_\mathbf{a}, P_\mathbf{a})$ of finite multisets, where $R_\mathbf{a}, P_\mathbf{a} \in S^\#$.

The multisets $R_\mathbf{a}$ and $P_\mathbf{a}$ are called the *reactant* of $\mathbf{a}$ and the *product* of $\mathbf{a}$, respectively. For convenience, a reaction $\mathbf{a} = (R, P)$ is denoted by $\mathbf{a} : R \to P$.

In [25], we considered two ways for applying reactions, i.e., sequential manner and maximally parallel manner. Intuitively, one reaction is applied to a multiset of objects in each step in sequential manner, while a multiset of reactions is exhaustively applied to a multiset in maximally parallel manner. However, in this paper, only sequential manner is considered.

Definition 2 Let A be a set of reactions in S. Then, for a finite multiset $T \in S^\#$, we say that
(1) a reaction $\mathbf{a}$ is *enabled by* T if $R_\mathbf{a} \subseteq T$,
(2) by $En_A^{sq}(T)$ we denote the set of reactions in A which are enabled by T,
(3) the *results of A on T*, denoted by $Res_A^{sq}(T)$, is defined as follows:

$$Res_A^{sq}(T) = \{T - R_\mathbf{a} + P_\mathbf{a} \mid \mathbf{a} \in En_A^{sq}(T)\}.$$

We note that $Res_A^{sq}(T) = \{T\}$ if $En_A^{sq}(T) = \emptyset$. Thus, if no reaction $\mathbf{a} \in A$ is enabled by T, then T remains unchanged.

It is obvious from the above definition that a reaction is applied to a multiset in *nondeterministic way*.

We are now in a position to introduce the notion of CRAs.

Definition 3 A *chemical reaction automaton* (CRA) $\mathcal{A}$ is a 5-tuple $\mathcal{A} = (S, \Sigma, A, D_0, F)$, where

- S is a finite set, called the *background set of* A,
- $\Sigma(\subseteq S)$ is called the *input alphabet of* A,
- A is a finite set of reactions in S,
- $D_0 \in S^{\#}$ is an *initial multiset*,
- $F \in 2^{S^{\#}}$ is a finite set of final multisets.

Definition 4 Let $A = (S, \Sigma, A, D_0, F)$ be a CRA and $w = a_1 \ldots a_n \in \Sigma^*$ be an input string. An *interactive process* in A by input w is a finite sequence $\pi = D_0, D_1, \ldots, D_n$, where for $0 \leq i \leq n - 1$,

$$D_{i+1} \in Res_A^{sq}(a_{i+1} + D_i),$$

In order to represent an interactive process π, we use the "arrow notation" for π: $D_0 \to^{a_1} D_1 \to^{a_2} D_2 \to^{a_3} \cdots \to^{a_{n-1}} D_{n-1} \to^{a_n} D_n$. By $IP_{sq}(A, w)$ we denote the set of all interactive processes in A with input w.

Let $\Sigma_\lambda = \Sigma \cup \{\lambda\}$. When $a_i = \lambda$ for several i's $(1 \leq i \leq n)$ in an input string $w = a_1 a_2 \ldots a_n$, an interactive process is said to be in *λ-input mode*. By $IP_{sq}^\lambda(A, w)$ we denote the set of all interactive processes in A with λ-input mode for the input $w \in \Sigma_\lambda^*$. We note that an ordinary interactive process, where no λ-input is used, is said to be *realtime* to distinguish from λ-input mode.

A CRA working with a realtime interactive process is called a realtime CRA, while a CRA working with a λ-input mode interactive process is called a CRA with λ-input mode.

A language accepted by a realtime CRA A is defined by

$$L_{sq}(A) = \{w \in \Sigma^* \mid \pi : D_0 \to^{a_1} D_1 \to^{a_2} \cdots \to^{a_n} D \in IP_{sq}(A, w), \, D \in F, \, w = a_1 a_2 \ldots a_n\}.$$

Similarly, a language accepted by a CRA $A = (S, \Sigma, A, D_0, F)$ with λ-input mode is defined by

$$L_{sq}^\lambda(A) = \{w \in \Sigma_\lambda^* \mid \pi : D_0 \to^{a_1} D_1 \to^{a_2} \cdots \to^{a_n} D \in IP_{sq}^\lambda(A, w), \, D \in F, \, w = a_1 a_2 \ldots a_n\}.$$

Example 1 Let $A = (S, \{a, b\}, A, p_0, \{f\})$ be a CRA, where

$$S = \{a, b\} \cup \{a', p_0, p_1, f, \natural\}$$
$$A = \{\mathbf{a_1} : p_0 + a \to p_0 + a', \, \mathbf{a_2} : p_0 + a' + b \to p_1, \, \mathbf{a_3} : p_0 + b \to \natural,$$
$$\mathbf{a_4} : p_0 \to f, \, \mathbf{a_5} : p_1 + a' + b \to p_1, \, \mathbf{a_6} : p_1 \to f,$$
$$\mathbf{a_7} : p_1 + a \to \natural, \, \mathbf{a_8} : f + a \to \natural, \, \mathbf{a_9} : f + a' \to \natural, \, \mathbf{a_{10}} : f + b \to \natural\}$$

In the graphic drawing in Fig. 1, each reaction $\mathbf{a}_i$ is applied to a multiset (a test tube) after receiving an input symbol (if any is provided) from the environment. In particular, applying $\mathbf{a_4}$ to $\{p_0\}$ leads to that the empty string is accepted by A. It is seen, for example, that reactions $\mathbf{a_1}$ and $\mathbf{a_2}$ are enabled by the multiset $D_1 = \{p_0, a', a'\}$

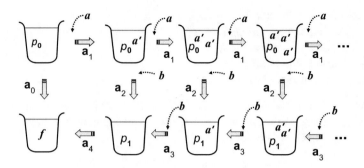

Fig. 1 A graphic illustration of interactive chemical reaction processes for accepting strings in the language $L = \{a^n b^n \mid n \geq 0\}$ by means of $\mathcal{A}$ in Example 1

only when inputs a and b, respectively, are received, which result in producing $D_2 = \{p_0, a', a', a'\}$ and $D_3 = \{p_1, a'\}$, respectively. Thus, we have that $Res_{\mathbf{a}_1}(D_1 \cup \{a\}) = D_2$ and $Res_{\mathbf{a}_2}(D_1 \cup \{b\}) = D_3$. Once applying $\mathbf{a}_2$ has brought about a change of p_0 into p_1, $\mathcal{A}$ has no possibility of accepting further inputs a's. Otherwise any possible application of rules leads to introducing the symbol ♮, eventually resulting in the failure of computations. (Thus, in the construction of A, the symbol ♮ plays a role of *trapdoor* for unsuccessful computations.) It is easily seen that

$$L_{sq}^{\lambda}(\mathcal{A}) = \{a^n b^n \mid n \geq 0\}.$$

2.3 Deterministic CRAs: DCRAs

We are now in a position to introduce the notions of determinism and reversibility of a CRA. Our goal is to define these two notions on not only realtime CRA but also CRA with λ-input mode, in an approvable fashion.

Unlike the determinism of conventional computation models such as pushdown automata, since a reactant is not divided into input part and memory (multiset) part, the determinism of CRAs cannot be decided only by a form of transition rules. This comes from the property of multiset memory, that is, from the current configuration alone, a CRA cannot identify a reactant of the next reaction to be applied. Therefore, the determinism of CRA has to be defined so as to exclude any branching computation, regardless of a non-empty input or empty input.

A CRA is said to be deterministic if for every input symbol $a \in \Sigma$ and every reachable configuration D, the result of reaction is a singleton. Similar to the definition of deterministic pushdown automata, this condition is extended to the case of λ-input mode.

Formally, a realtime CRA $\mathcal{A}$ is *deterministic* if for any $a \in \Sigma$ and any reachable configuration D, it holds that

(a)   **(b)**   **(c)**

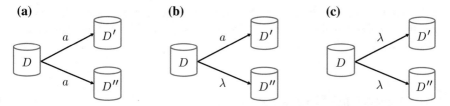

Fig. 2 There are three cases which are forbidden computations in a DCRA

$$Res_A^{sq}(D + a) = \{D'\}, \tag{i}$$

for some $D' \in S^\#$. A CRA $\mathcal{A} = (S, \Sigma, A, D_0, F)$ with λ-input mode is *deterministic* if for any $a \in \Sigma$ and any reachable configuration D, the followings hold

$$Res_A^{sq}(D + a) = \{D'\} \qquad\qquad \text{if } En_A^{sq}(D) = \emptyset \tag{ii}$$
$$Res_A^{sq}(D + a) \cup Res_A^{sq}(D) = \{D'\} \qquad\qquad \text{if } En_A^{sq}(D) \neq \emptyset, \tag{iii}$$

for some $D' \in S^\#$. We note that $En_A^{sq}(D) = \emptyset$ implies that without consuming an input, a configuration D remains unchanged. Hence, we exclude this "idling" case.

Computations which are not allowed in a deterministic CRA (abbrev. DCRA) are illustrated in Fig. 2, in which the case (a) is prohibited by (i) in a realtime DCRA and by (ii) in a DCRA with λ-input mode, while in addition, (b) and (c) are prohibited by (iii) in a DCRA with λ-input mode.

At first glance, it is not trivial, only from their definitions, to recognize the difference of computing powers of the determinism between realtime CRA and CRA with λ-input mode. In the following lemma, we show that an acceptance of a language L by a DCRA with λ-input mode implies that of L by a (realtime) DCRA.

Lemma 1 *If a language L is accepted by a DCRA with λ-input mode, then L is also accepted by a realtime DCRA.*

Proof Let $\mathcal{A} = (S, \Sigma, A, D_0, F)$ be a DCRA, $w = a_1 a_2 \ldots a_n$ be an input string, and $\pi : D_0 \rightarrow^{\hat{a}_1} D_1 \rightarrow^{\hat{a}_2} \cdots \rightarrow^{\hat{a}_m} D_m$ be an interactive process of M with λ-input mode, where $\hat{a}_1, \hat{a}_2, \ldots, \hat{a}_m \in \Sigma \cup \{\lambda\}$ and $w = \hat{a}_1 \hat{a}_2 \ldots \hat{a}_m$ for $0 \leq n \leq m$.

From the definition of a DCRA, for any $0 \leq i \leq m - 1$ and a part of $\pi : D_i \rightarrow^{\hat{a}_{i+1}} D_{i+1}$, an input $\hat{a}_{i+1}$ should be a symbol $a \in \Sigma$. This is because if $\hat{a}_{i+1} = \lambda$, then it is possible that $D_i \rightarrow^a D_{i+1} + a$ for any $a \in \Sigma$ using the same reaction as the one applied to D_i in π, leading to a contradiction to (iii) of the assumptions that $\mathcal{A}$ is deterministic.

This implies that any computations of DCRA $\mathcal{A}$ with λ-input mode is realized by using only a realtime computation of $\mathcal{A}$, that is, $L(= L(\mathcal{A}))$ is accepted by a realtime DCRA. $\square$

Remark As stated in the proof of Lemma 1, without receiving an input symbol, no configuration of a DCRA with λ-input mode can have an enable reaction. On the other hand, this is not the case for a realtime DCRA. In order to make a reaction, a realtime DCRA always requires an input symbol, even if its configuration has an enable reaction with no input symbol.

2.4 Reversible CRAs: RCRAs

Information preserving computations (forward and backward deterministic computations) are very important and considered as reversibility in many existing research papers [4, 10, 14, 19, 20]. However, in order to understand their properties in more details, we consider them apart, i.e., separate into forward and backward determinisms. Then, we take a new view of the "reversibility" which simply means that the previous configuration of computation can be uniquely determined (backward determinism). Hence, our definition requires an information preserving computation model to be both deterministic and reversible, and this paper investigates the computational capacity of each model.

A CRA is said to be reversible if for every input symbol $a \in \Sigma$ and every reachable configuration D, the set of configurations which directly reaches D with a is a singleton. For the case of λ-input mode, this condition is extended to the case of λ-input mode in a trivial manner.

Formally, a realtime CRA $\mathcal{A}$ is *reversible* if for any $a \in \Sigma$ and any two distinct reachable configurations D, D',

$$Res_A^{sq}(D + a) \cap Res_A^{sq}(D' + a) = \emptyset. \tag{iv}$$

A CRA $\mathcal{A} = (S, \Sigma, A, D_0, F)$ with λ-input mode is *reversible* if for any $a \in \Sigma$ and any two distinct reachable configurations D, D', the equation (iv) and the following equations (v), (vi) hold:

$$Res_A^{sq}(D + a) \cap Res_A^{sq}(D') = \emptyset, \qquad \text{if } En_A^{sq}(D') \neq \emptyset, \tag{v}$$
$$Res_A^{sq}(D) \cap Res_A^{sq}(D') = \emptyset \qquad \text{if } En_A^{sq}(D) \neq \emptyset \text{ and } En_A^{sq}(D') \neq \emptyset. \tag{vi}$$

Computations which are not allowed in a reversible CRA (abbrev. RCRA) are illustrated in Fig. 3, where (d), (e) and (f) are corresponding cases prohibited by (iv), (v) and (vi), respectively.

The following lemma holds true for a deterministic and reversible CRA (abbrev. DRCRA), whose proof is confirmed in a manner similar to Lemma 1.

Lemma 2 *If a language L is accepted by a DRCRA with λ-input mode, then L is also accepted by a realtime DRCRA.*

By $\mathcal{L}_{sq}(CRA)$, $\mathcal{L}_{sq}^{\lambda}(CRA)$, $\mathcal{L}_{sq}(DCRA)$, $\mathcal{L}_{sq}^{\lambda}(DCRA)$, $\mathcal{L}_{sq}(RCRA)$, $\mathcal{L}_{sq}^{\lambda}(RCRA)$, $\mathcal{L}_{sq}(DRCRA)$, and $\mathcal{L}_{sq}^{\lambda}(DRCRA)$, we denote the classes of languages accepted by realtime CRAs, CRAs with λ-input mode, realtime DCRAs, DCRAs with λ-input mode,

(d)   **(e)**   **(f)**

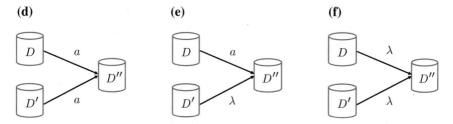

Fig. 3 There are three cases which are forbidden computations in an RCRA

realtime RCRAs, RCRAs with λ-input mode, realtime deterministic and reversible CRAs (DRCRAs), DRCRAs with λ-input mode, respectively.

3 Main Results

In this section, we will investigate the computational capabilities of variety of language classes introduced in the previous section.

3.1 Elementary Results

First, we consider the property of reversibility in general computing models. Let S be a set and $f : S \to 2^S$ be a mapping. An injective mapping can be defined as similar to the case of a function.

Definition 5 A mapping $f : S \to 2^S$ is an injection if for any $s_1, s_2 \in S$, $s_1 \neq s_2$ implies $f(s_1) \cap f(s_2) = \emptyset$.

A computing model is considered as reversible if a transition function (or mapping, for a nondeterministic model) is an injection. (It should be noted that in fact, for each $a \in \Sigma$ and a configuration D of an RCRA, if we define $f_a(D) = \{D' \mid D \to^a D' \text{ or } D \to D'\}$, then a transition mapping f_a is injective. Thus, a transition mapping of RCRA is injective.)

For each $i = 1, 2$, let $\mathcal{C}_i$ be the set of configurations of a computing model M_i, and let $f_i : \mathcal{C}_i \to 2^{\mathcal{C}_i}$ be a transition mapping (defined by M_i). Further, let $\phi : \mathcal{C}_1 \to \mathcal{C}_2$ be an injective function and we extend it to an injection from $2^{\mathcal{C}_1}$ to $2^{\mathcal{C}_2}$ in a natural manner.

Lemma 3 Assume that for any configuration $C \in \mathcal{C}_1$, it holds that $\phi(f_1(C)) = f_2(\phi(C))$. Then, f_2 is an injection implies that f_1 is an injection.

Proof Assume that f_1 is not an injection. Then, there exist $C_{11}, C_{12}, C_1 \in \mathcal{C}_1$ such that $C_{11} \neq C_{12}$ and $C_1 \in f_1(C_{11}) \cap f_1(C_{12})$. Let C_{21}, C_{22}, and C_2 be configurations in $\mathcal{C}_2$, where $C_{21} = \phi(C_{11})$, $C_{22} = \phi(C_{12})$, and $C_2 = \phi(C_1)$. Since ϕ is an injection, it holds that $C_{21} \neq C_{22}$. Here, it holds that $f_2(C_{21}) = f_2(\phi(C_{11})) = \phi(f_1(C_{11})) \ni \phi(C_1) = C_2$. Similarly, it holds that $f_2(C_{22}) \ni C_2$. This contradicts the assumption that f_2 is an injection $\qquad\square$

Lemma 4 (Normal form lemma for CRAs) *For any realtime CRA (RCRA) $\mathcal{A} = (S, \Sigma, A, D_0, F)$, there exists a realtime CRA (RCRA, resp.) $\mathcal{A}' = (S', \Sigma, A', D_0', F')$ such that for any input $w = a_1 a_2 \ldots a_n$ and an interactive process $\pi : D_0 \to^{a_1} D_1 \to^{a_2} \cdots \to^{a_n} D_n \in IP_{sq}(\mathcal{A}, w)$, there exists $\pi' : D_0' \to^{a_1} D_1' \to^{a_2} \cdots \to^{a_n} D_n' \in IP_{sq}(\mathcal{A}', w)$, where*

- *any reaction in A' is of the form $a\alpha' \to \beta'$ with $a \in \Sigma$ and $\alpha', \beta' \in (S' - \Sigma)^{\#}$,*
- *$D_0', D_1', \ldots, D_n' \in (S' - \Sigma)^{\#}$,*
- *$D_n \in F$ if and only if $D_n' \in F'$, i.e., $L_{sq}(\mathcal{A}) = L_{sq}(\mathcal{A}')$.*

Proof Consider a function $f : S \to \{a' \mid a \in S\}$ defined by $f(a) = a'$. The definition of f is naturally extended to a set of multisets. For a realtime CRA $\mathcal{A} = (S, \Sigma, A, D_0, F)$, construct $\mathcal{A}' = (S', \Sigma, A', D_0', F')$ as follows:

- $S' = \Sigma \cup f(S)$,
- A' consists of the following reactions:

type 1. $a \to a'$ for any $a \in \Sigma$,
type 2. $a + f(\alpha) \to f(\beta)$ for any $a + \alpha \to \beta \in A$ and any $a \in \Sigma$,
type 3. $f(a + \alpha) + b \to f(\beta) + b'$ for any $a + \alpha \to \beta \in A$ and any $b \in \Sigma$,
type 4. $f(\alpha) + b \to f(\beta) + b'$ for any $\alpha \to \beta \in A$ with $\alpha \in (S - \Sigma)^{\#}$ and any $b \in \Sigma$,

- $D_0' = f(D_0)$,
- $F' = f(F)$.

One step reaction $D_i \to^{a_{i+1}} D_{i+1}$ in π is simulated by $f(D_i) \to^{a_{i+1}} f(D_{i+1})$ in π' as follows:

- if there is no enable reaction to D_i, i.e. $D_{i+1} = a + D_i$ for $a = a_{i+1} \in \Sigma$, then by using a reaction of type 1, it holds that $f(D_i) \to^{a_{i+1}} a' + f(D_i)$ $(= f(D_{i+1}))$,
- if $D_i \to^{a_{i+1}} D_{i+1}$ is realized by a reaction $a_{i+1} + \alpha \to \beta \in A$, then the corresponding reaction of type 2, $a_{i+1} + f(\alpha) \to f(\beta)$, is applied,
- if $D_i \to^{a_{i+1}} D_{i+1}$ is realized by a reaction $a + \alpha \to \beta \in A$ with $a \neq a_{i+1}$, then the corresponding reaction of type 3, $f(a + \alpha) + a_{i+1} \to f(\beta) + a_{i+1}'$, is applied,
- if $D_i \to^{a_{i+1}} D_{i+1}$ is realized by a reaction $\alpha \to \beta \in A$ with $\alpha \in (S - \Sigma)^{\#}$, then the corresponding reaction of type 4, $f(\alpha) + a_{i+1} \to f(\beta) + a_{i+1}'$, is applied,

It is easily confirmed that $D \in F$ if and only if $D' \in F'$, i.e., $L_{sq}(\mathcal{A}) = L_{sq}(\mathcal{A}')$.

Considering an extended bijection f from the set of configurations of $\mathcal{A}'$ onto that of configurations of $\mathcal{A}$, it follows from Lemma 3 that if $\mathcal{A}$ is reversible, then $\mathcal{A}'$ is reversible. $\qquad\square$

In a computation of realtime CRA in normal form, each step is forced to require a non-empty input symbol and the symbol is consumed on the spot (in that step), which is not necessarily guaranteed by a realtime CRA in general case. This property of the normal form makes it possible to consider an input and a multiset separately, and easily to construct a CRA with λ-input mode that is equivalent to a given realtime CRA, as is seen below.

Lemma 5 *If a language L is accepted by a realtime CRA (RCRA), then L is also accepted by a CRA (RCRA, resp.) with λ-input mode.*

Proof We consider $\mathcal{A}'$ of Lemma 4 with λ-input mode. From the manner of constructing $\mathcal{A}'$, there is no chance to use λ-move in any computation of $\mathcal{A}'$, because any reachable configuration contains only elements in $f(S)$, and any reaction in A' inevitably requires $a \in \Sigma$ to be enabled. Hence, it holds that $L_{sq}(\mathcal{A}) = L_{sq}(\mathcal{A}') = L_{sq}^{\lambda}(\mathcal{A}')$. $\square$

The following relations are immediately derived from the definitions of determinism and reversibility of CRAs.

Lemma 6 *It holds that*

$$(i)\ \mathcal{L}_{sq}(DRCRA) \subseteq \mathcal{L}_{sq}(DCRA) \subseteq \mathcal{L}_{sq}(CRA),$$

$$(ii)\ \mathcal{L}_{sq}^{\lambda}(DRCRA) \subseteq \mathcal{L}_{sq}^{\lambda}(DCRA) \subseteq \mathcal{L}_{sq}^{\lambda}(CRA),$$

$$(iii)\ \mathcal{L}_{sq}(DRCRA) \subseteq \mathcal{L}_{sq}(RCRA) \subseteq \mathcal{L}_{sq}(CRA),$$

$$(iv)\ \mathcal{L}_{sq}^{\lambda}(DRCRA) \subseteq \mathcal{L}_{sq}^{\lambda}(RCRA) \subseteq \mathcal{L}_{sq}^{\lambda}(CRA).$$

3.2 Computational Capability of Various Classes of CRAs

In order to investigate a relationship among the various classes of languages accepted by *CRAs*, we use the following languages:

$$L_1 = \{a^n \mid n \geq 1\}, \qquad\qquad L_2 = \{a^n b^n, a^n b^{2n} \mid n \geq 0\},$$
$$L_3 = \{a^n \mid n \geq 0\} \cup \{b\}, \qquad L_4 = \{a^n d b^n e c^n \mid n \geq 0\}.$$

Lemma 7 *The class of languages $\mathcal{L}_{sq}(DRCRA)$ ($\mathcal{L}_{sq}^{\lambda}(DRCRA)$) is strictly included in $\mathcal{L}_{sq}(DCRA)$ ($\mathcal{L}_{sq}^{\lambda}(DCRA)$, resp.) and $\mathcal{L}_{sq}(RCRA)$ ($\mathcal{L}_{sq}^{\lambda}(RCRA)$, resp.).*

Proof We shall show by contradiction that L_1 is not in $\mathcal{L}_{sq}(DRCRA)$. Assume that there is a realtime DRCRA $M = (S, \Sigma, A, D_0, F)$ which accepts $L_1 = \{a^n \mid n \geq 1\}$.
Since F is finite, there exist i, j with $1 \geq i < j$ and interactive processes

$$\pi_i : D_0 \to^a D_1 \to^a \cdots \to^a D_i \in IP_{sq}(M, a^i),$$
$$\pi_j : D_0 \to^a D_1 \to^a \cdots \to^a D_i \to^a D_{i+1} \to^a \cdots \to^a D_j \in IP_{sq}(M, a^j),$$

where $D_i = D_j \in F$. Since M is reversible, it inductively holds that $D_{i-1} = D_{j-1}, D_{i-2} = D_{j-2}, \ldots, D_0 = D_{j-i}$. Moreover, since M is deterministic,

$$\pi_{j-i} : D_0 \to^a D_1 \to^a \cdots \to^a D_{j-i} = D_0 \in IP_{sq}(M, a^{j-i})$$

is the unique interactive process for input $a^{j-i} \in L_1$. Hence, the configuration $D_{j-i} = D_0$ must be in F, which contradicts that $\lambda \notin L_1$.

Let $M_{1D} = (\{a, p_0, p_1\}, \{a\}, \{p_0a \to p_1, p_1a \to p_1\}, p_0, \{p_1\})$ be a CRA. It is easily shown that $L_1 = L_{sq}^{\lambda}(M_{1D})$ and M_{1D} is deterministic. Similarly, for a CRA $M_{1R} = (\{a, p_0, p_1\}, \{a\}, \{p_0a \to p_0, p_0a \to p_1\}, p_0, \{p_1\})$, it can be shown that $L_1 = L_{sq}(M_{1R})$ and M_{1R} is reversible. Hence it holds that $L_1 \in \mathcal{L}_{sq}^{\lambda}(DCRA) \cap \mathcal{L}_{sq}(RCRA)$.

From Lemmas 1 and 6, the followings hold:

- $\mathcal{L}_{sq}^{\lambda}(DRCRA) \subseteq \mathcal{L}_{sq}(DRCRA) \subseteq \mathcal{L}_{sq}(DCRA)$,
- $\mathcal{L}_{sq}^{\lambda}(DRCRA) \subseteq \mathcal{L}_{sq}^{\lambda}(DCRA) \subseteq \mathcal{L}_{sq}(DCRA)$,
- $\mathcal{L}_{sq}^{\lambda}(DRCRA) \subseteq \mathcal{L}_{sq}(DRCRA) \subseteq \mathcal{L}_{sq}(RCRA) \subseteq \mathcal{L}_{sq}^{\lambda}(RCRA)$.

Hence, we can obtain that $\mathcal{L}_{sq}^{\lambda}(DRCRA) \subset \mathcal{L}_{sq}^{\lambda}(DCRA), \mathcal{L}_{sq}^{\lambda}(DRCRA) \subset \mathcal{L}_{sq}^{\lambda}(RCRA)$, $\mathcal{L}_{sq}(DRCRA) \subset \mathcal{L}_{sq}(DCRA)$, and $\mathcal{L}_{sq}(DRCRA) \subset \mathcal{L}_{sq}(RCRA)$. $\quad\square$

Lemma 8 *The class of languages $\mathcal{L}_{sq}(DCRA)$ is strictly included in $\mathcal{L}_{sq}(CRA)$.*

Proof It suffices to show the strictness of an inclusion relation. We can easily construct a realtime CRA accepting $L_2 = \{a^n b^n, a^n b^{2n} \mid n \geq 0\}$.

We shall show by contradiction that L_2 is not in $\mathcal{L}_{sq}(DCRA)$. Assume that there is a realtime DCRA $M = (S, \Sigma, A, D_0, F)$ which accepts L_2. Since the set of final multisets F is a finite set and M is deterministic, there exist i, j with $1 \leq i < j$ and four interactive processes such that

$$\pi_1 : D_0 \to^a D_1 \to^a \cdots \to^a D_i \to^b D_{i+1} \to^b \cdots \to^b D_{2i},$$
$$\pi_2 : D_0 \to^a D_1 \to^a \cdots \to^a D_i \to^b D_{i+1} \to^b \cdots \to^b D_{2i} \to^b D_{2i+1} \to^b \cdots \to^b D_{3i},$$
$$\pi_3 : D'_0 \to^a D'_1 \to^a \cdots \to^a D'_j \to^b D'_{j+1} \to^b \cdots \to^b D'_{2j},$$
$$\pi_4 : D'_0 \to^a D'_1 \to^a \cdots \to^a D'_j \to^b D'_{j+1} \to^b \cdots \to^b D'_{2j} \to^b D'_{2j+1} \to^b \cdots \to^b D'_{3j},$$

accepting $a^i b^i, a^i b^{2i}, a^j b^j$, and $a^j b^{2j}$, respectively, and these meet the following conditions:

- $D_{2i} = D'_{2j} \in F$,
- $D_0 = D'_0, D_1 = D'_1, \ldots, D_i = D'_i$,
- $D_{2i+1} = D'_{2j+1}, D_{2i+2} = D'_{2j+2}, \ldots, D_{3i} = D'_{2j+i}$.

Then, we can construct the interactive process π_5 accepting $a^j b^{j+i}$ as follows:

$$\pi_5 : D'_0 \to^a D'_1 \to^a \cdots \to^a D'_j \to^b D'_{j+1} \to^b \cdots \to^b D'_{2j} \to^b D'_{2j+1} \to^b \cdots \to^b D'_{2j+i}.$$

This contradicts that $L_{sq}(M) = L_2 = \{a^n b^n, a^n b^{2n} \mid n \geq 0\}$. $\quad\square$

Lemma 9 *The class of languages $\mathcal{L}_{sq}(RCRA)$ ($\mathcal{L}_{sq}^\lambda(RCRA)$) is strictly included in $\mathcal{L}_{sq}(CRA)$ (resp., $\mathcal{L}_{sq}^\lambda(CRA)$).*

Proof We can easily construct a realtime CRA accepting $L_3 = \{a^n \mid n \geq 0\} \cup \{b\}$.

Assume that there is an RCRA $M = (S, \Sigma, A, D_0, F)$ with λ-input mode which accepts L_3. Since the set of final multisets F is a finite set, there exist i, j with $1 \leq i < j$ and two interactive processes $\pi_i : D_0 \to^{\hat{a}_1} D_1 \to^{\hat{a}_2} \cdots \to^{\hat{a}_s} D_s$ and $\pi_j : D'_0 \to^{\hat{a}'_1} D'_1 \to^{\hat{a}'_2} \cdots \to^{\hat{a}'_s} D'_s \to^{\hat{a}'_{s+1}} D'_{s+1} \to^{\hat{a}'_{s+2}} \cdots \to^{\hat{a}'_t} D'_t$ (accepting a^i and a^j, respectively) such that $D_s = D'_t \in F$ and $D_0 = D'_0$ for some $t > s \geq 0$ and $\hat{a}_1, \hat{a}_2, \ldots, \hat{a}_s, \hat{a}'_1, \hat{a}'_2, \ldots, \hat{a}'_t \in \{\lambda, a\}$. From the definition of RCRA, it induc-tively holds that $D_{s-1} = D'_{t-1}, D_{s-2} = D'_{t-2}, \ldots, D_0 = D'_{t-s}$. Then, for the part of interactive process $D_0 = D'_0 \to^{\hat{a}'_1} D'_1 \to^{\hat{a}'_2} \cdots \to^{\hat{a}'_{t-s}} D'_{t-s} = D_0$, there is at least one a in $\{\hat{a}'_1, \hat{a}'_2, \ldots \hat{a}'_{t-s}\}$. This is because if this does not hold, it holds that $a^j = \hat{a}'_{t-s+1}\hat{a}'_{t-s+2} \ldots \hat{a}'_t = \hat{a}_1 \hat{a}_2 \ldots \hat{a}_s = a^i$, and this contradicts the assumption $i < j$.

Let $\pi_b : D_0 \to^{\hat{b}_1} D''_1 \to^{\hat{b}_2} \cdots \to^{\hat{b}_u} D''_u$ with $D''_u \in F$ be an interactive process accepting b. Then, we can consider an interactive process

$$\pi'_b : D'_0 \to^{\hat{a}'_1} D'_1 \to^{\hat{a}'_2} \cdots \to^{\hat{a}'_{t-s}} D'_{t-s} = D_0 \to^{\hat{b}_1} D''_1 \to^{\hat{b}_2} \cdots \to^{\hat{b}_u} D''_u$$

which accepts $a^v b$ for some $v > 0$. This contradicts that $L_{sq}^\lambda(M) = L_3 = \{a^n \mid n \geq 0\} \cup \{b\}$. $\qquad\square$

Lemma 10 *The language $L_4 = \{a^n db^n ec^n \mid n \geq 0\}$ is accepted by a DRCRA with λ-input mode.*

Proof Let $M = (S, \{a, b, c, d, e\}, A, p_0, \{p_2 d' e''\})$ be a CRA, where

- $S = \{a, b, c, d, e\} \cup \{a', b', c', d', e'\} \cup \{a'', b'', c'', d'', e''\} \cup \{a''', b''', d''', e'''\} \cup \{p_0, p_1, p_2\}$,
- A consists of the following sets of reactions A_0, A_1, A_2:

$$A_0 = \{r_1 : p_0 a \to p_0 a', \ r_2 : p_0 b \to p_0 b', \ r_3 : p_0 c \to p_0 c',$$
$$r_4 : p_0 d \to p_1 d', \ r_5 : p_0 e \to p_0 e'\},$$
$$A_1 = \{r_6 : p_1 a \to p_1 a'', \ r_7 : p_1 b \to p_1 b'', \ r_8 : p_1 c \to p_1 c'',$$
$$r_9 : p_1 d \to p_1 d'', \ r_{10} : p_1 e \to p_2 e''\},$$
$$A_2 = \{r_{11} : p_2 a \to p_2 a''', \ r_{12} : p_2 b \to p_2 b''', \ r_{13} : p_2 a' b'' c \to p_2,$$
$$r_{14} : p_2 d \to p_2 d''', \ r_{15} : p_2 e \to p_2 e'''\}.$$

We call an element of $\{p_0, p_1, p_2\}$ *state*. The following are important observations for computations of M:

(1) there is no chance of λ-move in any successful computation of M, because a reactant (left-hand side) of any reaction in A contains an element in Σ and any reachable configuration is in $(S - \Sigma)^\#$, where $\Sigma = \{a, b, c, d, e\}$,

(2) a computation of M proceeds through the three stages, i.e., the first stage containing state p_0, the second stage containing state p_1, and the final stage containing state p_2,

(3) A reaction in A_0, A_1, and A_2 is enabled if and only if the current configuration is in the first stage, the second stage, and the final stage, respectively.

[Proof for $L_4 = L(M)$]

Consider an input $a^\ell db^\ell ec^\ell$ with $\ell \geq 0$. First, when receiving "a^ℓ", an iterated ℓ-time applications of r_1 to p_0 produces $D_\ell = p_0 a'^\ell$. Then, when receiving "d", by applying r_4, $D_{\ell+1} = p_1 a'^\ell d'$ is produced. Further, when receiving "$b^\ell e$", by n-time applications of r_7 followed by r_{10}, $D_{2\ell+2} = p_2 a'^\ell b'''^\ell d' e''$ is produced. Lastly, on receiving "c^ℓ", by ℓ-time applications of r_{13}, the final multiset $p_2 d' e''$ is obtained. Hence, an input $a^\ell db^\ell ec^\ell$ is accepted by M with λ-input mode.

Note that only the reaction $r_{13} : p_2 a' b'' c \to p_2$ reduces the number of elements of a multiset (configuration) D. We also note that any successful configuration D must satisfy that (i) appearing "d" followed by "e" in an input string is necessary for D to reach the final state, (ii) only reaction r_{13} is effectively applicable on the final stage. Taking these into consideration, we conclude the only scenario for M to lead p_0 to the final multiset $(p_2 b' e'')$ as follows: when receiving "$a^\ell d$" on the first stage, M moves to the second stage with p_1. Then, after receiving "$b^\ell e$" on the stage, M produces $D_{2\ell+2}$ (mentioned above) on the final stage. Finally, with receiving 'c^ℓ", M brings $D_{2\ell+2}$ to the final multiset by using ℓ-time applications of r_{13}. Hence, M can accept only an input of the form $a^\ell db^\ell ec^\ell$.

[Proof for the determinism of M]

From the way of construction of A and the observation (1), it obviously follows that M is deterministic.

[Proof for the reversibility of M]

From the way of construction of A and the observation (2), (3), a reachable configuration is of the form

$$D = p_0 a'^{\ell_1} b'^{\ell_2} c'^{\ell_3} d' e'^{\ell_4} \text{ (on the first stage),}$$

$$D = p_1 a'^{\ell_5} b'^{\ell_6} c'^{\ell_7} d'^{\ell_8} e'^{\ell_9} a''^{\ell_{10}} b''^{\ell_{11}} c''^{\ell_{12}} d''^{\ell_{13}} e'' \text{ (on the second stage), or}$$

$$D = p_2 a'^{\ell_{14}} b'^{\ell_{15}} c'^{\ell_{16}} d'^{\ell_{17}} e'^{\ell_{18}} a''^{\ell_{19}} b''^{\ell_{20}} c''^{\ell_{21}} d''^{\ell_{22}} e''^{\ell_{23}} a'''^{\ell_{24}} b'''^{\ell_{25}} c'''^{\ell_{26}} d'''^{\ell_{27}} e'''^{\ell_{28}}$$

(on the final stage),

for some $\ell_i \geq 0$ with $1 \leq i \leq 28$. When reaching a configuration D with receiving "a", since the states in the products of r_1, r_6, r_{11} are different from each other, we can uniquely identify the applied reaction. This statement also holds a similar statement for the cases receiving "b" and "c". When reaching a configuration D' containing p_1 with receiving "d", r_4 and r_9 are candidates of the applied reaction. We can uniquely identify the applied reaction as follows: if D' contains d'', then r_9

is the applied reaction, otherwise r_4 is the applied reaction. Similarly, for the case receiving "e", the applied reaction can be uniquely identified.

Hence, L_4 is accepted by DRCRA M with λ-input mode. □

Lemma 11 *The class of regular languages $\mathcal{REG}$ is strictly included in $\mathcal{L}^\lambda_{sq}(DCRA)$.*

Proof For a deterministic finite state automaton $M = (Q, \Sigma, \delta, p_0, F)$, we construct a CRA $M' = (Q \cup \Sigma, \Sigma, A, p_0, F)$, where $A = \{pa \rightarrow q \mid \delta(p, a) = q\}$. Then, M' is obviously deterministic. Moreover, in any computation of M' with λ-input mode, no λ-move occurs. Hence, it holds that $L(M) = L^\lambda_{sq}(M')$.

From Lemma 10, L_4 belongs to $\mathcal{L}^\lambda_{sq}(DCRA) - \mathcal{REG}$. □

Lemma 12 *The class of language $\mathcal{L}^\lambda_{sq}(DCRA)$ ($\mathcal{L}_{sq}(DCRA)$) is incomparable with the class of language $\mathcal{L}_{sq}(RCRA)$ ($\mathcal{L}^\lambda_{sq}(RCRA)$, respectively).*

Proof For our purpose, it is sufficient to show $L_2 \in \mathcal{L}_{sq}(RCRA)$ and $L_3 \in \mathcal{L}^\lambda_{sq}(DCRA)$. From Lemma 11, it obviously holds that $L_3 \in \mathcal{L}^\lambda_{sq}(DCRA)$.

Let $M = (S, \{a, b\}, A, p_0, \{p_0, p_1, p_2\})$ be a CRA, where

- $S = \{a, b\} \cup \{a'\} \cup \{p_0, p_1, p_2, p_3\} \cup \{d_1, d_2, d_3, d_4, d_5, d_6\}$,
- A consists of the following sets of reactions A_0, A_1, A_2, A_3:

$$A_0 = \{r_1 : p_0a \rightarrow p_0a', \ r_2 : p_0a \rightarrow p_1a', \ r_3 : p_0a \rightarrow p_2a', \ r_4 : p_0b \rightarrow d_1\},$$
$$A_1 = \{r_5 : p_1a \rightarrow d_2, \ r_6 : p_1b \rightarrow d_3, \ r_7 : p_1a'b \rightarrow p_1\},$$
$$A_2 = \{r_8 : p_2a \rightarrow d_4, \ r_9 : p_2b \rightarrow d_5, \ r_{10} : p_2a'b \rightarrow p_3\},$$
$$A_3 = \{r_{11} : p_3a \rightarrow d_6, \ r_{12} : p_3b \rightarrow p_2\}.$$

We call an element of $\{p_0, p_1, p_2, p_3\} \cup \{d_1, d_2, d_3, d_4, d_5, d_6\}$ *state*. It is clear that a reachable configuration contains just one state. Especially, an element d_i ($1 \leq i \leq 6$) is a dead state, because once the configuration contains d_i, then the computation cannot reach the final multiset. Note that if the configuration contains a state in $\{p_0, p_1, p_2, p_3\}$, an input symbol have to be consumed by the next reaction.

For the input string λ, there exists a realtime interactive process $\pi : p_0 \in IP_{sq}(M, \lambda)$. This together with the fact that p_0 is a final multiset leads to $\lambda \in L(M)$. For the input string $a^\ell b^\ell$ with $\ell \geq 1$, by applying the sequence of reactions $r_1^{\ell-1} r_2 r_7^\ell$ to the initial multiset p_0, the computation reaches the final multiset p_1. Similarly,

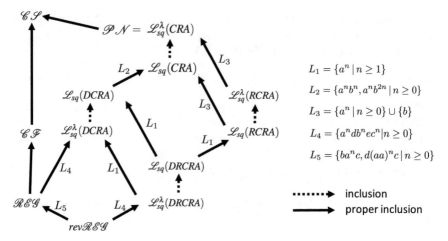

$$L_1 = \{a^n \mid n \geq 1\}$$

$$L_2 = \{a^n b^n, a^n b^{2n} \mid n \geq 0\}$$

$$L_3 = \{a^n \mid n \geq 0\} \cup \{b\}$$

$$L_4 = \{a^n db^n ec^n \mid n \geq 0\}$$

$$L_5 = \{ba^n c, d(aa)^n c \mid n \geq 0\}$$

$\cdots\cdots\blacktriangleright$ inclusion

$\longrightarrow$ proper inclusion

Fig. 4 Inclusion relations at large

for the input string $a^\ell b^{2\ell}$, by applying the sequence of reactions $r_1^{\ell-1} r_3 (r_{10} r_{12})^\ell$, the computation reaches the final multiset p_2. Moreover, from the way of constructing A, any other computation inevitably has to reach a configuration containing a dead state. Hence, M accepts $L_2 = \{a^n b^n, \; a^n b^{2n} \mid n \geq 0\}$.

Next, we confirm that M is reversible. We can easily check that any reachable configuration containing each state in $\{p_0, p_1, p_2, p_3\}$ cannot be reached by distinct computations with the same input symbol. The configuration D containing state d_1 is derived from either D' containing state p_0 by applying r_4 or D' containing state d_1 without applying any rule with input symbol b. In other words, D' is the previous configuration of D if and only if D' does not contain b. Thus, there is no configuration D containing state d_1 which is reached by distinct computations with the same input symbol. Similarly, any other dead state in $\{d_2, d_3, d_4, d_5, d_6\}$ cannot be reached by distinct computations with the same input symbol.

Hence, L_2 is accepted by realtime RCRA M.   □

3.3 Summary of Results

Figure 4 illustrates the inclusion relations among various classes of CRA languages introduced in this paper, where $\mathcal{PN}$ is the class of Petri net languages of L^λ-type [27] and $rev\mathcal{REG}$ is the class of zero-reversible regular languages [1, 28]. The equality result $\mathcal{PN} = \mathcal{L}_{sq}^\lambda(CRA)$ is shown in [25]. Further, $\mathcal{CS}$ and $\mathcal{CF}$ denote the classes of context-sensitive and context-free languages, respectively.

4 Discussion

4.1 Related Work

Aside from related literature of ours [22–25], here we would like to focus on and refer to the following articles classified into two themes which share common subjects with this paper.

- **Reversible computing models**: In his pioneering work [4], Bennett showed that any Turing machine can be simulated by a reversible Turing machine that leaves no garbage memory at the end of computation, suggesting a possibility of a physically reversible computer with arbitrary small amount of energy dissipation. The reality of this physically reversible computing with less energy dissipation was in principle supported by an interesting physical model called Billiard Ball Model [10]. Morita proved the universal computability of a reversible 2-counter machine [18], and studied the computational power of two-way reversible multi-head finite automata in [20]. A rather recent paper [14] investigated reversible (deterministic) pushdown automata, showing that their computing power lies properly in between regular and deterministic context-free languages.
  For cellular automata (CAs), [33] proved that 2-dimensional reversible CAs are computationally universal, which was further reinforced by the result on the universal computability of 1-dimensional reversible CAs in [17]. As for CRNs, it was shown in [32] that reversible CRNs can simulate in a space and energy efficient manner polynomially space-bounded Turing computation.
- **Multiset-based computing models**: Among many in the literature on computing models based on multiset rewriting, P systems have been intensively investigated [26] and reversible computation in P system models is investigated in several papers (e.g., [12, 16]). P systems with symport/antiport rules are considered in [21] which showed that the class of reversible P systems with four membranes is computationally universal. A paper [3] newly introduced strong versions of both determinism and reversibility into P systems and investigated in depth the computational powers of those P systems with not only sequential but also maximally parallel manners of rewriting. It should be noted that research on P system families mostly concerns computing subsets of *natural numbers* (not languages), which is in marked contrast to the classes of CRAs studied in this paper.
  As for other types of the multiset-based computing models that bear a certain resemblance to CRAs, P automata, a variant of P systems, were studied as language accepting devices (e.g., [6, 7]). On receiving an input (a multiset) from the environment at each step of computation, a P automaton changes its configuration by making region-wise applications of the equipped rules, and accepts a mapping image of the inputted sequence of multisets. In this sense, CRAs may be regarded as a simplified variant of P automata with symport/antiport rules but without membrane structure nor a mapping component.

CRAs are structurally akin to bag automata [8] and urn automata [2] in that the core structures of both are basically made up of finite automata with multiset data storages called "bag" and "urn", respectively. In the former paper, the computing power of bag automata is characterized in terms of AFL theory formulation, while the latter proved that the class of languages accepted by urn automata (with 2-way input tape) contains the class LOGSPACE.

However, we remark that, to the best of our knowledge, none of the results on reversibility has been published in any of the automaton-type computing models mentioned herein.

4.2 Conclusions and Future Work

By newly introducing the notions of determinism and reversibility into chemical reaction automata (CRAs), we have primarily investigated the computational capability of the following six subclasses of CRAs (in sequential rewriting manner): deterministic CRAs (DCRAs), reversible CRAs (RCRAs) and deterministic and reversible CRAs (DRCRAs) under two conditions of both realtime and λ-input modes. In this paper, we have shown the following.

1. The class of languages $\mathcal{L}_{sq}^{\lambda}(CRA)$ ($\mathcal{L}_{sq}(CRA)$) properly contains the class $\mathcal{L}_{sq}^{\lambda}(DCRA)$ ($\mathcal{L}_{sq}(DCRA)$, respectively) and the class $\mathcal{L}_{sq}^{\lambda}(RCRA)$ ($\mathcal{L}_{sq}(RCRA)$, respectively).
2. The class of languages $\mathcal{L}_{sq}(DCRA)$ is incomparable to the class $\mathcal{L}_{sq}^{\lambda}(RCRA)$.
3. The class of languages $\mathcal{L}_{sq}(DCRA)$ properly contains the class of regular languages $\mathcal{REG}$ and is incomparable to the class of context-free languages $\mathcal{CF}$.
4. The class of languages $\mathcal{L}_{sq}(DRCRA)$ properly contains the class of reversible regular languages $rev\mathcal{REG}$ and is incomparable to the class of regular languages $\mathcal{REG}$.

We remark again about the inclusion relations $\mathcal{L}_{sq}^{\lambda}(X) \subseteq \mathcal{L}_{sq}(X)$ in Fig. 4, where X is a deterministic subclass of CRAs, which might appear a little curious at a glance. However, as proved in Lemma 1, this is due to the fact that there is no chance for DCRAs to move with λ-input mode throughout any successful computation.

As mentioned previously, CRAs may be regarded as online computational models based on the chemical programming languages CRNs, and the class of CRAs offers a good formal model to explore the computational aspects of CRNs and DSD system implementation. From these, we believe that the obtained results on the computational powers of deterministic and/or reversible CRAs may shed new light on the feasibility analysis on biochemical programming paradigm. Further, considering that CRAs have a close resemblance to simpler version of some type of P automata, the obtained results may also offer some new insight into the reversible computations in a special variant of P automata theory.

There remain many subjects to be investigated in the direction suggested by CRAs and the related subjects.

- It is strongly encouraged to make further refinement of the inclusion relations in the CRA language hierarchy in Fig. 4.
- It is also intriguing to clarify the relationships between subclasses of CRAs studied here and others defined by reversible computing models such as reversible pushdown automata [14].
- Closure properties as well as decision problems for subclasses of deterministic/reversible CRAs are left open.
- The computing powers of CRA classes (introduced here) in maximally parallel manner are interesting to be studied. In fact, it remains open whether or not deterministic CRAs in maximally parallel manner are Turing universal.

Acknowledgements The authors are deeply indebted to referees for their useful comments which greatly improved the consistency and readability of an earlier version of this paper.

The work of F. Okubo was in part supported by JSPS KAKENHI Grant Number JP16K16008, Japan Society for the Promotion of Science. The work of T.Yokomori was in part supported by a Grant-in-Aid for Scientific Research on Innovative Areas "Molecular Robotics"(No. 24104003) and JSPS KAKENHI, Grant-in-Aid for Scientific Research (C) 17K00021 of The Ministry of Education, Culture, Sports, Science, and Technology, Japan, and by Waseda University grant for Special Research Projects: 2016B-067, 2016K-100 and 2017K-121.

References

1. Angluin, D.: Inference of reversible languages. J. ACM **29**(3), 741–765 (1982)
2. Angluin, D., Aspnes, J., Diamadi, Z., Fischer, M.J., Peralta, R.: Urn Automata. No.YLEU/DCS/TR-1280, Department of Computer Science, Yale University, New Haven CT, USA (2003)
3. Alhazov, A., Freund, R., Morita, K.: Sequential and maximally parallel multiset rewriting: reversibility and determinism. Nat. Comput. **11**, 95–106 (2012)
4. Bennett, C.H.: Logical reversibility of computation. IBM J. Res. Dev. **17**(6), 525–532 (1973)
5. Calude, C., Păun, Gh, Rozenberg, G.: In: Salomaa, A. (ed.) Multiset Processing. LNCS, vol. 2235. Springer, Heidelberg (2001)
6. Csuhaj-Varju, E., Vaszil, G.: P Automata or Purely Communicating Accepting P Systems. LNCS, vol. 2597, pp. 219–233. Springer, Berlin (2003)
7. Csuhaj-Varju, E., Vaszil, G.: P automata. In the Oxford Handbook of Membrane Computing, pp. 145–167. (2010)
8. Daley, M., Eramian, M., McQuillan, I.: The bag automaton: a model of nondeterministic storage. J. Autom. Lang. Comb. **13**, 185–206 (2008)
9. Ehrenfeucht, A., Rozenberg, G.: Reaction systems. Fundam. Inform. **75**, 263–280 (2007)
10. Fredkin, E., Toffoli, T.: Conservative logic. Int. J. Theor. Phys. **21**(3/4), 219–253 (1982)
11. Hopcroft, J.E., Motwani, T., Ullman, J.D.: Introduction to automata theory, language and computation, 2nd edn. Addison-Wesley, Reading (2003)
12. Ibarra, O.: On strong reversibility in P systems and related problems. Int. J. Found. Comput. Sci. **22**(1), 7–14 (2011)
13. Kudlek, M., Martin-Vide, C., Păun, Gh: Toward a formal macroset theory. In: Calude, C., Păun, Gh, Rozenberg, G., Salomaa, A. (eds.) Multiset Processing. LNCS, vol. 2235, pp. 123–133. Springer, Berlin (2001)
14. Kutrib, M., Malcher, A.: Reversible pushdown automata. J. Comput. Syst. Sci. **78**, 1814–1827 (2012)

15. Landauer, R.: Irreversibility and heat generation in the computing process. IBM J. Res. Dev. **5**(3), 183–191 (1961)
16. Leporati, A., Zandron, C., Mauri, G.: Reversible P systems to simulate Fredkin circuits. Fundam. Inform. **74**, 529–548 (2006)
17. Morita, K., Shirasaki, A., Gono, Y.: A 1-tape 2-symbol reversible turing machines. Trans. IEICE Japan **E72**(3), 223–228 (1989)
18. Morita, K.: Universality of a reversible two-counter machine. Theor. Comput. Sci. **168**, 303–320 (1996)
19. Morita, K.: Reversible computing and cellular automata - a survey. Theor. Comput. Sci. **395**, 101–131 (2008)
20. Morita, K.: Two-way reversible multi-head finite automata. Fundam. Inform. **110**(1–4), 241–254 (2011)
21. Nishida, T.Y.: Reversible P systems with symport/antiport Rules. In: Proceedings of the 10th Workshop on Membrane Computing, pp. 452–460 (2009)
22. Okubo, F.: Reaction automata working in sequential manner. RAIRO Theor. Inform. Appl. **48**, 23–38 (2014)
23. Okubo, F., Kobayashi, S., Yokomori, T.: Reaction automata. Theor. Comput. Sci. **429**, 247–257 (2012)
24. Okubo, F., Kobayashi, S., Yokomori, T.: On the properties of language classes defined by bounded reaction automata. Theor. Comput. Sci. **454**, 206–221 (2012)
25. Okubo, F., Yokomori, T.: The computational capability of chemical reaction automata. In: Murata, S., Kobayashi, S. (eds.) DNA20. LNCS, vol. 8727, pp. 53–66. Springer, Switzerland (2014). Also, in Natural Computing, vol. 15, pp. 215–224. (2016)
26. Paun, G., Rozenberg, G., Salomaa, A.: The Oxford Handbook of Membrane Computing. Oxford University Press, Inc., New York (2010)
27. Peterson, J.L.: Petri Net Theory and the Modelling of Systems. Prentice-Hall, Englewood Cliffs (1981)
28. Pin, J.E.: On Reversible Automata. In: Proceedings of LATIN '92, 1st Latin American Symposium on Theoretical Informatics, São Paulo, Brazil, April 6-10, pp. 401–416 (1992)
29. Qian, L., Winfree, E.: Scaling up digital circuit computation with DNA strand displacement cascades. Science **332**, 1196–1201 (2011)
30. Qian, L., Soloveichik, D., Winfree, E.: Efficient turing-universal computation with DNA polymers. In: Sakakibara, Y., Mi, Y. (eds.) DNA16. LNCS, vol. 6518, pp. 123–140. Springer, Heidelberg (2011)
31. Rozenberg, G., Back, T.: Section IV: molecular computation. In: Kok, J.N. (ed.) Handbook of Natural Computing, vol. 3, pp. 1071–1355. Springer, Berlin (2012)
32. Thachuk, C., Condon, A.: In: Stefanovic, D., Turberfield, A. (eds.) Space and energy efficient computation with DNA strand displacement systems. DNA 18. LNCS, vol. 7433, pp. 135–149. Springer, Heidelberg (2012)
33. Toffoli, T.: Computation and construction universality of reversible cellular automata. J. Comput. Syst. Sci. **15**, 213–231 (1977)

On Non-polar Token-Pass Brownian Circuits

Ferdinand Peper and Jia Lee

Abstract Brownian circuits are asynchronous circuits in which signals—represented as tokens—are able to fluctuate along wires. These fluctuations are used as a stochastic search mechanism to drive computations from a state of input to a state of output. Token-Pass circuits are a type of circuits in which wires will not merge or split. Rather, they are linear: each token remains on its wire during computation, and it will interact with other tokens only at points where they pass through circuit elements. The T-element, introduced in [Peper, Lee, Carmona, Cortadella, Morita, "Brownian Circuits: Fundamentals," 2013], is a circuit element in which three wires pass through, and it was shown to be universal, provided the circuit it is employed in is Brownian. The Brownian circuit designs based on the T-element in the above paper have in common that they implicitly assume a bias in the direction in which tokens flow on average, even though tokens may fluctuate forward and backward during the course of a computation. This chapter proposes a new type of Brownian Token-Pass circuit, called *Non-Polar Token-Pass Brownian Circuit*, in which no such directional bias is assumed. Though most wires in such circuits do have directional bias, the few wires that don't, allow for simpler circuit designs, as will be shown.

1 Introduction

Energy consumption has become a major issue for VLSI circuits, and this has led to an increased interest in designs that work near the thermal limit [7]. Noise and fluctuations play a major role in such a regime, and in order to realize low energy designs with acceptable reliability, researchers have started to devise strategies that not only cope with noise and fluctuations, but that can also exploit them. *Brownian circuits* [4, 6, 12] result from part of these efforts, and the central idea behind their

F. Peper (✉)
National Institute of Information and Communications Technology, Tokyo, Japan
e-mail: peper@nict.go.jp

J. Lee
ChongQing University, ChongQing, China
e-mail: lijia@cqu.edu.cn

© Springer International Publishing AG 2018
A. Adamatzky (ed.), *Reversibility and Universality*, Emergence, Complexity and Computation 30, https://doi.org/10.1007/978-3-319-73216-9_14

operation is rooted in biology. Molecules tend to move around in a fluid driven by Brownian motion, and its randomness allows them to interact with a wide range of other molecules. Those molecules that are compatible with each other will undergo biochemical reactions, but those that don't, continue on their way. This type of trial-and-error behavior does not appear to be explicitly designed, but it is actually a powerful control mechanism, since it does not need a supervising mechanism telling molecules where to go, yet molecules accomplish the goal of a system by finding their ways to each other through stochastic search.

Brownian circuits operate in a similar way. Their signals are modeled as tokens that move along wires and that can interact with each other in circuit elements, just like in token-based circuits, of which Petri-nets [8] are the most well-known. Unlike token-based circuits, however, Brownian circuits have tokens that have a fluctuation-driven choice to interact with each other or not, in a way resembling particles on molecular scales. In previous work [12] we analyzed Brownian circuits that are based on a particular type of circuit element, called the *T-element* (Fig. 3a). This element is noteworthy in the sense that it can be used to construct the set of circuits that are able to conduct all computable functions, but this feat comes with one condition, i.e., that tokens in the circuits are able to fluctuate. Absent this condition, the class of functions computable by this type of circuits based on the T-element is extremely limited, as has been proven in [12].

The T-element is an example of a so-called *token-pass circuit*. Getting their name from the fact that they basically bind together a bunch of wires through which tokens pass (Fig. 2), Token-Pass circuits lend themselves to a systematic analysis of circuit properties [12]. The wires can run from the input side of a circuit to the output side, or, alternatively, they can form loops on which one (or more) token will reside in eternity. Notwithstanding the fluctuating characteristics of tokens in the Brownian circuits in [12], there is a bias in the direction in which tokens move on average in time over a computation. For non-loop wires this bias is from input-side to output-side of the circuit, but for looping wires it is clockwise or counterclockwise, depending on how the loops are arranged. While the bias in directionality is conceptually in line with the way traditional circuits work, i.e., signals going from input to output, it runs somewhat counter to the basic idea of Brownian circuits that tokens can fluctuate forward and backward.

This chapter proposes a new type of Token-Pass Brownian circuit, in which such a directional bias does not necessarily exist for all wires. It focuses on Token-Pass circuits constructed from T-elements, but unlike in [12] the circuits may contain wires that neither run from input to output nor form loops, but rather are wires that have terminators at both sides. While one or more tokens reside on such terminated wires, they may move forward and backward without implicit directionality. Called *Non-Polar*, these circuits have less restrictions than conventional Token-Pass circuits, which pays off in terms of potentially simpler designs. In this chapter we illustrate this by a design for a 1-bit memory that is based on seven T-elements, which is one T-element less than we have up to now achieved with conventional Token-Pass circuits.

This chapter is organized as follows. Section 2 gives an intuitive explanation of Token-Based and Token-Pass circuits. A more formal treatment is given in [12]. Brownian circuits are introduced in Sect. 3, again in an intuitive way. This is followed in Sect. 4 with an explanation of Non-Polar Token-Pass Brownian circuits by introducing a simplified design for a 1-bit memory based on T-elements. We finish this chapter with a discussion.

2 Token-Based and Token-Pass Circuits

Tokens are discrete indivisible units that are used as signals in circuits or systems [12]. They are graphically represented by fat dots, like in the left of Fig. 2a, b. Token-based circuits are relevant for physical implementations in which signals have a discrete character, such as found in charge-state logic [2] or spintronics [1]. In such logic the behavior of devices and circuits depends on the movement and interaction of *individual* particles, like in single electron tunneling circuits [3, 9]. Charge-state logic is not yet common in current commercially available electronics and neither is spintronics; rather most circuits employ voltage-state logic, in which signals are represented by voltage levels.

The token-based circuits in this chapter are *Delay-Insensitive*: they allow arbitrary delays of signals in lines or in circuit elements, without this compromising the correctness of the circuit's operations. Delay-insensitive circuits are not governed by a clock, so they belong to the class of *asynchronous* circuits [13].

Token-based delay-insensitive circuits can be constructed from a fixed set of circuit primitives, like, for example, the NOT-gates and AND-gates in synchronous circuits. When a set of primitives offers sufficient functionality to construct any circuit possible in a class of circuits, this set is *universal* for that class. Figure 1a–c shows one such universal set for the class of delay-insensitive token-based circuits [5]: the *Merge* (called *P-Merge* in [5]), the *Fork*, and the *Tria* [10]. An important function of

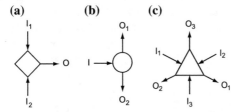

Fig. 1 Primitive modules for delay-insensitive token-based circuits. **a** Merge, which merges two streams of tokens on input lines I_1 and I_2 into one stream on output line O. **b** Fork, which duplicates every input token from input line I to its two output lines O_1 and O_2. **c** Tria, which joins two input tokens into a single output token as follows. Upon receiving one token each from input lines I_i ($i \in \{1, 2, 3\}$) and I_j ($j \in \{1, 2, 3\} \setminus \{i\}$), it outputs a token to the line O_{6-i-j}. If there is a token on only one of the input lines, it is kept pending, until a token on one more input line appears [12]

(a)   **(b)**

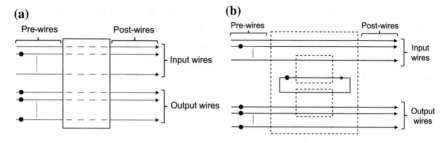

Fig. 2 a Scheme of a Token-Pass circuit. Input wires and output wires all pass through the circuit. **b** Detail of possible innards of a Token-Pass circuit. The dotted boxes denote modules containing circuit elements that act on tokens. Apart from wires passing through the circuit, there are also wires that are connected in loops. These loops contain one or more tokens and enforce an order on token interactions as well as store the state of the circuit. The tokens on loops interact with tokens on input wires and output wires, implementing the functionality of the circuit

the Tria is that it produces an output token only when it has at least two input tokens (*Join* functionality). The presence of merely one input token will keep that token pending on its input terminal until the second input token arrives at another input terminal. Join functionality synchronizes tokens on a local scale. It is of fundamental importance in delay-insensitive circuits, because it substitutes the clock that is used in synchronous circuits.

A *Token-Pass circuit* is a circuit through which tokens pass via linear wires (no merging or splitting of wires), on the way interacting with one other in circuit elements, but never veering from their wires (Fig. 2). Token-Pass circuits keep the number of tokens unchanged, be it on interconnection lines or inside circuit elements. A Token-Pass circuit can be characterized as a collection of wires that either are connected in loops, or that run through the circuit to form input wires or output wires (Fig. 2b). A token on the *pre-wire* of an input wire (left in Fig. 2a) denotes input to the circuit, and it passes to the corresponding *post-wire* (right) as part of the circuit operation. All output wires have tokens on their pre-wires before the circuit operation, and—depending on the input to the circuit—only part of these tokens are passed to the corresponding post-wires to signify output.

An example of a Token-Pass circuit is the *T-element* in Fig. 3a, which plays an important role in [12]. It has three wires passing through it, corresponding to the paths $\{a, a'\}$, $\{b, b'\}$, and $\{c, c'\}$ in the wire diagram in Fig. 3b. The wire c–c' passing through the base of the T-element is the *base wire*. The other two wires, each passing through an arm of the T-element, are the *choice wires*.

In a Token-Pass circuit the wires can be arranged either as an open path or as a loop. A token never moves from one wire to another. It will enter a wire as input if the wire is not connected as a loop, and it moves along a wire until it is output. The circuit constructed from two T-elements in Fig. 3c has five wires (Fig. 3d), one of which is connected as a loop. This loop contains one token, which will never leave the loop. By being at a certain location in the loop at a certain instance, this token enforces a certain order at which other tokens may pass through the net. Apart from

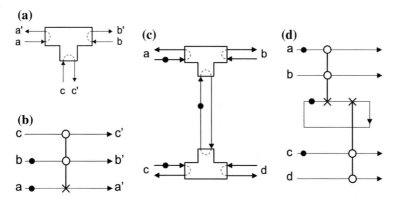

Fig. 3 **a** The T-element, which is a simple Token-Pass (TP) circuit. The wires a–a' and b–b' in the arms of the T are the *choice wires*, while the wire c–c' in the base of the T is the *base wire*. The T-element requires one token as input on its base wire and one token on one of its choice wires. The resulting transition moves the two tokens pairwise through the T-element to output them. If there are input tokens at the base wire as well as at both choice wires, then the T-element makes an arbitrary choice as to which of the choice wire tokens is processed, while leaving the token on the other choice wire pending. **b** Wire diagram of the T-element. **c** Circuit constructed from two T-elements and **d** its wire diagram. The loop corresponds to the pair of wires connecting the bases of the two T-elements [12]

enforcing an order of operations in a circuit, a loop may also be used to store a circuit state, as in the 1-bit memory in Fig. 7b or c. In general, loops are used for internal processing in circuits.

Token-based circuits are not necessarily Token-Pass. For example, the module in Fig. 4a is not Token-Pass, because it cannot be represented as a collection of wires. Rather, a token on b_0 can flow to either b_0' or b_1', depending on whether there is an input on a_0 or a_1, respectively, and the same holds for a token on b_1. For a more formal treatment of Token-Pass circuits see [12].

Theorem 1 *The class of Token-Pass circuits is closed under serial composition [12].*

This theorem implies that connecting two Token-Pass circuits to each other results in a structure that is also a Token-Pass circuit. Loops in the original two circuits to be joined stay loops, since they cannot be broken open by merely connecting wires, but non-looping wires in the original two circuits may become connected as loops, or they may become complete wires running through the circuit from input to output side.

Tokens in a Token-Pass circuit are processed according to the protocol in Fig. 5a–d [12]. A token on an input pre-wire indicates input for the module from that wire (top of Fig. 5a). When processing this input, the module moves the token to the corresponding input post-wire. While an input wire may lack a token on its pre-wire—when there is no input (bottom of Fig. 5a)—an output pre-wire always contains a token before an operation (top of Fig. 5c). To signify output of the operation, the module moves the token to the corresponding post-wire (bottom of Fig. 5c).

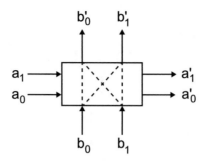

Fig. 4 A module that is not Token-Pass. The module assumes two tokens as input, one at one of the a-lines (left) and one at one of the b-lines (bottom). If a token is input to a_0, then the token at the b-lines passes through the module uncrossed, i.e. it moves from b_j to b'_j ($j \in \{0, 1\}$), while the other token moves to a'_0. If a token is input to a_1, then the module operates as if the b-lines were crossed, i.e., the token at b_j moves to b'_{1-j}, while the token at a_1 moves to a'_1. If there is only one input token, then it remains pending until a second input token arrives [12]

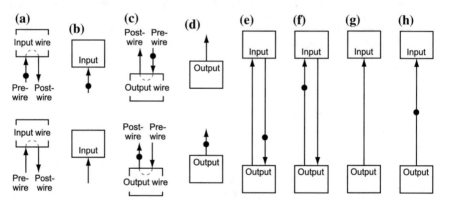

Fig. 5 Signaling protocol of Token-Pass circuit and its token-based equivalent. **a** Token-Pass module with input (top) and without input (bottom), and **b** the token-based equivalents. **c** Token-Pass module before producing output (top) and after output (bottom), and **d** the token-based equivalents. **e** Token on pair of wires connecting two modules is transferred through an operation of the bottom module from one wire to **f** the other wire. In terms of a conventional token-based circuit in which a bottom module produces output to a top module, the operations in **e** and **f** correspond to resp. **g** no token yet produced on a single interconnection wire, and **h** one token produced [12]

In other words, depending on the pattern of tokens present on the input pre-wires, a circuit will move certain tokens on the output pre-wires to the output post-wires.

While the protocol in Fig. 5a–d defines the external input/output behavior of Token-Pass circuits, it can also be used for interconnections between modules (Fig. 5e, f). Token-Pass modules are then connected to one other via a pair of wires that contains one token. The wire among this pair on which the token resides determines the state of the signaling between the modules. There are two cases. In the first case (Fig. 5e), the token is on the pre-wire of the bottom module, waiting to be let through and become output of the module. This results in the second case (Fig. 5f), in which the token is on the pre-wire of the top module, waiting to be accepted as input to this module. Once the token is accepted and let through, the first case applies again. In terms of a conventional token-based circuit, these two cases correspond to no tokens (Fig. 5g) resp. one token (Fig. 5h) on an interconnection wire between two modules.

The correspondence outlined above between the signaling protocols of token-based circuits and Token-Pass circuits can be carried over to the modules or elements from which circuits are made up. A module in a Token-Pass circuit then has a pair of wires wherever the equivalent token-based module has a single wire. When the Token-Pass equivalent of a token-based module M is denoted by *TP-M*, we then obtain the *TP-Merge* (Fig. 6a bottom), the *TP-Fork* (Fig. 6b bottom), and the *TP-Tria* (Fig. 6c bottom). The TP-Merge turns out to be identical to the T-element.

Token-Pass modules operate just like their token-based counterparts, though more wires and more tokens are involved. An operation of the TP-Tria is illustrated in Fig. 6e, together with its token-based counterpart (Fig. 6d). The equivalence of token-based circuits and Token-Pass circuits implies

Theorem 2 *The set {TP-Merge, TP-Fork, TP-Tria} is universal for the class of Token-Pass circuits [12].*

Proof Follows from the universality of the set {Merge, Fork, Tria} for token-based circuits [5].   □

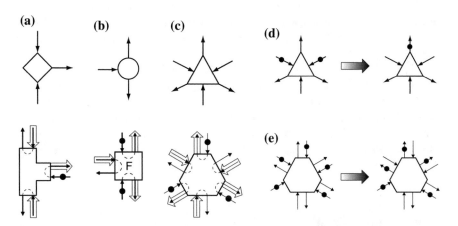

Fig. 6 **a** Merge (top) and TP-Merge (bottom); **b** Fork (top) and TP-Fork (bottom); **c** Tria (top) and TP-Tria (bottom). The hollow arrows of the Token-Pass modules at the bottom denote the corresponding input and output wires of the original modules at the top. The remaining wire ends carry "dummy" tokens, necessary to guarantee that the number of tokens remains unchanged from input to output. For example, the TP-Fork, displayed with the symbol T in it, consumes three tokens as input (of which two are dummies) and it produces three tokens as output (of which one is a dummy). **d** Transition describing an operation of a Tria, and **e** the equivalent operation of a TP-Tria [12]

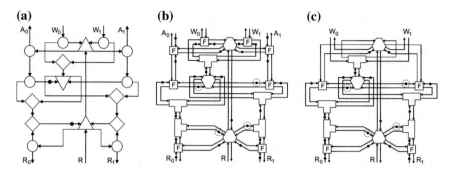

Fig. 7 a 1-Bit token-based memory. The state of the memory is stored by two tokens on the left of the two lower Trias (state 0, in this case), or on the right (state 1). State 0 or 1 is written into the memory by inputing a token to the input wire W_0 or W_1 respectively. The writing operation is acknowledged by a token from wire A_0 or A_1, respectively. By inputing a token to wire R, the memory is read out, and a token is produced on either wire R_0 or R_1, depending on the state of the memory. **b** Token-Pass equivalent of the 1-bit memory. Modules are connected to each other by pairs of wires, which effectively form loops, each containing a token. The memory's state is encoded by the positions of four tokens (encircled) in the loops on which they reside. **c** Simplified version of the Token-Pass 1-bit memory. Acknowledge wires are dropped, and their task is taken over by the post-wires of W_0 and W_1 [12]

    The circuit for the 1-bit memory [11] in Fig. 7a is token-based. The corresponding Token-Pass circuit is constructed by replacing the Merge, Fork, and Tria by their Token-Pass equivalents and connecting them by pairs of wires (Fig. 7b). The same circuit, but then with Acknowledge wires A_0 and A_1 left out, is shown in Fig. 7c; the function of the Acknowledge wires is taken over by the post-wires of the Writing wires W_0 and W_1. More efficient constructions for the 1-bit memory are given in Sect. 3.

    In [12] it is shown how a TP-Fork, TP-Merge, and TP-Tria can be constructed from T-elements. It is also shown that we cannot guarantee the absence of deadlocks in a TP-Tria constructed from T-elements, if no fluctuations of tokens are allowed. Such a deadlock occurs if a token waits in vain for another token to arrive in a Join functionality [12]. Fortunately, deadlocks do not occur if circuits are Brownian. This establishes the T-element as universal for the class of Token-Pass Brownian circuits [12].

3 Brownian Circuits

Circuits are *Brownian* when tokens fluctuate on wires and circuit elements (see [12] for a formal definition). Unlike with the monotonous behavior of tokens in the previous sections, operations in a Brownian circuit are done and undone repeatedly, as tokens move forward and backward. Delays of tokens due to fluctuations do not affect the correctness of a circuit's operation, since the circuit is delay-insensitive.

The added value of allowing tokens to fluctuate is that it is possible to undo operations of circuit elements, including those that lead to deadlocks. A circuit thus gains the ability to backtrack out of deadlocks, and this is why a TP-Tria constructed from Brownian T-elements according to the circuit design in [12] has no deadlocks. This implies

Theorem 3 *The Brownian T-element is universal for the class of Token-Pass circuits[12].*

When the 1-bit memories in Fig. 7b, c are built entirely from T-elements they are deadlock-free if fluctuations of tokens are allowed. Being designed without considerations of token fluctuations in mind, however, these memories are inefficient, since the searching behavior of fluctuations is only exploited inside their three TP-Trias, and not in other parts of the circuits. The 26 T-elements required to build the memory in Fig. 7c—five for the TP-Merges, three for each of the four TP-Forks, and three for each of the three TP-Trias—can be reduced to merely eight T-elements if the 1-bit memory is redesigned such that fluctuations are exploited to a fuller extent [12]. Figure 8 shows the resulting circuit design, which contains two parts. The top part is used to write a new value into the memory, while the bottom part can read out the contents of the memory. Though the design in Fig. 8 sharply reduces the number of required T-elements, it can be simplified even more as the next section shows.

4 Non-polar Token-Pass Brownian Circuits

The *polarity* of a T-element's wire terminals denotes the preferred direction of tokens through the element. For non-Brownian Token-Pass circuits this direction coincides with the flow of tokens from pre-wires to post-wires, and it is indicated by the arrows representing the wires in circuits. For the Brownian Token-Pass circuits in the previous sections the definition of polarity is similar, because tokens—even though they fluctuate—are still biased from pre-wires to post-wires.

When (part of) a Brownian circuit lacks a preferred direction of token flow, it is called *Non-Polar*. The traditional arrow-based notation of wires is abandoned, in favor of a labeling of wire terminals by a circle ○ on one side and a blank on the other side. Connecting terminals that have opposite labels results in the original polar circuits, but when wire terminals with identical labels are connected to each other, more flexible schemes arise, with less design complexity as a possible merit.

The non-polar 1-bit memory in Fig. 9 requires seven T-elements—which is one less than in the (polar) Brownian 1-bit memory in Fig. 8. Remarkable in this design is a wire passing through four T-elements that is neither connected in a loop, nor used for input or output. This wire is divided in segments labeled $0'$, 0, N, 1, and $1'$, one of which contains a token. The token is moved between the segments without any directional bias, and its location represents the state of the memory, which is:

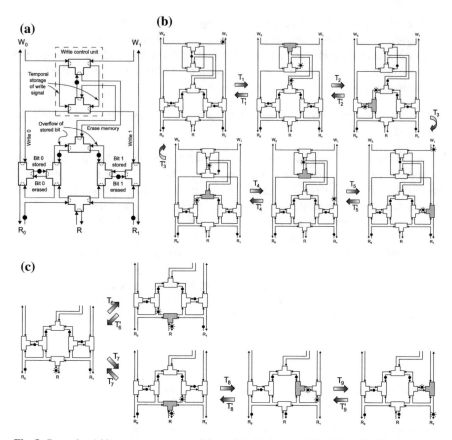

Fig. 8 Brownian 1-bit memory constructed from eight T-elements. The labels W_0, W_1, R, R_0, and R_1 have the same meanings as in Fig. 7. **a** The memory in its initial state 0. The state is stored by the positions of two tokens in two different loops, one token at position *Bit 0 stored* and the other token at position *Bit 1 erased*. A token in the bit-stored position may cause a token to flow into the corresponding *Overflow of stored bit* wire and back again, due to token fluctuations. This is an intermediate stage of erasing the memory during a writing operation. **b** Writing the state 1 in the memory. Tokens that have just moved to a new position are encircled by rays, and the T-element just used in the operation is colored gray. Tokens can fluctuate forward and backward as part of the writing operation, for example through steps T_2 and T_2' before bit erasure. The write control unit ensures that erasure takes place (step T_3) before a new bit is written (step T_5). The writing operation is acknowledged by a token output at the post-wire of W_1. **c** Reading out the memory (only lower half shown). Again, tokens fluctuate forward and backward, leading to dead ends in the search (via T_6), from which is backtracked (via T_6'), but eventually the memory's state is output to the post-wire of R_1 after step T_9 [12]

   0 State 0

   1 State 1

   N Neutral state, after temporarily erasing the memory to prepare for writing a new state

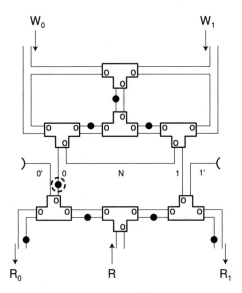

Fig. 9 Non-Polar 1-bit memory. The memory is in its initial state 0, which is stored by the encircled token on the wire passing through four T-elements. Inputs and outputs of the memory are indicated by arrows next to the corresponding wires. The wires W_0, W_1, R, R_0, and R_1 have the same meanings as in Fig. 7

0' Temporary state during read-out of state 0
1' Temporary state during read-out of state 1

The ends of the wire are *Terminators*, depicted as circle segments. Apart from preventing tokens from leaving the wire, terminators have no functionality. They do not accept input tokens nor produce output tokens. Though the T-elements in Fig. 9 are non-polar, there is still a rule governing their operations: tokens can only be processed by a T-element if they are on wire terminals with matching labels. So, if a T-element has a token at one of its choice wire terminals labeled ○, another token at its other choice wire terminal labeled blank, and a token at its base wire terminal labeled ○, the tokens at the terminals labeled ○ will pass through the T-element, while the token at the choice wire terminal labeled blank will not be processed.

Though the 1-bit memory in Fig. 9 is non-polar, there is still a polarity at the inputs and outputs of the circuit, which is denoted by the small arrows accompanying the wires W_0, W_1, R, R_0, and R_1. Seen from the outside, the non-polarity inside the circuit is invisible, and the pre-wires and post-wires of the circuit's input and output wires are well-defined. Figure 10a shows a sequence of steps the non-polar Brownian 1-bit memory undergoes when a token is input to its W_1 pre-wire, resulting in the memory's state to change to 1. During this operation the token representing the state of the memory is temporarily moved to the wire segment labeled N, after which it moves to segment 0 or 1, depending on whether there was input on wire W_0 or W_1,

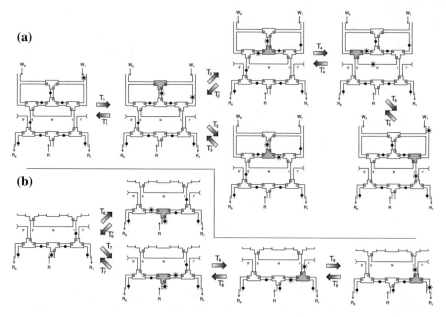

Fig. 10 Operations in a non-polar Brownian 1-bit memory. **a** Writing state 1 into the memory. After forward and backward steps, possibly getting into a dead end (through step T_3) and backtracking out of it (through step T_3'), the new state is written into the memory (T_5), while an Acknowledgement token is output to the post-wire W_1. **b** Reading out the memory when it is in state 1. Step T_6 leads to a dead end, but the memory can backtrack out of it via step T_6'. Eventually, the state of the memory is read out (T_9), with a token being output to the post-wire R_1

respectively. The memory is read out by inputing a token to the R wire (Fig. 10b). Since the state of the memory is 1, the token resides on the segment labeled 1, but during read-out it is temporarily stored on segment 1', before it returns to segment 1.

5 Discussion

Token-Pass Brownian circuits use fluctuations to explore their computational state space. This chapter revolves around a circuit primitive called T-element that is at the boundary of universality. Fluctuations provide tokens with the ability to search for computational paths in a circuit built from T-elements, allowing backtracking out of deadlocks. It is this ability that tilts the circuit in favor of universality.

The non-polar Token-Pass Brownian circuits introduced in this chapter operate in a similar way as their polar counterparts, except that some wires may lack a directional bias. Like the polar circuits, the non-polar circuits are amenable to the use of so-called *ratchets*. Unmentioned in this chapter, ratchets block tokens from fluctuating back on wires, thus making the wires effectively unidirectional, like in non-Brownian circuits. It is argued in [4, 6, 12] that ratchets speed up computation in a Brownian circuit, but their placement is limited to locations that do not require stochastic search. While some of such locations may be inside a circuit, ratchets are almost always suitable at the output wires of circuits [12]. One particular instance where ratchets are never suitable in non-polar circuits is on wires with terminators on their ends, because such wires are required to facilitate bidirectional traffic of tokens.

The searching and backtracking functionality afforded by fluctuations allows designs of circuits and circuit primitives to become simpler, since the functionality they supplement would otherwise need to be realized explicitly in the form of additional wires or states of circuit primitives, or in the form of more extensive circuitry. The Brownian 1-bit memory in Fig. 8, for example, requires less resources—in terms of number and complexity of primitives—than the non-Brownian 1-bit memory in Fig. 7a. Non-polar Brownian circuits do even better, by exploiting fluctuations to a fuller extent. We expect that further simplifications of the non-polar 1-bit memory in Fig. 9 are possible by exploiting non-polarity, but finding such improved designs remains a topic for future research.

References

1. Grollier, J., Querlioz, D., Stiles, M.D.: Spintronic nanodevices for bioinspired computing. Proc. IEEE **104**(10), 2024–2039 (2016)
2. Korotkov, A., Likharev, K.: Single-electron-parametron-based logic devices. J. Appl. Phys. **84**(11), 6114–6126 (1998)
3. Lageweg, C., Cotofana, S., Vassiliadis, S.: Single electron encoded latches and flip-flops. IEEE Trans. Nanotechnol. **3**(2), 237–248 (2004)
4. Lee, J., Peper, F.: On brownian cellular automata. In: Proceedings of Automata 2008, pp. 278–291. Luniver Press, UK (2008)
5. Lee, J., Peper, F., Adachi, S., Mashiko, S.: Universal delay-insensitive systems with buffering lines. IEEE Trans. Circuits Syst. I: Regul. Pap. **52**(4), 742–754 (2005)
6. Lee, J., Peper, F., Cotofana, S., Naruse, M., Ohtsu, M., Kawazoe, T., Takahashi, Y., Shimokawa, T., Kish, L., Kubota, T.: Brownian circuits: designs. Int. J. Unconv. Comput. **12**(5–6), 341–362 (2016)
7. Meindl, J., Chen, Q., Davis, J.: Limits on silicon nanoelectronics for terascale integration. Science **293**(5537), 2044–2049 (2001)
8. Murata, T.: Petri nets: properties, analysis and applications. Proc. IEEE **77**(4), 541–580 (1989)
9. Ono, Y., Fujiwara, A., Nishiguchi, K., Inokawa, H., Takahashi, Y.: Manipulation and detection of single electrons for future information processing. J. Appl. Phys. **97**, 031,101–1–031,101–19 (2005)
10. Patra, P., Fussell, D.: Conservative delay-insensitive circuits. In: Workshop on Physics and Computation, pp. 248–259 (1996)
11. Peper, F., Lee, J., Abo, F., Isokawa, T., Adachi, S., Matsui, N., Mashiko, S.: Fault-tolerance in nanocomputers: a cellular array approach. IEEE Trans. Nanotechnol. **3**(1), 187–201 (2004)
12. Peper, F., Lee, J., Carmona, J., Cortadella, J., Morita, K.: Brownian circuits: Fundamentals. ACM J. Emerg. Technol. Comput. Syst. **9**(1), 3:1–24 (2013)
13. Sparsø, J., Furber, S.: Principles of Asynchronous Circuit Design - A Systems Perspective. Kluwer Academic Publishers, Dordrecht (2001)

On the Reversibility of ECAs with Fully Asynchronous Updating: The Recurrence Point of View

Nazim Fatès, Biswanath Sethi and Sukanta Das

Abstract The reversibility of classical cellular automata is now a well-studied topic but what is reversibility when the evolution of the system is stochastic? In this context, we study a particular form of reversibility: the possibility of returning infinitely often to the initial condition after a random number of time steps. This corresponds to the *recurrence* property of the system. We analyse this property for the 256 elementary cellular automata with a finite size and a fully asynchronous updating, that is, we update only one cell, randomly chosen, at each time step. We show that there are 46 recurrent rules which almost surely come back to their initial condition. We analyse the structure of the communication graph of the system and find that the number of the communication classes may have different scaling laws, depending on the active transitions of the rules (those for which the state of the cell is modified when an update occurs).

1 Introduction

The question of reversibility is an important topic in the study of discrete dynamical systems because it is intimately linked to the question of time. For example, in physics a fundamental problem is to understand how rules which are reversible at the microscopic scale can generate macroscopic irreversible phenomena. In the context of cellular automata, a great amount of efforts have been deployed to understand the cases in which the local rule which defines the interaction between cells will

N. Fatès (✉)
Inria Nancy, LORIA, Université de Lorraine, 54000 Nancy, France
e-mail: nazim.fates@loria.fr

B. Sethi
Indira Gandhi Institute of Technology, Sarang, Dhenkanal, Odisha, India
e-mail: sethi.biswanath@gmail.com

S. Das
Indian Institute of Engineering Science and Technology, Shibpur, Howrah,
West Bengal, India
e-mail: sukanta@it.iiests.ac.in

© Springer International Publishing AG 2018
A. Adamatzky (ed.), *Reversibility and Universality*, Emergence, Complexity
and Computation 30, https://doi.org/10.1007/978-3-319-73216-9_15

generate a model which is reversible at the global scale [4]. In the majority of the cases, this reversibility has been interpreted as "backward determinism" [1, 5, 9], that is, the fact that each global state of the system has only a unique predecessor. In the specific case of deterministic *finite* cellular automata, this implies that the system will necessarily return to its initial state after some finite number of iterations.

The question can then be raised to know what is the equivalent of reversibility when the evolution of the cellular automaton is submitted to a form of noise. The non-determinism of the evolution implies that each configuration may have several successors. Consequently, the one-to-one correspondence between a configuration and its image can no longer be taken as a criterion of reversibility. Instead, the criterion which consists in looking if the system comes back to its initial condition may lead to interesting classifications of rules, depending on the conditions under which this "eternal return" phenomenon happens.

In this contribution, we consider the case of elementary cellular automata with fully asynchronous update. In other words, we study finite-size systems, with a local rule that involves nearest-neighbour interaction and binary states, and at each time step, we update one cell chosen randomly and uniformly in the set of cells.

So far, only a few results on the reversibility of asynchronous cellular automata have been obtained. In the case where one is allowed to choose freely the order of update of the cells, it was shown that there are some rules which allow one to return to the initial condition and some that do not always allow this [2, 6]. With a different perspective, Worsch and Wacker examined how to construct an "inverse" rule, in the sense that its transition graph would be the "inverse" of the transition graph of the original [10].

In a previous work, we analysed the set of elementary cellular automata with the aim of knowing in which case a cellular automaton may *not* return to its initial condition [7]. This resulted in the identification of the irreversible rules in the sense that such rules would not always return to their initial condition: out of the 256 rules, there are 132 strongly irreversible rules, which possess at least a configuration that is not reachable from itself, and 78 rules for which the system would return to its initial configuration a finite number of times and then stay away from it forever. It was conjectured that the rest of the rules would be *recurrent* in the sense that they would always return to their initial condition an infinite number of times. Our goal is to formally prove this recurrence property for the set of the remaining 46 candidate rules.

2 Notations

2.1 Basis

We consider one-dimensional binary cellular automata with periodic boundary conditions. The cells are arranged in a ring and the set of indices that represent each

Table 1 Table showing the transitions and their labels

A	B	C	D
000	001	100	101
010	011	110	111
E	F	G	H

cell is denoted by $\mathcal{L} = \mathbb{Z}/n\mathbb{Z}$, where $n > 1$ is the number of cells. We exclude the case $n = 1$ to avoid having a cell connected only to itself. At each time step $t \in \mathbb{N}$, a cell is assigned a state in $\{0, 1\}$. The collection of all states at given time is called a *configuration* and the set of configurations is denoted by $\mathcal{E}_n = \{0, 1\}^{\mathcal{L}}$. The two *homogeneous* configurations where all states are equal are denoted by $\mathbf{0} = 0^{\mathcal{L}}$ and $\mathbf{1} = 1^{\mathcal{L}}$.

We will often assimilate a configuration to a circular word. For a word w formed of letters of Q, we use $|x|_w$ to denote the number of occurrences of w in a configuration x. A 1-*region* (resp. 0-*region*) corresponds to a maximal set of contiguous cells with the state 1 (resp. 0). Note that for a configuration different from $\mathbf{0}$ and $\mathbf{1}$, the number of 0-regions and 1-regions are equal.

An elementary cellular automaton (ECA) is described by its local transition rule, a function $f : \{0, 1\}^3 \rightarrow \{0, 1\}$. Such a rule f describes the local transformations $(x, y, z) \rightarrow f(x, y, z)$ with $(x, y, z) \in \{0, 1\}^3$; we call each quadruplet $\big(x, y, z, f(x, y, z)\big)$, a *transition*. We say that a transition is *active* if it changes the state of a cell ($f(x, y, z) \neq y$); it is *passive* otherwise. The *decimal code* associated to f is the number: $W(f) = f(0, 0, 0) \cdot 2^0 + f(0, 0, 1) \cdot 2^1 + \cdots + f(1, 1, 1) \cdot 2^7$.

Since its introduction by Wolfram, this code has become a standard way to identify the ECAs. We will also use another code, called the *transition code*, to identify an ECA. This code consists of a concatenation of letters in $\{\mathsf{A}, \dots, \mathsf{H}\}$, where each letter denotes an active transition of the rule. The association of a letter to a transition follows the mapping of Table 1.

In the following, a rule will be denoted both by its decimal code and its transition code. For instance, majority rule is $232{:}\mathsf{DE}$ (only 101 and 010 are active; this corresponds to RMT 5 and 2 in the notation used by Das et al. [6]). The transition code of special rule that has no active transition is $204{:}\mathsf{I}$, where I stands for identity.

Given a local rule f and a ring size n, we define the global transition rule with fully asynchronous updating as $F : \mathcal{E}_n \times \mathcal{L} \rightarrow \mathcal{E}_n$ which maps a configuration $x \in \mathcal{E}_n$ and a cell to update $c \in \mathcal{L}$ to the configuration $y \in \mathcal{E}_n$ such that:

$$\forall i \in \mathcal{L}, \; y_i = \begin{cases} f(x_{i-1}, x_i, x_{i+1}) & \text{if } i = c, \\ x_i & \text{otherwise.} \end{cases}$$

Starting from an initial condition x, the evolution of the system can thus be described by the Markov chain $(X^t)_{t \in \mathbb{N}}$, where the random variables X^t are obtained recursively with $X^0 = x$ (with prob. 1) and $X^{t+1} = F(X^t, U^t)$ where $(U^t)_{t \in \mathbb{N}}$ is a series of uniformly and identically distributed random variables on $\mathcal{L}$.

2.2 Defining Recurrence

Recall that our aim is to know, given a cellular automaton f and a ring size n, if this system always returns to its initial condition. Since the updating is stochastic (one cell is chosen randomly and uniformly at each time step), one may believe at first sight that advanced tools from probability theory are needed to answer this question. In fact, because we are dealing with finite and memoryless systems, we only need some simple notions from the Markov chain theory. We now introduce these tools.

Definition 1 For two arbitrary configurations $x, y \in \mathcal{E}_n$, we say that y is a successor of x if there exists an update $u \in \mathcal{L}$ such that $F(x, u) = y$. For $x \in \mathcal{E}_n$, we denote by $\operatorname{succ}(x) = \{F(x, u), u \in \mathcal{L}\}$ the set of successors of x and by $\operatorname{succ}(S) = \cup_{x \in S} \operatorname{succ}(x)$ the set of successors of a set $S \subseteq \mathcal{E}_n$.

By iterating the successor relation, we obtain a new relation, "being reachable".

Definition 2 For two arbitrary configurations $x, y \in \mathcal{E}_n$, we say that y is reachable from x if y can be obtained from x by following a sequence of successor relations, that is, $\exists (x^i), i \in \{0, \ldots, k\}$ such that $x^0 = x$, $x^k = y$ and $x^{i+1} \in \operatorname{succ}(x^i)$ for $i \in \{0, \ldots, k - 1\}$.

We denote by $x \rightarrowtail y$ that y is reachable from x and the sequence of successors that leads from x to y is denoted by $[\![x = x^0, x^1, \ldots, x^k = y]\!]$. We call this sequence a chain of successors, or simply a chain.

We can now define the recurrence property:

Definition 3 A configuration $x \in \mathcal{E}_n$ is recurrent if for every configuration y that is reachable from x, x is also reachable from y. A configuration that is not recurrent is transient.

Intuitively, a transient configuration is such that there exists a particular sequence of updates which brings the system to a configuration from which it will never be possible to return back to the initial configuration.

More generally, if y is reachable from x and x is reachable from y, we say that x and y *communicate*. By convention, all states communicate with themselves. Clearly, the relationship "communicate" is an equivalence relation; this relation partitions the set of configurations into *communication classes*. For a ring size n, the number of communication classes is denoted by $\mathcal{C}(n)$.

Consequently, two major behaviours exist: for the transient configurations, the system remains a finite time in its communication class, then "escapes" this class and never returns back to it. In contrast, when the system starts from a recurrent configuration, it remains in the communication class of this configuration forever.

Definition 4 A rule is recurrent for size n if each configuration of $\mathcal{E}_n$ is recurrent; otherwise, it is irreversible (for this size). A rule is recurrent if it is recurrent for all the sizes $n > 1$.

Recall that our objective is to separate the 256 ECAs into recurrent and irreversible rules. By considering the conjugation symmetry, which inverts state 0 and 1 and the reflection symmetry, which exchanges left and right, one can reduce this set to 88 non-equivalent rules. In the rest of this text, we will only work with the *minimal representative* (or simply *minimal*) rules, that is, the rule which have the smallest Wolfram code when one considers these two symmetries and their composition.

We have previously established a list of irreversible rules. In order to give the sufficient conditions for a rule to be irreversible, we introduce the following notation: for a given rule, we will denote by A, B, ..., the fact that the respective transitions A, B are active and by $\overline{A}$, $\overline{B}$, ... the fact that they are passive.

Theorem 1 ([7]) *A rule is irreversible if it verifies one of the following conditions:*

- *1a: $A\overline{E}$ or 1b: $D\overline{H}$,*
- *2a: $\overline{A}E$ or 2b: $\overline{D}H$,*
- *3a: $\overline{ABCEF}$ or 3b:$\overline{ABCEG}$ or 3c:$\overline{BDFGH}$ or 3d:$\overline{CDFGH}$,*
- *4a: $\overline{ABEFG}$ or 4b:$\overline{ACEFG}$ or 4c:$\overline{BCDFH}$ or 4d:$\overline{BCDGH}$.*

The two or four conditions same line are equivalent up to the reflection or conjugation symmetries. The proof is simply obtained by finding examples of non-reversible evolutions. For example, consider rule BCE: it verifies condition 2a, as A is passive and E is active. If we take the configuration $x = 00100$ and update the third cell, we reach the fixed point 00000, and can not go back to the initial condition. Readers interested in the details may consult Ref. [7]. There are in total 70 minimal rules that verify the sufficient conditions expressed in the theorem above. Among these rules, 46 minimal rules were identified to be *strongly irreversible* in the sense that they have at least one initial condition for which it is *impossible* to go back, not even once. The 18 remaining rules were conjectured to be recurrent. Establishing this property will constitute our main result.

Theorem 2 *Among the 88 minimal ECA rules, the following 16 rules are recurrent:*

35:*ABDEFGH*	38:*BDFGH*	43:*ABDEGH*	46:*BDGH*
51:ABCDEFGH	**54:BCDFGH**	**57:ACDEGH**	**60:CDGH**
62:*BCDGH*	**105:ADEH**	**108:DH**	134:*BFG*
142:*BG*	**150:BCFG**	**156:CG**	**204:I**

and the rules 33:ADEFGH and 41:ADEGH are recurrent for $n \neq 3$.

The rules indicated bold font are directly recurrent (see below).

Let us now prove this theorem and analyse of how the number of communication classes $\mathcal{C}(n)$ scales with n.

2.3 Direct Recurrence

Definition 5 We say that a rule has the property of direct recurrence if for any two configurations $x, y \in \mathcal{E}_n$, if y is a successor of x, then x is also a successor of y. Formally: $\forall x \in \mathcal{E}_n$, $x \in \text{succ}(\text{succ}(x))$.

Proposition 1 *A rule that is directly recurrent is recurrent.*

Proof Consider a directly recurrent rule and assume that x and y are two configurations such that $x \rightarrowtail y$ with the chain of successors $[\![x^0 = x, \ldots, x^k = y]\!]$. The direct recurrence property gives $x^i \in \text{succ}(x^{i+1})$ for all $i \in \{0, \ldots, k-1\}$, which implies that $y \rightarrowtail x$ with the chain $[\![x^k = y, \ldots, x^0 = x]\!]$. □

Proposition 2 *A rule is directly recurrent if and only if it contains zero or two active transitions in each of the four sets of transitions:* $\{A, E\}, \{B, F\}, \{C, G\}, \{D, H\}$.

Proof The proof is clear by looking at Table 1. If the rule has no active transition, that is, if we have the identity rule, then the rule is recurrent because every configuration is a fixed point. If for example we assume that we go from x to y by applying transition A ($000 \rightarrow 1$), then, on the configuration y, it is now possible to apply transition E ($010 \rightarrow 0$) on the same cell to return to the previous configuration x. Symmetrically, if E was applied first, then it is possible to apply A to reverse the effect of the transition. The same arguments apply for the other couples of transitions $\{B, F\}, \{C, G\}$, and $\{D, H\}$. Reciprocally, if a rule is directly recurrent, one of these cases apply. □

We thus determine the minimal rules which are directly recurrent as:
51:ABCDEFGH, 54:BCDFGH, 57:ACDEGH, 60:CDGH, 105:ADEH, 108:DH, 150:BCFG, 156:CG, 204:I.

2.4 A Lemma for Determining the Communication Classes

Lemma 1 *For a given rule f and a given ring size n, let $(E_i)_{i \in \{1,\ldots,k\}}$ be a partition of the configuration space $\mathcal{E}_n$ and let $\{r_i\}$ be a set of configurations such that $\forall i \in \{1, \ldots, k\}$, $r_i \in E_i$. If for each $i \in \{1, \ldots, k\}$:*

(a) any configuration of E_i communicates with r_i,
(b) E_i is closed under the application of F (that is, $\text{succ}(E_i) \subseteq E_i$),

then f is recurrent and its communication classes are formed by the sets (E_i).

To verify the lemma, one simply needs to observe that for each $i \in \{1, \ldots, k\}$, the set E_i is recurrent and maximal. Consider $x \in E_i$, if we take a configuration $y \in E_i$, then x and y communicate through r_i and if we take a configuration y out of E_i, then

y is not reachable from x, otherwise this would contradict the second hypothesis which stipulates that the set E_i is closed.

Since communication classes are essentially equivalence classes, the configurations (r_i) are called the *representatives* of the partition (E_i). In our proofs, we will often show that two different representatives do not communicate, which is equivalent to the second property of the lemma.

3 Analysis of the Recurrent Rules

Let us now analyse the recurrent ECAs and their communication classes. We start with the rules that have a large number of communication classes and finish with the rules where the communication class is the whole set of configurations $\mathcal{E}_n$.

3.1 Rules with an Exponential Number of Communication Classes

Clearly, the identity rule $204 : \mathtt{I}$ is recurrent: each configuration is a fixed point and thus constitutes its own communication class.

Proposition 3 *Rule* $156 : \mathtt{GG}$ *is recurrent. The number of its communication classes* $\mathcal{C}_{\mathrm{CG}}$ *scales exponentially with the ring size* n. *More precisely, we have* $\mathcal{C}_{\mathrm{CG}}(n) \sim \phi^n$, *where* $\phi = (1 + \sqrt{5})/2$ *is the golden ratio.*

Proof This rule is directly recurrent.

To count the number of communication classes, let us define the set of configurations which do not contain the pattern 11 as $R = \{x \in \mathcal{E}_n; \nexists i \in \mathcal{L}; (x_i, x_{i+1}) = (1, 1)\}$ and show that R and the fixed point $\mathbf{1}$ constitute a set of representatives. Let us consider a configuration $x \notin R \cup \{\mathbf{1}\}$. This configuration necessarily contains the pattern 11 and, since it is different from $\mathbf{1}$, it contains the pattern 110. The transition G ($110 \to 0$) is active; x thus communicates with a configuration x' that has a 1-region which has been decreased by 1 (e.g., $x = 0011101$ and $x' = 0011001$). By transitivity x communicates with $x'' \in R$ which is such that all the 1-regions of x have been reduced to a size 1 (e.g., $x'' = 0010001$).

Moreover, two different configurations in R do not communicate. Indeed, the transitions C and G do not modify the positions of the 01-frontiers (see Fig. 1). The set $R \cup \{\mathbf{1}\}$ is thus a set of representatives and we have $\mathcal{C}_{\mathrm{CG}}(n) = card\ R + 1$.

Now, it is well-known that the number f_n of binary words of size n that do not contain the pattern 11 follows the Fibonacci sequence, defined by: $f_n = f_{n-1} + f_{n-2}$ with $f_1 = 2$ and $f_2 = 3$. This relation can be proven recursively by counting the number of words which end with a 0 or a 1, respectively. In order to obtain the size of R, we take into account the circularity of the configuration and suppress

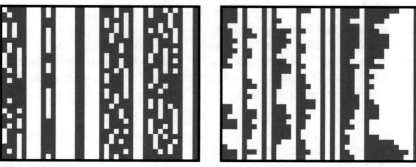

ECA 108:DH   ECA 156:CG

Fig. 1 Space-time diagrams of two ECA rules with an exponential number of communication classes. Blue and white squares represents cells in state 1 and 0, respectively. Time goes from bottom to top. Each row represents the state of the system after n random updates of cells. We have here $n = 40$ and the evolution is shown during $30 \cdot n$ time steps. This convention is kept

the words of length $n \geq 3$ which start with a 1 and end with a 1. The number δ_n of such words also verifies $\delta_n = \delta_{n-1} + \delta_{n-2}$ with $\delta_3=1$ (101) and $\delta_4 = 1$ (1001). With the convention $\delta_2 = 0$, $\delta_3 = 1$, for $n \geq 2$, the number of classes of communication of rule CG reads $\mathcal{C}_{CG}(n) = (f_n - \delta_n) + 1$. This implies that it obeys the recurrence equation $\mathcal{C}_{CG}(n) = \mathcal{C}_{CG}(n - 1) + \mathcal{C}_{CG}(n - 2) - 1$ with $\mathcal{C}(1) = 2$ and $\mathcal{C}(2) = 4$. We thus obtain $\mathcal{C}_{CG}(n) \sim \phi^n$, where $\phi = (1 + \sqrt{5})/2$ is the golden ratio.[1]   □

Proposition 4 *Rule* 108 : DH *is recurrent. The number of its communication classes scales exponentially with the ring size* n.

Proof This rule is directly recurrent. Let us evaluate the number of its communication classes. Let R be the set of configurations which do not have an isolated 0: We want to show that R constitutes a set of representatives.

Since transition D (101 → 1) is active, it is clear that any configuration x which does not belong to R communicates with a configuration x' with one less isolated 0, and, by transitivity, with a configuration which has no isolated 0, that is, with an element of R. Moreover, notice that for any configuration, a 0-region whose length is strictly greater than 1 is stable. Consequently, two different configurations of R cannot communicate because they necessarily have one 0-region for which they differ (otherwise they would be equal). The set R is thus a set of representatives.

Let us now estimate the size of R as a function of the ring size n. As for rule CG (see above), we can first evaluate the number g_n of (non-circular) words which do not contain the pattern 101. By looking at the recurrence relations that govern the number of (legal) words which end with the patterns 00, 01, 10, and 11, one

---

[1] The first elements of the sequence $\mathcal{C}_{CG}(n)$ are 2, 4, 5, 8, 12, 19, 30 and the sequence has index A001612 in the On-Line Encyclopedia of Integer Sequences (OEIS).

ECA 134:BFG   ECA 142:BG

Fig. 2 Space-time diagrams of two ECA rules with a linear number of communication classes

can prove that (g_n) verifies the recurrence equation[2]: $g_n = g_{n-1} + g_{n-2} + g_{n-4}$ with $g(0) = 1$, $g(1) = 2$, $g(2) = 4$ and $g(3) = 7$. One can then count the number δ_n of words which contain an isolated 0 which is exclusively created by the circularity of the configuration. This corresponds to the words of the form $01w1$ and $1w'10$ where w and w' are binary words of length $n - 3$ that do not contain an isolated 0. By applying the same techniques as above, one can prove that δ_n obeys the same recurrence equation as g_n, but with different initial conditions. More precisely, we have: $\delta_1 = 0$, $\delta_2 = \delta_3 = \delta_4 = 2$ and $\delta_n = 2g_{n-4}$ for $n > 4$.

As we have $\mathcal{C}_{DH}(n) = card\,R = g_n - \delta_n = g_n - 2.g_{n-4}$, we obtain the recurrence equation: $\mathcal{C}_{DH}(n) = \mathcal{C}_{DH}(n - 1) + \mathcal{C}_{DH}(n - 2) + \mathcal{C}_{DH}(n - 4)$ with $\mathcal{C}(1) = 2$, $\mathcal{C}(2) = 2$, $\mathcal{C}(1) = 5$, $\mathcal{C}(1) = 10$. It should be noted that the case $n = 1$, which was excluded from our study, does *not* correspond to a case where the rule is recurrent but it is here given only to initiate the recurrence.[3] To solve this linear recurrence equation, we need to find the roots of the polynomial $P(x) = x^4 - x^3 - x^2 - 1 = (x + 1)(x^3 - 2x^2 + x - 1)$. One can verify that the four roots of P are -1, $\lambda \approx 1.75488$, and two complex conjugate solutions whose module is strictly smaller than 1. As a result, we obtain that $\mathcal{C}_{DH}(n) \sim C.\lambda^n$ where C is a constant which is determined to be equal to 1 (see Appendix). □

3.2   Rules with a Linear Number of Communication Classes

Proposition 5 *Rules* 134:BFG, 142:BG, *and* 150:BCFG *are recurrent. Their communication classes correspond to the fixed points and to the configurations with an equal number of* 1*-regions.*

---

[2]This corresponds to the sequence A005251 of the OEIS (with different initial conditions since their sequence $(a(n))_{n \in \mathbb{N}}$ is given by $a(n) = g_{n-3}$).

[3]This corresponds to the sequence A259967 of the OEIS.

Table 2 Rule BG: sequence of configurations showing that the communication between an arbitrary configuration and a representative configuration. Below each configuration, we indicate which transition can be applied in each cell (a for transition A, b for transition B, etc.)

x	1100111010111 hgcbfhgdedchh	s_3	1010100000000 ededecaaaaaab
x'	0000100010100 aaabecabedeca	y''	0000001010100 aaaaabededeca
x''	0000101010000 aaabededecaaa	y'	0100001010000 becaaaabedeca
s_3	1010100000000 ededecaaaaaab	y	1100111010111 hgcbfhgdedchh
To a representative point		From a representative point	

The number of their communication classes is: $\mathcal{C}(n) = \begin{cases} 3 + n/2 & \text{if } n \in 2\mathbb{Z}, \\ 2 + \lfloor n/2 \rfloor & \text{otherwise.} \end{cases}$

Proof Let us first focus on rule BG (see Fig. 2). This rule has two fixed points: **0** and **1** and for n even, it has two additional fixed points: $(01)^{\frac{n}{2}}$ and $(10)^{\frac{n}{2}}$.

Let us now consider a configuration that is not a fixed point. For $k \in \{1, \ldots, \lfloor n/2 \rfloor\}$ and $k \neq n/2$, let E_k be the set of configurations with k 1-regions, that is, $E_k = \{x \in E, |x|_{01} = k\}$. We show that $s_k = (10)^k 0^{n-k}$ is a representative of E_k. Let us take a configuration $x \in E_k$ and examine if it communicates with s_k. To show that $x \rightarrowtail s_k$, we define the following intermediary configurations (see Table 2):

- x' is a transformation of x where the 1-regions are reduced to a size 1 from their right part.
- x'' is a transformation of x' where the 1-regions are "stacked" together.

We now show that: $x \rightarrowtail x' \rightarrowtail x'' \rightarrowtail s_k$. We advise the readers to follow the text below with an eye on Table 2.

Step (a): To see that $x \rightarrowtail x'$, we only need to remark that the transition G ($110 \rightarrow 0$) allows one to decrease the length of the 1-regions when this length is greater than or equal to 2.

Step (b): To show that $x' \rightarrowtail x''$, let us first observe the following property. If a configuration x contains the pattern 001, then the 1 can "move one cell to the left". Formally, if we have $z = w_1 001 w_2$, with w_1, w_2 two words of $\{0, 1\}$ such that $|w_1| + |w_2| = n - 3$, then $z \rightarrowtail z' = w_1 100 w_2$. Indeed, as the transition B ($001 \rightarrow 1$) and G are active, we have $z \rightarrowtail w_1 011 w_2 \rightarrowtail z'$.

By iterating this argument, we have that if x' has the form $x' = 0^i (10)^j 10^k w$ with $i, j \geq 0$ and $k > 1$ then $x' \rightarrowtail 0^i (10)^j 10^k w$. In words, this means that there are chains of successors which allow one to move a 1 to its leftmost position. Again, by considering each 1-region (from left to right) and moving the 1's to the left, we can build a chain such that $x' \rightarrowtail x''$.

Step (c): By considering the possibility of moving the 1's to the left as described above, we can show that for any configuration x that is not a fixed point, $x \rightarrowtail \sigma(x)$ where σ denotes the left shift. By repeating the shift operation as many times as needed, we can show that there exists a chain of successors such that $x'' \rightarrowtail s_k$.

In the same way, it can be shown that $s_k \rightarrowtail y$, where y is any configuration that is not a fixed point. We use the same arguments as above but follow a different path. First, we have that $s_k \rightarrowtail y''$, where y'' is the configuration obtained from s_k with a given number of shifts such that the *rightmost* 1 of y'' coincides with the rightmost 1 of s_k (see Table 2). Let y' be the configuration where all the ones are aligned with the rightmost 1's of the 1-regions y, then, clearly, $y'' \rightarrowtail y'$. Finally, the relationship $y' \rightarrowtail y$ can be obtained with the transition B to make the 1-regions grow to the appropriate size.

As the application of the rule BG does not modify the number of regions, the sets E_k are closed under the application of this rule, and, naturally, so are the fixed points. Consequently, by applying Lemma 1, the rule BG is recurrent.

The same conclusion holds for rules BFG and BCFG. Indeed, the presence of the active transitions C and F gives the chains of successors more possibilities of "movement" in the configuration space but does not change the communication classes.   □

We now turn our attention to what is probably the most interesting rule of our study. The number of its communication classes depends on the divisibility of n by 4 and in the general case the sizes of the classes of communication strongly vary depending on how much "freedom of movement" the system has.

Proposition 6 *Rule 105:$ADEH$ is recurrent. The number of its communication classes $\mathcal{C}$ depends on the ring size n according to:*

$$\mathcal{C}(n) = \begin{cases} 2 & \text{if } n \in 2\mathbb{Z} + 1 \\ n/2 + 3 & \text{if } n \in 4\mathbb{Z} \\ n/2 & \text{otherwise.} \end{cases}$$

Proof To understand the structure of the communication graph of rule ADEH, let us introduce the transformation T of the configuration from the states 0 and 1 to the binary alphabet $Q' = \{., \mathsf{X}\}$ where X and $.$ respectively denote the presence of absence of a change of state between a cell and its right neighbour. Formally, we associate to a configuration $x \in \mathcal{E}_n$ a word $x' = T(x) \in Q'^{\mathcal{L}}$ such that $\forall i \in \mathcal{L}$, $x'(i) = .$ if $x_i = x_{i+1}$ and $x'(i) = \mathsf{X}$ otherwise. Note that T is not injective since a configuration x and its inverse $\bar{x}$, that is, the configuration where the state of all cells has been inverted, have the same image. The function T is not surjective either because, since the number of 0-regions and 1-regions is equal on rings, only images with an even number of X's can be obtained (see Table 3). The two homogeneous fixed points $\mathbf{0}$ and $\mathbf{1}$ are mapped to the all-$.$ configuration $(.^{\mathcal{L}})$.

Now, remark that any application of the rule ADEH amounts to transforming a pair $..$ into a pair XX or a pair XX into a pair $..$ in the image of T. Moreover,

ECA 105:ADEH　　　　　　　　　　　　　　ECA 46:BDGH

Fig. 3 Space-time diagrams of ECA 105:ADEH (linear number of communication classes) and ECA 46:BDEH (two communication classes)

Table 3 Rule 105:ADEH:: two sequences of configurations showing the communication between an arbitrary configuration and a representative. The image according to the transformation T is shown on the right (see text for details). Time goes from top to bottom

0110110001	X.XX.X..XX	0110110001	X.XX.X..XX
0111110001	X....X..XX	0111110001	X....X..XX
0111110101	X....XXXXX	0001010001	X..XXX..XX
0111111101	X......XXX	0101010001	XXXXXX..XX
0111111111	X........X	(...)	(...)
1111111111	....	0000000000	....

remark that a class of communication in the image process corresponds to one or two communication classes of the original rule **ADEH**. Indeed, given a communication class of the image, there are two possibilities: either any two pre-images of this set communicate or they do not communicate and form two distinct classes (see examples below). These properties of the images allow us to determine the classes of communication of the original rule more easily. An example of evolution of the rule and its image is shown on Fig. 3 and Table 3. Moreover, an example of the communication graph is shown on Fig. 4 for $n = 6$.

(a) Let us start with the case where n is odd. Let us consider a configuration $x \notin \{\mathbf{0}, \mathbf{1}\}$, its image $T(x)$, and let us examine the regions of this image. Since n is odd, it can be remarked that there exists at least one region of . or X in $T(x)$ that has an *even* size, otherwise we would have an even number of regions of odd size and n would be even. This region can be made "to disappear" by changing the state of all its pairs. In other words, if for example $x = w_1 . X^{2k} . w_2$ and $y = w_1 . ^{2k+2}w_2$, where $k > 0$ and w_1 and w_2 are arbitrary words, then x and y communicate. By iterating this process, we obtain a homogeneous configuration, that is, an all-. or all-X configuration, but this latter possibility is excluded because the images necessarily

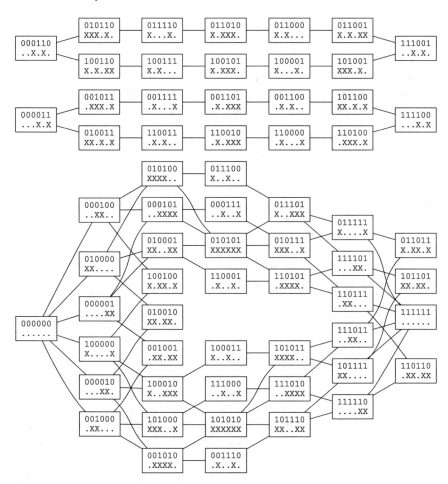

Fig. 4 Communication graph for rule 105 : ADEH and a ring of size $n = 6$. The symbols X and . represent the image by a transformation which displays the differences between a cell and its right neighbour (see text for details). Self-loops are not shown, links are bidirectional

have an even number of X's. Consequently, x communicates with at least one of the two homogeneous configurations **0** or **1** (the pre-images of all- .).

Now, consider the number of 00 pairs: by looking at the effect of the transitions A, D, E and H, it can be seen that its *parity* is a conserved quantity, that is, if $y \in \text{succ}(x)$, then $|x|_{00} \mod 2 = |y|_{00} \mod 2$. This implies that the two configurations **0** and **1** do not communicate and the space of configurations can be partitioned into two communication classes, each one containing the configuration with an even or an odd number of 00 pairs, respectively.

(b) We can now consider the case where n is even and write $n = 2k$. First let us examine what happens when n is not a multiple of four, that is, when $k = n/2$ is odd. We claim that there are k classes of communications, which representative images can be taken as the configurations $(\,.\,)^n$, $(\,.\,)^{n-4i}(X\,.\,)^{2i}$ and $(\,.\,)^{n-4i}(\,.\,X)^{2i}$, with $i \in \{1, \ldots, \lfloor k/2 \rfloor\}$. In words, we can always reach an image with an even number of X's which alternate with .'s.

Indeed, by reasoning on the images and by looking at the 1-regions, we can see that the transformation of pairs $XX \rightarrow \,.\,.$ makes an arbitrary configuration x communicate with a configuration x' whose image $T(x')$ is either all-., or where all the X's are isolated. Consequently, if $T(x')$ contains two 1's which are separated by an even number of cells, then x' communicates with a configuration x'' which is such that its image $T(x'')$ is obtained by replacing the two 1's of $T(x')$ by two .'s. By iterating this argument, we find that x'' communicates with y, which is such that either all the X's of $T(y)$ are on cells of even position or they are all on cells of odd positions. Last, we can verify that y communicates with the representative where the 1's have the same parity as in $T(y)$ (see Fig. 4).

We now examine whether two representatives communicate. Let us consider the following quantities: e and o, the number of even and odd cells with state X in the image. For a configuration $x \in \mathcal{E}_n$, we define: $e = card\,\{i \in \mathcal{L};\, i \in 2\mathbb{Z}$ and $T(x)_i = X\}$ and $o = card\,\{i \in \mathcal{L};\, i \in 2\mathbb{Z}+1$ and $T(x)_i = X\}$. It is then clear that $q = e - o$ is a conserved quantity, that is, if $y \in succ(x)$, then $q(T(y)) = q(T(x))$. Indeed, as the X's appear or disappear by pairs, the two functions e and o can only increase or decrease by one for each active update. Since all the representative have different values of q, they cannot communicate (see Fig. 4).

The last detail we need to settle is whether each representative image corresponds to one or two communication classes. Indeed, remember that each image corresponds to two pre-images, which are conjugate. We claim that each representative corresponds to a single communication class, as its two pre-images communicate. To see why, first note that the representative all-. corresponds the two pre-images $\mathbf{0}$ and $\mathbf{1}$, which both communicate with $(01)^k$, and thus communicate together. Then, if we take $s = (\,.\,)^{n-4i}(X\,.\,)^{2i}$ with $i \in \{1, \ldots, j\}$ as a representative, its two pre-images are $r = 0^{n-4i}(0011)^i$ and its conjugate $\bar{r} = 1^{n-4i}(1100)^i$. To show that r and $\bar{r}$ communicate, one may for example observe that r communicates with $r' = 1^{n-4i}(0011)^i$ (see Fig. 4). Then, by noting that r' and $\bar{r}$ have the same number of 1's, one may construct a chain of communication from r' to $\bar{r}$ by turning pairs of 0's into pairs of 1's (with two updates) to "shift" the pairs of 0's to their right position.

(c) We can now examine our last case: when n is a multiple of four. We write $n = 2k = 4j$. In fact, all the arguments stated above still apply, with an exception which concerns the two representative images $(01)^k$ and $(10)^k$. These two images have four pre-images, the four fixed points: $(0011)^j$, $(0110)^j$, $(1001)^j$ and $(1100)^j$. As each fixed point is its own class of communication, there are two more communication classes than in the previous case. There are thus $\mathcal{C}(n) = 1 + 2(j-1) + 4 = n/2 + 3$ communication classes, which respectively correspond to the following representatives: $\mathbf{0}$, $0^{n-2i}(01)^{2i}$, $0^{n-2i}(10)^{2i}$ with $i \in \{1, \ldots, j-1\}$, and the four fixed points. $\qquad\square$

3.3 Rules with Two Communication Classes

In the following, in many cases we will prove that **1** is a representative configuration. It is thus useful to know in which case **1** will be reachable and which are the configurations that can be reached from **1**.

Definition 6 For a given rule, we say that there is a density-decreasing path from x to y if $\exists (x^0, \ldots, x^k)$ such that $[\![x^0 = x, \ldots, x^k = y]\!]$ and: $|x^{i+1}|_1 < |x^i|_1$ for all $i \in \{0, \ldots, k-1\}$.

We say that **1** is an Olympic point of E if there is a density-decreasing path from **1** to any configuration of $E \setminus \mathbf{1}$.

We say that **0** is an Avernal point of E if there is a density-decreasing path from any configuration of $E \setminus \mathbf{0}$ to **0**.

It should be noted that the two definitions above are not exactly symmetric: in the first case we start from the Olympic point to reach other configurations, in the second case we start from a given configuration and reach the Avernal point.[4]

Proposition 7 *For a given rule, if (a) transition* H *is active and (b) transition* F *or* G *is active, then* **1** *is an Olympic point of* $\mathcal{E}_n \setminus \mathbf{0}$. *If conditions (a) and (b) are verified and transition* E *is active, then* **1** *is an Olympic point of the whole configuration space* $\mathcal{E}_n$ *and* **0** *is an Avernal point of* $\mathcal{E}_n$.

Proof The proof is straightforward. Let us assume that the conditions (a) and (b) are verified. Let us consider a configuration $x \notin \{\mathbf{0}, \mathbf{1}\}$. Then, if x does not contain a pair 00, that is, if all the 0's are isolated, then, x has a predecessor y where a 0 has been replaced by a 1 and such that transition H can be applied in this cell. Formally: $\exists i \in \mathcal{L}, y \in \text{succ}(x); \forall j \neq i, x_j = y_j$ and $x_i = 0$ and $(y_{i-1}, y_i, y_{i+1}) = (1, 1, 1)$. If x contains a pair 00, as it is different from **0**, it contains a triplet 001 and a triplet 100. In the same way as we did before, if transition F (resp. G) is active we can construct a configuration y where a 0 has been replaced by a 1 and such that transition F (resp. G) can be applied in this cell. Formally: $\exists i \in \mathcal{L}, y \in \text{succ}(x); x_i = 0, (y_{i-1}, y_i, y_{i+1}) = (0, 1, 1)$ or $(y_{i-1}, y_i, y_{i+1}) = (1, 1, 0)$, and $\forall j \neq i, x_j = y_j$.

As a consequence, every non-uniform configuration has a predecessor which has a higher number of 1's, which implies that this configuration can be reached from **1**.

In case E is active, it can easily be seen that **0** is reachable from the configurations with only one 1, which are themselves reachable from **1**. The uniform configuration **1** is thus an Olympic point for $\mathcal{E}_n$. It can also be verified that the number of 1's can always be decreased, which implies that **0** is an Avernal point. $\qquad\square$

---

[4]This has some logic: as expected one may go from the Olympus to any point, but the reverse is not granted, and one may also go from any point to the Avernus (the entry of the underworld in Roman mythology) but the other way round should be more adventurous.

Table 4 Rule 46:BDGH: sequence of configurations showing the communication between an arbitrary configuration and the Olympic point **1**

x	1100010100011
	hgcabedecabfh
x'	1111010111011
	hhhgdedfhgdfh
1	1111111111111
	hhhhhhhhhhhhh

Proposition 8 *The five rules 38:BDFGH, 46:BDGH, 54:BCDFGH, 60:CDGH, and 62:BCDGH are recurrent. Their communication classes are formed by the fixed point* **0** *and by the rest of the configuration space* $\mathcal{E}_n^* = \mathcal{E}_n \setminus \{0\}$.

Proof According to Proposition 7, as transitions G and H are active, we know that **1** is an Olympic point for $\mathcal{E}_n^*$ (any configuration of this set is reachable from **1**, see Table 4).

Now, let us take $x \in \mathcal{E}_n^*$ and show $x \rightarrowtail \mathbf{1}$. It can be observed that if $x \neq \mathbf{1}$, for all these rules, as either B and D are active or C and D are active, it is always possible to increase the number of 1's. Indeed, if x contains the pattern 00, then it also necessarily contains the pattern 001 and 100, and transition B or C can be applied. If x does not contain the pattern 00, this means that all the 0's are isolated, that is, in the form 101, and transition D can be applied. The configuration **1** is thus reachable from x (see Table 4).

Consequently, **1** is a representative configuration of $\mathcal{E}_n^*$ and, given that **0** is a fixed point that cannot be reached from $\mathcal{E}_n^*$, we have $\mathrm{succ}(\mathcal{E}_n^*) \subseteq \mathcal{E}_n^*$, which allows us to use Lemma 1 to conclude that the four rules are recurrent.   □

3.4 Rules with a Single Communication Class

Proposition 9 *For a ring size* $n \neq 3$, *the rules 33:ADEFGH and 41:ADEGH are recurrent. Their communication class is the whole configuration space.*

The case of these two rules is interesting: their proof is less direct than the previous rules. In particular, the rules are not recurrent for $n = 3$. Indeed, if one starts from 011, the configuration 001 is reachable, but it is not possible to go back to the initial condition from this configuration as transitions B and C are inactive. The rule is however recurrent for $n > 3$ (see Fig. 5).

Proof The proof of recurrence is decomposed into three steps: (a) **0** is an Avernal point; (b) **1** is reachable from **0**; (c) **1** is an Olympic point.

Steps (a) and (c) are directly obtained from Proposition 7.

To prove step (b) for the two rules, let us first assume that the ring size is even and write $n = 2k$. The sequence of configurations $(x_i)_{i \in \{1, \ldots, k\}}$ with $x_i = (10)^i 0^{2k-i}$ forms a chain of successors since the active transition A can always be applied in

ECA 33:ADEFGH   ECA 43:ABDEGH

Fig. 5 Space-time diagrams of two ECA rules for which all the configurations communicate

the cells in state 0 surrounded by two 0's (odd positions). We obtain $\mathbf{0} \rightarrowtail (01)^k$. Similarly, we have a chain of successors (y_i) with $y_i = (11)^i (10)k - i$ because the transition D is active. We obtain $(01)^k \rightarrowtail \mathbf{1}$ and thus $\mathbf{0} \rightarrowtail \mathbf{1}$.

Now let us consider an odd ring size $n = 2k + 1$. Using the same operations as above, we can show that $\mathbf{0} \rightarrowtail z = 1^{2k-1} 00$ (e.g., for $z = 11111100$ for $n = 7$). Now, if we want to reach **1**, it is necessary to "go backwards". With the active transition G, we have $z \rightarrowtail z' = 11111000$ and then, by applying transition A on the penultimate 0, a successor $z \rightarrowtail z'' = 11111000$ with two isolated 0's (e.g., 11111010). Applying twice transition D, we then have $z'' \rightarrowtail \mathbf{1}$ which gives us $\mathbf{0} \rightarrowtail \mathbf{1}$.

The three properties (a), (b) and (c) show the recurrence property of the two rules 33 : ADEFGH and 41 : ADEGH. Indeed for any two configurations $x, y \in \mathcal{E}_n$, we have $x \rightarrowtail \mathbf{0} \rightarrowtail \mathbf{1} \rightarrowtail y$. □

Proposition 10 *The four rules:* 35 : ABDEFGH, 43 : ABDEGH, 51 : ABCDEFGH, *and* 57 : ACDEGH *are recurrent. Their communication class is the whole configuration space.*

Proof For these three rules, by Proposition 7 we know **1** is an Olympic point. Moreover, as either the three transitions A, B, and D or the three transitions A, C, and D are active, the number of 1's can always be increased from any configuration, which implies that **1** can be reached from any configuration. It is thus a representative point of the whole configuration space $\mathcal{E}_n$ and the rules are recurrent by Lemma 1. □

4   Discussion

We have analysed the complete set of elementary cellular automata under fully asynchronous updating. We have asked if the behaviour of the system was recurrent and used the Markov chain theory to partition the set of rules between 46 recurrent rules (or 18 equivalence classes), and the rest of the rules, for which the return to the initial condition either never happens, or happens a finite number of times. The main tool to make this partition was to analyse the reachability relationships in the communication graph of the system. Moreover, the analysis of the communication classes showed various scaling relation of the number of communication classes as a function of the number of cells: we found the presence of an exponential number, a linear number and a constant number (two or one) of classes. The rule that displayed the greatest complexity in the analysis of the communication graph was rule $105 : \mathtt{ADEH}$, which is rather surprising given that this rule can be written $f(x, y, z) = 1 \oplus x \oplus y \oplus z$ (affine rule). It would be interesting to relate this complexity of analysis to some formal properties of the communication graph. The fact that there are fixed points of length 4 but no fixed points of smaller size may give some hints on why this rule has such a peculiar behaviour.

We can note that determining the recurrence properties of the system seems somehow an easier task than, for example, distinguishing classes of asynchronous rules as to whether they converge rapidly or not to a fixed point [3]. The situation is similar with synchronous systems: only recently was achieved the classification of the 256 elementary rules according to their dynamical properties [8]. One may thus legitimately ask to which extent the results on the recurrence obtained above can extend to a larger class of rules, radius-2 rules or two-dimensional rules for instance.

Another path of research is to estimate the average time for returning to the initial configuration. For the "flip" rule (ECA 51), one may verify easily that the symmetry properties of the rule imply that the stationary distribution, that is, the distribution one approaches as time goes to infinity, is uniform over the set of configurations. As a result, the average return time of a n-sized system is 2^n. This result can be generalised to the directly reversible rules: the stationary distribution is uniform on the configurations of the communication class and the average return time from an initial configuration is exactly the size of the communication class of this configuration. For the other rules, where a non-uniform stationary distribution exists, a more detailed analysis would be required.

Finally, we ask what definitions of the recurrence property can be given for an infinite-size system. In this context, one may either use interacting particle systems with continuous time or α-asynchronous systems, where the cells are updated or not according to a (global) probability (see Ref. [3] for more details). One may then think about defining recurrence as the ability to come back infinitely often arbitrarily close to the initial condition. However, if we take the asynchronous shift as an example (ECA 170), even though this rule has a tendency to make the configurations more and more uniform, it is always possible that the information that is far away influences the central zone to make it "close" to the initial configuration. In the infinite context, contrarily to what we have in the finite case, it seems that the property to erase information locally [7] would not be sufficient to make a rule non-recurrent. To sum up, the notion of reversibility in the context of stochastic cellular automata is a rich problem and its exploration is only at its beginning.

Acknowledgements The authors have benefitted from the careful reading of the manuscript by Jordina Francès de Mas and from helpful remarks of the reviewers.

Appendix

We calculate here an equivalent the number of communication classes of rule $105\!:\!\text{ADEH}$ for a ring of size n. For simplicity, let us denote this number by $u_n = \mathcal{C}_{DH}(n)$.

This quantity verifies the following linear recurrence equation:

$$u_n = u_{n-1} + u_{n-2} + u_{n-4} \text{ with } u_0 = 3, u_1 = 2, u_2 = 2, u_3 = 5.$$

The values u_0 and u_1 are defined only for convenience and do not correspond to $\mathcal{C}_{DH}(0)$ and $\mathcal{C}_{DH}(1)$, which are excluded form our study. To solve this equation, we need to find the roots of the polynomial

$$P(x) = x^4 - x^3 - x^2 - 1 = (x + 1)(x^3 - 2x^2 + x - 1).$$

There are two real roots, λ and -1, and two complex roots, which are conjugate, and which we denote by h and $\bar{h}$. The general solution can thus be written:

$$u_n = A\lambda^n + Bh^n + C\bar{h}^n + D(-1)^n, \tag{1}$$

where A, B, C, D are four constants which belong to the set of complex numbers.

Looking at the first terms, we have:

$$
\begin{aligned}
u_0 &= & A + & B + & C + D = 3 \\
u_1 &= \lambda\ & A + h\ & B + \bar{h}\ & C - D = 2 \\
u_2 &= \lambda^2\ & A + h^2\ & B + \bar{h}^2\ & C - D = 2 \\
u_3 &= \lambda^3\ & A + h^3\ & B + \bar{h}^3\ & C - D = 5
\end{aligned}
$$

Now let us combine these equations with the evaluation of $q = u_3 - 2u_2 + u_1 - u_0$. We find that $q = Q(\lambda)A + Q(h)B + Q(\bar{h})C - (1 + 2 + 1 + 1)D = 5 - 2 \times 2 + 2 - 3 = 0$, where $Q(x) = x^3 - 2x^2 + x - 1$. Since by definition $Q(\lambda) = Q(h) = Q(\bar{h}) = 0$, we find $D = 0$.

In a second step, we evaluate

$$q' = u_2 - \bar{h}u_1 = (\lambda^2 - \lambda\bar{h})A + (h^2 - h\bar{h})B = 2 - 2\bar{h}$$

and

$$q'' = h(u_1 - \bar{h}u_0) = (\lambda - \bar{h})hA + h(h - \bar{h})B = h(2 - 3\bar{h}).$$

The evaluation of $q' - q''$ allows one to get rid of B and we have:

$$\left(\lambda^2 - \lambda\bar{h} - h(\lambda - \bar{h})\right)A = 2 - 2\bar{h} - h(2 - 3\bar{h}). \tag{2}$$

It is now useful to write:

$$Q(x) = x^3 - 2x^2 + x - 1$$
$$= (x - \lambda)(x - h)(x - \bar{h})$$
$$= x^3 - (\lambda + h + \bar{h})x^2 + (\lambda h + \lambda \bar{h} + h \bar{h})x + \lambda h \bar{h},$$

from which we obtain by identification $\lambda + h + \bar{h} = 2$ and $\lambda h \bar{h} = 1$.

Now, by using the relationships $h + \bar{h} = 2 - \lambda$ and $h\bar{h} = 1/\lambda$ in Eq. 2, we obtain: $(2\lambda^2 - 2\lambda + 3)A = (2\lambda^2 - 2\lambda + 3)$, which leads to $A = 1$.

The algebraic or numerical estimation of the roots of Q gives: $\lambda \approx 1.75488$ and $h = 0.12256 + 0.74486i$, whose module is strictly smaller than 1.

We thus obtain $u_n \sim \lambda^n$, which was the desired result.   □

References

1. Amoroso, S., Patt, Y.N.: Decision procedures for surjectivity and injectivity of parallel maps for tesselation structures. J. Comput. Syst. Sci. **6**, 448–464 (1972)
2. Das, S., Sarkar, A., Sikdar, B.K.: Synthesis of reversible asynchronous cellular automata for pattern generation with specific hamming distance. In: Proceedings of ACRI'12, pp. 643–652. Springer (2012)
3. Fatès, N.: A guided tour of asynchronous cellular automata. J. Cell. Autom. **9**, 387–416 (2014)
4. Morita, K.: Reversible computing and cellular automata – a survey. Theor. Comput. Sci. **395**(1), 101–131 (2008)
5. Richardson, D.: Tessellations with local transformations. J. Comput. Syst. Sci. **6**, 373–388 (1972)
6. Sarkar, A., Mukherjee, A., Das, S.: Reversibility in asynchronous cellular automata. Complex Syst. **21**(1), 71–84 (2012)
7. Sethi, B., Fatès, N., Das, S.: Reversibility of elementary cellular automata under fully asynchronous update. In: Gopal, T.V., Agrawal, M., Li, A., Barry Cooper, S. (eds.) Proceedings of TAMC 2014. Lecture Notes in Computer Science, vol. 8402, pp. 39–49. Springer (2014)
8. Schüle, M., Stoop, R.: A full computation-relevant topological dynamics classification of elementary cellular automata. Chaos **22**(4), 043143 (2012)
9. Toffoli, T.: Computation and construction universality of reversible cellular automata. J. Comput. Syst. Sci. **15**, 213–231 (1977)
10. Wacker, S., Worsch, T.: On completeness and decidability of phase space invertible asynchronous cellular automata. Fundamenta Informaticae **126**(2–3), 157–181 (2013)

An Overview of 2D Picture Array Generating Models Based on Membrane Computing

K. G. Subramanian, Sastha Sriram, Bosheng Song and Linqiang Pan

Abstract A variety of two-dimensional array grammar models generating picture array languages have been introduced and investigated, utilizing and extending the well-established notions and techniques of formal string language theory. On the other hand the versatile computing model with a generic name of P system in the area of membrane computing, has turned out to be a rich framework for different kinds of problems in a variety of fields. Picture array generation in the field of two-dimensional (2D) languages is one such area where P systems with array objects and array rewriting, referred to as array P systems, have been fruitfully employed in increasing the generating power of the 2D grammar models. A variety of array P systems have been proposed in the literature. The objective of this survey is to review and describe the salient features of the major types of array P systems, which have served as the basis for developing other kinds of array P systems. Applications of these array P systems are also briefly described besides indicating possible new directions of investigation.

K. G. Subramanian
Honorary Visiting Professor (March 2017–February 2021), Department of Mathematics and Computer Science, Faculty of Science, Liverpool Hope University,
Liverpool L16 9JD, UK
e-mail: kgsmani1948@gmail.com

S. Sriram
Department of Mathematics, KL University, Vaddeswaram, Guntur 522502,
Andhra Pradesh, India
e-mail: sriram.discrete@gmail.com

B. Song · L. Pan (✉)
School of Automation, Huazhong University of Science and Technology,
Wuhan 430074, China
e-mail: lqpan@mail.hust.edu.cn

B. Song
e-mail: boshengsong@hust.edu.cn

© Springer International Publishing AG 2018
A. Adamatzky (ed.), *Reversibility and Universality*, Emergence, Complexity and Computation 30, https://doi.org/10.1007/978-3-319-73216-9_16

1 Introduction

The field of membrane computing [28, 30, 55] originated with Gh. Păun developing a new computing model around the year 2000, inspired by the structure and functioning of living cells. The basic version of this new computing model, with a generic name of P system (named in honour of the originator of this system) involves a hierarchical arrangement of membranes, one within another but all of them within one membrane, called the skin membrane and the membrane with no other membrane inside, being referred to as an elementary membrane. The regions delimited by the membranes can have objects and evolution rules. The minimal activity in a P system involves processing at the same time, the objects in all regions of the system by a nondeterministic and maximally parallel manner of application of the rules to the objects, thereby allowing the objects to evolve. The objects evolved can continue to remain in the same region or go to an adjacent region, with the communication being specified by a target indication. A computation halts when no object in all the regions can further evolve and the result of a computation is the number of objects in a specified membrane. Very many modifications and variants of the basic model of a P system have been proposed and studied but we do not enter into these details here. Instead our interest here is in a variant known as rewriting P system, initially introduced in [27] for the string case. The basic idea in a string rewriting P system is to consider the objects in the regions to be finite (structured) strings over an alphabet and the evolution rules as rewriting rules transforming a finite string in a region into another string. When the transformed string is passed through a membrane, it is sent as a whole in this kind of P system. There has been a number of investigations on rewriting P systems by several researchers (see, for example, [2–4, 9, 21, 22, 31]) introducing other features.

On the other hand, motivated by problems in the framework of image analysis and picture processing, several kinds of two-dimensional grammar models (see for example [17, 25, 33, 34, 45, 51] and references therein) have been introduced for picture array generation, with many of these models extending the rewriting feature in string grammars in formal language theory [35, 36]. Utilizing the rewriting rules in the two-dimensional grammar models, the string-rewriting P systems have been extended to arrays, resulting in a variety of array P systems (see for example [40]).

In this survey, some of the major types of array P systems are reviewed bringing out their constructions and their picture array generative power. Applications of the array P systems in the description of picture patterns are also indicated besides pointing out possible directions of study in future.

2 Preliminaries

We refer to [35, 36] for notions related to formal string grammars and languages. For picture array grammars and languages we refer to [15, 17] and for concepts relating to P systems to [28]. We briefly recall here certain needed notions.

An alphabet Σ is a finite set of symbols. A word or a string α over Σ is a finite sequence of symbols taken from Σ. The empty word with no symbols is denoted by λ. The set of all words over Σ including λ, is denoted by Σ^*. For any word $\alpha = a_1 a_2 \ldots a_n$, we denote by $^t\alpha$ the word α written vertically, so that $^t\alpha = {}^t(\alpha)$.

For example, if $w = abb$ over $\{a, b\}$, then ^tw is $\begin{matrix} a \\ b \\ b \end{matrix}$.

Interpreting the two-dimensional digital plane as a set of unit squares, a picture array in the two-dimensional plane (also called, simply as an array) over Σ, is composed of a finite number of labelled unit squares (also called pixels), with the labels taken from the alphabet Σ. The set of all picture arrays over Σ will be denoted by Σ^{**}. The empty picture array is also denoted by λ and $\Sigma^{++} = \Sigma^{**} - \{\lambda\}$. An empty unit square in the plane is indicated by the blank symbol $\# \notin \Sigma$.

A pictorial way of representing a picture array is done by showing the labels of the non-blank unit squares that constitute the picture array. For example, a picture array representing the digitized Chinese character "center" [52, p. 228] is shown in Fig. 1. Sometimes, if needed, the blank symbol is shown in some empty square but in general, we assume that the empty unit square in the plane contains this symbol even if the blank symbol is not shown. A picture array can be given in a formal manner by listing the coordinates of the non-blank unit squares of the picture array along with the corresponding labels of the unit squares. Note that a translation of the picture array in the two-dimensional plane changes only the coordinates of the unit squares of a picture array and hence only relative positions of the symbols in the non-blank unit squares are essential for describing a picture array. For example, for the picture array in Fig. 1, taking the origin $(0, 0)$ at the lowermost non-blank unit square of the leftmost vertical line of $x's$, the coordinates of the non-blank unit squares of the picture array can be specified as follows:

$$\{((p, 0), x), ((p, 3), x) \mid 0 \le p \le 10\} \cup \{((q, r), x) \mid q \in \{0, 5, 10\}, r \in \{1, 2\}\}$$

$$\cup\{((5, s), x) \mid s \in \{-1, -2, -3, 4, 5\}\}.$$

```
              x
              x
x x x x x x x x x x x
x           x           x
x           x           x
x x x x x x x x x x x
              x
              x
              x
```

Fig. 1 A picture array representing the chinese character "center" in digitized form

3   Array Grammars and Languages

Basically there are two types of array grammars, referred to as isometric array grammars [10–12, 15, 33] and non-isometric array grammars [34]. We first describe isometric array grammars. In an isometric array grammar, the geometric shape of an array is preserved while an array grammar rule is used in generating an array from another array.

3.1   Isometric Array Grammars

Analogous to the Chomsky hierarchy [35] in string grammars, there is a corresponding hierarchy [15] in the isometric array grammars. Here our interest is in the context-free type of rules which we recall now.

Definition 1 [14, 15] A *context-free (CF) array grammar* $G = (N, T, S, P, \#)$, where N and T are finite sets of symbols, respectively called nonterminals and terminals with $V = N \cup T$ and N and T are disjoint. The symbol $S \in N$ is the start symbol. The set P is a finite set of array rewriting rules of the form $r : \alpha \to \beta$ where α and β are arrays over $V \cup \#$ (# is the blank symbol) satisfying the following conditions:

1. the arrays α and β have geometrically identical shapes;
2. exactly one square in α is labelled by a nonterminal in N while the remaining squares contain the blank symbol # but β contains no blank symbol;
3. in rewriting α by β by the rule $r : \alpha \to \beta$, the rule should be such that its application does not erase the non-blank symbols in α *i.e* a unit square with a non-blank symbol should not get replaced by the blank symbol #;
4. the symbols of T that occur in α should be retained in their corresponding squares in β while applying the rule;
5. the application of the rule $r : \alpha \to \beta$ should preserve the connectivity of the array in which rewriting is done.

For two arrays γ, δ over V and a rule r as above, we write $\gamma \Rightarrow_r \delta$ if δ can be obtained by replacing with β, a subarray of γ identical to α. The reflexive and transitive closure of the relation $\Rightarrow$ is denoted by $\Rightarrow^*$.

A *CF* array grammar is called *regular*, if the rules are of the following forms:

$$A \# \to a\, B, \ \# A \to B\, a, \ \begin{matrix} \# \\ A \end{matrix} \to \begin{matrix} B \\ a \end{matrix}, \ \begin{matrix} A \\ \# \end{matrix} \to \begin{matrix} a \\ B \end{matrix}, \ A \to B, \ A \to a,$$

where A, B are nonterminals and a is a terminal.

The picture array language generated by G is

$$L(G) = \{\mathcal{A} \mid S \Rightarrow^* \mathcal{A} \in T^{++}\}.$$

Note that the start array is indeed $\{((0, 0), S)\}$ and it is understood that this square labelled S is surrounded by #, denoting empty squares with no labels. We denote by $L(AREG)$ and $L(ACF)$ respectively the families of array languages generated by array grammars with regular and context-free array rewriting rules.

We illustrate with an example.

Example 1 Consider the context-free array grammar G_p with rules

$$
(1)\ \begin{matrix} \# \\ \# \ S \ \# \end{matrix} \rightarrow \begin{matrix} A \\ C \ a \ B \end{matrix}, \ (2)\ \begin{matrix} \# \\ A \end{matrix} \rightarrow \begin{matrix} A \\ a \end{matrix},
$$

$$
(3)\ B\ \# \rightarrow a\ B,\ (4)\ \#\ C \rightarrow C\ a,\ (5)\ A \rightarrow a,\ (6)\ B \rightarrow a,\ (7)\ C \rightarrow a,
$$

where S, A, B, C are nonterminals and a is a terminal. This grammar generates arrays with three arms over $\{a\}$, but with the arms not necessarily of equal length where the length equals the number of $a's$ in an arm. In a derivation starting with S, the rule (1) is applied once. In other words, the first step of the derivation is as follows:

$$
\begin{matrix} \# \\ \# \ S \ \# \end{matrix} \Rightarrow \begin{matrix} A \\ C \ a \ B \end{matrix}
$$

This can then be followed, for example, by the application of the rule (2) as many times as needed, thus growing the vertical upper arm and the growth can be terminated with an application of the rule (5). For example, if the rule (2) is applied twice, then the derivation takes place as follows:

$$
\begin{matrix} \# \\ \# \ S \ \# \end{matrix} \Rightarrow \begin{matrix} A \\ C \ a \ B \end{matrix} \Rightarrow \begin{matrix} A \\ a \\ C \ a \ B \end{matrix}
$$

$$
\Rightarrow \begin{matrix} A \\ a \\ a \\ C \ a \ B \end{matrix} \Rightarrow \begin{matrix} a \\ a \\ a \\ C \ a \ B \end{matrix}
$$

Likewise, the horizontal right or left arm can be grown by the respective application of the rules (3), (4). Again the growth in these arms can be terminated by the respective application of the rules (6), (7), thus yielding an array in the form of the digitized Math symbol of perpendicularity (Fig. 2).

It is known that analogous to the string case, CF array grammars are more powerful than regular array grammars in generating picture array languages.

Theorem 1 [5, 15] $L(AREG) \subset L(ACF)$.

$$a$$
$$a$$
$$a$$
$$a$$
$$a$$
$$a\ a\ a\ a\ a\ a\ a\ a$$

Fig. 2 A picture array representing the math symbol for perpendicularity in digitized form

The inclusion is straightforward from the Definition 1 while the strict inclusion is seen by observing that the picture array language in Example 1 cannot be generated by regular array grammar rules as the rewriting in a regular array grammar cannot handle all three arms at the "junction" in the Fig. 2. In fact only two of the three arms can be generated by regular array grammar rules due to the nature of the rules.

Remark 1 Generation of geometric figures such as rectangles and squares by array grammars has been a problem of interest. One of the earliest studies in this direction has been done in [50]. Although regular array grammars have the simplest kind of rules and hence cannot have high generating power, it is interesting to note that Yamamoto, Morita and Sugata [50] have constructed regular array grammars for generating picture languages of rectangles and squares, utilizing the ability of the regular array grammars in sensing the blank symbol #.

While context-free array grammar is an extension to two-dimensions of context-free string grammar of the Chomsky hierarchy [35], contextual array grammar [16] is an extension of the contextual string grammars [26] introduced by Marcus [24]. We now recall the definition of a two-dimensional contextual array grammar [16]. For our purposes of later relating this study to array P systems we consider the restriction as in [13], with the "selector" and the "context" being connected and labelled by only symbols from an alphabet and not the blank symbol # amounting to the case of the selector and the context having no empty unit squares.

Definition 2 [26] A contextual array grammar (CAG) is a construct $G = (V, P, A)$ where V is an alphabet, A is a finite set of axioms which are two-dimensional arrays in V^{++} and P is a finite set of array contextual rules of the form (α, β) where

(*i*) α is a function defined on $U_\alpha \subset \mathcal{Z} \times \mathcal{Z}$ with values in V and (U_α, α) is called the *selector* and U_α, the *selector area* of the production (α, β); (Here $\mathcal{Z}$ is the set of all integers);

(*ii*) β is a function defined on $U_\beta \subset \mathcal{Z} \times \mathcal{Z}$ with values in V where (U_β, β) is called the *context* and U_β, the *context area* of the production (α, β);

(*iii*) U_α and U_β are finite and disjoint.

For arrays $\mathcal{C}_1, \mathcal{C}_2 \in V^{++}$, and a contextual rule $p : (\alpha, \beta)$, we have a derivation, denoted by $\mathcal{C}_1 \Longrightarrow_p \mathcal{C}_2$, if in $\mathcal{C}_1$ we can find a sub-array that corresponds to the selector (U_α, α), and if the positions corresponding to (U_β, β) are labelled only by the blank symbol #, so that we can add the context (U_β, β), thus obtaining $\mathcal{C}_2$. We then

$$
\begin{array}{ccccccc}
a & a & a & a & a & a & a \\
a & c & c & c & c & c & a \\
a & c & a & a & a & c & a \\
a & c & a & c & a & c & a \\
a & c & a & a & a & c & a \\
a & c & c & c & c & c & a \\
a & a & a & a & a & a & a
\end{array}
$$

Fig. 3 A 5×5 solid square with c in its central position

say that C_2 is derivable from C_1 and we write $C_1 \Longrightarrow_G C_2$. We denote the reflexive transitive closure of $\Longrightarrow_G$ by $\Longrightarrow_G^*$. A t-mode of derivation, denoted by $\Longrightarrow_G^t$, is defined for arbitrary arrays $A, B \in V^{++}$ in the following manner: $A \Longrightarrow_G^t B$ if and only if $A \Longrightarrow_G^* B$ and there is no $C \in V^{++}$ such that $B \Longrightarrow_G C$. The picture array language generated by a *CAG* G in the t-mode is defined as follows:

$$
L_t(G) = \left\{ B \in V^{++} \mid A \Longrightarrow_G^t B \text{ for some } A \in A \right\}.
$$

For a given *CAG* G, the t-mode (also called maximal mode) corresponds to collecting only the arrays produced by derivations which cannot be continued at some stage. On the other hand, in the *∗-mode* of derivation, all pictures derivable from an axiom are taken in the picture language generated. We consider here mainly the t-mode of derivation. The family of picture languages generated by contextual array grammars of the form $G = (V, P, A)$ in the t-mode will be denoted by $\mathcal{L}(cont, t)$.

We give an example illustrating the working of a contextual array grammar in t-mode, which generates an array language L consisting of picture arrays of solid squares of odd side length $4n - 1, n \geq 1$, in the form as shown in Fig. 3 for $n = 2$, with the central square labelled c, surrounded by a single "layer" of a's followed by a single "layer" of c's and again followed by a single "layer" of a's.

Example 2 Let $G = (\{a, c\}, P, A)$ be a contextual array grammar with A containing two axiom picture arrays A_1, A_2, which are given in pictorial form.

$$
A_1 := \begin{array}{ccc} a & a & a \\ a & c & a \\ a & a & a \end{array}, \quad A_2 := \begin{array}{ccc} & c & \\ a & a & a \\ a & c & a \\ a & a & a \end{array}
$$

Since the selector area U_α and the context area U_β are disjoint in a contextual array production, the rules can be represented by patterns where the unit squares and symbols of the selector are indicated by enclosing these in boxes. The rules are defined as follows:

$$p_1 := \boxed{\begin{smallmatrix} \boxed{c} & c \\ \boxed{a} & \boxed{a} & \boxed{a} \end{smallmatrix}}, \quad p_2 := \boxed{\begin{smallmatrix} \boxed{c} & c & c \\ \boxed{a} & \boxed{a} & c \end{smallmatrix}}, \quad p_3 := \begin{smallmatrix} \boxed{a} & \boxed{c} \\ \boxed{a} & c \\ \boxed{a} \end{smallmatrix}, \quad p_4 := \begin{smallmatrix} \boxed{a} & \boxed{c} \\ \boxed{a} & c \\ c & c \end{smallmatrix},$$

$$p_5 := \boxed{\begin{smallmatrix} \boxed{a} & \boxed{a} & \boxed{a} \\ c & \boxed{c} \end{smallmatrix}}, \quad p_6 := \begin{smallmatrix} c & \boxed{a} & \boxed{a} \\ c & c & \boxed{c} \end{smallmatrix}, \quad p_7 := \begin{smallmatrix} & \boxed{a} & \\ c & \boxed{a} & \\ & \boxed{a} & \end{smallmatrix}, \quad p_7' := \begin{smallmatrix} & \boxed{c} \\ c & \boxed{a} \\ & \boxed{a} \end{smallmatrix}, \quad p_8 := \begin{smallmatrix} & & a \\ c & \boxed{c} \\ & \boxed{c} & \boxed{a} \end{smallmatrix},$$

$$r_1 := \boxed{\begin{smallmatrix} \boxed{a} & a \\ \boxed{c} & \boxed{c} & \boxed{c} \end{smallmatrix}}, \quad r_2 := \boxed{\begin{smallmatrix} \boxed{a} & a & a \\ \boxed{c} & \boxed{c} & a \end{smallmatrix}}, \quad r_3 := \begin{smallmatrix} \boxed{c} & \boxed{a} \\ \boxed{c} & a \\ \boxed{c} \end{smallmatrix}, \quad r_4 := \begin{smallmatrix} \boxed{c} & \boxed{a} \\ \boxed{c} & a \\ a & a \end{smallmatrix}, ,\quad r_5 := \boxed{\begin{smallmatrix} \boxed{c} & \boxed{c} & \boxed{c} \\ a & \boxed{a} \end{smallmatrix}},$$

$$r_6 := \begin{smallmatrix} a & \boxed{c} & \boxed{c} \\ a & a & \boxed{a} \end{smallmatrix}, \quad r_7 := \begin{smallmatrix} & \boxed{c} & \\ a & \boxed{c} & \\ \boxed{a} & \boxed{c} \end{smallmatrix}, \quad r_7' := \begin{smallmatrix} & \boxed{a} \\ a & \boxed{c} \\ \boxed{a} & \boxed{c} \end{smallmatrix}, \quad r_8 := \begin{smallmatrix} & a & \\ a & \boxed{a} \\ \boxed{a} & \boxed{c} \end{smallmatrix}, \quad r_9 := \begin{smallmatrix} c & a & a \\ \boxed{a} & \boxed{c} \end{smallmatrix}.$$

A maximal (that is, t-mode) derivation in G generating a 7×7 picture array of the language L_1 is shown below:

```
       c                 c c              c c c c           c c c c          c c c c
     a a a             a a a            a a a c           a a a c          a a a c
     a c a    ⟹p₁     a c a   ⟹p₂    a c a    ⟹p₃    a c a c  ⟹p₄   a c a c
     a a a             a a a            a a a             a a a            a a a c
                                                                            c c
```

```
       c c c c           c c c c           c c c c          c c c c           a
       a a a c           a a a c           a a a c          c a a a c       c c c c c
⟹p₅  a c a c  ⟹p₆   a c a c  ⟹p₇   c a c a c  ⟹p₇'  c a c a c  ⟹p₈  c a a a c
       a a a c           c a a a c         c a a a c        c a a a c         c a c a c
       c c c             c c c c c         c c c c c        c c c c c         c a a a c
                                                                              c c c c c
```

```
        a a                 a a a              a a a a a a         a a a a a a
      c c c c c           c c c c c           c c c c c a        c c c c c a
⟹r₁  c a a a c   ⟹r₁    c a a a c   ⟹r₂    c a a a c   ⟹r₃   c a a a c a
      c a c a c           c a c a c           c a c a c          c a c a c
      c a a a c           c a a a c           c a a a c          c a a a c
      c c c c c           c c c c c           c c c c c          c c c c c
```

The derivation proceeds in this way due to the t-mode of derivation and halts when no rule is applicable. The resulting array is collected in the picture language of the CAG.

3.2 Non-isometric Array Grammars

We review here a basic non-isometric array grammar model introduced in [37] and extensively studied by different researchers and another comparatively recent model, which has been introduced in [41] and investigated in [1]. In these models, the picture arrays considered are rectangular arrays with the symbols defining a picture array being arranged in rows and columns. We first recall the two-dimensional grammar model (originally named as 2D matrix grammar [37]), which we call here as two-dimensional context-free grammar (2CFG), consistent with the terminology used in [17].

Definition 3 [17] A two-dimensional context-free grammar (2CFG) is $G = (V_h, V_v, V_i, T, S, R_h, R_v)$, where V_h, V_v, V_i are finite sets of symbols, respectively called as horizontal nonterminals, vertical nonterminals and intermediate symbols; $V_i \subset V_v$; T is a finite set of terminals; $S \in V_h$ is the start symbol; R_h is a finite set of context-free rules of the form $X \rightarrow \alpha, X \in V_h, \alpha \in (V_h \cup V_i)^*$; R_v is a finite set of right-linear rules of the form $X \rightarrow aY$ or $X \rightarrow a, X, Y \in V_i, a \in T \cup \{\lambda\}$.

When the rules in R_h are regular grammar rules, then the 2CFG is called a two-dimensional right-linear grammar (2RLG).

There are two phases of derivation in a 2CFG. Starting with S, the context-free rules are applied in the first horizontal phase, generating strings over intermediates, which act as terminals of this first phase. The intermediate symbols in a string generated in the first phase, serve as the start symbols for the second vertical phase. The right-linear rules are applied in parallel to this string of intermediates in the second phase for generating the columns of the rectangular arrays over terminals. In this phase at a derivation step, either all the rules applied are of the form $X \rightarrow aY$, $a \in T$ or all the rules are of the form $X \rightarrow Y$ or all the rules are the terminating vertical rules of the form $X \rightarrow a$, $a \in T$ or all are of the form $X \rightarrow \lambda$. When the vertical generation halts, the picture array obtained over T, is collected in the picture language generated by the 2CFG. Note that the picture language generated by a 2CFG consists of rectangular arrays of symbols. We denote by $L(2CFG)$, the family of picture array languages generated by two-dimensional context-free grammars and by $L(2RLG)$, the family of picture array languages generated by two-dimensional right-linear grammars.

We illustrate the working of 2RLG with an example.

Example 3 We consider the 2RLG G_1 having in the first phase, the horizontal regular rules $S \rightarrow AX, X \rightarrow BY, Y \rightarrow AZ, Z \rightarrow B$ where A, B are the intermediate symbols and in the second phase the vertical right-linear rules $1 : A \rightarrow aC, 2 : C \rightarrow bC, 3 : C \rightarrow bD, 4 : D \rightarrow a, 5 : B \rightarrow aB, 6 : B \rightarrow a$ where a, b are the terminal symbols. The derivation in the first phase is as in a string grammar. The horizontal rules generate words of the form ABA^nB, $n \geq 1$. In the second phase every symbol in such a word ABA^nB is rewritten in parallel either by using the right-linear nonterminal rules rules 1 and 5 adding a row of the form a^{n+3} followed by using the rules 2 or 3 and 5 adding a row of the form bab^na. When the terminal rules 4 and 6 are applied in parallel, the derivation terminates adding a row of the form a^{n+3} thus generating a rectangular picture array. Such a picture array for $n = 3$ is shown in Fig. 4. In fact, for this picture array, the derivation in the second phase is as follows:

$$
\begin{array}{llllll}
a\ a\ a\ \ a\ \ a\ a & a\ a\ a\ \ a\ \ a\ a \\
b\ a\ b\ \ b\ \ b\ a & a \qquad\qquad a \\
b\ a\ b\ \ b\ \ b\ a & a \qquad\qquad a \\
b\ a\ b\ \ b\ \ b\ a & a \qquad\qquad a \\
a\ a\ a\ \ a\ \ a\ a & a\ a\ a\ \ a\ \ a\ a \\
\quad (i) & \quad\ (ii)
\end{array}
$$

Fig. 4 **i** A picture array generated in Example 3 **ii** Digitized form of symbol D

$$
A\,B\,A\,A\,A\,B \Rightarrow \begin{array}{c} a\ a\ a\ a\ a\ a \\ C\ B\ C\ C\ C\ B \end{array} \Rightarrow \begin{array}{c} a\ a\ a\ a\ a\ a \\ b\ a\ b\ b\ b\ a \\ C\ B\ C\ C\ C\ B \end{array}
$$

$$
\Rightarrow \begin{array}{c} a\ a\ a\ a\ a\ a \\ b\ a\ b\ b\ b\ a \\ b\ a\ b\ b\ b\ a \\ C\ B\ C\ C\ C\ B \end{array} \Rightarrow \begin{array}{c} a\ a\ a\ a\ a\ a \\ b\ a\ b\ b\ b\ a \\ b\ a\ b\ b\ b\ a \\ b\ a\ b\ b\ b\ a \\ D\ B\ D\ D\ D\ B \end{array} \Rightarrow \begin{array}{c} a\ a\ a\ a\ a\ a \\ b\ a\ b\ b\ b\ a \\ b\ a\ b\ b\ b\ a \\ a\ a\ a\ a\ a\ a \end{array}
$$

The rules used are 1, 5 once, followed by 2, 5 two times, followed by 3, 5 once and finally 4, 6 once. Note that, if we interpret the symbol b as the blank symbol, then the picture array in Fig. 4a corresponds to the digitized form of the letter D over the symbol a Fig. 4b.

An extension of the $2CFG$ (resp., $2RLG$) which we call here as the two-dimensional tabled context-free grammar ($2TCFG$) (resp., tabled right-linear grammar ($2TRLG$)) was introduced in [39] (originally called as 2D tabled matrix grammar in [39]) to generate picture languages that cannot be generated by any $2CFG$ (resp., $2RLG$). This kind of two-dimensional grammar is defined similar to $2CFG$ (resp., $2RLG$) except that in the second phase, the right-linear rules are grouped into different sets, called tables such that (*i*) all the rules in a table are nonterminal rules of the form either $A \rightarrow aB$ only or $A \rightarrow B$ only or (*ii*) all are terminal rules of the form either $A \rightarrow a$ only or $A \rightarrow \lambda$ only and in the parallel derivation in the vertical direction, at a time rules in a single table are used. We denote by $L(2TCFG)$ (resp., $L(2TRLG)$) the family of picture array languages generated by two-dimensional tabled context-free (resp., right-linear) grammars.

Another non-isometric array grammar model which is simple to handle yet expressive, is pure 2D context-free grammar introduced in [41] and subsequently investigated in [1]. We recall this 2D model now.

Definition 4 [41] A pure 2D context-free grammar ($P2DCFG$) is a 4-tuple

$$
G = (\Sigma, P_1, P_2, \mathcal{M}_0)
$$

where

(i) Σ is a finite set of symbols;

(ii) $P_1 = \{c_i \mid 1 \le i \le p\}$, where c_i, called a column rule table, is a finite set of context-free rules of the form $a \to \alpha, a \in \Sigma, \alpha \in \Sigma^*$ satisfying the condition that for any two rules $a \to \alpha, b \to \beta$ in c_i, α and β have equal length;

(iii) $P_2 = \{r_j \mid 1 \le j \le q\}$, where r_j, called a row rule table, is a finite set of rules of the form $c \to {}^t\gamma, c \in \Sigma, \gamma \in \Sigma^*$ such that for any two rules $c \to {}^t\gamma, d \to {}^t\delta$ in r_j, γ and δ have equal length;

(iv) $\mathcal{M}_0 \subseteq \Sigma^{**} - \{\lambda\}$ is a finite set of axiom arrays.

A derivation in a $P2DCFG$ G is defined as follows: Let $A, B \in \Sigma^{**}$. The picture array B is derived in G from the picture array A, denoted by $A \Rightarrow B$, if B is obtained from A either (*i*) by rewriting in parallel all the symbols in a column of A, by rules in some column rule table or (*ii*) rewriting in parallel all the symbols in a row of A, by rules in some row rule table. All the rules used to rewrite a column (or a row) at a time should belong to the same table.

The picture language generated by G is the set of picture arrays $L(G) = \{M \in \Sigma^{**} \mid M_0 \Rightarrow^* M \text{ for some } M_0 \in \mathcal{M}_0\}$. The family of picture languages generated by $P2DCFGs$ is denoted by $L(P2DCFG)$.

Example 4 Consider the $P2DCFG$ $G_2 = (\Sigma, P_1, P_2, \{M_0\})$ where $\Sigma = \{a, b, c, d\}$,

$$P_1 = \{c_t\}, P_2 = \{r_t\}, \text{ where } c_t = \{a \to bab, c \to ada\}, r_t = \left\{ \begin{matrix} & b & & a \\ a \to & a, d \to & c \\ & b & & a \end{matrix} \right\},$$

$$\text{and } M_0 = \begin{matrix} b\ a\ b \\ a\ c\ a \\ b\ a\ b \end{matrix}.$$ It can be seen that starting with the axiom array, the column table c_t alone is applicable,which can be followed by an application of the row table r_t only, giving a derivation as shown below:

$$\begin{matrix} & & & & b\ b\ a\ b\ b \\ b\ a\ b & & b\ b\ a\ b\ b & & b\ b\ a\ b\ b \\ a\ c\ a & \Rightarrow & a\ a\ d\ a\ a & \Rightarrow & a\ a\ c\ a\ a \\ b\ a\ b & & b\ b\ a\ b\ b & & b\ b\ a\ b\ b \\ & & & & b\ b\ a\ b\ b \end{matrix}$$

Remark 2 It has been shown in [41] that the family $L(P2DCFG)$ is incomparable with the family $L(2CFG)$ (Definition 3) but is not disjoint with it.

4 P Systems for Picture Arrays

We now review the array P system models that have been proposed for handling picture array languages. In recent years, construction of rewriting P systems for the generation of two-dimensional picture languages has received more attention (see, for example, [40]). We consider here mainly the array-rewriting P systems that are based on the array grammar models reviewed in Sect. 3.

4.1 Array-Rewriting P System

The array P system introduced in [5] linked the two areas of membrane computing and picture grammars and is one of the earliest models in this direction, which has stimulated further research in the area of array P systems. We now review this system.

Definition 5 [5] An array-rewriting P system of degree $m \geq 1$ is a construct

$$\Pi = (V, T, \#, \mu, F_1, \ldots, F_m, R_1, \ldots, R_m, i_o),$$

where

(i) V is an alphabet;
(ii) $T \subseteq V$ is the terminal alphabet;
(iii) $\# \notin V$ is the blank symbol;
(iv) μ is a membrane structure with m membranes labelled in a one-to-one correspondence with $1, 2, \ldots, m$;
(v) F_i, $1 \leq i \leq m$ is a finite set of axiom arrays over V in the region i, $1 \leq i \leq m$;
(vi) R_i, $1 \leq i \leq m$ is a finite set of context-free array rewriting rules over V in the region i, $1 \leq i \leq m$ with the rules having attached targets *here, out, in*; and
(vii) i_o is the label of an elementary membrane of μ, called the output membrane.

Here we consider array-rewriting P systems with context-free (CF) or regular (REG) array-rewriting rules.

A computation in an array-rewriting P system is defined analogous to the computation in a string rewriting P system [27]. But halting computations are considered as successful computations. Each array in every region of the system, that can be rewritten by a rule in that region, should be rewritten and the rewriting is sequential at the level of arrays in the sense that only one rule is applied to an array. The target associated with the rule used in rewriting an array decides to which region (immediately inner or immediately outer or the same region) the generated array is sent and these are indicated by the targets *in, out, here*. In fact *here* means that the array remains in the same region, *out* means that the array exits the current membrane (but if the rewriting was done in the skin membrane, then it exits the system and is "lost" in the environment) and *in* means that the array is immediately sent to one of the directly lower membranes, nondeterministically chosen if several there exist and if no internal membrane exists, then a rule with the target indication *in* cannot be used. A computation is successful only if it stops in the sense that no rule can be applied to the existing arrays in any of the regions. The result of an halting computation consists of the picture arrays composed only of symbols from T collected in the membrane with label i_o in the halting configuration. The set of all such picture arrays generated or computed by an array-rewriting P system Π is denoted by $AL(\Pi)$.

The family of all picture array languages $AL(\Pi)$ generated by such systems Π, with at most m membranes and with rules of type $\alpha \in \{REG, CF\}$ is denoted by $EAP_m(\alpha)$; if we have $V = T$ (referred to as non-extended systems), we ignore the condition to have at least one nonterminal unit square in the left hand side of rules and the family of picture languages in this case is written as $AP_m(\alpha)$.

We give an example of an array-rewriting P system with regular array grammar rules in the regions, generating a picture array language L_p consisting of picture arrays representing the Math symbol of perpendicularity as in Fig. 2 but with all three "arms" of equal length, i.e., the number of symbols a in the middle vertical line equals the number of symbols in the horizontal bottom line in the portion to the right or to the left counting from the symbol in the "junction".

Example 5 Consider the array-rewriting P system

$$\Pi_p = (\{A, B, B', C, D, a\}, \{a\}, \#, [_1[_2[_3 \ ]_3]_2]_1, F_1, \emptyset, \emptyset, R_1, R_2, R_3, 3),$$

where $F_1 = \left\{ \begin{array}{c} A \\ C \ a \ B \end{array} \right\}$,

$$R_1 = \{ \begin{array}{c} \# \\ A \end{array} \to \begin{array}{c} A \\ a \end{array} (in)\}, R_2 = \{B \ \# \to a \ B'(in), \ B' \ \to B(out)\},$$

$$R_3 = \{\# \ C \to C \ a(out), \# \ C \to D \ a, A \to a, B' \to a, D \to a\}.$$

In a computation in Π_p, only region 1 has an axiom array $\left\{ \begin{array}{c} A \\ C \ a \ B \end{array} \right\}$ and so an application of the rewriting rule in region 1, to this array grows the middle column vertically up by one symbol a and the generated array is sent to region 2 due to the target indication *in*. In region 2, an application of the rule for B in this region, grows the bottommost horizontal arm to the right by one symbol and the array is sent to region 3. An application of the rule $\# \ C \to C \ a(out)$, grows the bottommost horizontal arm by one symbol to the left and the array is sent back to region 2 where the primed symbol B' is changed into B after which the array is sent back to region 1. The process can be repeated. If in region 3, the rule applied is $C \to D \ a$, then the array remains in the same region so that the application of the terminal rules $A \to a, B' \to a, D \to a$ generates a picture array over a in the output region 3. The picture array is then collected in the language of the system. Note that in region 3, if A, B' are changed into a prior to an application of a rule for C and the nonterminal rule $\# \ C \to C \ a(out)$, is applied, then the array is sent to region 2 and remains stuck there. Note also that the lengths of the vertical arm, the left horizontal arm and the right horizontal arm of the picture array generated are of equal length.

In [5], the generative power of the array-rewriting P system is investigated in a very general sense. Since we are concerned here with array-rewriting P system mainly with CF or regular array grammar rules, we state a result on the generative power of the array-rewriting P system with only regular array grammar (Definition 1) rules in order to show the power of such array-rewriting P systems.

Theorem 2 [5] $EAP_3(REG) - ACF \neq \emptyset$.

The picture array language in Example 5 is in $EAP_3(REG)$ but it can be seen that this language L_p cannot be generated by any context-free array grammar as in Definition 1 in view of the fact that the growth in the arms of the picture arrays of the language L_p can take place independently on using context-free array grammar rules and cannot be controlled.

Remark 3 Recently, in [29], rewriting in parallel mode was employed on the arrays in the regions of the array-rewriting P system (Definition 5) and the constructions in [42] were improved in [29] resulting in a reduction in the number of membranes used.

4.2 Contextual Array P System

We now review an array P system which is based on contextual array grammar kind of rules and the method of application of the rules as described in Definition 2.

Definition 6 [13] A contextual array P system with $m \geq 1$ membranes is a construct

$$\Pi = (V, \#, \mu, A_1, \ldots, A_m, P_1, \ldots, P_m, i_o),$$

where

  (i) V is an alphabet;
 (ii) $\# \notin V$ is the blank symbol;
(iii) μ is a membrane structure with m membranes labelled in a one-to-one correspondence with $1, 2, \ldots, m$;
 (iv) A_i, $1 \leq i \leq m$ is a finite set of axiom arrays over V in the region i, $1 \leq i \leq m$;
  (v) P_i, $1 \leq i \leq m$ is a finite set of array contextual rules over V as in Definition 2; the rules have attached targets *here, out, in_j*, $1 \leq j \leq m$ or *in* and
 (vi) i_o is the label of an elementary membrane of μ, called the output membrane.

A computation in a contextual array P system is done analogous to the array-rewriting P system (Definition 5) again with the halting computations defined as successful computations. For each array $\mathcal{A}$ in each region of the system, if an array contextual rule p in the region, nondeterministically chosen can be applied to $\mathcal{A}$, then it should be applied and the application of a rule is sequential at the level of arrays. The resulting array, if any, is sent to the region indicated by the target associated with the rule used interpreting the attached target *here, in, out* in the usual manner as described in Definition 5. Also, in_j is a target indication which means that the array is immediately sent to the directly lower membrane with label j. A computation is successful only if it stops such that a configuration is reached where no rule can be applied to the existing arrays. The result of a halting computation consists of the arrays collected in the membrane with label i_o in the halting configuration. The set of all such arrays computed or *generated* by a system Π is denoted by $CAL(\Pi)$.

$$b$$
$$a$$
$$a$$
$$a$$
$$a\ a\ a\ a\ a$$

Fig. 5 A picture array describing the digitized shape L

The family of all picture array languages $CAL(\Pi)$ generated by systems Π as mentioned above, with at most m membranes, is denoted by $AP_m(cont)$.

Example 6 [13] Let L_l be a picture language consisting of arrays describing the digitized shape of the letter L with each square labelled by a except for the one in the uppermost position of the vertical arm which is labelled by b and both the arms having equal length *i.e.* equal number (at least three) of unit squares. A member of L_l is shown in Fig. 5. A contextual array P system Π_l with 2 membranes can generate L_l and is given by $\Pi_l = (\{a, b\}, \#, \mu, A_1, A_2, P_1, P_2, 2)$ where $\mu = [_1 \, [_2 \,]_2 \,]_1$ and $A_1 := \{^a_{a\,a}\}, A_2 := \emptyset$. The rules are as follows:

$$P_1 := \{p_1\} := \left\{ \left(\boxed{a}\,\boxed{a}\, a, in \right) \right\},$$

$$P_2 := \{p_{2,1}, p_{2,2}\} := \left\{ \left(\begin{matrix} a \\ \boxed{a} \\ \boxed{a} \end{matrix}, out \right), \left(\begin{matrix} b \\ \boxed{a} \\ \boxed{a} \end{matrix}, here \right) \right\}.$$

Starting from the axiom array $^a_{a\,a}$ in membrane 1, the rule p_1 is applied adjoining the context a to the selector a a and then the resulting array $^{a}_{a\,a\,a}$ is sent to membrane 2 due to the target indication *in* attached to p_1. Note that there is no initial array in membrane 2. If the rule $p_{2,1}$ is applied in membrane 2, then the context a will be adjoined resulting in the array $^{a}_{a}_{a\,a\,a}$ which is sent back to membrane 1, due to the target indication *out* in rule $p_{2,1}$. The rule p_1 in membrane 1, can now be applied and the resulting array is sent again to membrane 2. The process can be repeated. If rule $p_{2,2}$ is applied in membrane 2, then the uppermost square is "filled" with b and the picture remains in membrane 2 due to the target indication *here*. Note that the number of symbols in both the horizontal and vertical arms of the resulting L shaped picture will be the same so that the arms are of equal length (of at least three). Hence, only arrays from L_l can be produced and all arrays of L_l can be produced in this way.

We now state comparison results established in [13] for small number of membranes in contextual array P systems.

Theorem 3 $\mathcal{L}\,(cont, t) \subset AP_2(cont)$ *where* $\mathcal{L}\,(cont, t)$ *is defined in Definition 2.*

Theorem 4 $AP_2(cont) \subset AP_3(cont).$

An interesting result established in [13] is now stated which shows the existence of a proper infinite hierarchy with respect to the classes of languages described by contextual array P systems.

Theorem 5 *For every* $k \geq 1$, $AP_k(cont) \subset AP_{3k}(cont)$.

5 Sequential/Parallel Array P Systems

Here we recall array P systems, called sequential / parallel array P systems [46], having the objects in the membranes as rectangular arrays and the rules as 2CFG or 2RLG type of rules as described in Definition 3.

Definition 7 [46] A sequential / parallel array P system of degree $m \geq 1$ is a construct $\pi = (V_1 \cup V_2, I, T, \mu, F_1, \cdots, F_M, R_1, \cdots, R_M, i_0)$ where

(i) $V = V_1 \cup V_2$ is an alphabet, $V_1 - I$ is the set of horizontal nonterminals, $I \subseteq V_1$ is the set of intermediates, $V_2 - T$ is the set of vertical non-terminals, $T \subseteq V_2$ is the set of terminals, $V_2 - T$ includes the elements of I;

(ii) μ is a membrane structure with m membranes labeled in a one-to-one manner with $1, 2, \cdots, m$;

(iii) $F_1, \cdots, F_m$ are finite sets (can be empty) of axiom strings in the regions of μ where horizontal rules are present; the regions where vertical rules are present, are initially empty;

(iv) $R_1, \cdots, R_m$ are finite sets of rules as in Definition 3 associated with the m regions of μ; the rules can be either context free rules (called horizontal rules) of the form $A \to \alpha$, $A \in V_1 - I$, $\alpha \in V_1^*$ or sets of right-linear nonterminal rules (called vertical nonterminal rules) of the form $X \to aY, X, Y \in V_2 - T$, $a \in T \cup \{\lambda\}$ or sets of right-linear terminal rules (called vertical terminal rules) of the form $X \to a$, $X \in V_2 - T$, $a \in T \cup \{\lambda\}$. The horizontal CF rules can be, in particular, regular rules of the form $A \to aB$, $A \to a$, $A \in V_1 - I$, $w \in I^*$. Horizontal rules and sets of vertical rules have attached targets, *here, out, in* (in general, *here* is omitted). A membrane has either horizontal rules or sets of vertical rules; horizontal rules are applied in a sequential manner; the vertical rules in a parallel manner in the vertical direction as in a $2CFG$ grammar;

(v) Finally, i_o is the label of an elementary membrane of μ which is the output membrane.

A computation in a sequential / parallel array P system is similar to a string rewriting P system; the successful computations are the halting ones; each object (either a string or a rectangular array), from each region of the system, which can be rewritten by suitable rules (a horizontal rule or a set of vertical rules) associated with that region, should be rewritten; The target (*here, in, out*) associated with the horizontal rule or the set of vertical rules, have the usual meaning as in other rewriting P systems (described earlier). A computation is successful only if it stops by reaching a configuration where no rule can be applied to the existing arrays. The result of a halting computation consists of the rectangular arrays composed only of symbols from T placed in the membrane with label i_o in the halting configuration.

The set of all such arrays computed or generated by the system π is denoted by $RAL(\pi)$. The family of all array languages $RAL(\pi)$ generated by systems π as above, with at most m membranes, with horizontal rules of type $\alpha \in \{REG, CF\}$ is denoted by $S/PAP_m(\alpha)$ where REG, CF stand respectively for regular or context-free type of rules.

Example 7 Consider the sequential / parallel array P system in the class S/PAP_4 (*REG*)

$$\Pi = (V_1 \cup V_2, I, T, [_4[_3[_2[_1 \ ]_1]_2]_3]_4, S_1BS_2, \emptyset, \emptyset, \emptyset, R_1, R_2, R_3, R_4, 4).$$

$V_1 = \{S_1, S_2, A, B\}, V_2 = \{A, B, C, a, b\}, I = \{A, B\}, T = \{a, b\}$

$R_1 = \{S_1 \to AS_1(in), S_1 \to A(in)\}$ $\qquad$ $R_2 = \{S_2 \to AS_2(out), S_2 \to A(in)\}$,

$R_3 = \left\{ \left\{ A \to \dfrac{a}{C}, B \to \dfrac{a}{B}, C \to \dfrac{b}{C} \right\}, \{C \to b, B \to a\} \right\}, R_4 = \emptyset.$

Region 1 has the initial axiom word S_1BS_2. There are no objects initially in the other regions. Starting with S_1BS_2 an application of the rule $S_1 \to AS_1$ generates the word AS_1BS_2 which is sent to region 2 where an application of the rule $S_2 \to AS_2$ sends the string AS_1BAS_2 generated back to region 1. When the terminating rule $S_1 \to A$ is applied in region 1, the string $AABAS_2$ generated is sent to region 2 where an application of the rule $S_2 \to A$ generates $AABAA$ and sends it to region 3. An improper application of the rules in regions 1 and 2 will make the string get stuck in one of the regions. In general, the string sent to region 3 is of the form A^nBA^n. In region 3, application in parallel of the vertical rules in the table $\left\{ A \to \dfrac{a}{C}, B \to \dfrac{a}{B}, C \to \dfrac{b}{C} \right\}$ allows growth in the vertical direction with an array generated being of the form

$$
\begin{array}{ccc}
a^n & a & a^n \\
b^n & a & b^n \\
& \cdots & \\
& \cdots & \\
b^n & a & b^n \\
C^n & B & C^n
\end{array}
$$

Application of the rules of the table $\{C \to b, B \to a\}$ in parallel terminates the derivation, yielding arrays, one member of which is shown in Fig. 6. The picture language generated by Π consists of rectangular arrays describing the letter T with equal "horizontal arms" over a (Fig. 6) to the left and right of the middle vertical arm, interpreting b as blank.

Now the generative power of the sequential/parallel array P systems is given in the following Theorem 6.

	(a)						(b)				
a	a a	a	a a a	a a a	a	a a a
b b b	a	b b b					a
b b b	a	b b b					a
b b b	a	b b b					a
b b b	a	b b b					a

Fig. 6 **a** A 5×7 array representing Letter T over a with equal "horizontal arms" **b** Digitized form of Letter T

Theorem 6 *(i)* $S/PRAP_3(REG) \supset L(2RLG)$;

(ii) $S/PRAP_3(CF) \supset L(2CFG)$;

(iii) $S/PRAP_4(REG) \subset S/PRAP_4(CF)$.

5.1 Pure 2D Context-Free Grammar Based P System

We now recall an array P system that involves rules of pure 2D context-free grammars as in Definition 3.

Definition 8 [44] An array-rewriting P system (of degree $m \geq 1$) with pure 2D context-free rules is a construct

$$\Pi = (V, \mu, F_1, \ldots, F_m, R_1, \ldots, R_m, i_o),$$

where V is an alphabet, μ is a membrane structure with m membranes labelled in a one-to-one way with $1, 2, \ldots, m$; F_i is a finite set of rectangular arrays over V associated with the region i, $1 \leq i \leq m$, of μ; R_i is a finite set of column tables or row tables of context-free rules over V (as in a $P2DCFG$) associated with region i, $1 \leq i \leq m$, of μ; the tables have attached targets *here,out, in* (in general, *here* is omitted), finally, i_o is the label of an elementary membrane of μ, which is the output membrane.

A computation in Π is defined in the same way as in an array-rewriting P system with the successful computations being the halting ones: each rectangular array in each region of the system, which can be rewritten by a column/row table of rules (with the rewriting as in a $P2DCFG$) associated with that region, should be rewritten; this means that one table of rules is applied; the array obtained by rewriting is placed in the region indicated by the target associated with the table used (*here* means that the array remains in the same region, *out* means that the array exits the current membrane; and *in* means that the array is immediately sent to one of the directly lower membranes, nondeterministically chosen if several exist there; if no internal membrane exists, then a table with the target indication *in* cannot be used). A computation is successful only if it stops, a configuration is reached where no table of rules can be applied to the existing arrays. The result of a halting computation consists of rectangular arrays over V placed in the membrane with label i_o in the halting configuration.

The set of all such arrays computed or *generated* by a system Π is denoted by $AL(\Pi)$. The family of all array languages $AL(\Pi)$ generated by systems Π as above, with at most m membranes, is denoted by $AP_m(P2DCFG)$.

We illustrate with an example.

Example 8 [44] Let

$$\Pi_1 = (V, \mu, F_1, F_2, F_3, F_4, R_1, R_2, R_3, R_4, i_o),$$

where $V = \{u, u_t, u_b, u_l, u_l', v, v_t, v_b, v_r, v_r', x, y, z, w, s, s_1, s_2\}, \mu = [_1[_2[_3[_4]_4]_3]_2]_1$, indicating that the system has four regions, one within another, i.e. region 1 is the 'skin' membrane which contains region 2, which in turn contains region 3, and which in turn contains region 4. $i_0 = 4$ indicates that the region 4 is the output region.

$$R_1 = \{t_{c1}\}, \quad R_2 = \{t_{c2}, t_{r3}\}, \quad R_3 = \{t_{r1}, t_{r2}\}, \quad R_4 = \emptyset.$$

The tables of rules are given by

$$t_{c1} = \{u_t \to zu_tx, u \to yux, u_b \to wu_bx, s_1 \to ss_1x, s \to ssx\}(in),$$

$$t_{c2} = \{v_t \to v_tz, v \to vy, v_b \to v_bw, s_2 \to s_2s, s \to ss\}(in),$$

$$t_{r1} = \left\{ \begin{matrix} u_l \\ u_l \to y \\ u \end{matrix} , \begin{matrix} s \\ s \to y \\ s \end{matrix} , \begin{matrix} s_1 \\ s_1 \to u \\ s \end{matrix} \right\} \bigcup \left\{ \begin{matrix} x \\ x \to x \\ x \end{matrix} , \begin{matrix} s_2 \\ s_2 \to v \\ s \end{matrix} , \begin{matrix} v_r \\ v_r \to y \\ v \end{matrix} \right\} (in),$$

$$t_{r2} = \{u_l \to u_l', s \to s, s_1 \to s_1, x \to x, s_2 \to s_2, v_r \to v_r'\}(out),$$

$$t_{r3} = \{u_l' \to u_l, s \to s, s_1 \to s_1, x \to x, s_2 \to s_2, v_r' \to v_r\}(out),$$

$$F_1 = \{M_0\}, F_2 = F_3 = F_4 = \emptyset, M_0 = \begin{matrix} z & z & u_t & v_t & z & z \\ u_l & s & s_1 & s_2 & s & v_r \\ w & w & u_b & v_b & w & w. \end{matrix}$$

Starting with M_0 in the region 1, the rules of the column table t_{c1} are applied and the array is sent to region 2 wherein the rules of the column table t_{c2} are applied and the array is again sent to region 3. If in region 3, the rules of the row table t_{r1} are applied, then the array is sent to region 4 wherein it remains and gets collected in the language generated. On the other hand, if in region 3, the rules of the row table t_{r2} are applied changing u_l into u_l' and v_r into v_r', then the array is sent back to region 2 wherein the symbols u_l' and v_r' are respectively changed back into v_l and v_r and the array is sent to region 1. The process then repeats. An array M generated by Π_1 is shown in Fig. 7.

$$
M = \begin{array}{ccccccccccccc}
z & z & z & z & z & u_t & x & x & x & v_t & z & z & z & z & z \\
u_l & s & s & s & s & s_1 & x & x & x & s_2 & s & s & s & s & v_r \\
y & y & y & y & y & u & x & x & x & v & y & y & y & y & y \\
u & s & s & s & s & s & x & x & x & s & s & s & s & s & v \\
y & y & y & y & y & u & x & x & x & v & y & y & y & y & y \\
u & s & s & s & s & s & x & x & x & s & s & s & s & s & v \\
w & w & w & w & w & u_b & x & x & x & v_b & w & w & w & w & w
\end{array}
$$

Fig. 7 An array M generated by Π_1

Results on comparison of generative power of array-rewriting P systems based on Pure 2D context-free grammars have been established in [44]. We state some of these here (Theorem 7).

Theorem 7 [44] (*i*) $L(P2DCFG) = AP_1(P2DCFG)$;
(*ii*) $L(P2DCFG) \subset AP_2(P2DCFG)$;
(*iii*) $L(2RLG) \subset L(T2RLG) \subseteq AP_2(P2DCFG)$;
(*iv*) $L(2CFG) \subset L(T2CFG) \subseteq AP_2(P2DCFG)$;
(*v*) $AP_2(P2DCFG) \setminus L(T2CFG) \neq \emptyset$.

6 Application to Picture Pattern Generation

Interesting classes of picture patterns can be generated using the array P systems. We illustrate by considering one of the systems discussed in the earlier sections, namely the array-rewriting P system based on *P2DCFG* type of rules. In the arrays generated by the array-rewriting P system with *P2DCFG* type of rules in the Example 8, we use a well-known technique of replacing the letter symbols in the generated picture arrays by 'primitive patterns' (Fig. 8). Each symbol of a rectangular array M over an alphabet Σ is considered to occupy a unit square in the rectangular grid so that a row of symbols or a column of symbols in the array respectively occupies a horizontal or a vertical sequence of adjacent unit squares. A mapping i, called an interpretation, from the alphabet $\Sigma = \{a_1, a_2, \ldots, a_n\}$ to a set of primitive picture patterns $\{p_1, p_2, \ldots p_m\}$ is defined such that for $1 \leq i \leq n, i(a_i) = p_j$, for some $1 \leq j \leq m$. Then $i(M)$ is obtained by replacing every symbol $a \in M$ by the corresponding picture pattern $i(a)$. In the Example 8, the interpretation i is given by $i(u_l) = i(u_t) = i(u_b) = i(u) = u$, $i(z) = z$, $i(x) = x$, $i(s_1) = i(s_2) = i(s) = s$, $i(v_r) = i(v_t) = i(v_b) = i(v) = v$, $i(w) = w$, $i(y) = y$. The interpretation i applied to the array M gives a pattern commonly called a "kolam" pattern [38] (or a floor design) as in Fig. 9. The primitive kolam patterns are shown in Fig. 8.

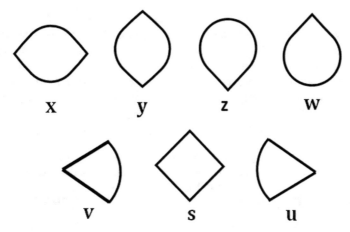

Fig. 8 Primitive patterns of a "floor design": x, y are called pupil, u, v fan, s diamond and z, w drop

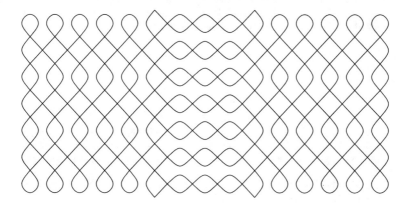

Fig. 9 A "floor design" based on a "kolam" pattern

7 Conclusion and Discussion

We have reviewed here certain main types of array-rewriting P systems, with array objects and with two of these systems based on isometric type of array grammar rules and another two based on non-isometric type of array generating grammar rules. We have indicated an application of one of these systems in the description of floor-designs, commonly called "kolam" patterns [38].

Based on and following the types of array generating P systems reviewed here, several other kinds of array P systems have been proposed in the literature (see, for example, [6, 8, 18, 19, 43, 47]). The interesting application problem of generation of patterns of "floor-designs" has been considered in some of these array P systems (see, for example [19]). On the other hand string language generating models based on spiking neural P systems [7, 20, 53, 54, 56] have been proposed. It will be interesting to examine the possibility of picture array generation based on spiking neural P systems, which is an intensively investigated research area in recent times.

P systems with controlled computations have been introduced and investigated in [23]. In [32], study of the family of control languages of spiking neural P systems in comparison with other families of string languages, is undertaken. Recently, in [48, 49], besides considering 8-directional array P systems, control language study is also done. It will be interesting to investigate the effect of such a control language on the array P systems reviewed here, like the contextual array P system.

Acknowledgements The authors are grateful to Prof. Andrew Adamatzky for extending the authors a valuable opportunity to have this paper in a volume celebrating the life-time achievements of Prof. Kenichi Morita, a wonderful researcher with great contributions in many areas including two-dimensional languages. The authors thank the referees for their very useful comments which enabled them to provide a better presentation. The work was supported by National Natural Science Foundation of China (61320106005, 61602192, and 61772214), China Postdoctoral Science Foundation (2016M600592 and 2017T100554). The first and second authors, K.G. Subramanian and Sastha Sriram, gratefully acknowledge the support extended to them by Prof. Linqiang Pan for an academic visit to Huazhong University of Science and Technology, Wuhan, China from May 03, 2017 to June 02, 2017, where part of this research was done.

References

1. Bersani, M.M., Frigeri, A., Cherubini, A.: Expressiveness and complexity of regular pure two-dimensional context-free languages. Int. J. Comput. Math. **90**(8), 1708–1733 (2013)
2. Besozzi, D., Ferretti, C., Mauri, G., Zandron, C.: Parallel rewriting P systems with deadlock. DNA Computing. Lecture Notes Computer Science, vol. 2568, pp. 302–314. Springer, Berlin (2003)
3. Besozzi, D., Mauri, G., Zandron, C.: Hierarchies of parallel rewriting P systems - a survey. New Gener. Comput. **22**(4), 331–347 (2004)
4. Bottoni, P., Labella, A., Martin-Vide, C., Păun, Gh: Rewriting P systems with conditional communication. Formal and Natural Computing. Lecture Notes in Computer Science, pp. 325–353. Springer, Berlin (2002)
5. Ceterchi, R., Mutyam, M., Păun, Gh, Subramanian, K.G.: Array-rewriting P systems. Nat. Comput. **2**, 229–249 (2003)
6. Ceterchi, R., Subramanian, K.G., Venkat, I.: P systems with parallel rewriting for chain code picture languages. In: Proceeding 11th Conference on Computability in Europe (CiE), pp. 45–155 (2015)
7. Chen, H., Freund, R., Ionescu, M., Păun, G., Pérez-Jiménez, M.J.: On string languages generated by spiking neural P systems. Fundam. Inf. **75**(1–4), 141–162 (2007)
8. Dersanambika, K.S., Krithivasan, K.: Contextual array P systems. Int. J. Comput. Math. **81**(8), 955–969 (2004)
9. Ferretti, C., Mauri, G., Păun, Gh, Zandron, C.: On three variants of rewriting P systems. Theor. Comput. Sci. **301**, 201–215 (2003)
10. Fernau, H., Freund, R ., Holzer, M.: The generative power of d-dimensional #-context-free array grammars. In: Proceedings Intenational Colloquium Universal Machines and Computations, MCU'98, Vol. II, pp. 43–56 (1998)
11. Fernau, H., Freund, R., Holzer, M.: Regulated array grammars of finite index. Part I: theoretical investigations. Grammatical Models of Multi-Agent Systems, pp. 157–181. Gordon and Breach, Reading (1999)

12. Fernau, H., Freund, R., Holzer, M.: Regulated array grammars of finite index. Part II: syntactic pattern recognition. Grammatical Models of Multi-Agent Systems, pp. 284–296. Gordon and Breach, Reading (1999)

13. Fernau, H., Freund, R., Schmid, M.L., Subramanian, K.G., Wiederhold, P.: Contextual array grammars and array P systems. Ann. Math. Artif. Intell. **75**(1–2), 5–26 (2015)

14. Freund, R.: Control mechanisms on #-context-free array grammars. In: Păun, Gh (ed.) Mathematical Aspects of Natural and Formal Languages, pp. 97–137. World Scientific Publishing, Singapore (1994)

15. Freund, R.: Array grammars. Technical Rep. 15/00, Research Group on Mathematical Linguistics, Rovira i Virgili University, Tarragona, 164 pages (2000)

16. Freund, R., Păun, Gh, Rozenberg, G.: Contextual array grammars. In: Subramanian, K.G., et al. (eds.) Formal Models, Languages and Applications. Series in Machine Perception and Artificial Intelligence, vol. 66, pp. 112–136. World Scientific, Signapore (2007)

17. Giammarresi, D., Restivo, A.: Two-dimensional languages. In: Rozenberg, G., Salomaa, G. (eds.) Handbook of Formal Languages, pp. 215–267. Springer, Berlin (1997)

18. Isawasan, P., Muniyandi, R.C., Venkat, I., Subramanian, K.G.: Array-rewriting P systems with basic puzzle grammar rules and permitting features. International Conference on Membrane Computing. Lecture Notes in Computer Science, vol. 10105, pp. 272–285. Springer, Berlin (2017)

19. Isawasan, P., Venkat, I., Muniyandi, R.C., Subramanian, K.G.: A membrane computing model for generation of picture arrays. Advances in Visual Informatics. Lecture Notes in Computer Science, pp. 155–165. Springer, Berlin (2015)

20. Jiang, K., Chen, W., Zhang, Y., Pan, L.: On string languages generated by sequential spiking neural P systems based on the number of spikes. Nat. Comput. **15**(1), 87–96 (2016)

21. Krishna, S.N., Rama, R.: On the power of P systems based on sequential/parallel rewriting. Int. J. Comput. Math. **77**(1–2), 1–14 (2000)

22. Krishna, S.N., Rama, R.: A note on parallel rewriting in P systems. Bull. EATCS **73**, 147–151 (2001)

23. Krithivasan, K., Păun, Gh, Ramanujan, A.: On controlled P systems. Fundam. Inf. **131**(3–4), 451–464 (2014)

24. Marcus, S.: Contextual grammars. Revue Roumaine de Mathématiques Pures et Appliquées. **14**, 1525–1534 (1969)

25. Morita, K.: Two-dimensional languages. In: Martin-Vide, C., Mitrana, V., Păun, Gh (eds.) Formal Languages and Applications. Series in Fuzziness and Soft Computing, vol. 148, pp. 426–437. Springer, Berlin (2004)

26. Păun, Gh: Marcus contextual grammars. Studies in Linguistics and Philosophy, vol. 67. Springer, Dordrecht (1997)

27. Păun, Gh: Computing with membranes. J. Comput. Syst. Sci. **61**, 108–143 (2000)

28. Păun, Gh: Membrane Computing: An Introduction. Springer, Berlin (2000)

29. Pan, L., Păun, Gh: On parallel array P systems. Automata, Universality, Computation. Emergence, Complexity and Computation, vol. 12, pp. 171–181. Springer, Berlin (2015)

30. Păun, Gh, Rozenberg, G., Salomaa, A.: The Oxford Handbook of Membrane Computing. Oxford University Press, New York (2010)

31. Pan, L., Song, B., Subramanian, K.G.: Rewriting P systems with flat-splicing rules. Proceeding International Conference on Membrane Computing, vol. 10105, pp. 340–345. Springer, Berlin (2016)

32. Ramanujan, A., Krithivasan, K.: Control languages associated with spiking neural P systems. Rom. J. Inf. Sci. Technol. **15**(4), 301–318 (2012)

33. Rosenfeld, A.: Picture Languages. Academic Press, Reading (1979)

34. Rosenfeld, A., Siromoney, R.: Picture languages - a survey. Lang. Des. **1**(3), 229–245 (1993)

35. Rozenberg, G., Salomaa, A. (eds.): Handbook of Formal Languages, vol. 3. Springer, Berlin (1997)

36. Salomaa, A.: Formal Languages. Academic Press, Reading (1973)

37. Siromoney, G., Siromoney, R., Krithivasan, K.: Abstract families of matrices and picture languages. Comput. Gr. Image Process. **1**, 284–307 (1972)
38. Siromoney, G., Siromoney, R., Krithivasan, K.: Array grammars and Kolam. Comput. Gr. Image Process. **3**(1), 63–82 (1974)
39. Siromoney, R., Subramanian, K.G., Rangarajan, K.: Parallel/sequential rectangular arrays with tables. Int. J. Comput. Math. **6**(2), 143–158 (1977)
40. Subramanian, K.G.: P Systems and picture languages. Machines, Computations, and Universality. Lecture Notes in Computer Science, pp. 99–109. Springer, Berlin (2007)
41. Subramanian, K.G., Ali, R.M., Geethalakshmi, M., Nagar, A.K.: Pure 2D picture grammars and languages. Discret. Appl. Math. **157**(16), 3401–3411 (2009)
42. Subramanian, K.G., Isawasan, P., Venkat, I., Pan, L.: Parallel array-rewriting P systems. Rom. J. Inf. Sci. Technol. **17**(1), 103–116 (2014)
43. Subramanian, K.G., Isawasan, P., Venkat, I., Pan, L., Nagar, A.K.: Array P systems with permitting features. J. Comput. Sci. **5**(2), 243–250 (2014)
44. Subramanian, K.G., Pan, L., Lee, S.K., Nagar, A.K.: A P system model with pure context-free rules for picture array generation. Math. Comput. Model. **52**, 1901–1909 (2010)
45. Subramanian, K.G., Rangarajan, K., Mukund, M. (eds.): Formal Models, Languages and Applications. Series in Machine Perception and Artificial Intelligence, vol. 66. World Scientific, Singapore (2007)
46. Subramanian, K.G., Saravanan, R., Robinson, T.: P system for array generation and application to kolam patterns. Forma **22**, 47–54 (2007)
47. Subramanian, K.G., Saravanan, R., Geethalakshmi, M., Helen Chandra, P., Margenstern, M.: P systems with array objects and array rewriting rules. Prog. Nat. Sci. **17**(4), 479–485 (2007)
48. Sureshkumar, W., Rama, R.: Chomsky hierarchy control on isotonic array P systems. Int. J. Pattern Recogit. Artif. Intell. **30**(2), 1–20 (2016)
49. Sureshkumar, W., Rama, R., Krishna, S.N.: 8-directional array P systems: power and hierarchy. In: Gheorghe, M., et al. (eds.) Multidisciplinary Creativity, pp. 150–169. Spandugino Publishing House, Bucharest (2015)
50. Yamamoto, Y., Morita, K., Sugata, K.: Context-sensitivity of two-dimensional regular array grammars. Array Grammars, Patterns and Recognizers. WSP Series in Computer Science, vol. 18, pp. 17–41. World Scientific, Singapore (1989)
51. Wang, P.S.P. (ed.): Array Grammars Patterns and Recognizers. World Scientific, Singapore (1989)
52. Wang, P.S.P.: A Formal Parallel Model for Three-Dimensional Object Pattern Representation. In: Chen, C.H., et al. (eds.) Handbook of Pattern Recognition and Computer Vision, pp. 211–231. World Scientific, Singapore (2010)
53. Wu, T., Zhang, Z., Pan, L.: On languages generated by cell-like spiking neural P systems. IEEE Trans. NanoBiosci. **15**(5), 455–467 (2016)
54. Zeng, X., Xu, L., Liu, X., Pan, L.: On languages generated by spiking neural P systems with weights. Inf. Sci. **278**, 423–433 (2014)
55. Zhang, G., Pan, L.: A survey of membrane computing as a new branch of natural computing. Chin. J. Comput. **33**(2), 208–214 (2010)
56. Zhang, X., Zeng, X., Pan, L.: On string languages generated by spiking neural P systems with exhaustive use of rules. Nat. Comput. **7**(4), 535–549 (2008)

Input-Position-Restricted Models of Language Acceptors

Oscar H. Ibarra and Ian McQuillan

Abstract Machines of various types are studied with some restriction on the moves that can be made either on or before the end of the input. For example, for machine models such as deterministic reversal-bounded multicounter machines, one restriction is the class of all machines that do not subtract from any counters before the end the input. Similar restrictions are defined on different combinations of stores with many machine models (nondeterministic and deterministic), and their families studied.

1 Introduction

In this paper, various one-way machine models are studied where there is a restriction on the instructions that are permitted on or before hitting the end of the one-way input (on a right input end-marker $\lhd$). For example, one such model is a pushdown automaton that cannot pop until hitting the end of the input, or a pushdown automaton that cannot pop until hitting the end of the input and also cannot push after hitting the end of the input. A preliminary investigation started regarding such concepts on reversal-bounded multicounter machines (**NCM**) in [13]. This model consists of an **NFA** augmented by some number of reversal-bounded counters (each counter stores a non-negative integer that can be increased or decreased by one, or zero, and tested for being zero or non-zero, and there is a bound on the number of changes between non-decreasing and non-increasing), and **DCM** is the same type of machine that is deterministic. For example, in [13], it was shown that every **NCM** can be converted to another that does not decrease before hitting the end-marker. For the deterministic

O. H. Ibarra (✉)
Department of Computer Science,
University of California, Santa Barbara, CA 93106, USA
e-mail: ibarra@cs.ucsb.edu

I. McQuillan
Department of Computer Science,
University of Saskatchewan, Saskatoon, SK S7N 5A9, Canada
e-mail: mcquillan@cs.usask.ca

© Springer International Publishing AG 2018
A. Adamatzky (ed.), *Reversibility and Universality*, Emergence, Complexity and Computation 30, https://doi.org/10.1007/978-3-319-73216-9_17

case however, this is not true, as there is a **DCM** with only one counter that cannot be accepted by any **DCM** machine that does not decrease before hitting the end of the input. Furthermore, the languages accepted by this same restricted model of **DCM** were shown to coincide with the languages accepted by deterministic Parikh automata [3].

A key tool used in this paper for studying many of these machine models with input-position-based restrictions is the *store language* of a machine. The store language of a machine M is an encoding of the set of all configurations (state plus concatenated store contents) that can appear in an accepting computation. For example, the store language of a pushdown automaton M is the set of all words $q\gamma$ where there is an accepting computation of M that passes through state q with pushdown contents γ. It is known that the store language of every pushdown automaton is in fact a regular language [1], and the store language of the more general stack automata (similar to pushdown automata with the additional ability to read the contents of the pushdown in read-only mode) are all regular languages [2]. The store languages of several other models were also recently studied in [14, 15]. For example, the store languages of reversal-bounded queue automata (this is an **NFA** augmented by a queue data structure with a bound on the number of switches between enqueueing and dequeueing) and the store language of r-flip pushdown automata (pushdown automata with the additional ability to reverse their pushdown contents at most r times) are also all regular [15].

Here, at first nondeterministic machines are investigated. The restriction that a store cannot decrease in size until the end-marker is a major focus. When restricting the pushdown of a nondeterministic pushdown automaton in this fashion, the machines coincide with the regular languages, and similarly for reversal-bounded queue automata and r-flip nondeterministic pushdown automata. When augmenting any of these three models with reversal-bounded counters, and the decreasing property is only applied to the pushdown or queue, then these machines coincide with **NCM**. For any of these models, also enforcing that the reversal-bounded counters cannot decrease until the end-marker does not even reduce the capacity. For deterministic machines, the situation is more complicated. In particular, **DCM** and **DCM** augmented by an unrestricted pushdown (**DPCM**) are studied with the decreasing restriction placed on the counters and pushdown separately. Several witnesses are found to separate deterministic classes. Also, a bridging technique is created to determine that languages are not in **DCM** and **DPCM** from the decreasing-restricted restriction.

2   Preliminaries

We assume an introductory background in the area of formal language and automata theory; see e.g. [11].

Let $\mathbb{N}_0$ be the set of non-negative integers. Given a set X and $k \in \mathbb{N}_0$, let $[X]^k$ be the set of k-tuples over X.

An *alphabet* is a finite set of symbols, with Σ^* being the set of all words over Σ^*. The empty word is denoted by λ. A *language* is any $L \subseteq \Sigma^*$. Given a word w, w^R is the word obtained by reversing the letters of w, which can be extended to languages L^R in the natural way, and also to families of languages. Given $w \in \Sigma^*$, then u is a prefix (resp. suffix) of w if $w = ux, x \in \Sigma^*$ (resp. $w = xu, x \in \Sigma^*$). The length of w is denoted by $|w|$, and the number of a's in w, $a \in \Sigma$, is $|w|_a$.

A one-way nondeterministic k-pushdown automaton is a tuple $M = (Q, \Sigma, \Gamma, \delta, q_0, F)$, where Q is a finite set of states, Σ is the input alphabet, Γ is the pushdown alphabet, containing the bottom-of-stack marker Z_0, $q_0 \in Q$ is the initial state, $F \subseteq Q$ is the set of final states, and δ is a finite relation from $Q \times (\Sigma \cup \{\lambda, \lhd\})^* \times [\Gamma]^k$ to $Q \times [\Gamma^*]^k$, where $\lhd$ is the right input end-marker. A configuration of M is a tuple $(q, w, \gamma_1, \ldots, \gamma_k)$, where $q \in Q$ is the current state, $w \in \Sigma^* \lhd \cup \{\lambda\}$ is the remaining input, and $\gamma_i \in Z_0(\Gamma - \{Z_0\})^*$ is the contents of the i'th pushdown, for $1 \leq i \leq k$. The derivation relation between configurations, $\vdash_M$, is such that $(q, aw, v_1 b_1, \ldots, v_k b_k) \vdash_M (p, w, v_1 u_1, \ldots, v_k u_k)$, where $(p, u_1, \ldots, u_k) \in \delta(q, a, b_1, \ldots, b_k)$. Then $\vdash_M^*$ is the reflexive and transitive closure of $\vdash_M$.

The language accepted by M,

$$L(M) = \{w \mid (q_0, w\lhd, Z_0, \ldots, Z_0) \vdash_M^* (q_f, \lambda, \gamma_1, \ldots, \gamma_k), q_f \in F, \gamma_i \in \Gamma^*, 1 \leq i \leq k\}.$$

Furthermore, the store language of M,

$$S(M) = \{q\gamma_1 \cdots \gamma_k \mid (q_0, w\lhd, Z_0, \ldots, Z_0)$$
$$\vdash_M^* (q, v, \gamma_1, \ldots, \gamma_k) \vdash_M^* (q_f, \lambda, \gamma_1', \ldots, \gamma_k'), q_f \in F\}.$$

Also, M is deterministic if $|\delta(q, a, b_1, \ldots, b_k) \cup \delta(q, \lambda, b_1, \ldots, b_k)| \leq 1$, for all $q \in Q, a \in \Sigma \cup \{\lhd\}, b_i \in \Gamma, 1 \leq i \leq k$.

Instead of a pushdown, other data structures, such as a queue can be attached to such a machine.

Often in the literature, one-way machines are defined without the right input end-marker $\lhd$. For nondeterministic machines, $\lhd$ is not necessary as machines can guess when they have hit the end of the input. And for some types of deterministic machines, such as DPDAs, it is not needed [8], but for others, such as DCM, this is not the case (see [13] for further discussion). For consistency, we will define all machines using $\lhd$. This is also useful as we will study restrictions of machines based on whether $\lhd$ has been scanned yet or not.

A *counter* is a pushdown stack that uses only one stack symbol, in addition to a distinguished bottom-of-stack symbol, Z_0, which is never altered. It is known that deterministic machines with only two pushdowns that are both counters have the same power as Turing machines [11]. When there is only one pushdown, this is the well-known nondeterministic pushdown automaton, denoted by NPDA (DPDA for the deterministic variant). Also, a counter is l-reversal-bounded if the machine makes at most l changes between non-decreasing and non-increasing (and vice versa) on the counter. Nondeterministic (resp. deterministic) finite automata with some number of reversal-bounded counters are denoted by **NCM** (resp. **DCM**). These machines have been extensively investigated in the literature, e.g. [17], and they are closed under intersection and have a decidable emptiness problem (resp. containment problem for deterministic machines). Furthermore, machines with one unrestricted pushdown plus some number of reversal-bounded counters also have a decidable emptiness problem, and are denoted by **NPCM** (**DPCM** for the deterministic variant).

By a slight abuse of notation, we will denote each family of machines synonymously with the family of languages they accept. Therefore, **DCM** will denote both the reversal-bounded counter machines, and also the family of languages they accept.

3 Restrictions On/Before the End of Input

In this section, we will restrict the operation of different classes of machines so that any instruction that reduces the size of the store can only occur when the input tape has read the right input end-marker $\lhd$. Notice that this can be studied separately for each store. For example, **NPCM**s could be studied where the pushdown cannot decrease in size before the end-marker, and separately, they could be studied where the counters cannot be decreased before hitting the end-marker, and lastly, they can be studied where both stores have this restriction. This restriction will be denoted on the family by placing a "bar" on top of the letter denoting the store with this restriction. For example, **DP$\overline{\text{C}}$M** are machines where the pushdown is unrestricted, but the counters cannot decrease before hitting the end-marker, and **D$\overline{\text{PC}}$M** is where the restriction is on both the counters and the pushdown.

In addition, a stronger notion whereby no decreasing of storage before the end-marker, plus no increasing once the end-marker is hit is studied, and for this notion, two bars are placed over the appropriate store letter, such as **DP$\overline{\overline{\text{C}}}$M**.

We first recall a result from [13], where the families **N$\overline{\text{C}}$M** and **D$\overline{\text{C}}$M** were introduced (eNCM and eDCM was the notation used in [13]). For **NCM**, it was found that restricting the counters to not decrease until the end-marker did not change the family accepted. It was also found that **D$\overline{\text{C}}$M** coincides with the family of languages accepted by deterministic Parikh automata [3], and that this family is a strict subset of **DCM**. We will extend this proof by adding in **N$\overline{\text{C}}$M**.

Proposition 1 $D\overline{C}M \subsetneq DCM \subsetneq NCM = N\overline{C}M = N\overline{\overline{C}}M.$

Proof All but the last equality were shown in [13]. But we will briefly describe the construction of **NCM** = **N$\overline{\overline{\text{C}}}$M** here for use later in the paper.

Let M be a k counter **NCM** over Σ, with the counters labelled by $c_1, \ldots, c_k$. It can be assumed without loss of generality that all counters are 1-reversal-bounded [17].

Then, construct a **N$\overline{\overline{\text{C}}}$M** M' with counters labelled by $c_1, d_1, \ldots, c_k, d_k$. On input $w \in \Sigma^*$ followed by the end-marker, M' simulates M exactly using counters $c_1, \ldots, c_k$ so long as they are non-decreasing. If a counter c_i attempts to decrease before hitting the end-marker (nothing is required to be changed on the end-marker), counter d_i is instead increased and thus records the number of decrements of c_i in M. At some nondeterministically guessed point after a counter decrease, M' verifies that the contents of d_i and c_i are equal, and then continues the simulation, simulating transitions on the counter being zero. Then $L(M') = L(M)$, and M' does not decrease any counter before hitting the end-marker.

This proof can be modified slightly to create a machine M'' in $\overline{\overline{\mathsf{NCM}}}$. M' uses a third set of counters $e_1, \ldots, e_k$. Then M'' similarly simulates M exactly using $c_1, \ldots, c_k$ as long as they are non-decreasing and before the end-marker. Then, at some nondeterministically guessed spot before the end-marker, on λ-transitions, for all counters c_i that have not already started decrementing in M, M'' increases e_i and d_i to the same arbitrary number. Then, M'' verifies that the next character is the end-marker. On the end-marker, for every increase of c_i in M, M'' instead decreases e_i, verifying that the simulated increasing ends when counter e_i hits zero, and then when simulating the decreases of c_i in M, it decreases d_i until empty, then decreases c_i until empty. Essentially then, M' guesses right before it reaches $\lhd$, and does all the remaining additions nondeterministically instead. Then, it decreases instead of increases at the end-marker. But it therefore needs two identical copies, d_i and e_i, one to simulate the increases of M and one to simulate decreases of M. □

Next, we study these restrictions on standard NPDAs. Both restrictions induce a large collapse in contrast to NCM.

Proposition 2 $\overline{\mathsf{NPDA}} = \overline{\overline{\mathsf{NPDA}}} = \mathsf{REG}.$

Proof It is enough to show it for $\overline{\mathsf{NPDA}}$.

Let $M = (Q, \Sigma, \Gamma, \delta, q_0, F)$ be an $\overline{\mathsf{NPDA}}$. Assume without loss of generality that M empties the pushdown in every computation on the end-marker. Assume also that Q is partitioned into $Q^{\leftarrow}$, $Q^{\lhd}$, and $Q^{\rightarrow}$, whereby $Q^{\leftarrow}$ are all states that are used before the end-marker, $Q^{\lhd}$ are the states that can be used on the end-marker (which immediately read the end-marker), and then $Q^{\rightarrow}$ are all states that can be used after the end-marker has been read. Lastly, assume without loss of generality that all transitions before the end-marker (on states of $Q^{\leftarrow}$), either replace the top of the pushdown X with XY, where $X, Y \in \Gamma$, or replaces X with X. Indeed, we know that the pushdown does not decrease before the end-marker, it is easy to see that the machine only needs to push one symbol at a time, and also it never needs to replace the top symbol by storing the topmost symbol in the finite control and not pushing it until another symbol is pushed.

From [1], for each state $q \in Q$, the set $co\text{-}Acc(q) = \{\beta \in \Gamma^* \mid (q, v\lhd, \beta) \vdash^*_M (q_f, \lambda, p), q_f \in F\}$ is a regular language. Let $R = \bigcup_{q \in Q^{\lhd}} (co\text{-}Acc(q) \cdot q)$ (over the alphabet of $\Gamma \cup Q^{\lhd}$), which also must be regular. Let $M_R = (Q_R, \Gamma \cup Q^{\lhd}, \delta_R, q_0^R, F_R)$ be an NFA accepting R.

Next, let $M' = (Q', \Sigma, \delta', q_0', F')$ be an NFA with λ transitions with state set $Q' = (Q^{\leftarrow} \cup Q^{\lhd}) \times Q_R$, $q_0' = (q_0, q_0^R)$, $F' = Q^{\lhd} \times F_R$ that operates as follows.

If M has a transition $(p, \gamma) \in \delta(q, a, X), q \in Q^{\leftarrow}, p \in Q^{\leftarrow} \cup Q^{\lhd}, a \in \Sigma \cup \{\lambda\}, X \in \Gamma$, then create:

$$(p, p_R) \in \delta'((q, q_R), a) \text{ if } \gamma = XY, \text{ and } p_R \in \delta_R(q_R, Y),$$
$$(p, q_R) \in \delta'((q, q_R), a) \text{ if } \gamma = X.$$

Also, create $((q, p_R) \in \delta'((q, q_R), \lambda)$ if $q \in Q^{\lhd}$, and $p_R \in \delta_R(q_R, q)$.

Let $w \in L(M)$. Then

$$(q_0, w \lhd, Z_0) \vdash_M^* (q', \lhd, \beta) \vdash_M^* (q_f, \lambda, Z_0),$$

where $q' \in Q^{\lhd}, q_f \in F$. Then, $\beta \in co\text{-}Acc(q')$, and so $\beta q' \in L(M_R)$, and there exists $p \in F_R$ such that $p \in \hat{\delta}_R(q_0^R, \beta q')$. Then, by the construction, $(q', p) \in \hat{\delta}'((q_0, q_0^R), w)$, and $(q', p) \in Q^{\lhd} \times F_R = F'$, and hence $w \in L(M')$.

Let $w \in L(M')$. Then $(q', p) \in \hat{\delta}'((q_0, q_0^R), w)$, where $q' \in Q^{\lhd}, p \in F_R$. By the construction, $(q_0, w \lhd, Z_0) \vdash_M^* (q', \lhd, \beta)$, where $q' \in Q^{\lhd}, \beta q' \in L(M_R)$. Hence, there exists $q_f \in F$ such that $(q', \lhd, \beta) \vdash_M^* (q_f, \lambda, Z_0)$. Hence, $w \in L(M)$.

Then $L(M') = L(M)$. □

Similarly, a reversal-bounded queue automaton, NQA, is an NFA augmented with a queue store, with a bound on the number of switches between enqueuing and dequeueing, and let NQCM be the same system augmented by reversal-bounded counters. Although it is known that queue automata without a reversal-bound on the queue has the same power as a Turing machine, it is known that both NQA and NQCM are more limited, and indeed only accept semilinear languages [10]. It was shown in [15] that the store languages of all NQA are regular languages, and the store languages of all NQCM are all in NCM.

Essentially the same proof of Proposition 2 can be used for NQA as well, ie. when enqueueing before the end-marker, verify in parallel that whatever would be enqueued is in the store language. Therefore,

Corollary 1 $N\overline{Q}A = N\overline{\overline{Q}}A = REG$.

This same proof technique can be used for NPCM and NQCM, showing the resulting languages are all NCM languages. Indeed, take an input NPCM M with k counters. The store language of every NPCM is an NCM language [14]. Let $M_s \in$ NCM with l counters be the store language of M. And in the store language of an NPCM, the word on the pushdown comes first, and then the counters (such as $xc_1^{i_1} \cdots c_k^{i_k}$ where x is the contents of the pushdown, and i_j is the contents of counter j). So, build an NCM M' machine accepting $L(M)$ with $k + l$ counters as follows: M' simulates M with k counters, but if M pushes y onto the pushdown, M' runs it through the store language in parallel using the other l counters. Then, when M' reaches the end of the input, it just needs to verify that the pushdown contents read this far, concatenated with the counter contents are in the store language. So, it subtracts from each counter from c_1 to c_k, while still simulating the machine accepting the store language. Hence,

Proposition 3 $\overline{NP}CM = \overline{\overline{NP}}CM = \overline{NQ}CM = \overline{\overline{NQ}}CM = NCM = \overline{NC}M = \overline{\overline{NC}}M$.

Next, consider r-flip pushdown automata. These machines are similar to pushdown automata with the additional ability to "flip" the pushdown stack at most r times (more precisely, they flip everything above the bottom-of-stack marker Z_0, transforming pushdown contents $Z_0\gamma$, with γ over the pushdown alphabet, to $Z_0\gamma^R$). Let r-NPDA be this family, and let r-NPCM be the same type of machines augmented by reversal-bounded counters. In [15], it was shown that the store languages of all r-NPDA are regular, and in [14], it was shown that the store languages of r-NPCM are all in NCM. When restricting it to not decrease before the end-marker, these machines can therefore only push or flip before the end-marker, but not pop.

Proposition 4 r-$\overline{NP}DA = r$-$\overline{\overline{NP}}DA = REG$, and r-$\overline{NP}CM = r$-$\overline{\overline{NP}}CM = NCM$.

Proof First, for r-$\overline{NP}DA$, let M be such a machine with input alphabet Σ, and stack alphabet Γ. Then build a 2NFA (a two-way NFA [11] where there is a left and right end-marker on the input) M' over alphabet $\Sigma \cup \Gamma \cup \{\$_1, \ldots, \$_r\}$ such that M' reads $a \in \Gamma$ of M as input instead of pushing it, and also reads a new character $\$_i$ when flipping the pushdown for the ith time. When M' reaches the end of the input, it now needs to verify that the pushdown letters are a "representation" of a word in the store language of M. Let $Z_0 v_0 \$_1 \cdots \$_i v_i$, $v_j \in \Gamma^*$, $0 \le j \le i$, be the sequence of letters read on the input from $\Gamma \cup \{\$_i \mid 1 \le i \le r\}$ (ignoring letters of Σ). Then, in M, the pushdown contents would be $x = Z_0(\cdots((v_0)^R v_1)^R \cdots v_{i-1})^R v_i$. By using the two-way input, M' can verify that x is in the store language of M since $S(M)$ is a regular language. Then, creating a homomorphism h that erases all letters not in Σ, and fixes all others, leaves $L(M) = h(L(M'))$, and the regular languages are closed under homomorphisms.

Similarly for r-$\overline{NP}CM$, build a 2NCM (a two-way machine with reversal-bounded counters [9]) M' that simulates M while reading symbols of $\Gamma \cup \{\$_i \mid 1 \le i \le r\}$ but uses counters of M' to do so faithfully. And then, at the end of the input, it needs to verify that the representation of the pushdown letters is in the store language. For this, it uses the two-way input as above together with additional counters as the store language of M is in NCM. Further, M' makes a bounded number of turns on the input since r is fixed, and therefore M' can be converted to a one-way NCM [9]. Similarly again, since NCM is closed under homomorphism [17], the proposition follows. $\square$

We can similarly study the increasing and decreasing restrictions when applied to the counters. Notice that in the proof of Proposition 1, this same proof would hold to show that $NPCM = NP\overline{C}M = NP\overline{\overline{C}}M$ since the pushdown is not used. In the same way, for example, $N\overline{P}CM = N\overline{\overline{P}}CM$ since the counters are independent of the other stores. Similarly for all other types of stores. Hence:

Proposition 5 $NPCM = N\overline{P}CM = N\overline{\overline{P}}CM$, and $\overline{NP}CM = \overline{N}\overline{P}CM = \overline{N}\overline{\overline{P}}CM = NCM$. Similarly for **NQCM** and r-**NPCM**.

Next, we will turn our attention to deterministic machines. First, a result on reversal is needed. Although closure of **DCM** under reversal has not been formally studied to the best of our knowledge, it follows relatively easily that **DCM** is not closed under reversal from a recent paper.

Proposition 6 *DCM is not closed under reversal. Hence, 2DCM that makes one turn on the input is strictly more powerful than DCM.*

Proof Assume that **DCM** is closed under reversal. In [7], it was shown that **DCM** is closed under the prefix operator, but not closed under the suffix operator. Let $L \in$ **DCM** such that the suffix closure of L is not in **DCM**. By assumption, and by the closure of **DCM** under prefix, $(\mathrm{pref}(L^R))^R \in$ **DCM** (where pref is the prefix operator). But this is equal to the suffix closure of L, a contradiction.

From this it follows that 2**DCM** that makes one turn on the input is strictly more powerful than **DCM**. □

The above proposition is quite interesting as the previous candidate witness language conjectured to separate 2**DCM** from **DCM** was significantly more complex, being accepted by a 5-crossing 2**DCM** (ie. the boundary of each input cell is crossed at most five times, a more general notion than turns on the input) [16], and the proof that it is not in **DCM** did not appear in the text.

Next, it will be shown that $\overline{DCM}$ is no more general than $\overline{\overline{DCM}}$.

Lemma 1 $\overline{DCM} = \overline{\overline{DCM}}$.

Proof Let $M \in \overline{DCM}$ with k counters called $c_1, \ldots, c_k$, and with m states. First, on an input of size n, each counter can increase until it is at most $m \cdot n$ by the time the right input end-marker is hit, otherwise M enters an infinite loop and does not accept. Then at the end-marker, if one counter is increasing, another must be decreasing after m transitions, otherwise an infinite loop is entered. Then for every decrease, there is at most m increases of some other counter. Thus, other counters reach a value of at most $m^2 n$. Continuing across all counters, the most M can store in any counter of an accepting computation is $f(n) = m^k n$.

Then, a $3k$-counter $\overline{\overline{DCM}}$ M' machine will be built accepting $L(M)$. M' simulates M using counters $c_1, \ldots, c_k$, but in parallel, M' increases counters d_i, e_i, for each $1 \leq i \leq k$, to $f(n)$ by the end of the input, which is possible by adding additional states. Then, at the right input end-marker, instead of increasing counter c_i say, it instead decreases counter d_i. Then, when M would start decreasing counter c_i, say M has increased c_i by x_i since hitting the end-marker. Then d_i holds $f(n) - x_i$ (and indeed, $f(n) \geq x_i$ in any accepting computation by the calculation of $f(n)$). At this point, M' subtracts both d_i and e_i in parallel until both are zero. Then e_i holds $f(n) - (f(n) - x_i) = x_i$. Then M' can continue to simulate c_i using counters e_i until empty, then c_i, as their combined length is the same as counter c_i in M at the point where counter c_i starts to decrease. □

Thus, every language by a deterministic Parikh automaton (equal to $D\overline{C}M$) can be accepted by a DCM where the machine only adds until the end-marker, and then at the end-marker, only subtracts.

Next, the families of $D\overline{P}CM$ and $D\overline{P}CM$ will be studied.

Lemma 2 $DCM^R \subseteq D\overline{P}CM \subseteq D\overline{P}CM$.

Proof The second inclusion is immediate.

For the first, given $M \in DCM$, a machine $M' \in D\overline{P}CM$ can be built, that on input $x \in \Sigma^*$, pushes x on the pushdown, and then it simulates M on x^R by popping from the pushdown instead of reading from the input, while simulating the counters exactly (all counter operations are performed after the end-marker of M' has been reached). Thus, $L(M)^R \in D\overline{P}CM$. □

From this, the following can be shown:

Proposition 7 $D\overline{\overline{C}}M = D\overline{C}M \subsetneq DCM \subsetneq D\overline{P}CM$.

Proof The first equality is from Lemma 1, and the first strict inclusion follows from Proposition 1. The second inclusion follows trivially by not using the pushdown. Strictness follows since DCM is not closed under reversal by Proposition 6, and since $DCM^R \subseteq D\overline{P}CM$ by Lemma 2. □

Next, it will be shown that $D\overline{C}M$ and its reverse are in $D\overline{P}CM$.

Lemma 3 $D\overline{C}M \cup D\overline{C}M^R \subseteq D\overline{P}CM$.

Proof $D\overline{C}M \subseteq D\overline{P}CM$ follows by simulating the counters verbatim without using the pushdown. And $D\overline{C}M^R$ follows from Lemma 2. □

Next, we will show that this containment is strict. A technique in [5] was used to find languages that could not be accepted by deterministic reversal-bounded multi-counter machines, and also deterministic machines with a pushdown augmented by counters. Essentially, it was shown that if L is a $DPCM$, then there exists $w \in \Sigma^*$ such that $L \cap w\Sigma^*$ is in $DPDA$, and similarly if L is a DCM, then there exists $w \in \Sigma^*$ such that $L \cap w\Sigma^*$ is a regular language. Then this property can be used to find languages not in DCM or $DPCM$. However, a close reading of this paper shows that the definition of $DPCM$ and DCM used in this paper does not have an end-marker on the right end of the input. However, at least for DCM, it is known that one-way deterministic machines with reversal-bounded counters, accept strictly less languages when an end-marker is not used (unlike deterministic pushdown automata) [13]. And, a careful reading of the proof technique used to find languages outside of DCM and $DPCM$, illustrates that the technique only works on machines without the end-marker (as defined in the paper). Here though, we are using the more general definition with an end-marker. The same technique though can be used to show that if $L \in DCM$, then there exists $w \in \Sigma^*$ such that $L \cap w\Sigma^*$ is in $D\overline{C}M$. Similarly with $DPCM$ and $D\overline{P}CM$. This can be used as a type of "bridge" to show languages are not in

DCM. And indeed, $\overline{\text{DCM}}$ coincides with deterministic Parikh automata where witness languages are known [4] and can be used with this property. This will be shown next.

First, a definition is needed. Let $M \in$ DPCM (resp. DCM) with k 1-reversal-bounded counters (without loss of generality [6]). For an integer $1 \leq i \leq k$, then w is i-decreasing if, while M is reading w, then the ith counter is decreased before reading the right end-marker.

The key here is that, if there is an i-decreasing word w, then for all $y \in \Sigma^*$, then counter i decreases on wy as well. If instead a word w decreases counter i on the end-marker, then there is no guarantee that wy decreases for all $y \in \Sigma^*$ before the end-marker. The first lemma is immediate.

Lemma 4 *Let $M \in$ DPCM (resp. DCM) with k-counters. If there exists $1 \leq i \leq k$ such that no word in $L(M)$ is i-decreasing, then counter i only decreases on the end-marker in an accepting computation, and another machine M' of the same type can be created where all transitions that decrease counter i before the end-marker are removed, and $L(M) = L(M')$.*

Lemma 5 *Let $M \in$ DPCM (resp. DCM) with k counters and $L(M) \neq \emptyset$, such that, by Lemma 4, for all counters $c_1, \ldots, c_l$ ($l \leq k$) that decrease before the end-marker, there is some $w \in L(M)$ that is i-decreasing, for each $1 \leq i \leq l$. Then, for each such w, there is a machine of the same type accepting $L(M) \cap w\Sigma^* \neq \emptyset$ that has at most $l - 1$ counters that decrease before the end-marker.*

Proof By applying Lemma 4 on all counters i whereby there is no i-decreasing word, the only counters that can decrease before the end-marker are those that can do so on at least one word in the language, in an accepting computation. Let i be such a counter, and $w \in L(M)$ be an i-decreasing word. Then, build a DPCM M' that simulates M but enforces using the states that the input must start with w. Also, it does not need to include counter i as this counter has already started decreasing for any $L(M) \cap w\Sigma^*$ in M by the time w is read, and therefore, the counter can be stored in the finite control. Therefore, $L(M') \neq \emptyset$. By another application of Lemma 4, another M'' can be created whereby at most $l - 1$ counters can decrease before the end-marker. $\square$

By applying Lemma 5 iteratively, the following is true:

Proposition 8 *Let $L \in$ DPCM (resp. DCM) be non-empty. Then there exists $w \in \Sigma^*$ such that $L \cap w\Sigma^*$ is a non-empty $\overline{\text{DPCM}}$ (resp. $\overline{\text{DCM}}$).*

This same technique also clearly works for all other deterministic machine models augmented by reversal-bounded counters considered in this paper. Then this can serve as a "bridge" where witnesses known for versions of machines with counters that only decrease on the end-marker can possibly be used to show a witness in the more general model where the counter restriction does not occur.

Furthermore, it is known that $\overline{\text{DCM}}$ coincides with deterministic Parikh automata [13]. And, it is known that

$$L = \{v \in \{a, b\}^* \mid v[|v|_a] = b\},$$

where $v[j]$ is the jth letter of v, cannot be accepted by deterministic Parikh automata [4]. Assume by way of contradiction that $L \in \mathsf{DCM}$. Then, by Proposition 8, there exists $w \in \Sigma^*$ such that $L \cap w\Sigma^* \in \overline{\mathsf{DCM}}$. Let w be such a word, let i be the length of w, and let j be the number of a's in w. Then $L \cap w\Sigma^* = \{wv \in \{a, b\}^* \mid (wv)[|wv|_a] = b\} \in \overline{\mathsf{DCM}}$. Since $\overline{\mathsf{DCM}}$ is closed under left quotient with a fixed word, it follows that $L' = \{v \in \{a, b\}^* \mid (wv)[|v|_a + j] = b\} \in \overline{\mathsf{DCM}}$. Let $x = a^{i-j}$. Let $L'' = (L')(x)^{-1} = \{v \in \{a, b\}^* \mid (wvx)[|v|_a + j + i - j] = b\} = \{v \in \{a, b\}^* \mid (wvx)[|v|_a + i] = b\} = \{v \in \{a, b\}^* \mid v[|v|_a] = b\}$. But $\overline{\mathsf{DCM}}$ is closed under right quotient with words [13], and $L'' = L$, a contradiction.

Lemma 6 $L = \{v \in \{a, b\}^* \mid v[|v|_a] = b\} \notin \mathsf{DCM}$.

Next, we will show that $L \notin \mathsf{DCM}^R$. This is equivalent to showing that $L^R \notin \mathsf{DCM}$. Then $L^R = \{v \in \{a, b\}^* \mid v[|v| - |v|_a + 1] = b\} \notin \mathsf{DCM}$. Assume, by contradiction that $L^R \in \mathsf{DCM}$. Given $v \in L^R$, notice that $|v| - |v|_a$ is equal to $|v|_b$. Therefore, $L^R = \{v \in \{a, b\} \mid v[|v|_b + 1] = b\}$. But, this language can also be shown to not be accepted by a deterministic Parikh automaton, similar to the proof that L cannot in [4], a contradiction. Hence:

Proposition 9 *There exists* $L \in \overline{\mathsf{DPCM}}$ *such that* $L \notin (\mathsf{DCM} \cup \mathsf{DCM}^R)$.

Proof It has been shown already that L above is not in $\mathsf{DCM} \cup \mathsf{DCM}^R$.

Also, $L \in \overline{\mathsf{DPCM}}$, as a $\overline{\mathsf{DPCM}}$ M can be built with 2 counters c_1, c_2, that on input v, pushes v to the pushdown while in parallel, recording $|v|_a$ in counter c_1, and recording $|v|$ in counter c_2. Then at the end of the input, M subtracts the value of c_1 from c_2, so that c_2 now contains $|v| - |v|_a$. Then, M pops $|v| - |v|_a$ characters from the pushdown, and then verifies that the next character on the pushdown is a b, which then has the effect of verifying that position $|v|_a$ of v contains a b. □

Corollary 2 $\overline{\mathsf{DCM}} \cup \overline{\mathsf{DCM}}^R \subsetneq \overline{\mathsf{DPCM}}$.

It is still open as to whether DCM is a subset of $\overline{\mathsf{DPCM}}$, and whether $\overline{\mathsf{DPCM}}$ is a strict subset of $\overline{\mathsf{DPCM}}$. Though, the language $L = \{c^i\$wa^jb^jv \mid w, v \in \{a, b\}^*, j > 0, |w| = i\}$ is in DCM and $\overline{\mathsf{DPCM}}$. But we conjecture that L is not in $\overline{\mathsf{DPCM}}$, which would resolve both open problems. Even though L can be accepted by a DCM with one counter and three reversals, we also conjecture that all languages accepted by DCM with one counter and one reversal are in $\overline{\mathsf{DPCM}}$.

4 Restrictions When Reading/Not Reading Input Letters

In this section, we generalize the concept of machines that can only decrease the store on the end-marker. For example, in an NPCM, the first stack reversal only occurs on the end-marker. Here, we will create a more general model that restricts what can happen when non-λ transitions are used on the input.

Definition 1 An sNPCM M is an NPCM with the restriction that the pushdown stack can only pop on a λ transition.

An sNPCM is an slNPCM if all transitions that keep the same size of pushdown (i.e. the top of the pushdown symbol X is replaced with a symbol Y with potentially $X = Y$) are λ transitions.

For both types, M is reversal-bounded if the pushdown makes at most k alternations between non-increasing and non-decreasing the size of its pushdown, for some odd k.

Definition 2 An NPCM M is in simple normal form if at every step, M can only do one of the following:

1. reads an input symbol and pushes exactly one symbol on the stack,
2. reads λ and pops one symbol (i.e. the top symbol) from the stack,
3. reads λ and does not change the stack.

Note that for any machine in simple normal form, any transition that does not change the contents of the pushdown must be a λ transition. Also, note that a simple normal form NPCM machine is an slNPCM.

Lemma 7 *An NPCM M over Σ can be converted to a slNPCM M' over $\Sigma \cup \{\#\}$ in simple normal form such that $L(M) = h(L(M'))$, where h is a homomorphism that erases $\#$ and fixes all letters of Σ.*

Proof First, from M with pushdown alphabet Γ, create an intermediate M_1 as follows: For all transitions that replaces A on the stack with $B\gamma$, $A, B \in \Gamma$, replace this with transitions that replace A with B, then pushes each symbol of γ, one at a time. Then, all transitions where a letter is replaced with some $\gamma \in \Gamma^*$ on the pushdown has $|\gamma| \le 2$.

Next, from M_1, create M_2 such that the only transitions that replaces A with B on the stack, $A, B \in \Gamma$ satisfy $A = B$, and are λ transitions. Indeed, M_2 simulates M_1, but for all transitions that replaces A with B on the top of the pushdown, M_2 instead pushes a primed symbol B' (leaving A on the stack). Then, when eventually decreasing, if a primed symbol is seen, M_2 removes all primed symbols plus one more, and continues the simulation as if the topmost primed symbol is the top of the pushdown. Next, if there is a transition that reads a letter and pops A, M_2 instead pushes A' on the letter, then on a λ, pops A' then pops A

Lastly, let $\#$ be a new input symbol. From M_2, create M' such that, if a transition pushes on a λ transition, it instead reads $\#$.

Then M' is in simple normal form, $L(M) = h(L(M'))$, where h is a homomorphism that erases $\#$ and leaves the other symbols unchanged. $\square$

Notice that the normal form NPCM can have a non-reversal-bounded pushdown even if the original machine has a reversal-bounded pushdown. Since NPCM is closed under homomorphism, the following is obtained:

Corollary 3 $NPCM = sl NPCM = s NPCM.$

We will show that reversal-bounded slNPCMs are equivalent to NCMs. We will need two lemmas first.

Lemma 8 *Every reversal-bounded NPCM M in simple normal form can be converted to $M' \in$ NCM such that $L(M) = L(M')$.*

Proof Let M be an NPCM which makes at most k reversals on the stack for some k. Since the family of NCM languages is closed under union, it is sufficient to assume that M makes exactly k-reversals for some odd k.

We will show that we can construct a finite-crossing 2NCM M' such that $L(M) = h(L(M'))$ for some homomorphism h. The result would then follow, since finite-crossing 2NCMs are equivalent to NCMs [9], and by closure of NCM under homomorphism [17].

We illustrate the construction for $k = 3$. Thus, M makes exactly 3 reversals: pushing, popping, pushing, and popping. The input w to M' has two "tracks":

- Track 1 contains an encoding of the input x to M.
- Track 2 contains the string which represents the entire string that M pushes on its stack (from the start to accepting) where the positions when the first popping started and ended and when the second popping started and ended are marked.

Let P_1, P_2, E_1, E_2 be new symbols not in the input alphabet and stack alphabet of M. The input to M' would have two tracks. The first track would look like:

$$x_1 E_2 x_2 E_1 x_3 P_1 x_4 P_2$$

The second track would represent contents of the stack:

$$y_1 E_2 y_2 E_1 y_3 P_1 y_4 P_2$$

where $|x_i| = |y_i|$ for $1 \le i \le 4$ (so, each symbol of each x_i is in the same position of track 1 as the corresponding symbol of y_i on track 2). The 2-track input w to M' indicates that the input to M is $x = x_1 x_2 x_3 x_4$, and M performs the following processes:

1. M reads input segments $x_1 x_2 x_3$ while pushing $y_1 y_2 y_3$ on the stack.
2. Then, M pops the stack content y_3 on λ, leaving $y_1 y_2$ on the stack. (These are indicated by the markers P_1 and E_1.)
3. Then M reads input segment x_4 while pushing y_4 on the stack.
4. Finally, on λ, M pops the stack content y_4 then y_2.

The finite-crossing 2NCM M', when given the two-track input w (provided with left and right end-markers), simulates M and confirms the processes above. Then, M' makes three turns on the input w. We also note that because the number of reversals the stack of M makes exactly is k, the NCM M' can remember the relative positions of P_1, P_2, E_1, E_2.

The 2-track input could have other forms, depending on the relative positions of the P_i's and the E_i's (e.g., $E_1 P_1 E_2 P_2$ is another such form), and the processes of when to read/push and start/end the popping is modified accordingly.

The above construction can easily be generalized for any k. The finite-crossing 2NCMs will need markers $P_1, \ldots, P_{(k+1)/2}$ and markers $E_1, \ldots, E_{(k+1)/2}$.   □

Lemma 9 *Every reversal-bounded slNPCM M can be converted to a reversal-bounded slNPCM M' in simple normal form such that $h(L(M')) = L(M)$.*

Proof First, from M with stack alphabet Γ, create M_1 as follows: For all transitions that replaces A on the stack with $B\gamma$, A, $B \in \Gamma$, $\gamma \in \Gamma^+$, replace this with transitions that replace A with B that does not read input, then pushes each symbol of γ, one at a time (reading the input on the first letter of γ). Then, all transitions where any $A \in \Gamma$ is replaced with γ on the pushdown has $|\gamma| \leq 2$, and all transitions that keep the same size of pushdown are λ transitions. (By the definition of slNPCMs, all those transitions of M that replace A with B are on λ, and all those that pop are on λ.)

Next, from M_1, we will create M_2 such that the only transitions that replaces A with B on the stack, A, $B \in \Gamma$ satisfy $A = B$. Then, for all transitions that replace A with B on the top of the stack, A, $B \in \Gamma$, $A \neq B$ (then these must be λ transitions), then M_2 simulates this as follows: M_1 keeps track of whether it is in "non-decreasing mode" or "non-increasing mode".

- If it is in "non-decreasing mode", then M_2 instead pushes a primed symbol B' (leaving A on the stack) and continues the simulation as if B were the top of the stack. Then, when eventually in decreasing mode, if a primed symbol is seen, M_2 removes all primed symbols plus one more, and continues the simulation as if the topmost primed symbol is the top of the pushdown.
- If the machine is "non-increasing mode", then M_2 leaves A on the top of the pushdown and remembers B in the state and continues the simulation as if B was the top of the pushdown. If this eventually pops B, then M_2 pops A and continues. If the simulation does pop B, but undergoes a reversal (so the simulation reverses), M_2 pops A on λ transition, pushes B on a λ transition, then continues the simulation.

It is clear then that $L(M_2) = L(M_1)$, and M_2 is reversal-bounded.

Lastly, introduce a new input letter #, and create M' from M_2. Then, M' simulates M_2 and for all transitions of M that pushes a symbol on a λ transition, instead it will read input #.

Then $L(M) = h(L(M'))$, where h is a homomorphism that erases # and leaves the other symbols unchanged. Furthermore M' is in simple normal form, and is reversal-bounded if M is reversal-bounded.   □

From this lemma, and since every reversal-bounded NPCM in simple normal form is in NCM, and from closure of NCM under homomorphism [17], we can conclude:

Proposition 10 *Every reversal-bounded slNPCM M can be converted to a NCM M' such that $L(M) = L(M')$.*

But this is not true with only sNPCM, as we see next.

Proposition 11 *Every reversal-bounded NPCM can be accepted by a reversal-bounded sNPCM.*

Proof Let M be a reversal-bounded **NPCM**. Then, for every pop transition that moves right on input letter d, replace it with a transition that moves right on d that keeps the pushdown the same, followed by a transition that pops without reading any input letter. $\square$

In this construction, the number of counters, and the reversal-bounds on the pushdown remain unchanged.

Then, if an *sl***NPDA** is an *sl***NPCM** without reversal-bounded counters:

Corollary 4 *If M is a reversal-bounded sl**NPDA**, then $L(M)$ is regular.*

Proof This is true as the constructions in the proofs above do not introduce any new counters. $\square$

We now show that Proposition 11 and Corollary 4 are not true when the pushdown stack is not reversal-bounded.

Proposition 12 *There is a non-regular language L such that:*

1. *L can be accepted by a **DPDA** in simple normal form (but the stack is not reversal-bounded).*
2. *L cannot be accepted by an **NCM**.*

Proof Let $L = \{x\#x^R \mid x \in \{0,1\}^*\}$. Clearly, L is non-regular. For Part 1, we construct a **DPDA** M which operates as follows, when given input w. M reads the symbols and pushes them in the stack until it sees #, which it pushes on the stack, Then M pops the top of the stack (which is #), and repeats the following process: It reads the next input symbol, say a, and remembers it in the finite control and pushes a fixed dummy symbol D on top of the stack. Then it makes two consecutive λ-moves, where on the first λ move, it pops D, and on the second, it verifies that the symbol on top of the stack is the same as the symbol a remembered in the finite control. M accepts if it finds no discrepancy during the process. Note that M is not reversal-bounded.

Part 2 follows from the fact that L cannot be accepted by an NCM [18]. $\square$

We note that restriction (2) in Definition 2 of an **NPCM** M in simple normal form is essential, since if we remove this restriction, i.e., we allow M to read a symbol on the input while popping the top of the stack, a DPDA whose stack makes only 1 reversal can clearly accept $L = \{x\#x^R \mid x \in \{0,1\}^*\}$, which cannot be accepted by an **NCM**. Hence, Proposition 10 and Corollary 4 are not valid without restriction (2).

Now **NPCM**s are closed under union. But, they are not closed under intersection. In fact, it can be shown using the proof of Theorem 4.2 in [12] that there are languages L_1 and L_2 accepted by 1-reversal **DPDA**s such that $L_1 \cap L_2$ cannot be accepted by any **NPCM**. However, from Lemma 8, **NPCM**s in simple normal form are effectively closed under union and intersection since **NCM**s are clearly closed under these operations.

Acknowledgements The research of O. H. Ibarra was supported, in part, by NSF Grant CCF-1117708. The research of I. McQuillan was supported, in part, by Natural Sciences and Engineering Research Council of Canada Grant 2016-06172.

References

1. Autebert, J.M., Berstel, J., Boasson, L.: Context-free languages and pushdown automata. Handbook of Formal Languages, vol. 1. Springer, Berlin (1997)
2. Bensch, S., Björklund, J., Kutrib, M.: Deterministic stack transducers. In: Han, Y.-S., Salomaa, K. (eds.) Implementation and Application of Automata: 21st International Conference, CIAA 2016, Seoul, South Korea, 19–22 July 2016, Proceedings. Lecture Notes in Computer Science, vol. 9705, pp. 27–38. Springer International Publishing, Berlin (2016)
3. Cadilhac, M., Finkel, A., McKenzie, P.: Bounded Parikh automata. Int. J. Found. Comput. Sci. **23**(08), 1691–1709 (2012)
4. Cadilhac, M., Finkel, A., McKenzie, P.: Affine Parikh automata. RAIRO - Theor. Inf. Appl. **46**, 511–545 (2012)
5. Chiniforooshan, E., Daley, M., Ibarra, O.H., Kari, L., Seki, S.: One-reversal counter machines and multihead automata: revisited. In: Proceedings of the 37th International Conference on Current Trends in Theory and Practice of Computer Science, SOFSEM'11, pp. 166–177. Springer, Berlin (2011)
6. Chiniforooshan, E., Daley, M., Ibarra, O.H., Kari, L., Seki, S.: One-reversal counter machines and multihead automata: revisited. Theor. Comput. Sci. **454**, 81–87 (2012)
7. Eremondi, J., Ibarra, O.H., McQuillan, I.: Deletion operations on deterministic families of automata. Inf. Comput. **TBA**, 1–20 (2017)
8. Ginsburg, S., Greibach, S.: Deterministic context free languages. Inf. Control **9**(6), 620–648 (1966)
9. Gurari, E.M., Ibarra, O.H.: The complexity of decision problems for finite-turn multicounter machines. J. Comput. Syst. Sci. **22**(2), 220–229 (1981)
10. Harju, T., Ibarra, O., Karhumki, J., Salomaa, A.: Some decision problems concerning semilinearity and commutation. J. Comput. Syst. Sci. **65**(2), 278–294 (2002)
11. Hopcroft, J.E., Ullman, J.D.: Introduction to Automata Theory, Languages, and Computation. Addison-Wesley, Reading (1979)
12. Ibarra, O.H.: Visibly pushdown automata and transducers with counters. Fundam. Inform. **148**(3–4), 291–308 (2016)
13. Ibarra, O.H., McQuillan, I.: The effect of end-markers on counter machines and commutativity. Theor. Comput. Sci. **627**, 71–81 (2016)
14. Ibarra, O.H., McQuillan, I.: Applications of store languages to reachability (submitted, 2017)
15. Ibarra, O.H., McQuillan, I.: On store languages of language acceptors (submitted, 2017). https://arxiv.org/abs/1702.07388
16. Ibarra, O.H., Yen, H.C.: On the containment and equivalence problems for two-way transducers. Theor. Comput. Sci. **429**, 155–163 (2012)
17. Ibarra, O.H.: Reversal-bounded multicounter machines and their decision problems. J. ACM **25**(1), 116–133 (1978)
18. Ibarra, O.H., Seki, S.: Characterizations of bounded semilinear languages by one-way and two-way deterministic machines. Int. J. Found. Comput. Sci. **23**(6), 1291–1306 (2012)

On the Persistency of Gellular Automata

Masami Hagiya and Katsunobu Imai

Abstract Gellular automata have been proposed as a new model of cellular automata that are intended to be implemented with gel materials. Computational universality has been investigated and has been shown with unidirectional signal propagation through Moritas rotary elements. A way to realize a Margolus neighborhood has also been proposed, so that block cellular automata can be realized directly in the model. In this chapter, the persistency of those results is examined with numerical simulations. It is shown that block cellular automata can undergo an infinite number of state transitions, and unidirectional signals can be transmitted repeatedly over a circuit. To show persistency, slight modifications of the proposed reactions are needed.

1 Introduction

Within the molecular robotics project [2, 7], we have been working on a model of cellular automata, called *gellular automata* [1, 8–10], with the intention of implementing the model using gel materials and chemical reactions. Cells in the model are separated by a gel material, as a wall. Two kinds of molecules are assigned to each wall. The first is called a *decomposer* that gradually decomposes the wall and eventually makes a hole in it. When a hole is made, the cells separated by the wall are merged instantaneously, i.e., their solutions are mixed. In this rather crude manner, two cells communicate with each other. However, molecules called *composers* fill the hole and separate the cells again. They can then communicate with other neighbors by decomposing other walls.

In the molecular robotics project, more realistic models, based on the diffusion of molecules through gel materials, have also been investigated [3] with actual wet

M. Hagiya (✉)
University of Tokyo, Tokyo, Japan
e-mail: hagiya@is.s.u-tokyo.ac.jp

K. Imai
Hiroshima University, Higashihiroshima, Japan
e-mail: imai@hiroshima-u.ac.jp

© Springer International Publishing AG 2018
A. Adamatzky (ed.), *Reversibility and Universality*, Emergence, Complexity
and Computation 30, https://doi.org/10.1007/978-3-319-73216-9_18

implementations [4]. They assume two kinds of molecules, where one can diffuse through gel walls while the other remains within a cell.

We have also examined possible wet implementations of the model above with decomposers and composers. For example, a gel chunk can be used as a valve in a pipe connecting cells; decomposers shrink the valve to open the pipe while composers swell the valve to close the pipe. Some preliminary experiments were reported [1], but actual implementation requires gel materials with suitable properties and more research needs to be done.

However, the model with decomposers and composers is theoretically interesting because the way in which the cells communicate is severely restricted. We first examined the computational universality of the model [1]. To show universality, we used unidirectional signal propagation through Moritas rotary elements [6]. We then showed a direct implementation of block cellular automata in the model [10]. This implementation was based on the observation that the evolution of merged cells can be regarded naturally as a state transition by a block rule. We thus tried to implement a Margolus neighborhood in the model [5].

In this chapter, we focus on the persistency of those implementations in the model. Unfortunately, we found that both implementations lacked persistency in the sense that Turing machines or block cellular automata cannot execute forever. We examined this situation and looked into how to correct the implementations.

The chapter is organized as follows. We first introduce a simplified version of the model of gellular automata. We then examine the persistency of the implementation of a Margolus neighborhood, and then that of unidirectional signal propagation.

2   The Model

For simplicity, we deal only with one-dimensional cellular automata in this chapter. The model is also simplified in some other aspects, compared with [8], in that we only allowed one decomposer and one composer for each wall.

Each cell is identified by an integer. The wall between cell $i - 1$ and cell i is denoted by w_i. Each wall w_i has a decomposer and a composer, which are chemical species, indicated as $D(w_i)$ and $C(w_i)$, respectively. Each cell i contains a solution consisting of some molecular species. For molecular species A, its concentration in cell i is denoted by $c_i(A)$, which changes over time by chemical reactions defined globally and by the change of w_i and w_i.

Each wall w_i has two discrete states, OPEN and CLOSE, denoted by $\sigma(w_i)$, and a continuous parameter ranging in the interval [0,1], denoted as $\tau(w_i)$. If $\sigma(w_i) =$ CLOSE, $\tau(w_i)$ changes as follows:

$$\frac{d\tau(w_i)}{dt} = k_d(c_{i-1}(D(w_i)) + c_i(D(w_i))),$$

where k_d is a parameter, defined globally, i.e., not depending on walls. When $\tau(w_i)$ reaches 0, $\sigma(w_i)$ is changed to OPEN and, at the same time, for each molecular species A, $c_{i-1}(A)$ and $c_i(A)$ both become $(c_{i-1}(A) + c_i(A))/2$ instantaneously.

If $\sigma(w_i) = $ OPEN, $\tau(w_i)$ changes as follows:

$$\frac{d\tau(w_i)}{dt} = k_c(c_{i-1}(C(w_i)) + c_i(C(w_i))),$$

where k_c is also a parameter defined globally, and $c_{i-1}(C(w_i))$ is equal to $c_i(C(w_i))$ in this case. When $\tau(w_i)$ reaches 1, $\sigma(w_i)$ is changed to CLOSE.

3 Margolus Neighbourhood

In [8, 10], a Margolus neighbourhood was implemented quite elegantly. Six molecular species A_i ($i = 0, 1, 2, 3, 4, 5$) were introduced with the following chemical reactions:

$$A_i + A_{i\oplus 1} + A_{i\oplus 2} + A_{i\oplus 3} \rightarrow 2A_{i\oplus 1} + 2A_{i\oplus 2} \ (k_a),$$

where i ranges over 0, 1, 2, 3, 4, and 5, and denotes addition mod 6. The rate of these reactions is denoted by k_a. The decomposer and composer of wall w_i are defined as follows:

$$D(w_i) = A_{i\oplus 1}$$
$$C(w_i) = A_{i\oplus 0}.$$

In the initial configuration,

$$c_i(A_j) = 1.0 \text{ if } i \equiv j \ (\text{mod } 6)$$
$$c_i(A_j) = 0.0 \text{ if } i \not\equiv j \ (\text{mod } 6)$$
$$\sigma(w_i) = \text{CLOSE and } \tau(w_i) = 1 \text{ if } i \text{ is even}$$
$$\sigma(w_i) = \text{OPEN and } \tau(w_i) = 0 \text{ if } i \text{ is odd.}$$

Because odd walls are open, each even cell is immediately merged with the next odd cell.

At the beginning of Fig. 1, cells 0 and 1 consist of A_0 and A_1, cells 2 and 3 consist of A_2 and A_3, and cells 4 and 5 consist of A_4 and A_5, all with concentration 0.5. Because A_1 is the composer of w_1 while A_3 is the decomposer of w_2, if $k_c > k_d$, w_1 is closed first and then w_2 is opened, so that cells 1 and 2 are merged to contain A_0, A_1, A_2, and A_3. They soon react according to the rule above: they turn into A_1 and A_2.

The naïve observation was verified by a simulation that numerically integrates the ordinary differential equations obtained from the reaction rules among the species.

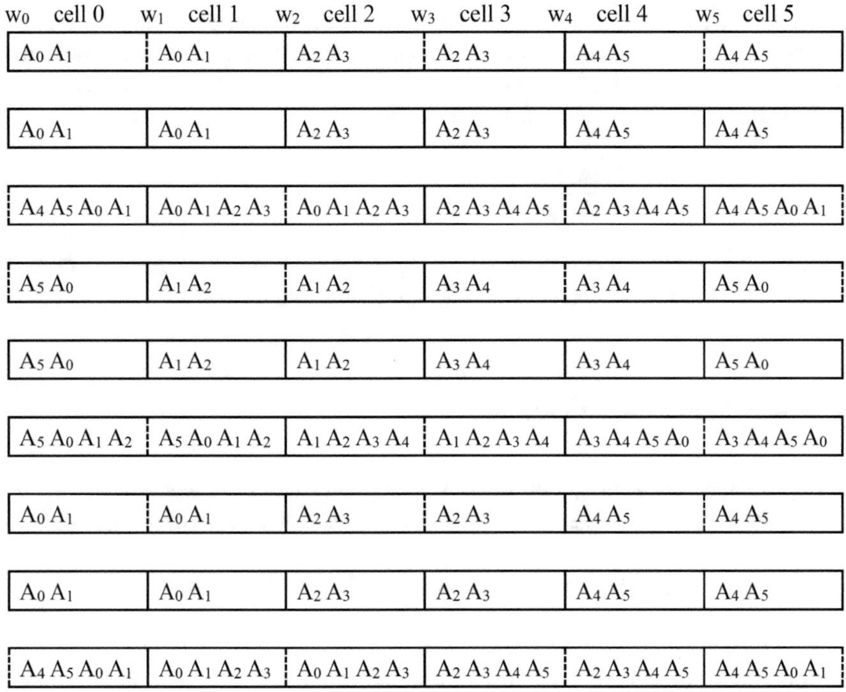

w_0 cell 0	w_1 cell 1	w_2 cell 2	w_3 cell 3	w_4 cell 4	w_5 cell 5
$A_0 A_1$	$A_0 A_1$	$A_2 A_3$	$A_2 A_3$	$A_4 A_5$	$A_4 A_5$
$A_0 A_1$	$A_0 A_1$	$A_2 A_3$	$A_2 A_3$	$A_4 A_5$	$A_4 A_5$
$A_4 A_5 A_0 A_1$	$A_0 A_1 A_2 A_3$	$A_0 A_1 A_2 A_3$	$A_2 A_3 A_4 A_5$	$A_2 A_3 A_4 A_5$	$A_4 A_5 A_0 A_1$
$A_5 A_0$	$A_1 A_2$	$A_1 A_2$	$A_3 A_4$	$A_3 A_4$	$A_5 A_0$
$A_5 A_0$	$A_1 A_2$	$A_1 A_2$	$A_3 A_4$	$A_3 A_4$	$A_5 A_0$
$A_5 A_0 A_1 A_2$	$A_5 A_0 A_1 A_2$	$A_1 A_2 A_3 A_4$	$A_1 A_2 A_3 A_4$	$A_3 A_4 A_5 A_0$	$A_3 A_4 A_5 A_0$
$A_0 A_1$	$A_0 A_1$	$A_2 A_3$	$A_2 A_3$	$A_4 A_5$	$A_4 A_5$
$A_0 A_1$	$A_0 A_1$	$A_2 A_3$	$A_2 A_3$	$A_4 A_5$	$A_4 A_5$
$A_4 A_5 A_0 A_1$	$A_0 A_1 A_2 A_3$	$A_0 A_1 A_2 A_3$	$A_2 A_3 A_4 A_5$	$A_2 A_3 A_4 A_5$	$A_4 A_5 A_0 A_1$

Fig. 1 Evolution of a Margolus neighbourhood. Straight lines denote closed walls while broken lines denote opened walls

We used the following parameters, from [8].

$$k_c = 0.02$$
$$k_d = 0.01$$
$$k_a = 15.0$$

In Fig. 2, each iterated pattern in the graph shows the state of two adjacent cells (e.g., 0 and 1) when the wall between them (w_1) is opened and until the other wall (w_0 or w_2) is opened (after w_1 is closed). This process corresponds to the last four steps in Fig. 1. The thin straight line denotes $\tau(w_0)$, which is equal to $\tau(w_2)$, and decreases (almost) constantly from (almost) 1 to 0, whereas the thin broken line indicates $\tau(w_1)$, which quickly increases from 0 to 1 when w_1 is closed, and then decreases slightly. The thick black line denotes the concentration of A_0 (or A_1), which increases quickly from 0.25 to 0.5, whereas the thick gray line indicates that of A_2 (or A_5), which decreases, from 0.25, and approaches 0. Note that the two cells soon consist only of A_0 and A_1. This pattern is iterated indefinitely and, thus, a Margolus neighborhood is persistently realized.

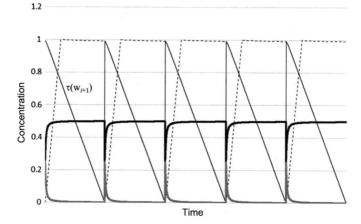

Fig. 2 Evolution of walls and A_i species. The x-axis corresponds to time in abstract unit, and the y-axis concentration in abstract unit or the value of $\tau(w_{i+1})$. The thick black line denotes the concentration of $A_{i\oplus1}$ (or $A_{i\oplus2}$), and the thick gray line is that of $A_{i\oplus0}$ (or $A_{i\oplus3}$). The thin straight line indicates $\tau(w_i)$ (or $\tau(w_{i+2})$), and the thin broken line $\tau(w_{i+1})$. Index i is incremented in each iteration

In [8, 10], we further showed how to implement block rules as in Fig. 3. States of cells and block rules were implemented with two kinds of molecular species, collectively denoted by X and U here. Two U species are assumed to exist in two adjacent cells (e.g., 0 and 1) when the wall between them (w_1) is open. They interact with A_0 and A_1, and are transformed into two X species, which denote the states of cells 0 and 1, respectively. The two X species then interact cooperatively with A_0 and A_1, and produce another U species as a result of state transition by a block rule. This U species will produce X species after the state transition in the next Margolus neighbourhood.

The above construct is realised by reaction rules with the following forms:

$$U_1 + A_0 \rightarrow A_0 + 2X_1 \ (k_r)$$
$$U_0 + A_1 \rightarrow A_1 + 2X_0 \ (k_r)$$
$$X_0 + X_1 + A_0 + A_1 \rightarrow A_0 + A_1 + 2U \ (k_m)$$

In addition to these rules, the X species are assumed to degrade, to eliminate the effect caused by those copies of the X species that are not consumed by the rules above and spread into adjacent cells.

$$X_0 \rightarrow \emptyset \ (k_e)$$
$$X_1 \rightarrow \emptyset \ (k_e)$$

However, with only these rules, the amount of the U or X species increases or decreases exponentially, depending on the reaction rates. In Fig. 4, we use the following parameters in addition to those listed above.

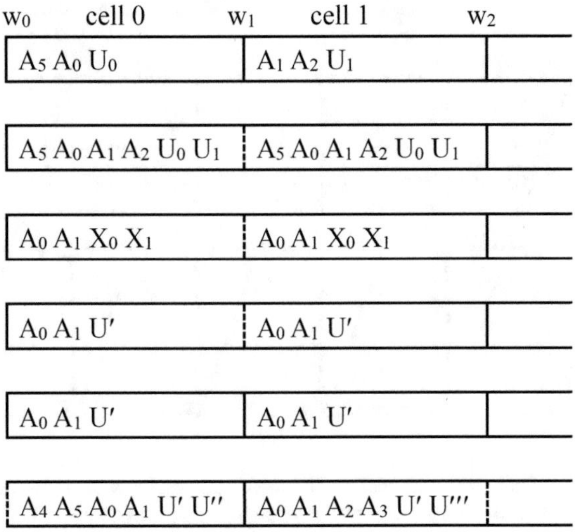

w_0 cell 0 w_1 cell 1 w_2
$A_5\ A_0\ U_0$
$A_1\ A_2\ U_1$
$A_5\ A_0\ A_1\ A_2\ U_0\ U_1$
$A_5\ A_0\ A_1\ A_2\ U_0\ U_1$
$A_0\ A_1\ X_0\ X_1$
$A_0\ A_1\ X_0\ X_1$
$A_0\ A_1\ U'$
$A_0\ A_1\ U'$
$A_0\ A_1\ U'$
$A_0\ A_1\ U'$
$A_4\ A_5\ A_0\ A_1\ U'\ U''$
$A_0\ A_1\ A_2\ A_3\ U'\ U'''$

Fig. 3 State transition in a Margolus neighbourhood. X_0 and X_1 denote the current state of cells 0 and 1. U is the result of state transition and will turn into the next state of the cells

18
16
14
12
10
8
6
4
2
0
Concentration
Time

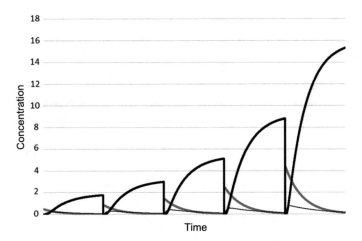

Fig. 4 Evolution of the U species. The x-axis corresponds to time in abstract unit, and the y-axis concentration in abstract unit. The thick grey line denotes the concentration of the U species before state transition, and the thick black line that of the U species after state transition. The thin straight line indicates the concentration of the X species

$$k_m = 0.8$$
$$k_r = 0.035$$
$$k_e = 0.0005$$

In each iteration, the thick gray line denotes the concentration of the U species before the state transition whereas the thick black line denotes that of the U species after the state transition. Its final value is halved and used as the initial value of the U species in the next iteration. The thin straight line denotes the concentration of the intermediate X species.

To address this problem, we introduce a higher-order degradation rule for the U species, as follows.

$$U + U + U \rightarrow \emptyset \ (k_e)$$

This rule means that trimers of the U species are decomposed.

Figure 5 shows the concentrations of U and X species at the steady limit where the same pattern is iterated. The lines, except for the thin broken line, indicate the same entities as in Fig. 4. The thin broken line denotes the accumulated error, as explained below. For the higher-order degradation rule, we use the same parameter, k_e.

The main sources of error are the U and X species that remain at the end of iteration; they interact erroneously and produce incorrect X and U. To reduce the errors, we add reactions of the following form:

$$U + U' \rightarrow \emptyset \ (k_f),$$

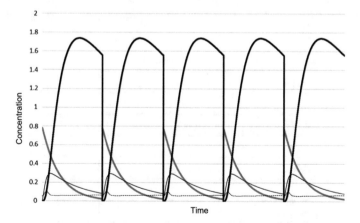

Fig. 5 Evolution of the U species with the modified implementation of block rules. The x-axis corresponds to time in abstract unit, and the y-axis concentration in abstract unit. The thick gray line denotes the concentration of the U species before the state transition, and the thick black line that of the U species after the state transition. The thin straight line indicates the concentration of the X species. The thin broken line indicates the total concentration of erroneous U species

where U and U′ are two conflicting U species. We call U and U′ conflicting if they can appear in the same cell (mod 6) either in an even step or in an odd step. In Fig. 5, we set

$$k_f = 1.0.$$

It shows the result of reactions including the above one. The total concentration of erroneous U species is estimated as the thin broken line. Due to the rules above, this amount is kept to a relatively small value. The estimation of errors depends on the assumption of the U and X species, but with the rules above, only the main U species can remain at the end of each iteration.

4   Unidirectional Signal Propagation

In [1], we showed a way to realise unidirectional signal propagation by gellular automata, as in Fig. 6.

If there exists A_i in cell i, it decomposes a wall between cells i and $i + 1$, the solutions of which are mixed. A_i and X_{i+1} then react and turn into A_{i+1} and X_i by the following rules:

$$A_i + X_{i+1} \rightarrow 2A_{i+1} \; (k_1)$$
$$A_{i+1} \rightarrow X_i \; (k_3).$$

A_{i+1} then reconstructs the wall between cells i and $i + 1$. In cell i, A_{i+1} continues to turn into X_i, and cell i eventually consists of only X_i. In this way, cell i returns to its original state. However, A_{i+1} in cell $i + 1$ decomposes the wall between cells $i + 1$

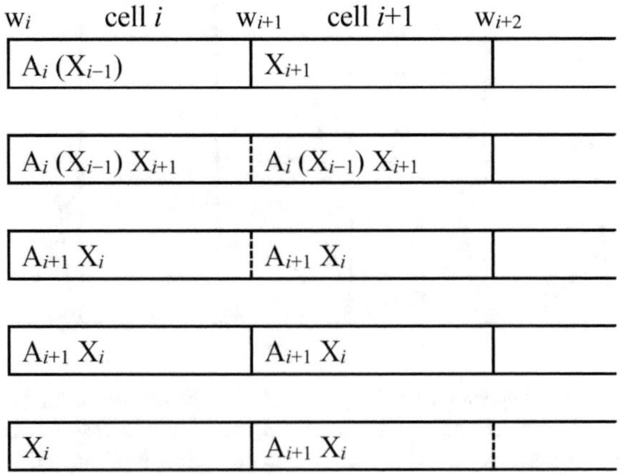

Fig. 6 Unidirectional signal propagation. At the beginning, cell i consists of A_i and X_{i+1}. Cell $i + 1$ then consists of A_{i+1} and X_i, while cell i eventually returns to its original state, consisting of only X_i

and $i + 2$, so the signal continues to propagate. Because X_i also exists in cell $i + 1$, we should have assumed that X_{i-1} also exists in cell i at the beginning. We thus add the following rule:

$$X_{i-1} + X_{i+1} \to 2A_{i+1} \; (k_2).$$

Two conditions should be examined in this approach. First, after a signal is transmitted, cell i should become stable and consist of only X_i. Second, the wall between cells i and $i + 1$ should remain closed.

The second condition may be violated as follows. After the wall between cells i and $i + 1$ is reconstructed by A_{i+1}, there still remains a certain amount of A_i, which continues to decompose the reconstructed wall. Even when cell i becomes stable, consisting of only X_i, the wall between cells i and $i + 1$ is partially broken. This damage may accumulate each time a signal is transmitted, and the wall may eventually disappear.

By numerical simulation, it turns out that with five kinds of cells, walls, and species X_i and species A_i, as in [1], the two conditions above are both violated due to some unexpected interactions among A_i and X_i. The parameters used in the simulation were as follows and were taken from [8]:

$$k_1 = 0.8$$
$$k_2 = 0.8$$
$$k_3 = 0.01$$
$$k_c = 0.02$$
$$k_d = 0.02.$$

Fortunately, if we prepare six kinds of cells, walls, and species, the first condition is almost satisfied and the damage in the second condition is limited. Figure 7 shows how cell i and the wall between cells i and $i + 1$ become stable after a signal is transmitted. The thick line shows the concentration of X_i, which approaches almost 1 at the limit, the bright gray line is that of A_i, and the dark gray line is that of A_{i+1}. The thin line denotes $\tau(w_{i+1})$, which approaches a constant at the limit.

Figure 8 shows the limit of $\tau(w_{i+1})$ in successive signal transmissions. As the graph shows, this limit also approaches a constant at the limit. Consequently, signal propagation can be iterated infinitely often.

5 Concluding Remarks

We examined whether the proposed model of gellular automata can execute two applications - a Margolus neighborhood and unidirectional signal propagation – in a persistent manner. We used numerical integration of ordinary differential equations, which we considered to correctly compute the limits of functions defined by the differential equations and the limits of those limits. In future work, a mathematical

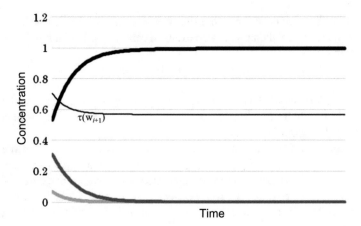

Fig. 7 Cell i after signal transmission. The x-axis corresponds to time in abstract unit, and the y-axis concentration in abstract unit or the value of $\tau(w_{i+1})$. The thick line shows the concentration of X_i, the bright gray line that of A_i, and the dark gray line that of A_{i+1}. The thin line denotes $\tau(w_{i+1})$

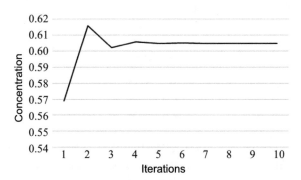

Fig. 8 The limit of $\tau(w_{i+1})$ in successive signal transmissions. The x-axis corresponds to time in abstract unit, and the y-axis the limit value of $\tau(w_{i+1})$.It approaches a constant value

justification of the numerical simulation should be provided, based on an analysis of the differential equations.

Another issue that should be examined is how non-uniformity or perturbation of cells affects the computations. The Margolus neighborhood, in particular, may be affected because some cells may change the neighborhood faster than others, and this difference may accumulate and eventually break the synchronization.

Acknowledgements This research is supported by Grant-in-Aid for Scientific Research on Innovative Areas "Molecular Robotic" (No. 24104003 and No. 24104005) of The Ministry of Education, Culture, Sports, Science, and Technology, Japan.

References

1. Hagiya, M., Wang, S., Kawamata, I., Murata, S., Isokawa, T., Peper, F., Imai, K.: On DNA-based gellular automata. In: Unconventional Computation and Natural Computation, 13th International Conference, UCNC 2014. Lecture Notes in Computer Science, vol. 8553, pp. 177–18 (2014). https://doi.org/10.1007/978-3-319-08123-6_15
2. Hagiya, M., Konagaya, A., Kobayashi, S., Saito, H., Murata, S.: Molecular robots with sensors and intelligence. Acc. Chem. Res. ACS **47**(6), 1681–1690 (2014). https://doi.org/10.1021/ar400318d
3. Isokawa, T., Peper, F., Kawamata, I., Matsui, N., Murata, S., Hagiya, M.: Universal totalistic asynchronous cellular automaton and its possible implementation by DNA. In: Unconventional Computation and Natural Computation, 15th International Conference, UCNC 2016. Lecture Notes in Computer Science, vol. 9726, pp. 182–195 (2016). https://doi.org/10.1007/978-3-319-41312-9_15
4. Kawamata, I., Yoshizawa, S., Takabatake, F., Sugawara, K., Murata, S.: Discrete DNA reaction-diffusion model for implementing simple cellular automaton. In: Unconventional Computation and Natural Computation, 15th International Conference, UCNC 2016. Lecture Notes in Computer Science, vol. 9726, pp. 168–181 (2016)
5. Margolus, N.: Physics-like models of computation. Physica D **10**, 81–95 (1984). https://doi.org/10.1016/0167-2789(84)90252-5
6. Morita, K.: A simple universal logic element and cellular automata for reversible computing. Machines, Computations, and Universality, MCU 2001. Lecture Notes in Computer Science, vol. 2055, pp. 102–113 (2001)
7. Murata, S., Konagaya, A., Kobayashi, S., Saito, H., Hagiya, M.: Molecular robotics: a new paradigm for artifacts. New Gener. Comput. **31**, 27–45 (2013)
8. Wang, S.: An approach to constructing and simulating block cellular automata by gellular automata. Master Thesis (2016)
9. Wang, S., Imai, K., Hagiya, M.: On the composition of signals in gellular automata. In: Second International Symposium on Computing and Networking (CANDAR-AFCA), pp. 499–502 (2014). https://doi.org/10.1109/CANDAR.2014.71
10. Wang, S., Imai, K., Hagiya, M.: An approach to constructing and simulating block cellular automata by gellular automata. In: Third International Symposium on Computing and Networking (CANDAR-AFCA), pp. 442–448 (2015). https://doi.org/10.1109/CANDAR.2015.97

Queue Automata: Foundations and Developments

Martin Kutrib, Andreas Malcher and Matthias Wendlandt

Abstract A queue automaton is basically a finite automaton equipped with a storage obeying the first-in-first-out principle, a queue. The power of queue automata has been studied from several perspectives. One of the classical results frequently cited in the literature is that a machine equipped with a queue storage can be capable of universal computations. This result has been discovered several times. At least implicitly it has already been mentioned by Post in 1943. In connection with formal languages, Vollmar studied in 1970 queue automata for the first time. Despite their versatility queue automata received only occasional attention, probably due to their high computational power with consequent low manageability. These facts triggered the study of subclasses and restricted variants of queue automata which documents the importance of these devices also from a practical point of view. In the present paper, we tour a fragment of the literature on queue automata to give a comprehensive overview of fundamental results and recent developments.

1 Introduction

The investigation of finite automata was probably one fundamental starting point in the theory of formal languages and automata. This model has widely been investigated from different vantage points both theoretical and practical. Finite automata feature many positive and desirable properties such as, for example, the equivalence of nondeterministic and deterministic models, the existence of minimization algorithms, the closure under many operations, and decidable questions such as emptiness, inclusion, or equivalence (see, for example, [28]). But on the other hand, their computational capacity is rather limited since only regular languages are accepted.

M. Kutrib (✉) · A. Malcher · M. Wendlandt
Institut für Informatik, Universität Giessen, Arndtstr. 2, 35392 Giessen, Germany
e-mail: kutrib@informatik.uni-giessen.de

A. Malcher
e-mail: malcher@informatik.uni-giessen.de

M. Wendlandt
e-mail: matthias.wendlandt@informatik.uni-giessen.de

© Springer International Publishing AG 2018
A. Adamatzky (ed.), *Reversibility and Universality*, Emergence, Complexity
and Computation 30, https://doi.org/10.1007/978-3-319-73216-9_19

Natural and well-studied extensions of finite automata are additional storage media such as, for example, pushdown stores [16] or stacks [20]. The model of a *queue* automaton where in comparison with pushdown automata the data structure of a pushdown store is replaced by a queue obeying a *first-in-first-out* principle and its investigation from a formal language point of view has not gained much attention in the literature yet. Some contributions relate these devices to other well-known concepts in formal language theory and theoretical computer science. For instance, in [10], queue automata (there called Post machines) with certain features are shown to characterize the family of languages accepted by multi-reset machines [9], as well as some classes of languages defined by equality sets. In [15] a variant of context-free grammars, called breadth-first grammars, is provided as a generating system for certain classes of languages accepted by queue automata. In [13] queue automata are considered as a theoretical model to investigate scheduling problems.

The fact that, despite their versatility, queue automata received only occasional attention, is probably due to their high computational power with consequent low manageability. In fact, it can easily be seen that the computational power of their deterministic version (DQA), considered for example in [11, 12, 15, 60], equals that of Turing machines if no time restriction is required. Thus, it is natural to define and investigate restricted versions of queue automata. In [60] realtime DQA are considered, where an input symbol is consumed at any step. It is shown that they define a family of languages which is strictly included in the family of context-sensitive languages, includes the regular languages, and is incomparable with the family of context-free languages. A further possibility to limit the computational power is studied in [15], where the realtime condition is slightly relaxed and queue automata are considered that work in quasi realtime, that is, the number of consecutive λ-transitions is bounded by a constant. It is shown in [15] that quasi-realtime queue automata are still a powerful model, since the emptiness problem for such automata is shown to be undecidable. Another restriction bounds the number of alternating enqueuing and dequeuing phases to a finite number and yields so-called finite-turn queue automata. It is shown in [36] that emptiness and finiteness are decidable for such automata.

In [36] *input-driven* queue automata are considered in which the input symbols govern the behavior on the queue and which work by definition in realtime. Nevertheless, emptiness remains undecidable even in this restricted model. The first references to input-driven models date back to [48, 61], where input-driven pushdown automata are introduced. The first results show that the membership problem for input-driven pushdown automata can be solved in logarithmic space, and that nondeterministic and deterministic models are equivalent. Later, it has been shown in [17] that their membership problem belongs to the parallel complexity class NC^1. The investigation of input-driven pushdown automata has been renewed in [3, 4], where such devices are called visibly pushdown automata or nested word automata. The main result is that input-driven pushdown automata describe a language family lying properly in between the regular and the deterministic context-free languages, and sharing with regular languages many desirable features such as many closure properties and decidable questions. Moreover, a tight bound of $2^{\Theta(n^2)}$ is known for the size increase when removing nondeterminism in input-driven pushdown automata. The upper bound has been shown in [61], whereas the lower bound has been derived in [4].

The edge between languages that are accepted by input-driven queue automata (IDQA) or not is very small. For example, language $\{a^n\$b^n \mid n \geq 1\}$ is accepted by an IDQA where an a means an enter operation, b means a remove operation, and a $\$$ leaves the queue unchanged. On the other hand, the very similar language $\{a^n\$a^n \mid n \geq 1\}$ is not accepted by any IDQA. Similarly, the language $\{w\$w \mid w \in \{a, b\}^+\}$ is not accepted by any IDQA, but if the second w is written down with some marked alphabet $\{\hat{a}, \hat{b}\}$, then language $\{w\$\hat{w} \mid w \in \{a, b\}^+\}$ is accepted by an IDQA. To overcome these obstacles, in [37] input-driven queue automata are equipped with internal sequential transducers that preprocess the input. An IDQA with internal transducer is said to be *tinput-driven* (TDQA). While an internal transducer does not affect the computational capacity of general queue automata, it clearly does for *input-driven* versions. To implement the idea without giving the transducers too much power for the overall computation, essentially, only deterministic injective and length-preserving transducers are considered.

Reversibility is a practically motivated property that has been investigated for several automata models. The reversibility of a computation means in essence that every configuration has a unique successor configuration and a unique predecessor configuration or, in other words, the computations are forward and backward deterministic. The main motivation to study reversible computations is the observation that information is lost in irreversible computations, and this loss results in heat dissipation [39]. In one of the first studies Bennett [8] investigated reversible Turing machines. It turned out that every Turing machine can be simulated by a reversible Turing machine or, in other words, any recursively enumerable language can be accepted in a reversible way. Since due to the Theorem of Rice almost nothing is decidable for (reversible) Turing machines, it is reasonable from a practical perspective to study reversibility in devices of lower computational capacity. Reversible regular languages are studied in [5, 54]. It turns out that reversible deterministic finite automata are less powerful than general (possibly irreversible) finite automata. An example is the regular language a^*b^* which is shown [54] to be not acceptable by any reversible deterministic finite automaton. In the same paper it is proved that for a given deterministic finite automaton the existence of an equivalent reversible finite automaton can be decided in polynomial time. Reversible queue automata are introduced and studied in [38].

The difference between pushdown stores and *stack stores* is that the latter model possesses a pushdown head which may move inside the pushdown store in read-only mode and may compare pushdown contents with the input several times without deleting them. Pushdown automata are often used in the context of compilers checking the syntax of programming code and the generalized model of stack automata is also applied to compiling questions in [21]. In [7] the extension from pushdown automata to stack automata is translated to queue automata with the additional constraints of quasi-realtime and finite-turn boundedness. In analogy to stack automata, two heads are provided, one at the front of the queue and one at the tail of the queue, which may move inside the queue in read-only mode. Such automata are called *diving queue automata* for which it is then possible to compare queue contents with the input several times without deleting them and, since there are two heads, to compare some part of the queue with some other part of the queue. Also the restricted model is considered where only one head is available.

In the present paper we tour a fragment of the literature on these issues of queue automata.

2 Definitions and Preliminaries

We write Σ^* for the *set of all words* over the finite alphabet Σ. The *empty word* is denoted by λ, and we set $\Sigma^+ = \Sigma^* \setminus \{\lambda\}$. The set of words of length at most $n \geq 0$ is denoted by $\Sigma^{\leq n}$. For convenience, we use Σ_x for $\Sigma \cup \{x\}$, and similarly for other sets. The *reversal* of a word w is denoted by w^R, and for the *length* of w we write $|w|$. The number of *occurrences* of a symbol $a \in \Sigma$ in $w \in \Sigma^*$ is written as $|w|_a$. We use $\subseteq$ for *inclusions* and $\subset$ for *strict inclusions*. The *complement* of a language $L \subseteq \Sigma^*$ is defined as $\overline{L} = \Sigma^* \setminus L$.

A classical *queue automaton* is a system consisting of a finite state control, a one-way input tape, and a data structure *queue*. At each time step, it is possible to remove or keep the symbol at the front and to possibly enter a symbol at the end of the queue. The transition depends on the current state, the current input symbol or λ, and the symbol which is currently at the front of the queue (see Fig. 1). A formal definition is:

A *nondeterministic queue automaton* (NQA) is a system $\langle Q, \Sigma, \Gamma, \delta, q_0, \bot, F \rangle$, where Q is the finite set of *internal states*, Σ is the finite set of *input symbols*, Γ is the finite set of *queue symbols*, $q_0 \in Q$ is the *initial state*, $\bot \notin \Gamma$ is the *empty-queue symbol*, $F \subseteq Q$ is the set of *accepting states*, δ is a partial *transition function* mapping from $Q \times \Sigma_\lambda \times \Gamma_\bot$ to the subsets of $Q \times \Gamma_\lambda \times \{\text{keep}, \text{remove}\}$.

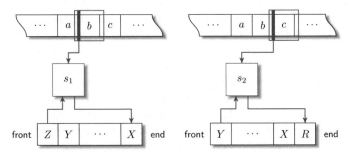

Fig. 1 A queue automaton applying the transition $(s_2, R, \text{remove}) \in \delta(s_1, b, Z)$

A *configuration* of an NQA $M = \langle Q, \Sigma, \Gamma, \delta, q_0, \bot, F \rangle$ is a quadruple (u, q, v, s), where $q \in Q$ is the current state, $u \in \Sigma^*$ is the already read and $v \in \Sigma^*$ is the unread part of the input, and $s \in \Gamma^*$ denotes the current queue content, where the leftmost symbol is at the front. Thus, the *initial configuration* for an input string w is set to $(\lambda, q_0, w, \lambda)$. During the course of its computation, M runs through a sequence of configurations. One step from a configuration to a *successor configuration* is denoted by $\vdash$. Let $a \in \Sigma_\lambda, u, v \in \Sigma^*, z' \in \Gamma_\lambda$ and $z_1, z_2, \ldots, z_m \in \Gamma, m \geq 1$. We set

1. $(u, q, av, z_1 z_2 \cdots z_m) \vdash (ua, q', v, z_1 z_2 \cdots z_m z')$, if $(q', z', \mathtt{keep}) \in \delta(q, a, z_1)$,
2. $(u, q, av, \lambda) \vdash (ua, q', v, z')$, if $(q', z', \mathtt{keep}) \in \delta(q, a, \bot)$, and
3. $(u, q, av, z_1 z_2 \cdots z_m) \vdash (ua, q', v, z_2 \cdots z_m z')$, if $(q', z', \mathtt{remove}) \in \delta(q, a, z_1)$.

So, whenever the queue is empty, a successor configuration is computed by the transition function with the special empty-queue symbol $\bot$. We denote the reflexive and transitive (resp., transitive) closure of $\vdash$ by $\vdash^*$ (resp., $\vdash^+$). The language accepted by the NQA M is the set $L(M)$ of words for which the computation beginning in the initial configuration halts in a configuration in which the whole input is read and an accepting state is entered. Formally:

$$L(M) = \{ w \in \Sigma^* \mid (\lambda, q_0, w, \lambda) \vdash^* (w, q, \lambda, s) \text{ with } q \in F, s \in \Gamma^* \}.$$

If in any case δ is either undefined or a singleton, then the queue automaton is said to be *deterministic*. In particular, there must never be a choice of using an input symbol or of using λ input. So, it is required that for all q in Q and z in $\Gamma_\bot$: if $\delta(q, \lambda, z)$ is defined, then $\delta(q, a, z)$ is undefined for all a in Σ. Deterministic queue automata are denoted by DQA.

Apart from general queue automata that turn out to be computational universal, in the sequel we also consider restricted variants. In particular we are interested to limit the maximal number of consecutive λ-steps that may appear in computations. A queue automaton is said to be *quasi-realtime* if there is a constant that bounds this number for all computations. The queue automaton is said to be *realtime* if this constant is 0, that is, if there are no λ-steps at all.

In order to clarify the notions we continue with an example.

Example 1 The non-semilinear language $\{ a^n \mid n \text{ is a Fibonacci number} \}$ is accepted by a deterministic queue automaton with a number of steps that is linear in the length of the input (see [15]). Formally, the queue automaton is constructed as $M = \langle \{q_0, q_0', q_1, q_1', q_2, q_2'\}, \{a\}, \{A, B, \$\}, \delta, q_0, \bot, \{q_0', q_1\} \rangle$, where δ is given as follows.

(1) $\delta(q_0, a, \perp) =$	$(q_0', \$, \text{keep})$	
(2) $\delta(q_0', a, \$) =$	(q_1, A, keep)	
(3) $\delta(q_1, a, \$) =$	$(q_1', \$, \text{keep})$	
(4) $\delta(q_1', \lambda, \$) =$	(q_2, A, remove)	
(5) $\delta(q_2, \lambda, A) =$	(q_2', A, keep)	
(6) $\delta(q_2', \lambda, A) =$	(q_2, B, remove)	
(7) $\delta(q_2, \lambda, \$) =$	$(q_1, \lambda, \text{keep})$	
(8) $\delta(q_2, a, B) =$	(q_2, A, remove)	

The idea of the construction is as follows. Let F_j, for $j \geq 1$, denote the jth Fibonacci number. It is easily verified that M is in state q_1 storing $\$A$ in the queue after having read the second input symbol. In general, assume that the queue contains a word of the form $\$s$ with $s \in \{A, B\}^*$ such that $|s|_A = F_i - 1$ and $|s|_B = F_{i-1} - 1$, when M has reached state q_1 after having read the F_ith, $i \geq 2$, input symbol. If there is no further input symbol, M halts and accepts. Otherwise the next input symbol is read, M enters state q_1', and starts to revolve the queue until the $\$$ is again at the front, which eventually drives M into state q_1 again. During this process, for every A an AB is enqueued, the $\$$ is replaced by $\$A$ (in two steps), and for every B an A is enqueued. So after this process, the queue content is a word of the form $\$s$ with $s \in \{A, B\}^*$ such that $|s|_A = F_{i+1} - 1$ and $|s|_B = F_i - 1$. Moreover, for every B or $\$$ in the queue M reads one input symbol. So, during the process it reads F_{i-1} further symbols. Together with the F_i symbols read before, M has read F_{i+1} input symbols. Since the $\$$ in front of the queue eventually drives M into the accepting state q_1, M accepts if and only if the input length is a Fibonacci number.   ∎

3 The Power of the Queue

The power of queue automata has been studied from several perspectives. One of the classical results frequently cited in the literature is that a machine equipped with a queue storage can be capable of universal computations. This result has been discovered several times. At least implicitly it has already been mentioned by Post in 1943. In [55] he considered normal productions and proved their universality. A normal production $u \to v$ applied to a string ux yields the string xv. So, the u at the front of the string is removed and the v at the end of the string is added. To some extent this is the behavior of a queue. After Vollmar studied automata with buffer storage [60], which are essentially queue automata, Manna [46] took up the ideas of Post and defined an automaton with queue storage that he called *Post machine*.

Theorem 1 *Deterministic queue automata are computationally universal.*

However, this statement needs some justification. Often it is given in terms as "a queue automaton can simulate a (one-tape, one-head) Turing machine". Basically, the idea of such a simulation is as follows. The queue automaton starts its computation by reading the input entirely, whereby the initial configuration of the Turing machine is stored into the queue with a separating symbol at the end. Then every step of the Turing machine is simulated by revolving the queue, whereby the successor configuration is stored. So, after reading the input, the queue automaton performs only λ-steps. In order to implement such simulation, the queue automaton has to switch from steps in which the input is read to λ-steps. However, for deterministic devices this is impossible. Either a state is used that may read another input symbol, that is, a state for which λ-steps are undefined, or a state is used for which only λ-steps are defined. In the former case, the queue automaton would halt immediately after reading the last input symbol. In the latter case, further input symbols would be ignored. An incremental simulation on the prefixes of the input is, in general, impossible either, since the Turing machine could halt on the input but loop on a prefix.

To overcome these problems either queue automata could be fed with inputs having an endmarker, or a weak encoding of the input is used for the simulation. The latter means that the queue automaton always gets the input $w\#$ when the Turing machine gets some input w. Here $\#$ is a new symbol. While these arrangements are acceptable for *simulations*, they cannot be accepted for *language acceptance*, because it is a difference whether language L or language $L\#$ is accepted. So, in the sequel we will tacitly assume the encoding when we speak about simulations, but will make a distinction when we speak about language acceptance.

Theorem 2 *Let L be a recursively enumerable language over some alphabet Σ and $\# \notin \Sigma$. Then $L\#$ is accepted by a deterministic queue automaton.*

More generally, the simulation sketched above shows the following theorem, where in the deterministic case the endmarker $\#$ can be omitted if the Turing machine is always halting. The latter property should be achieved, for example, for constructible complexity bounds. Since a nondeterministic queue automaton can simply guess and verify whether or not it reads the last input symbol, the endmarker is not needed for nondeterministic computations.

Theorem 3 *Let T be a nondeterministic one-tape, one-head Turing machine with space complexity s and time complexity t. Then there exists a nondeterministic queue automaton M with the same space complexity s and time complexity $s \cdot t$ accepting $L(T)$. If T is deterministic, then M is deterministic and accepts $L(T)\#$.*

In particular, the complexity classes P and NP are equal to the families of languages accepted by deterministic and nondeterministic queue automata, respectively, in polynomial time.

Simulations

Once the results on the simulations of Turing machines are known, the question for simulations by and of other types of machines is raised immediately. The complexity of mutual simulations of devices with different storage types has been studied in several papers. These storage types include multidimensional tapes, dequeues, queues, and pushdowns as well as devices with combinations thereof, even with multiple storages of the same type. The details of these simulations are beyond this survey, where we focus on machines with one storage only. A valuable source of references and results of this type is [52]. So, here we turn to consider devices with one push-down and one tape. It is straightforward to see that an upper bound for the simulation of a queue by a tape is quadratic in the deterministic and nondeterministic case. In the deterministic case this bound is tight in the order of magnitude, since the lower bound shown in [43] is quadratic as well. Let n denote the number of steps simulated.

Theorem 4 *Let M be a deterministic queue automaton. Then $O(n^2)$ time is sufficient to simulate M by a deterministic one-tape, one-head Turing machine. A lower bound for the simulation is $\Omega(n^2)$.*

In the nondeterministic case, the trivial upper bound can be lowered [42]. Nevertheless, there is still a gap between the currently best known lower [43] and the upper bound.

Theorem 5 *Let M be a nondeterministic queue automaton. Then $O(n^{3/2} \log^{1/2} n)$ time is sufficient to simulate M by a nondeterministic one-tape, one-head Turing machine. A lower bound for the simulation is $\Omega(n^{4/3}/(\log n)^{2/3})$.*

For the converse simulation, according to current knowledge there is no difference between simulating a tape or a pushdown. For deterministic and nondeterministic computations only the trivial quadratic upper bound is known. Also the currently best known lower bound coincides for both cases [45] (cf. Theorem 3).

Theorem 6 *Let M be a deterministic (nondeterministic) one-tape, one-head Turing machine or pushdown automaton. Then $O(n^2)$ time is sufficient to simulate M by a deterministic (nondeterministic) queue automaton. A lower bound for the deterministic (nondeterministic) simulation is $\Omega(n^{4/3}/\log n)$.*

Clearly, there are non-context-free languages accepted by queue automata (see Example 1). So, in general a pushdown automaton cannot simulate queues or tapes. However, the last theorem sheds first light on the relations between queue automata and the language families of the Chomsky hierarchy. Since any context-free language is accepted by some nondeterministic pushdown automaton in realtime, it is accepted by a nondeterministic queue automaton in $O(n^2)$ time.

It is well known that the deterministic context-free languages are accepted by deterministic pushdown automata and that, to this end, the pushdown automata have to be able to perform λ-steps. On the other hand, each deterministic pushdown automaton can be normalized is such a way that it pops a symbol from the pushdown whenever it performs a λ-step. This implies that every deterministic pushdown automaton can be made running in linear time. So, any deterministic context-free language is accepted by some deterministic queue automaton in $O(n^2)$ time.

For linear deterministic context-free languages, which are precisely the languages accepted by one-turn deterministic pushdown automata, the upper bound could be improved in [53].

Theorem 7 *Let M be a one-turn deterministic pushdown automaton. Then $O(n^{3/2})$ time is sufficient to simulate M by a deterministic queue automaton.*

So, queue automata and Turing machines are polynomially related. This and applications of queue automata attract notice to restricted variants. In [60] queue automata are studied that are restricted such that they are not allowed to enter more than a constant number of symbols to the queue while keeping the input head stationary, that is, with consecutive λ-steps. By now, this property is called *quasi-realtime* and has formally been defined for queue automata in [15] for the first time. In the same paper several languages are reported to be accepted by quasi-realtime queue automata. In particular, the non-semilinear language $\{a^n \mid n$ is a Fibonacci number$\}$ of Example 1 belongs to that class. A closer look at the DQA constructed in the example reveals that any queue content in a computation never contains more than two consecutive symbols A. Since λ-steps are only performed with an A at the front of the queue which is removed in the next step, the DQA performs no more than four consecutive λ-steps. So, it works in quasi-realtime. Further examples reported in [15] are the languages $\{wcw \mid w \in \{a, b\}^*\}$, $\{a^n b^n c^n \mid n \geq 0\}$, $\{(a^n b)^m \mid n \geq 0, m \geq 0\}$, and $\{a^n b^{k \cdot n} \mid n \geq 0, k \geq 0\}$, which are accepted by quasi-realtime DQA and the languages $\{ww \mid w \in \{a, b\}^*\}$ and $\{a^n \mid n$ is not a prime number$\}$, which are accepted by quasi-realtime NQA.

The role played by λ-steps in quasi-realtime computations has been solved in [15] as follows.

Lemma 1 *For every quasi-realtime nondeterministic queue automaton M an equivalent realtime queue automaton M' can effectively be constructed. If M is deterministic, so is M'.*

Basically, the idea of the construction that shows the previous lemma is to encode a constant number of queue symbols into one, and to perform the consecutive λ-steps together with a step that reads an input symbol at once. In the following we will focus on properties of (quasi-)realtime computations.

Here the acceptance mode of queue automata is by accepting states. Inspired by pushdown automata also acceptance by empty queue has been considered [15]. For nondeterministic pushdown automata it is well known that acceptance by accepting states and acceptance by empty pushdown are equally powerful. On the other hand, acceptance by empty pushdown is strictly weaker than acceptance by accepting states for deterministic pushdown automata. This situation is nicely complemented by the results for queue automata.

Lemma 2 *For every realtime NQA an equivalent realtime NQA accepting by empty queue can effectively be constructed, and vice versa.*

However, in the deterministic case, acceptance by empty queue is not as powerful as acceptance by accepting state [15]. For example, the language $\{ba^n ca^m \mid m \leq n\}$ is a witness for that fact: It is accepted by some realtime DQA but cannot be accepted by any realtime DQA accepting with empty queue.

Lemma 3 *The family of languages accepted by realtime DQA with empty queue is strictly included in the family of languages accepted by realtime DQA.*

The difference between nondeterministic and deterministic queue automata with respect to the acceptance modes suggests that the nondeterministic devices are stronger than the deterministic ones. A corresponding result is shown in [60], but the acceptance mode considered there is only by empty queue. So, the result follows immediately from Lemma 3. In order to separate nondeterminism from determinism also for realtime queue automata accepting with accepting states another result from [15] can be used. It says that the language $L_k = \bigcup_{i=1}^{k} L_i'$, where $k \geq 2$ and $L_i' = \{ ba^{n_1} ba^{n_2} \cdots ba^{n_i} ca^{n_i} b \mid 0 \leq n_1, n_2, \ldots, n_i \}$, is not accepted by any realtime DQA M, even if M is equipped with $k-1$ queues. On the other hand, L_k is accepted by some realtime DQA with at least k queues. So, in particular a deterministic queue hierarchy follows. Since L_2 is clearly accepted by a realtime NQA, the separation follows as well.

Theorem 8 *The family of languages accepted by realtime DQA is strictly included in the family of languages accepted by realtime NQA.*

Next we turn to compare the language families in question with well-known families of the Chomsky hierarchy (cf. [60]). Clearly, every regular language is accepted by some realtime DQA. The properness of the inclusion follows by Example 1.

Climbing up the hierarchy we arrive next at the family of linear realtime deterministic context-free languages. Such a language is the mirror language with centermarker: $L = \{ wcw^R \mid w \in \{a, b\}^* \}$. In [45] it has been shown that any NQA accepting L needs at least $\Omega(n^{4/3} / \log n)$ steps, where n is the length of the input. So, in particular, L is not accepted by any realtime NQA. Conversely, Example 1 provides a non-context-free language accepted by a realtime DQA. So, we have:

Theorem 9 *Neither the family of realtime DQA languages nor the family of realtime NQA languages is comparable with any of the families of (linear) (realtime) (deterministic) context-free languages.*

Next we turn to the family of context-sensitive languages. In general, for realtime queue automata the length of the queue contents that appear in computations is linearly bounded (with respect to the length of the input). On the other hand, any context-sensitive language is accepted by some linear bounded automaton. So for any realtime queue automaton an equivalent linear bounded automaton can be constructed in a straightforward way and we derive that any queue automaton language is context sensitive. It is an open problem whether deterministic and nondeterministic linear bounded automata accept the same language family. Therefore, we now sketch a construction of a *deterministic* linear bounded automaton from a given *nondeterministic* realtime queue automaton that accepts the same language.

Let $M = \langle Q, \Sigma, \Gamma, \delta, q_0, \bot, F \rangle$ be a realtime NQA and $c \geq 0$ be the maximal number of alternative moves that appear in a situation, that is,

$$c = \max\{ |\delta(q, a, z)| \mid q \in Q, a \in \Sigma, z \in \Gamma_\bot \}.$$

So, there are at most c^n different computation paths on a given input of length n. Each of these paths can be encoded by the input word and a c-ary number of length n. A deterministic linear bounded automaton M' that simulates M keeps the original input on one track of its tape. A second track is used as a c-ary counter that produces successively all numbers up to length n. For each of these numbers, M' tries to simulate the computation path encoded by the input and the number. If a number does not encode a valid computation of M, the next number is tested. If it does encode a valid but rejecting computation of M, the next number is tested as well. If it does encode a valid and accepting computation of M, the input is accepted. So, we have the following inclusion.

Lemma 4 *The family of languages accepted by realtime NQA is included in the family of deterministic context-sensitive languages.*

The next step is to show that the inclusion of Lemma 4 is strict. To this end and for further purposes, we present a device that can simulate realtime NQA. Then we can utilize witness languages that are not accepted by that device and, thus, not accepted by realtime NQA.

In particular, we consider one-way multi-head finite automata whose heads may write on the tape [58]. Let $k \geq 1$ be an integer. A *nondeterministic k-head writing finite automaton* (NWFA(k)) is a finite automaton having a single input tape with endmarker processed by k one-way read-write heads. At each step, the device may move each of its heads to the right or keep it stationary, may write a symbol on each tape square scanned by some head, and may change its internal state.

Let M be some realtime NQA. Then an NWFA(2) M' simulates M as follows. The first head of M' simulates the input head of M. So, whenever M reads a symbol and enters z to the queue, M' reads the same symbol with the first head that rewrites the current square with z. If M reads a symbol without entering something to the queue, the first head of M' reads the same symbol, whereby the current square is rewritten by a blank symbol. The second head of M' is located at the square containing the symbol at the front of the queue. Whenever M removes that symbol, M' moves the second head to the next non-blank square which contains the new symbol at the front of the queue or an original input symbol if the queue gets empty.

From the (sketch of the) simulation we derive the next lemma.

Lemma 5 *The family of languages accepted by realtime NQA is included in the family of languages accepted by nondeterministic 2-head writing finite automata.*

In [58] it is stated without proof that both families of Lemma 5 coincide. However, the inclusion is sufficient for our purposes, since in the same paper it is proved that language

$$L_{xy} = \{ xycyx \mid x \in \{0, 1\}^*, y \in \{a, b\}^* \}$$

is not accepted by any NWFA(2). So, it is not accepted by any NQA either. Since L is deterministic context sensitive, the inclusion of Lemma 4 is strict.

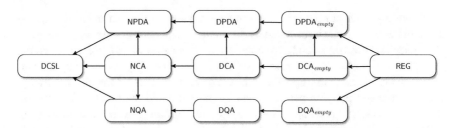

Fig. 2 Inclusion structure of language families described by different types of automata. The arrows indicate strict inclusions. PDA means pushdown automata, CA counter automata, REG denotes the family of regular languages, DCSL the family of deterministic context-sensitive languages, and the index *empty* acceptance by empty storage. All families not connected by a path are incomparable. All queue automata depicted work in realtime

Theorem 10 *The family of languages accepted by realtime NQA is strictly included in the family of deterministic context-sensitive languages.*

So, we obtain the inclusion structure of language families in consideration as depicted in Fig. 2.

An obvious observation is that pushdown automata and realtime queue automata are the same (apart from negligible technical details) when the set of queue and pushdown symbols is a singleton. In this case, both automata degenerate to *counter automata*. Therefore, the family of counter automata languages is a subset of the intersection of the families of realtime queue and pushdown automata languages. An immediate question in this context is whether this inclusion is strict or whether it is an equality, such that a counter would be the intersection of a pushdown and a realtime queue. This and related questions have been solved negatively in [12] for nondeterministic machines. In order to show that counters are strictly weaker than the intersection of pushdowns and realtime queues, the witness language

$$L = \{\, a^i b^j c^m d^n e^p f^q \mid i \neq n \text{ or } j \neq m \text{ or } (i < p \text{ and } j > q) \,\}$$

is used. A realtime NQA M accepting L has the three choices $i \neq n$, $j \neq m$, and (if $i = n$ and $j = m$ then $i < p$ and $j > q$) for a guess. These guesses are handled and verified independently where M accepts if a guess is verified. The verification of $i \neq n$ and $j \neq m$ is obvious. For the third case, M stores $a^i b^j$ into the queue. Then it checks that $i < p$ and $j > q$, in which case it accepts. Now assume that M accepts a correctly formatted input. Then at least one of $i \neq n$ or $j \neq m$ or $i < p$ and $j > q$ is true. In the first two cases the input belongs to L. In the latter case, if $i = n$ and $j = m$ then the input belongs to L, and if $i \neq n$ or $j \neq m$ the input belongs to L as well. On the other hand, any input from L is accepted by at least one of the guesses.

A pushdown automaton accepting L works similarly. It pushes $c^m d^n$ to its pushdown store instead of $a^i b^j$. The proof in [12] is completed by showing that L is not accepted by any nondeterministic counter automaton.

Table 1 Summary of closure properties of realtime queue automata languages

	$-$	$\cap$	$\cup$	$\cdot$	$*$	R	h_λ	h	h^{-1}
Realtime NQA	$-$	$-$	$+$	$+$	$+$	$+$	$+$	$-$	$+$
Realtime DQA	$+$	$-$	$-$	$-$	$-$	$-$	$-$	$-$	$+$

Closure Properties

Now we turn to the closure properties of the families of languages accepted by deterministic and nondeterministic realtime queue automata. The properties are summarized in Table 1. They have been obtained in [15, 58, 60], where in the latter paper the weaker acceptance mode by empty queue is used, but most proofs also work for the more general accepting mode by accepting state. We briefly discuss the witness languages and basic proof ideas and define

$$L_x = \{ xy_1cy_2x \mid x \in \{0, 1\}^*, y_1, y_2 \in \{a, b\}^* \},$$
$$L_y = \{ x_1ycyx_2 \mid x_1, x_2 \in \{0, 1\}^*, y \in \{a, b\}^* \}.$$

The closure of the realtime DQA languages under complementation follows in the same way as the same closure of deterministic pushdown automata languages [15]. The non-closure of the realtime NQA languages follows since the complement of the language L_{xy} from above is clearly accepted by some realtime NQA, but L_{xy} is not.

The intersection of L_x and L_y is the language L_{xy} from above which is not accepted by any realtime NQA. This implies both non-closures under intersection.

Now the closure under complementation and the non-closure under intersection shows by De Morgan's law the non-closure of the realtime DQA languages under union.

In terms of nondeterministic 2-head writing finite automata in [58] the following result has been stated. Let $R \subseteq \Sigma^*$ be a regular language and, for each $a \in \Sigma$, σ be a substitution such that $\sigma(a)$ is a realtime NQA language, then so is $\sigma(R)$.

Since the family of regular languages is the smallest family that includes all finite languages and is closed under union, concatenation, and Kleene star, the result shows the closure of the family of realtime NQA languages under union, concatenation, and Kleene star.

Recall from above that the language $L_k = \bigcup_{i=1}^{k} L_i'$, where $k \geq 2$ and

$$L_i' = \{ ba^{n_1}ba^{n_2} \cdots ba^{n_i}ca^{n_i}b \mid 0 \leq n_1, n_2, \ldots, n_i \},$$

is not accepted by any realtime DQA M. Since $L_k = (ba^*)^{\leq k-1} \cdot L_1$ and L_1 as well as $(ba^*)^{\leq k-1}$ are accepted by some realtime DQA, the non-closure of that family under concatenation follows. Similarly, the non-closure under Kleene star can be shown by the witness language $\{ ba^nca^m \mid m \leq n \}$, whose star is not accepted by any realtime DQA [15].

The closure of the realtime NQA languages under reversal is reported in [58], where a reference to [57] is given. The non-closure of the deterministic languages is seen by language L_k whose reversal is clearly accepted by a realtime DQA.

If the languages L_i' are slightly modified to

$$L_i'' = \{ ba^{n_1} ba^{n_2} \cdots da^{n_i} ca^{n_i} b \mid 0 \le n_1, n_2, \ldots, n_i \},$$

then $L_k' = \bigcup_{i=1}^{k} L_i''$, for $k \ge 2$, is accepted by some realtime DQA. With the λ-free homomorphism $h(a) = a, h(b) = b, h(c) = c$, and $h(d) = b$ we obtain $h(L_k') = L_k$ and, thus, the non-closure of the realtime deterministic queue automata languages under (λ-free) homomorphisms. According to a construction in [57] the nondeterministic family is closed under λ-free homomorphisms, but it is not under arbitrary homomorphisms (otherwise it would include the family of recursively enumerable languages).

The final operation we consider here is the inverse homomorphism. The construction that shows the closure is fairly standard. Let $h : \Sigma \to \Delta$ be a homomorphism. If $L \subseteq \Delta^*$ is accepted by some realtime queue automaton M, then a realtime queue automaton of the same type is constructed that reads symbols from the input and simulates M on the homomorphic image of these symbols. The image may be stored in the finite control. If M accepts the input then M' accepts. So, M' accepts $h^{-1}(L)$.

Decidability

We conclude this section with some remarks on decidability problems. Clearly, the membership problem is decidable for realtime NQA languages since the family is effectively included in the deterministic context-sensitive languages. However, the emptiness problem is not even semidecidable for realtime DQA languages. This implies that many other classical problems are also undecidable. Since this is true also for more restricted variants of queue automata, we refer to the following sections for more on undecidability.

4 Input-Driven Queues and Finite-Turn Queues

In the previous section we have seen that queue automata are a very powerful model. Their capacities have consequences on their computational complexity, in particular, certain closure properties do not hold, minimization algorithms do not exist, and many problems and properties are undecidable. This is at least disconcerting from an applied perspective, where we are much more interested in manageable devices. So, the question arises under which restrictions "good" properties are regained. One such restriction is represented by input-driven automata. Basically, for such devices the operations on the storage medium are dictated by the input symbols. The first references date back to [48, 61], where input-driven pushdown automata (PDA) are introduced as classical PDA in which the input symbols define whether a push operation, a pop operation, or no operation on the pushdown store has to be performed.

The first results show that the membership problem for input-driven PDA can be solved in logarithmic space, and that nondeterministic and deterministic models are equivalent. The investigation of input-driven PDA has been renewed in [3, 4], where such devices are called visibly PDA or nested word automata. The main result is that input-driven PDA describe a language family lying properly in between the regular and the deterministic context-free languages, and sharing with regular languages many desirable features such as many closure properties and decidable questions.

In [36] the input-driven paradigm has been imposed on queue automata which yields input-driven queue automata (IDQA). In the same paper another more "manageable" restriction is considered, namely, queue automata performing only a finite number of turns. This means queue automata in which the number of alternations between enqueuing and dequeuing is bounded by some fixed finite number. The property of finite turns is well known and has been studied in many papers. It has been introduced and studied for pushdown automata and counter automata in [19, 22, 29]. The results reported in this section are from [36].

A classical deterministic queue automaton is called *input-driven* if the next input symbol defines the next action on the queue, that is, entering a symbol at the end of the queue, removing a symbol from the front of the queue, or changing the internal state without modifying the queue content. To this end, we assume that the input alphabet Σ is partitioned into the sets Σ_D, Σ_R, and Σ_N, that control the actions enter (D), remove (R), and state change only (N). Such a partition is called a *signature*.

According to the partition of the input alphabet the transition function of a *deterministic input-driven queue automaton* (IDQA) is split into δ_D mapping $Q \times \Sigma_D \times \Gamma_\perp$ to $Q \times \Gamma$, δ_R mapping $Q \times \Sigma_R \times \Gamma_\perp$ to Q, and δ_N mapping $Q \times \Sigma_N \times \Gamma_\perp$ to Q. Let $a \in \Sigma, u, v \in \Sigma^*, z' \in \Gamma$, and $z_1, z_2, \ldots, z_m \in \Gamma, m \geq 1$. We set

1. $(u, q, av, z_1 z_2 \cdots z_m) \vdash (ua, q', v, z_1 z_2 \cdots z_m z')$,
   if $a \in \Sigma_D$ and $(q', z') = \delta_D(q, a, z_1)$,
2. $(u, q, av, \lambda) \vdash (ua, q', v, z')$, if $a \in \Sigma_D$ and $(q', z') = \delta_D(q, a, \perp)$,
3. $(u, q, av, z_1 z_2 \cdots z_m) \vdash (ua, q', v, z_2 \cdots z_m)$, if $a \in \Sigma_R$ and $q' = \delta_R(q, a, z_1)$,
4. $(u, q, av, \lambda) \vdash (ua, q', v, \lambda)$, if $a \in \Sigma_R$ and $q' = \delta_R(q, a, \perp)$,
5. $(u, q, av, z_1 z_2 \cdots z_m) \vdash (ua, q', v, z_1 z_2 \cdots z_m)$, if $a \in \Sigma_N$ and $q' = \delta_N(q, a, z_1)$,
6. $(u, q, av, \lambda) \vdash (ua, q', v, \lambda)$, if $a \in \Sigma_N$ and $q' = \delta_N(q, a, \perp)$.

The difference between an IDQA and a classical DQA is that the latter makes no distinction of the types of the input symbols, and may perform λ-moves. However, even comparing IDQA with realtime DQA shows that the property of being input driven is a strict weakening of realtime DQA. For example, the language $L = \{ wcw \mid w \in \{a, b\}^* \}$ is clearly accepted by some realtime DQA, but it is not hard to see that it cannot be accepted by any IDQA since each of the symbols a and b either does belong to Σ_D or to Σ_R. So inputs of the form $a^n c a^n$ or $b^n c b^n$ have to be accepted without utilizing the queue, which is impossible since they do not describe regular languages.

Theorem 11 *The family of languages accepted by deterministic input-driven queue automata is strictly included in the family of languages accepted by realtime DQA.*

The next example shows that even deterministic input-driven queue automata accept non-semilinear languages.

Example 2 The non-semilinear language

$$L = \{\, \$_0 \$_1 abb \$_2 \$_1 (abb)^2 \$_2 \$_1 (abb)^4 \$_2 \cdots \$_1 (abb)^{(2^n)} \$_2 \mid n \geq 0 \,\}$$

is accepted by the IDQA $M = \langle \{q_0, q_1, \ldots, q_6\}, \Sigma, \{A, \$\}, \delta, q_0, \bot, \{q_6\} \rangle$, where $\Sigma_N = \emptyset$, $\Sigma_D = \{b, \$_0, \$_1\}$, $\Sigma_R = \{a, \$_2\}$, and the transition functions are as follows:

(1) $\delta_D(q_0, \$_0, \bot) = (q_1, A)$,	
(2) $\delta_D(q_1, \$_1, A) = (q_2, \$)$,	
(3) $\delta_D(q_3, b, Z) = (q_4, A)$,	for $Z \in \{A, \$\}$,
(4) $\delta_D(q_4, b, Z) = (q_5, A)$,	for $Z \in \{A, \$\}$,
(5) $\delta_D(q_6, \$_1, A) = (q_2, \$)$,	
(6) $\delta_R(q_2, a, A) = q_3$,	
(7) $\delta_R(q_5, a, A) = q_3$,	
(8) $\delta_R(q_5, \$_2, \$) = q_6$.	

In an accepting computation of M, the Transitions (1) and (2) are used to read the initial symbols $\$_0$ and $\$_1$ and, moreover, to enqueue the string $A\$$. Then, the computation iterates through phases. In each phase the number of A in the queue is doubled while sequences of adjacent abb blocks are read. For each block, by Transition (6) the a drives M to remove one symbol A from the queue. Then, by Transitions (3) and (4) the following two b are read, whereby two symbols A are enqueued. If symbol $\$_2$ appears in the input exactly when all abb blocks have been read, and the separating symbol $\$$ is at the front of the queue, Transition (8) removes the separating symbol from the queue and switches to the accepting state. If there are no further input symbols, the input is accepted as it belongs to L. If there follows an input symbol $\$_1$, another phase is started by Transition (5) enqueuing the separating symbol again.

It is easy to see that the IDQA M accepts L since: (i) M accepts only at the end of a phase, (ii) in each phase as many abb blocks as A symbols before the separating symbol in the queue are processed, and (iii) in each phase the number of A symbols in the queue are doubled. ∎

Example 2 enables us to prove an interesting property, which makes a difference between pushdown automata and *input-driven* queue automata, though the latter are not allowed to perform λ-moves. It is well known (see, for example, [6, 48]) that for pushdown automata the *pushdown store language* of all words appearing in the pushdown in any accepting computation is always regular. Instead, the corresponding language of queue words for M is not. Contrarily, assume it is regular. Then its intersection with A^* is regular as well. However, this intersection is the non-semilinear language $\{\, A^{2^n} \mid n \geq 0 \,\}$, a contradiction.

It is interesting to compare the class of languages accepted by IDQA with that accepted by input-driven PDA. It turns out that these two classes are incomparable. Let $h_p : \{a, b\}^* \to \{a', b'\}^*$ be the homomorphism defined as $h_p(a) = a'$ and $h_p(b) = b'$. The language $L = \{ wh_p(w)\# \mid w \in \{a, b\}^* \}$ is clearly not context free and thus cannot be accepted by any input-driven PDA. On the other hand, it is not hard to see that L is accepted by an IDQA. Let us now consider the language $L' = \{ wh_p(w^R)\# \mid w \in \{a, b\}^* \}$ which is accepted by an input-driven PDA. Instead, no IDQA for L' can exist. With a little work, this can be obtained by the fact that any NQA accepting palindromes needs at least $\Omega(n^{4/3}/\log n)$ steps, where n is the length of the input [45].

Finite-Turn Queue Automata

In order to study the computational capacities of queue automata in more detail, we now consider turn-bounded devices. For a computation of a queue automaton, a turn is a phase in which the length of the queue first increases and then decreases. This may happen in one or at least two steps. Formally, either a step of the form $(u_1, q_1, v_1, z_1 z_2 \cdots z_m) \vdash (u_2, q_2, v_2, z_2 z_3 \cdots z_m z)$, where δ removes the first symbol from the queue and enters the symbol z into the queue at the same time, or a sequence of at least three configurations

$$(u_1, q_1, v_1, s_1) \vdash (u_2, q_2, v_2, s_2) \vdash \cdots \vdash (u_m, q_m, v_m, s_m)$$

where $|s_1| < |s_2| = \cdots = |s_{m-1}| > |s_m|$ and no step in between removes from and enters into the queue at the same time is a *turn*. For any given $k \geq 0$, a *k-turn* computation is any computation containing exactly k turns.

An IDQA performing *at most k* turns in *any* computation is called k-turn IDQA and will be denoted by $IDQA_k$. We use a similar notation for other devices as well.

Next, we are going to discuss a proper hierarchy on the number of turns for $IDQA_k$, DQA_k, and NQA_k. In order to establish such hierarchies, some results for flip-pushdown automata are used [24, 25]. Basically, a flip-pushdown automaton is an ordinary pushdown automaton with the additional ability to flip its pushdown during the computation. In [36] it has been shown that any realtime queue automaton can effectively be simulated by a flip-pushdown automaton. The direct simulation is straightforward. The idea of the construction is to use one end of the pushdown store as the front and the other end as the tail of the queue to be simulated. So, whenever the queue automaton performs a turn, that is, changes from increasing to decreasing mode, the flip-pushdown automaton flips the front end of the pushdown store to the top and pops. Similarly, whenever the queue automaton changes from decreasing to increasing mode, the tail end has to be flipped to the top and a push is performed. So, for any turn of the queue, one flip is necessary, plus one flip to change from one turn to the next, plus possibly another one to change from the decreasing phase following the last turn to a final increasing phase. Altogether, this makes $2k$ flips.

Since any language accepted by some nondeterministic flip-pushdown automaton (NFPDA) whose number of pushdown flips is bounded by a constant is semilinear [24], we have the following result.

Theorem 12 *Let $k \geq 0$ be a constant and M be a k-turn NQA. Then $L(M)$ is semilinear. In particular, if $L(M)$ is a unary language then it is regular.*

To prove tight turn hierarchies, the straightforward simulation from above is too weak. Therefore, we consider the homomorphism $h_p : \{a, b\}^* \to \{a', b'\}^*$ defined as $h_p(a) = a'$ and $h_p(b) = b'$ again. For all $j \geq 0$, we set

$$C_j = \{ \#w\#h_p(w) \mid w \in \{a, b\}^* \}^j \cdot \#,$$

and define for all $k \geq 0$ the language $L_k = \bigcup_{j=0}^{k} C_j$. The next example shows that L_k is accepted by an IDQA_k.

Example 3 Let $k \geq 0$ be a constant. The language L_k is accepted by the IDQA_k $M = \langle Q, \Sigma, \Gamma, \delta, (q_0, 0), \perp, F \rangle$, where

$$Q = (\{q_0, q_1, q_2\} \times \{1, 2, \ldots, k\}) \cup \{(q_0, 0), (q_1, k+1)\},$$
$$\Sigma_N = \{\#\}, \Sigma_D = \{a, b\}, \Sigma_R = \{a', b'\},$$
$$\Gamma = \{A, B\},$$
$$F = \{ (q_1, j) \mid 1 \leq j \leq k+1 \},$$

and the transition functions are as follows:

(1) $\delta_N((q_0, j), \#, \perp) = (q_1, j+1)$	for $0 \leq j \leq k$,
(2) $\delta_N((q_1, j), \#, Z) = (q_0, j)$	for $1 \leq j \leq k, Z \in \{A, B, \perp\}$,
(3) $\delta_N((q_2, j), \#, Z) = (q_0, j)$	for $1 \leq j \leq k, Z \in \{A, B\}$,
(4) $\delta_D((q_1, j), a, Z) = ((q_2, j), A)$	for $1 \leq j \leq k, Z \in \{A, B, \perp\}$,
(5) $\delta_D((q_1, j), b, Z) = ((q_2, j), B)$	for $1 \leq j \leq k, Z \in \{A, B, \perp\}$,
(6) $\delta_D((q_2, j), a, Z) = ((q_2, j), A)$	for $1 \leq j \leq k, Z \in \{A, B\}$,
(7) $\delta_D((q_2, j), b, Z) = ((q_2, j), B)$	for $1 \leq j \leq k, Z \in \{A, B\}$,
(8) $\delta_R((q_0, j), a', A) = (q_0, j)$	for $1 \leq j \leq k$,
(9) $\delta_R((q_0, j), b', B) = (q_0, j)$	for $1 \leq j \leq k$

The computation of M starts in the state $(q_0, 0)$, necessarily with Transition (1). So, only inputs beginning with $\#$ are accepted. If this is the whole input, it is accepted by the state $(q_1, 1)$. Otherwise, the computation may continue with Transitions (4)–(7), which are used together with the state $(q_2, 1)$ to store in the queue the subsequent word w over $\{a, b\}$ in capitalized form. The input is rejected by blocking if a symbol a' or b' appears before symbol $\#$. When $\#$ is read, M enters the state $(q_0, 1)$ which is used with Transitions (8) and (9) to match the subsequent input factor $h_p(w)$ with the queue content, that is, with w. Only if the queue is empty exactly when the next $\#$ appears in the input, the computation can be accepting. In this case, Transition (1) is used to move to the accepting state $(q_1, 2)$. The 2 indicates that the second infix of the form $w\#h_p(w)\#$ is processed. Finally, when $i \leq k$ infixes have successfully been checked, the automaton enters the accepting state $(q_1, i+1)$. Since no transition is defined for the state $(q_1, k+1)$, the input is accepted only if there are no further input symbols. Clearly, every turn increases the second component of the states. Therefore, M never performs more than k turns. ∎

However, the language L_k cannot be accepted by any queue automaton with less than k turns. This result has been shown in [36] for deterministic queue automata. Literally the same proof shows that L_k is even not accepted by any *nondeterministic* $(k-1)$-turn queue automaton.

Lemma 6 *Let $k \geq 1$ be a constant. The language L_k is not accepted by any $(k-1)$-turn NQA.*

Example 3 and Lemma 6 prove the following proper turn hierarchies for general queue automata as well as for input-driven queue automata.

Theorem 13 *For all $k \geq 0$, the family of languages accepted by deterministic k-turn input-driven queue automata is properly included in the family of languages accepted by deterministic $(k+1)$-turn input-driven queue automata. The same inclusion holds for deterministic and nondeterministic k-turn queue automata.*

Closure Properties

Let us now turn to the closure properties of language families defined by deterministic (finite-turn) input-driven queue automata. For input-driven pushdown automata, strong closure properties have been derived in [3] *provided that* all automata involved share the same partition of the input alphabet into enter, remove, and state-change symbols. Here we distinguish this important special case from the general one. For easier writing, we call the partition of an input alphabet a *signature*, and say that two signatures $\Sigma = \Sigma_D \cup \Sigma_R \cup \Sigma_N$ and $\Sigma' = \Sigma'_D \cup \Sigma'_R \cup \Sigma'_N$ are *compatible* if both $\bigcup_{j \in \{D,R,N\}} (\Sigma_j \setminus \Sigma'_j) \cap \Sigma' = \emptyset$ and $\bigcup_{j \in \{D,R,N\}} (\Sigma'_j \setminus \Sigma_j) \cap \Sigma = \emptyset$ hold.

The closure properties are summarized in Table 2. Let $k \geq 1$ be the number of turns. The closure under intersection with regular sets can be shown by the usual cross-product construction.

Basically, the closure under union and intersection with compatible signatures is shown in the same way. Since it may happen that both alphabets are different, a little extra care has to be taken when one of the automata simulated encounters an input symbol not belonging to its alphabet.

On the contrary, due to turn restrictions, $IDQA_k$ are not closed under complementation despite their determinism. As witness for this fact, language L_k of Example 3 can be used. The example shows that L_k is accepted by some $IDQA_k$. One the other

Table 2 Closure properties of the classes of languages accepted by IDQA and $IDQA_k$. Symbols $\cup_c$ and $\cap_c$ denote union and intersection with compatible signatures

	$-$	$\cup$	$\cap$	$\cap_{REG}$	$\cdot$	$*$	h	h^{-1}	$\cup_c$	$\cap_c$
IDQA	$+$	$-$	$-$	$+$	$-$	$-$	$-$	$-$	$+$	$+$
$IDQA_k$	$-$	$-$	$-$	$+$	$-$	$-$	$-$	$-$	$+$	$+$

hand, one can show that the complement of L_k is not accepted by any IDQA_k. Basically, the idea of the proof is to consider words in the complement which are of the form $\#a^m\#a'^n\#$, for sufficiently large m, n satisfying $m \neq n$. Then one can derive that $a \in \Sigma_D$ and $a' \in \Sigma_R$. The input $(aa')^{k+1}$ must be accepted. However, since $a \in \Sigma_D$ and $a' \in \Sigma_R$, an accepting IDQA_k has to perform one turn on each aa' pair. Thus, it performs at least $k + 1$ turns.

If the requirement of compatible signatures is dropped, several non-closure properties for finite-turn devices are obtained.

In order to disprove the closure under union, consider the languages

$$L_1 = \{\, a^m b^m c^n \# \mid m, n \geq 0 \,\} \text{ and } L_2 = \{\, a^m b^n c^n \# \mid m, n \geq 0 \,\}$$

that are accepted by some IDQA_1. On the other hand, their union $L_1 \cup L_2$ is not even accepted by some IDQA with unbounded turns [36].

For the non-closure under concatenation and Kleene star, take the languages

$$L_1 = \{\lambda\} \cup \{\, a^n b \mid n \geq 1 \,\} \text{ and } L_2 = \{a^n c^n \# \mid n \geq 1\}$$

as witnesses that are accepted by some IDQA_1 as is their union. The star $(L_1 \cup L_2)^*$ contains $L = L_1 L_2 = \{a^n c^n \# \mid n \geq 1\} \cup \{a^m b a^n c^n \# \mid n, m \geq 1\}$. But L cannot be accepted by any IDQA since, for sufficiently large n, the word $a^n c^n \# \in L$ implies that $a \in \Sigma_D$ and $c \in \Sigma_R$. So, an accepting computation on input $a^m b a^n c^n \# \in L$, for sufficiently large m, n satisfying $m > n$, would imply an accepting computation even for words $a^m b a^{n+kp} c^n \# \notin L$ for some $p \geq 1$ and any $k \geq 1$.

For the non-closure under homomorphisms, consider the language

$$L = \{a^n b c^n \# \mid n \geq 0\}$$

and the homomorphism $h : \{a, b, c, \#\}^* \to \{a, b, \#\}^*$ with $h(\sigma) = \sigma$ for $\sigma \in \{a, b, \#\}$ and $h(c) = a$. By similar arguments as before, the language $h(L) = \{a^n b a^n \# \mid n \geq 0\}$ cannot be accepted by any IDQA.

For inverse homomorphism, consider the language $L = \{a^n b^n \# \mid \text{even } n \geq 0\}$ and the homomorphism $h : \{a, b, \#\}^* \to \{a, b, \#\}^*$ defined as $h(a) = a$ and $h(b) = bb$. The language $h^{-1}(L) = \{\, a^{2n} b^n \# \mid n \geq 0 \,\}$ cannot be accepted by any IDQA. Again, one can derive that $a \in \Sigma_D$ and $b \in \Sigma_R$. Then, for sufficiently large n, from an accepting computation on $a^{2n} b^n \# \in h^{-1}(L)$ accepting computations on words $a^n a^{n+kp} b^n \# \notin h^{-1}(L)$ for some $p \geq 1$ and any $k \geq 1$, can be derived.

For incompatible signatures the non-closure under intersection has been shown in [36] as well.

In contrast to the finite-turn case, the general model of IDQA turns out to be closed under complementation. Moreover, the general model of IDQA as well as the turn-bounded variants are clearly closed under union and intersection for compatible signatures and intersection with regular languages.

Decidability

We recall (see, for example, [28]) that a decidability problem is *semidecidable* (resp., *decidable*) if and only if the set of all instances for which the answer is 'yes' is recursively enumerable (resp., recursive). Clearly, any decidable problem is also semidecidable, while the converse does not generally hold. In the remainder of the section we report on several decidability problems for (input-driven) (finite-turn) queue automata. Details of the proofs can be found in [36].

We have seen that a k-turn queue automaton can be simulated by a flip-pushdown automaton making $2k$ flips. An essential technique for flip-pushdown automata is the so-called "flip-pushdown input-reversal" technique, which is developed and proved in [24]. It allows to simulate flipping the pushdown by reversing the (remaining) input. So, given an NFPDA M there is a language accepted by another NFPDA with one flip less. The words of the new language are obtained by reversing the part of the input that is still to be read when M flips its pushdown for the last time. Therefore, the new language is letter equivalent to $L(M)$. In particular, given a $2k$-flip NFPDA M, the $2k$-fold application of the technique yields a language accepted by some NFPDA without any flip, that is, the language is context free. But it is still letter equivalent to $L(M)$. Since the construction is effective and emptiness and finiteness are decidable for context-free languages [28], we derive the first decidability result.

Theorem 14 *Let $k \geq 0$ be a constant and M be a k-turn NQA. Then the emptiness and finiteness of M are decidable.*

Next we turn to equivalence with regular languages and universality. Let R be a regular language. Testing $L(M) = R$ amounts to test $L(M) \cap \overline{R} = \emptyset$ and $R \cap \overline{L(M)} = \emptyset$. Since the family of languages accepted by k-turn DQA is closed under intersection with regular languages, $L(M) \cap \overline{R}$ is accepted by some DQA_k, whose emptiness can be tested by Theorem 14. For testing $R \cap \overline{L(M)} = \emptyset$, we first construct a DQA_k M' accepting $\overline{L(M)}$. To this end, basically, we switch the role of accepting and rejecting states and introduce an additional accepting state f to which undefined transitions are directed. From f, the input head is shifted to the end of the input. Additionally, we have to make sure that M' accepts if M rejects by looping with λ-moves. This can be achieved by inspecting the transition function of M and redirecting in M' λ-moves that would enter a loop to the already added state f. Clearly, M' is a DQA_k accepting $\overline{L(M)}$. Due to the closure under intersection with regular languages, we again obtain that $R \cap \overline{L(M)}$ is accepted by some DQA_k, whose emptiness can be tested.

Theorem 15 *Let $k \geq 0$ be a constant and M be a k-turn DQA. Then the equivalence with regular languages and, in particular, universality is decidable for M.*

So, while emptiness and finiteness are decidable even for nondeterministic finite-turn queue automata, the decidability of equivalence with regular languages and universality has been derived for deterministic devices only. For the general equivalence and inclusion we now consider deterministic input-driven k-turn queue automata. The inclusion $L(M) \subseteq L(M')$ is equivalent to $L(M) \cap \overline{L(M')} = \emptyset$. Since the family of languages accepted by $IDQA_k$ is not closed under complementation, but the family of languages accepted by IDQA is, we obtain that $\overline{L(M')}$ is accepted by some IDQA M'' possibly performing more than k, even an unbounded number of turns but having the same signature as M'. Since the family of languages accepted by IDQA is closed under intersection with compatible signatures, we obtain an IDQA N which accepts $L(M) \cap \overline{L(M')}$ by simulating M with at most k turns in its first component and simulating M'' in its second component. Since M' and M'' have compatible signatures, it is not hard to see that N cannot perform more than k turns: any attempt of M'' to use more than k turns would lead M to halt. Thus, N is an $IDQA_k$ and its emptiness can be tested.

Theorem 16 *Let $k \geq 0$ be a constant and M, M' be k-turn IDQA with compatible signatures. Then inclusion and equivalence of M and M' are decidable.*

Let us now turn to undecidable problems. Basically the proofs rely on reductions of undecidable problems for variants of Turing machines. The method of choice is to encode histories of Turing machine computations into single words [23]. These encodings and variants thereof are of tangible advantage for our purposes. Basically, we consider *valid computations of Turing machines*. Roughly speaking, these are words built from a sequence of configurations passed through during an accepting computation. In [36] valid computations (VALC(M)) of deterministic one-sided one-tape one-head Turing machines M have been encoded such that they become acceptable by deterministic input-driven queue automata.

Lemma 7 *Let M be a deterministic one-sided one-tape one-head Turing machine. Then an IDQA accepting VALC(M) can effectively be constructed.*

Since the emptiness problem for deterministic one-sided one-tape one-head Turing machines M is not semidecidable (see, for example, [28]), and VALC(M) is empty if and only if $L(M)$ is empty, the next theorem follows.

Theorem 17 *Emptiness is not semidecidable for IDQA.*

From this result further non-semidecidability results are obtained by standard methods. It is easy to construct an IDQA for the empty language. Thus, equivalence and inclusion are not semidecidable since emptiness is not semidecidable. Since the family of languages accepted by IDQA is closed under complementation, universality is not semidecidable as well. For any given Turing machine M, VALC(M) is finite (resp., infinite) if and only if $L(M)$ is finite (resp., infinite). So, finiteness and infiniteness are not semidecidable for IDQA. Finally, in [23] it is shown that, given a Turing machine M, VALC(M) is regular if and only if $L(M)$ is finite. Similarly, one obtains that VALC(M) is context free if and only if $L(M)$ is finite (see also [26]). This implies that regularity and context-freeness are not semidecidable for IDQA.

Theorem 18 *Finiteness, infiniteness, universality, inclusion, equivalence, regularity, and context-freeness are not semidecidable for IDQA.*

In Theorem 16 the inclusion and equivalence problem for k-turn IDQA with compatible signatures has been settled. However, the compatibility of signatures is an essential ingredient since inclusion is not semidecidable for two $IDQA_k$ with incompatible signatures. In order to show this assertion, in [36] a special variant of valid computations ($VALC'$) of linear bounded automata (LBA) is considered, whose emptiness problem is also not semidecidable (see, for example, [28]). The key tool is the following lemma.

Lemma 8 *Let M be an LBA. Then 1-turn IDQA M_1 and M_2 can effectively be constructed such that $VALC'(M) = L(M_1) \cap L(M_2)$.*

By utilizing this lemma the non-semidecidability follows.

Theorem 19 *Let $k \geq 1$ be a constant and M, M' be two k-turn IDQA with incompatible signatures. Then the inclusion $L(M) \subseteq L(M')$ is not semidecidable.*

We conclude this section by reporting two further results from [36], where the second result is shown by valid computations and the reduction of the emptiness problem for Turing machines.

Theorem 20 *Let M be an IDQA. It is decidable whether M is k-turn, for some fixed $k \geq 0$. It is semidecidable, but undecidable whether M is finite-turn.*

It turned out that equivalence is not semidecidable for IDQA and, thus, for DQA and NQA, it is currently an open problem whether or not equivalence is decidable for $IDQA_k$ with not necessarily compatible signatures. Taking into account that equivalence is decidable for deterministic pushdown automata [56] and, in particular, for finite-turn deterministic pushdown automata [59], it might be true that similar approaches could prove the decidability of equivalence for DQA_k.

5 Boosting Input-Driven Queues by Preprocessing

The edge between languages that are accepted by input-driven queue automata or not is very small. For example, language $\{\, a^n \$ b^n \mid n \geq 1 \,\}$ is accepted by an IDQA where an a means an enter operation, b means a remove operation, and a $\$$ leaves the queue unchanged. On the other hand, the very similar language $\{\, a^n \$ a^n \mid n \geq 1 \,\}$ is not accepted by any IDQA. Similarly, the language $\{\, w \$ w \mid w \in \{a, b\}^+ \,\}$ is not accepted by any IDQA, but if the second w is written down with some marked alphabet $\{\hat{a}, \hat{b}\}$, then language $\{\, w \$ \hat{w} \mid w \in \{a, b\}^+ \,\}$ is accepted by an IDQA. To overcome these obstacles we provide the input-driven queue automaton with an internal sequential transducer that preprocesses the input.

In the first example above such a transducer translates every a before reading $\$$ to a and after reading $\$$ to b. An IDQA with internal transducer is said to be *tinput-driven*. While an internal transducer does not affect the computational capacity of general queue automata, it clearly does for *input-driven* versions. To implement the idea without giving the transducers too much power for the overall computation, essentially, only deterministic injective and length-preserving transducers are considered. The results reported in this section are from [37].

For the definition of tinput-driven queue automata we need the notion of *deterministic one-way sequential transducers* (DST) which are basically deterministic finite automata equipped with an initially empty output tape. In every transition a DST appends a string over the output alphabet to the output tape. The transduction defined by a DST is the set of all pairs (w, v), where w is the input and v is the output produced after having read w completely. Formally, a DST is a system $T = \langle Q, \Sigma, \Delta, q_0, \delta \rangle$, where Q is the finite set of internal states, Σ is the finite set of input symbols, Δ is the finite set of output symbols, $q_0 \in Q$ is the initial state, and δ is the total transition function mapping $Q \times \Sigma$ to $Q \times \Delta^*$. By $T(w) \in \Delta^*$ we denote the output produced by T on input $w \in \Sigma^*$. Here we will consider only injective and length-preserving DST which are also known as injective Mealy machines.

Let M be an IDQA and T be an injective and length-preserving DST so that the output alphabet of T is the input alphabet of M. The pair (M, T) is called a *tinput-driven queue automaton* (TDQA) and the language accepted by (M, T) is $L(M, T) = \{ w \in \Sigma^* \mid T(w) \in L(M) \}$.

As before, a TDQA performing *at most* k turns in *any* computation is called k-turn TDQA and will be denoted by TDQA$_k$.

Example 4 Language $L_1 = \{ a^n \$ a^n \mid n \geq 1 \}$ is accepted by a TDQA$_1$. Before reading symbol $\$$ the transducer maps an a to an a, and after reading $\$$ it maps an a to a b. Thus, L_1 is translated to $\{ a^n \$ b^n \mid n \geq 1 \}$ which is accepted by some IDQA$_1$.

Similarly, $L_2 = \{ w \$ w \mid w \in \{a, b\}^* \}$ can be accepted by some TDQA$_1$. Here, the transducer maps any a, b to a, b before reading $\$$ and to $\hat{a}, \hat{b}$ after reading $\$$. This gives the language $\{ w \$ \hat{w} \mid w \in \{a, b\}^* \}$ which clearly is accepted by some IDQA$_1$.

Finally, consider $L_3 = \{ a^n b^{2n} \mid n \geq 1 \}$. Here, the transducer maps an a to a and every b alternately to b and c. This gives language $\{ a^n (bc)^n \mid n \geq 1 \}$ which is accepted by some IDQA$_1$: every a implies an enter-operation, every b implies a remove, and every c leaves the queue unchanged.   ∎

The following technical lemma allows to earn several results [37].

Lemma 9 *The concatenation of the input-driven queue automata languages*

$$\{ a^n b^n \mid n \geq 1 \} \text{ and } \{ b^m a^m \mid m \geq 1 \}$$

is not accepted by any TDQA.

The languages given in Example 4 show that even the computational capacity of TDQA with weak, that is, deterministic injective and length-preserving, internal transducers, is larger than that of IDQA. By Lemma 9 the language $\{ a^n b^n b^m a^m \mid n, m \geq 1 \}$ is not accepted by any TDQA. However, this language can easily be accepted by a realtime deterministic queue automaton. So we have the following inclusions.

Table 3 Closure properties of the family of languages accepted by TDQA. Symbols $\cup_c$ and $\cap_c$ denote union and intersection with compatible signatures, $h_{l.p.}$ denotes length-preserving homomorphisms

	—	∪	∩	$\cap_{\text{REG}}$	·	*	R	$h_{l.p.}$	$\cup_c$	$\cap_c$
TDQA	+	−	−	+	−	−	−	−	+	+

Theorem 21 *The family of languages accepted by deterministic input-driven queue automata is strictly included in the family of languages accepted by deterministic tinput-driven queue automata which, in turn, is strictly included in the family of languages accepted by realtime deterministic queue automata.*

Closure Properties

Closure properties of the family of languages accepted by TDQA are summarized in Table 3.

As an immediate consequence of Lemma 9 we obtain the non-closure under concatenation.

It is not hard to construct a tinput-driven queue automaton that accepts language $L = \{a^n b^n c b^{m-1} a^m \mid m, n \geq 1\}$. Let $h : \{a, b, c\}^* \to \{a, b\}^*$ be the length-preserving homomorphism defined by $h(a) = a$, $h(b) = b$, $h(c) = b$. Then the homomorphic image $h(L) = \{a^n b^n b^m a^m \mid m, n \geq 1\}$ is the language of Lemma 9 which is not accepted by any TDQA. So, the non-closure under length-preserving homomorphisms follows.

The language $L = \{a^n b^n \mid n \geq 1\} \cup \{b^n a^n \mid n \geq 1\}$ is clearly accepted by some tinput-driven queue automaton. Words of the form $a^+ b^+ a^+$ that belong to L^* must be from the concatenation $\{a^n b^n \mid n \geq 1\} \cdot \{b^n a^n \mid n \geq 1\}$, that is, from the two-fold iteration of L. However, it can be shown that if these words are accepted by some TDQA, then also some words not belonging to L^* are accepted. This implies the non-closure under Kleene star.

Now we turn to the Boolean operations and obtain a first positive closure under complementation. Let (M, T) be a tinput-driven queue automaton. For the construction of a TDQA (M', T') accepting the complement of $L(M, T)$ the transducer can be used as it is, that is, $T' = T$. An input can be rejected by (M, T) if the computation of M ends in a non-accepting state or if the computation of M blocks since the transition function is undefined for the current situation. Recall that M may perform remove operations on the empty queue. So, for the construction of M' it is sufficient to add transitions to M whenever a transition is undefined. These new transitions let M enter a new non-accepting state s_-. Once in state s_-, M obeys the signature and stays in state s_-. In this way, now all non-accepting computations of M end in a non-accepting state after having read the input entirely. Finally, interchanging accepting and non-accepting states concludes the construction of M'. So, the closure under complementation follows.

For the remaining Boolean operations union and intersection one has to distinguish whether or not the given TDQA have identical or at least compatible signatures. Clearly, two TDQA M and M' have identical signatures if they are defined over the same input alphabet and the behavior of the queue of M and M' is identical for all input symbols. In case of *compatible* signatures, the input alphabets of M and M' may differ, but the behavior of the queue of M and M' is identical for all input symbols belonging to the intersection of both input alphabets. We consider first TDQA having compatible signatures and identical translations.

The basic idea for the construction that shows the closure under intersection for compatible signatures is the well-known cross-product construction. The closure under union follows from the closure under intersection and complementation.

In contrast, the family of languages accepted by TDQA is not closed under union and intersection in case of incompatible signatures. As witnesses for the non-closure consider $L_1 = \{ ba^n ca^n b \mid n \geq 0 \}$ and $L_2 = \{ ba^n ba^m ca^m b \mid m, n \geq 0 \}$. Both languages are accepted by some TDQA, but it is shown in [15] that the union $L_1 \cup L_2$ is not even accepted by any realtime deterministic queue automaton. Since the family of languages accepted by TDQA is closed under complementation, we obtain non-closure under intersection as well.

Similarly, the non-closure under reversal can be derived. The reversal of the union $L = \{ ba^n ca^n b \mid n \geq 0 \} \cup \{ ba^n ba^m ca^m b \mid m, n \geq 0 \}$ is accepted by some TDQA, but L itself is not.

Determinization

Each language accepted by a nondeterministic input-driven pushdown automaton is also accepted by a deterministic input-driven pushdown automaton. The simulation costs for an n-state nondeterministic input-driven pushdown automaton are $2^{\Theta(n^2)}$ states [61]. It is shown in [4] that this size is necessary in the worst case.

The basic idea of the proof is that the automaton stores applicable transition rules onto the pushdown store when a push operation should be done, instead of pushing the appropriate symbol. The actual push operation is simulated at the time at which the symbol to be pushed is popped. In this way, it may happen that some transition rule pushed does not belong to a valid computation. However, the construction allows to distinguish these cases and, thus, only valid transitions will be evaluated. This technique cannot be assigned to input-driven queue automata. The reason is that on input-driven pushdown automata the last symbol pushed is used first. Thus it can be determined whether it belongs to a valid computation or not. In contrast, input-driven queue automata work according to the FIFO principle. Therefore, there could be many symbols in between the queue symbol currently to be removed and the symbol that has been entered last. So, the technique does not allow to verify that the remove operation simulates only enter operations belonging to a valid computation.

This brings us to the question whether the nondeterministic version of a tinput-driven queue automaton can be determinized as well. There are four different working modes for a tinput-driven queue automaton. The sequential transducer can be deterministic or nondeterministic and also the input-driven queue automaton may be deterministic or nondeterministic. We use the notation $\text{TDQA}_{x,y}$ with $x, y \in \{d, n\}$ where x stands for the working mode of the transducer and y for the mode of the input-driven queue automaton. For example, $\text{TDQA}_{n,d}$ is a tinput-driven queue automaton with a nondeterministic sequential transducer and a deterministic input-driven queue automaton. We require here that nondeterministic sequential transducers are injective and length-preserving as their deterministic variant. The next result reveals that, for nondeterministic transducers, the nondeterministic IDQA can be determinized.

Theorem 22 *The family of languages accepted by* $TDQA_{n,n}$ *and* $TDQA_{n,d}$ *coincide.*

So, the nondeterminism of the transducer is a powerful resource. Once it is available, it does not matter whether the IDQA is nondeterministic or not. In both cases the same language family is accepted. However, the next result shows that the absence of nondeterminism of the transducer strictly weakens the TDQA, regardless whether or not nondeterminism is provided for the IDQA.

The inclusion of the next theorem follows immediately from the equality of Theorem 22. The properness of the inclusion is witnessed by language

$$\{\, u\$v\#_1 u \mid u, v \in \{a, b\}^* \,\} \cup \{\, u\$v\#_2 v \mid u, v \in \{a, b\}^* \,\}.$$

It is accepted by a $TDQA_{n,d}$ in such a way that the transducer nondeterministically guesses whether subword u or v has to be matched. Moreover, it can be shown that it cannot be accepted by any $TDQA_{d,n}$.

Theorem 23 *The family of languages accepted by* $TDQA_{d,n}$ *is properly included in the family of languages accepted by* $TDQA_{n,d}$.

The previous theorem showed once more that the nondeterminism of the transducer is a powerful resource. Even if the associated IDQA is deterministic, the nondeterminism of the transducer cannot be compensated by a nondeterministic IDQA. On the other hand, the next result reveals that nondeterminism for the IDQA is better than determinism if the transducers are deterministic. The inclusion follows immediately for structural reasons. The properness of the inclusion is witnessed by language

$$\{\, a^n \$h(w_1)\$h(w_2)\$ \cdots \$h(w_m) \mid m, n \geq 1, w_k \in \{a, b\}^n, 1 \leq k \leq m,$$
$$\text{and there exist } 1 \leq i < j \leq m \text{ such that } w_i = w_j \,\},$$

where h is the homomorphism that maps a to $\#a$ and b to $\#b$.

Theorem 24 *The family of languages accepted by* $TDQA_{d,d}$ *is properly included in the family of languages accepted by* $TDQA_{d,n}$.

Summarizing the results regarding determinization, we end up with the following hierarchy:

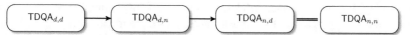

Decidability

We have already seen that IDQA with an unbounded number of turns are a powerful model, since they can accept encodings of computations of linear bounded automata. So, it is possible to reduce undecidability questions for linear bounded automata to undecidability questions for IDQA. Since every IDQA can be considered as a TDQA such that its corresponding injective and length-preserving deterministic sequential transducer simply realizes the identity map, we immediately obtain that all non-semidecidable questions are not semidecidable for TDQA as well.

Corollary 1 *Emptiness, finiteness, infiniteness, universality, inclusion, equivalence, regularity, and context-freeness are not semidecidable for TDQA.*

On the other hand, the questions of emptiness, finiteness, and equivalence with regular languages are decidable for k-turn DQA which are not necessarily input-driven. These results can be transferred to TDQA. Let (M, T) be a k-turn TDQA. Then the DST T is in particular length-preserving and maps every letter of the input alphabet Σ to another letter of M's input alphabet Δ. Now, we construct a DQA M' which translates in its state set every input symbol $a \in \Sigma$ to $T(a) \in \Delta$ and simulates M on input $T(a)$. Thus, $w \in L(M')$ if and only if $w \in L(M, T)$ and we obtain that $L(M') = L(M, T)$. Moreover, M' is k-turn since M is k-turn. In this way, the next theorem follows.

Theorem 25 *Let $k \geq 0$ be a constant and (M, T) be a k-turn TDQA. Then emptiness and finiteness of $L(M, T)$ is decidable. Furthermore, equivalence with regular sets and, in particular, universality are decidable for (M, T) as well.*

Now, we turn to further decidable questions for k-turn TDQA. The questions of inclusion and equivalence are decidable as long as the signatures are compatible [37].

Theorem 26 *Let $k \geq 0$ be a constant and (M, T) as well as (M', T) be k-turn TDQA with compatible signatures. Then the inclusion and the equivalence of $L(M, T)$ and $L(M', T)$ are decidable.*

The positive decidability of inclusion gets lost if the signatures are no longer compatible. From above we know that the inclusion problem for IDQA with incompatible signatures is not semidecidable. Clearly, this result holds for TDQA as well. Thus, we obtain the following corollary.

Corollary 2 *Let $k \geq 1$ be a constant and (M, T) as well as (M', T) be two k-turn TDQA. Then the inclusion $L(M, T) \subseteq L(M', T)$ is not semidecidable.*

It should be noted once more that it is currently an open question whether the equivalence of k-turn IDQA or k-turn TDQA becomes undecidable in case of incompatible signatures. Now, we turn back to TDQA with an unbounded number of turns. Here, we get the interesting result that inclusion and equivalence remain non-semidecidable even if compatible signatures are provided. To this end, let (M, T) be a TDQA. We take a copy of (M, T) and modify M to M' by setting its set of accepting states to $\emptyset$.

Clearly, $L(M', T) = \emptyset$ and (M, T) as well as (M', T) have compatible signatures. If we could semidecide the inclusion or equivalence of $L(M, T)$ and $L(M', T)$, we could semidecide the emptiness of (M, T) as well. This is a contradiction to Corollary 1.

Theorem 27 *Let (M, T) and (M', T) be two TDQA with compatible signatures. Then the inclusion $L(M, T) \subseteq L(M', T)$ and the equivalence $L(M, T) = L(M', T)$ is not semidecidable.*

6 Reversibility

Reversibility is a practically motivated property that has been investigated for several automata models. The reversibility of a computation means in essence that every configuration has a unique successor configuration and a unique predecessor configuration or, in other words, the computations are forward and backward deterministic. The main motivation to study reversible computations is the observation that information is lost in irreversible computations and this loss results in heat dissipation [39]. In one of the first studies Bennett [8] investigated reversible Turing machines. It turned out that every Turing machine can be simulated by a reversible Turing machine or, in other words, any recursively enumerable language can be accepted in a reversible way. Reversible regular languages have been studied in [5, 54], where it turned out that reversible deterministic finite automata are less powerful than general (possibly irreversible) finite automata. As extension of finite automata, reversible pushdown automata have been studied in [34], where it was shown that their corresponding language family lies properly in between the regular and the deterministic context-free languages. Reversibility has been studied also in other computational devices such as, for example, space-bounded Turing machines [41], two-way multi-head finite automata [50], one-way multi-head finite automata [35], and the massively parallel model of cellular automata [33, 49]. Different aspects of reversibility for classical automata are discussed in [32].

Reversible queue automata have been introduced and investigated in [38], where the results reported in this section are from.

With an eye towards backward steps in reversible computations, the transition function of reversible queue automata depends not only on the current state, the current input symbol or λ, and the symbol currently at the front of the queue, but *also on the symbol at the end of the queue*. It is worth mentioning that the additional knowledge of the last queue symbol does not increase the computational power of queue automata. A device can simply store the symbol lastly entered at the end of the queue in its state to simulate the behavior of the extended automaton.

Formally, the transition function of a DQA $\langle Q, \Sigma, \Gamma, \delta, q_0, \bot, F \rangle$ is now a partial function mapping from

$$Q \times \Sigma_\lambda \times ((\Gamma \times \Gamma) \cup (\{\bot\} \times \{\bot\})) \text{ to } Q \times \Gamma_\lambda \times \{\texttt{keep}, \texttt{remove}\}.$$

As before, there must never be a choice of using an input symbol or of using λ input. So, it is required that for all q in Q and z_1, z_2 in $\Gamma \cup \{\bot\}$: if $\delta(q, \lambda, z_1, z_2)$ is defined, then $\delta(q, a, z_1, z_2)$ is undefined for all a in Σ.

Now we turn to reversible queue automata. Reversibility is meant with respect to the possibility of stepping the computation back and forth. To this end, the queue automata have to be also backward deterministic. That is, any configuration must have at most one predecessor which, in addition, is computable by a DQA. For reverse computation steps the head of the input tape is moved to the *left*. Therefore, the automaton rereads the input symbol which has been read in a preceding forward step. Moreover, the roles played by the front and end of the queue are interchanged. That is, in reversed computation steps the symbols are removed from the end and added to the front of the queue. So, for reversible queue automata there must exist a *reverse transition function* $\delta^{\leftarrow}$ mapping from $Q \times \Sigma_\lambda \times ((\Gamma \times \Gamma) \cup (\{\bot\} \times \{\bot\}))$ to $Q \times \Gamma_\lambda \times \{\texttt{keep}, \texttt{remove}\}$ that maps a configuration to its *predecessor configuration*. One step from a configuration to its *predecessor configuration* is denoted by $\vdash^{\leftarrow}$. Let $a \in \Sigma_\lambda, u, v \in \Sigma^*, z' \in \Gamma_\lambda$ and $z_1, z_2, \ldots, z_m \in \Gamma, m \geq 1$. We set

1. $(ua, q, v, z_1 z_2 \cdots z_m) \vdash^{\leftarrow} (u, q', av, z' z_1 z_2 \cdots z_m)$,
   if $(q', z', \texttt{keep}) = \delta^{\leftarrow}(q, a, z_1, z_m)$,
2. $(ua, q, v, \lambda) \vdash^{\leftarrow} (u, q', av, z')$, if $(q', z', \texttt{keep}) = \delta^{\leftarrow}(q, a, \bot, \bot)$, and
3. $(ua, q, v, z_1 z_2 \cdots z_m) \vdash^{\leftarrow} (u, q', av, z' z_1 z_2 \cdots z_{m-1})$,
   if $(q', z', \texttt{remove}) = \delta^{\leftarrow}(q, a, z_1, z_m)$.

A deterministic queue automaton M is said to be *reversible* (REV-DQA), if there exists a reverse transition function $\delta^{\leftarrow}$ inducing a relation $\vdash^{\leftarrow}$ from one configuration to the next, so that

$$(u, p, v, s) \vdash^{\leftarrow} (u', p', v', s') \text{ if and only if } (u', p', v', s') \vdash (u, p, v, s)$$

(see Fig. 3).

In order to clarify our notion we continue with examples of realtime and quasi-realtime REV-DQA.

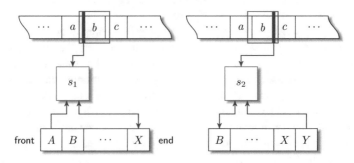

Fig. 3 Successive configurations of a REV-DQA, where $\delta(s_1, b, A, X) = (s_2, Y, \texttt{remove})$ (left to right) and $\delta^{\leftarrow}(s_2, b, B, Y) = (s_1, A, \texttt{remove})$ (right to left)

Example 5 The deterministic linear context-free language $\{a^n b^n \mid n \geq 1\}$ is accepted by the quasi-realtime REV-DQA

$$M = \langle \{q_0, q_1, q_2\}, \{a, b\}, \{A_0, B_0, B_1\}, \delta, q_0, \perp, \{q_2\} \rangle$$

where the transition functions δ and $\delta^{\leftarrow}$ are as follows. For all $X \in \Gamma$ set:

Transition function δ
(1) $\delta(q_0, a, \perp, \perp) = (q_0, A_0, \text{keep})$
(2) $\delta(q_0, a, A_0, X) = (q_0, A_0, \text{keep})$
(3) $\delta(q_0, b, A_0, X) = (q_1, B_0, \text{remove})$
(4) $\delta(q_1, b, A_0, X) = (q_1, B_1, \text{remove})$
(5) $\delta(q_1, \lambda, B_0, X) = (q_2, \lambda, \text{keep})$

Reverse transition function $\delta^{\leftarrow}$
(1) $\delta^{\leftarrow}(q_0, a, A_0, A_0) = (q_0, \lambda, \text{remove})$
(2) $\delta^{\leftarrow}(q_1, b, X, B_0) = (q_0, A_0, \text{remove})$
(3) $\delta^{\leftarrow}(q_1, b, X, B_1) = (q_1, A_0, \text{remove})$
(4) $\delta^{\leftarrow}(q_2, \lambda, B_0, X) = (q_1, \lambda, \text{keep})$

The enqueued symbols have two flags, namely the letter and its index. The letters characterize whether the machine is in mode A or mode B. The machine is in mode A while reading a's from the input and is in mode B when reading b's and checking whether the number of a's and b's coincide. The index describes in which state the machine was in its last step.

The Transitions (1) and (2) of δ are used by M to store for a's in the input an A_0 in the queue to keep in mind that the last state was state q_0 and an a has passed in the input. When the first b appears in the input, Transition (3) is used to change to state q_1 and put a B_0 in the queue. Now the machine enters a cycle with Transition (4) by putting for every b a B_1 in the queue as long as there is an A_0 left at the front of the queue. This ensures that M reads as many b's as a's. If M sees B_0 at the front of the queue, it has read as many b's as a's and it enters the accepting state q_2 with Transition (5). If the whole input is read, the input is accepted and otherwise rejected. ∎

Example 6 Only slight modifications of the construction given in Example 5 show that the languages $\{a^n b^n c^n \mid n \geq 1\}$ and $\{a^m b^n c^m d^n \mid m, n \geq 1\}$ are accepted by quasi-realtime REV-DQA as well. ∎

Example 7 The language $L = \{w \# w \mid w \in \{a, b\}^*\}$ is accepted by a quasi-realtime REV-DQA

$$M = \langle \{q_0, q_1, q_2\}, \{a, b, \#\}, \{A_0, B_0, Z_{1, A_0}, Z_{1, B_0}, \#_0\}, \delta, q_0, \perp, \{q_2\} \rangle$$

where the transition function δ is as follows. For all $X \in \Gamma$ and $Z \in \{A_0, B_0\}$ set:

Transition function δ
(1) $\delta(q_0, a, \bot, \bot)$ = (q_0, A_0, keep)
(2) $\delta(q_0, b, \bot, \bot)$ = (q_0, B_0, keep)
(3) $\delta(q_0, a, Z, X)$ = (q_0, A_0, keep)
(4) $\delta(q_0, b, Z, X)$ = (q_0, B_0, keep)
(5) $\delta(q_0, \#, Z, X)$ = $(q_1, \#_0, \text{keep})$
(6) $\delta(q_1, a, A_0, X)$ = $(q_1, Z_{1,A_0}, \text{remove})$
(7) $\delta(q_1, b, B_0, X)$ = $(q_1, Z_{1,B_0}, \text{remove})$
(8) $\delta(q_1, \lambda, \#_0, X)$ = $(q_2, \lambda, \text{keep})$

State q_0 enqueues for every symbol read a special symbol which stores the input symbol and state q_0. When symbol # is read, symbol $\#_0$ is enqueued and state q_1 is entered. From state q_1 it is checked whether the input symbols coincide with the information stored in the queue. Additionally, the history is stored by enqueuing symbols Z_{1,A_0} and Z_{1,B_0}, respectively. If the whole input is read and M sees symbols $\#_0$ at the front of the queue, the input is accepted and otherwise rejected. As in Example 5 the history of the computation is stored in the queue. Thus, the reversibility of M can easily be checked.   ∎

Example 8 Language $L = \{a^m b^n \mid m, n \geq 0\}$ is accepted by a realtime REV-DQA having two accepting states q_0 and q_1. In the initial state q_0, symbol A_0 is enqueued for every input symbol a. When the first b is read, another A_0 is enqueued and state q_1 is entered. Now, for every input symbol b symbol B_1 is put into the queue. Clearly, in this way language L is accepted. Since the history of the computation is stored in the queue, the computation is reversible. Note that L is not accepted by any reversible DFA [54].   ∎

Example 9 The language $\{a, b\}^*$ is accepted by a one-state reversible DFA, which can directly be simulated by a realtime REV-DQA without using the queue.   ∎

The first result on reversible queue automata reported here is that without any time restriction each deterministic queue automaton can be simulated by a reversible queue automaton. In particular this implies that general REV-DQA are as powerful as Turing machines. The rough idea of the simulation is as follows. The queue of the DQA is simulated in the queue of the REV-DQA. In addition, the latter one stores the whole history of the computation "at the end" of the queue. To this end, for the simulation of one step, the queue content has to be revolved entirely.

Theorem 28 *Let M be a deterministic queue automaton. Then there exists a reversible queue automaton accepting the language $L(M)$.*

Due to the computational universality of reversible queue automata it is of natural interest to impose time restrictions to the computations. As above, here we consider quasi-realtime and realtime computations. We start to examine the role played by λ-steps. As for general quasi-realtime queue automata, also reversible quasi-realtime queue automata can be sped up to realtime. The basic idea of the effective proof is to simulate a possibly empty sequence of λ-moves, the subsequent non-λ-step, and another possibly empty sequence of λ-moves following the non-λ-step at once. Special attention has to be paid for computations where a sequence of λ-steps appears after reading the last input symbol. These λ-steps may drive the queue automaton through accepting and rejecting states. So, one cannot simply simulate a sequence entirely, since the last state could be rejecting while predecessor states are accepting. A detailed proof of the following theorem is given in [38].

Theorem 29 *For every quasi-realtime REV-DQA an equivalent realtime REV-DQA can effectively be constructed.*

So, the family of languages accepted by quasi-realtime REV-DQA equals the family of languages accepted by realtime REV-DQA. This result is particularly used to show the closure of the family of accepted languages under inverse homomorphism. We continue with a more or less immediate result. In [54] it has been shown that there are regular languages which are not accepted by any reversible finite automaton. However, any regular language is accepted by some realtime REV-DQA. The apparent idea is to simulate a deterministic finite automaton in the finite control and to store its state history in the queue. The strictness of the next inclusion is witnessed, for example, by the non-regular language $\{\,a^n b^n \mid n \geq 1\,\}$.

Theorem 30 *The family of regular languages is strictly included in the family of languages accepted by reversible realtime queue automata.*

While the principal statement of the next result is obvious, one has to take care about the reversibility. In fact, irreversible regular languages cannot be accepted by REV-DQA with constantly bounded queue lengths. The construction of the next lemma is similar to a construction in [30], where general queue automata whose queue lengths are bounded by a constant are studied in detail.

Lemma 10 *The family of languages accepted by REV-DQA whose queue lengths are bounded by a constant is equal to the family of languages accepted by reversible deterministic finite automata.*

In the next part of this section, we compare reversible to general realtime queue automata. In particular it turns out that there are languages accepted by realtime DQA which cannot be accepted by realtime REV-DQA. To this end, a language L_{mcp} is exhibited which is accepted by a realtime DQA. Moreover, any REV-DQA accepting L_{mcp} needs $\Omega\left(\frac{n^2}{\log(n)}\right)$ time steps.

Example 10 We consider the regular language $L_{bin} = ((aa + a)(bb + b))^+$. Then the language

$$L_{mcp} = \{\, p\$w_1\$w_1\$w_2\$w_2\$ \cdots \$w_n\$w_n \mid p \in L_{bin}, n \geq 0, w_i \in \{a, b\}^* \,\}$$

is accepted by a realtime DQA. Informally, a DQA works as follows. The prefix up to the first $ can be tested without using the queue, because it belongs to a regular language. Then the first copy of each w_i is stored in the queue and subsequently compared and removed from the queue while reading the second copy. Whenever the second copy matches the first copy, the automaton enters an accepting state. Whenever a mismatch is detected, a non-accepting state is entered so that the input is rejected.   ∎

The proof of the lower bound on the time necessary for any REV-DQA to accept L_{mcp} uses the next lemma which shows that the queue length of a REV-DQA processing language L_{bin} has to be of linear size. The lemma is proven by means of Kolmogorov complexity and incompressibility arguments. General information on Kolmogorov complexity and the incompressibility method can be found, for example, in [44]. Let $w \in \{0, 1\}^+$ be an arbitrary binary string. The Kolmogorov complexity $C(w)$ of w is defined to be the minimal size of a binary program describing w. The following key argument for the incompressibility method is well known. There are binary strings w of *any* length such that $|w| \leq C(w)$.

Next, we encode words $w \in \{0, 1\}^+$ as follows. From left to right the digits are alternately represented by a's and b's such that a 0 is represented by a single letter and a 1 by a double letter. For example, the word 010110 is encoded as *abbabbaab*. Let $t(w)$ denote the code of w. Clearly, for any word $w \in \{0, 1\}^+$ of even length, its code $t(w)$ belongs to the regular language L_{bin}.

The idea of the proof of the next lemma is to choose a word $w \in \{0, 1\}^+$ of even length and long enough such that $C(w) \geq |w|$, consider an accepting computation on $t(w)$, and show that w can be compressed if L_{bin} is accepted by some REV-DQA that has a maximal queue length of order $o(|w|)$.

Lemma 11 *Any REV-DQA accepting L_{bin} has a queue length of $\Omega(|w|)$ for infinitely many inputs w.*

Now, the lower bound result can be derived with the help of Lemma 11.

Lemma 12 *Any REV-DQA accepting L_{mcp} has a time complexity of $\Omega\left(\frac{n^2}{\log(n)}\right)$.*

With these results the separation of the language families according to reversible and irreversible realtime DQA follows since language L_{mcp} from Example 10 is accepted by a realtime DQA, but Lemma 12 shows that any equivalent REV-DQA needs more than realtime.

Theorem 31 *The family of languages accepted by realtime REV-DQA is strictly included in the family of languages accepted by realtime DQA.*

Thus, under realtime conditions reversible queue automata are less powerful than general queue automata. Moreover, the lower bound result raises the question for the costs of simulating any realtime DQA, not necessarily reversible, by an equivalent REV-DQA. It turns out that quadratic time is sufficient for such simulations. The rough idea of the construction is to store information for each computation step of M in a history at the end of the queue. For the simulation of one step the queue content has to be revolved entirely.

Theorem 32 *Every realtime DQA can be simulated by a REV-DQA that needs at most quadratic time.*

Finally, we report two incomparability results showing that the language accepting capabilities of reversible realtime DQA are different from those of reversible pushdown automata [34] and input-driven queue automata. The following witness language $L_{rww} = \{a^m b^n \$ w \# w \mid m, n \geq 0, w \in \{a, b\}^* \}$ is also used below to derive closure properties.

It is shown in [54] that the regular language $\{a^m b^n \mid m, n \geq 0\}$ is not accepted by any reversible deterministic finite automaton. The same follows for the regular language $\{a^m b^n \$ \# \mid m, n \geq 0\}$. So, Lemma 10 implies that the length of the queue of some realtime reversible queue automaton accepting L_{rww} exceeds any given constant while processing some input prefixes of the form $a^* b^*$ that are long enough. Now the idea of the proof of the following lemma is that after processing such input prefix, some realtime REV-DQA cannot access the queue content stored while reading the first instance of w, when it has to read the second instance.

Lemma 13 *Language L_{rww} is not accepted by any realtime REV-DQA.*

Since the mirror language $\{w \# w^R \mid w \in \{a, b\}^+\}$ is accepted by a reversible pushdown automaton [34], but not by any even irreversible quasi-realtime queue automaton, and on the other hand, the non-context-free language $\{a^n b^n c^n \mid n \geq 1\}$ is accepted by some realtime REV-DQA by Example 6, we derive the following incomparability.

Theorem 33 *The family of languages accepted by realtime REV-DQA is incomparable with the family of languages accepted by reversible pushdown automata.*

A straightforward modification of Example 7 shows that $L = \{a^n \# a^n \mid n \geq 1\}$ is accepted by some realtime REV-DQA. Since for input-driven devices the input letter determines the operation on the storage media, one derives immediately a contradiction to the assumption that L is accepted by some input-driven queue automaton. Conversely, the latter can accept a slight modification of the witness language $L_{rww} = \{a^m b^n \$ w \# w \mid m, n \geq 0, w \in \{a, b\}^*\}$ of Lemma 13. The modification is to use a copy of the alphabet for the subword w following the #. Roughly speaking, using the copies of the symbols allows the input-driven queue automaton to store the first w and retrieve the second w. However, even the modified language cannot be accepted by any realtime REV-DQA.

Theorem 34 *The family of languages accepted by realtime REV-DQA is incomparable with the family of languages accepted by input-driven queue automata.*

Table 4 Closure properties of the classes of languages accepted by realtime REV-DQA, realtime DQA, reversible pushdown automata (REV-PDA), and deterministic pushdown automata (DPDA)

	$-$	$\cup$	$\cap$	$\cdot$	R	h_λ	h^{-1}	$\cap_{\text{REG}}$	$\cup_{\text{REG}}$
Realtime REV-DQA	+	−	−	−	−	−	+	−	−
Realtime DQA	+	−	−	−	−	−	+	+	+
REV-PDA	+	−	−	−	−	−	+	−	−
DPDA	+	−	−	−	−	−	+	+	+

Closure Properties

Here we compare the closure properties of the families of languages accepted by realtime REV-DQA, realtime DQA, reversible pushdown automata (REV-PDA), and deterministic pushdown automata (DPDA). The properties for pushdown automata have been shown in [34]. It turns out that realtime REV-DQA and REV-PDA as well as realtime DQA and DPDA have similar closure properties which are summarized in Table 4.

In order to show the closure of the family of languages accepted by realtime REV-DQA under complementation, first a given realtime REV-DQA M is modified so that it always reads the input entirely. The modification has to preserve the reversibility. To this end, M is simulated reversibly until it halts. Then the transition function is modified such that it maps to a special state instead of halting, whereby the current state is enqueued. Once in this special state, the remaining input is read successively while in each step a special symbol is enqueued. This modification preserves reversibility and, finally, interchanging accepting and non-accepting states yields a realtime REV-DQA that accepts the complement of $L(M)$.

Given a realtime REV-DQA M and a homomorphism $h : \Delta^* \to \Sigma^*$, basically, a realtime REV-DQA M' accepting $h^{-1}(L(M))$ works as follows. In every step it reads an input symbol, say a, and simulates M on $h(a)$. In particular, if $|h(a)| = 0$ then M leaves the queue unchanged. If $|h(a)| = 1$, it simulates the step of M on input symbol $h(a)$. The crucial part is the simulation for $|h(a)| > 1$. Let k be max $\{ |h(a)| \mid a \in \Delta \}$. Then M' has to simulate at most k steps of M at once. The construction now is as usual, where the queue alphabet of M' is grouped and some special registers of the state set are used as buffers. So, the family of languages accepted by realtime REV-DQA is closed under inverse homomorphism.

Next, we turn to non-closure results. The languages $L_1 = \{ ba^n ca^n b \mid n \geq 0 \}$ and $L_2 = \{ ba^n ba^m ca^m b \mid m, n \geq 0 \}$ are each accepted by some realtime REV-DQA. But it is shown in [15] that the union $L_1 \cup L_2$ is not even accepted by any realtime deterministic queue automaton. Since the family of languages accepted by REV-DQA is closed under complementation, we obtain non-closure under intersection as well.

The languages $L_1 = \{ a^m b^n \$ \mid m, n \geq 0 \}$ and $L_2 = \{ w\#w \mid w \in \{a, b\}^* \}$ are both accepted by some realtime REV-DQA due to Examples 8 and 7. But their concatenation gives the language $L_{rww} = \{ a^m b^n \$ w \# w \mid m, n \geq 0, w \in \{a, b\}^* \}$ which is not accepted by any realtime REV-DQA due to Lemma 13. Thus, we obtain non-closure under left concatenation with regular languages.

A combination and slight extension of the constructions of Examples 7 and 8 shows that the language $L_3 = \{ w \# w \$ b^n a^m \mid m, n \geq 0, w \in \{a, b\}^* \}$ is accepted by some REV-DQA in realtime. On the other hand, the reversal of L_3 gives the language L_{rww}. Thus, we obtain non-closure under reversal.

We consider the language $L_4 = \{ \{a, b\}^m \$ w \# w \mid m \geq 0, w \in \{a, b\}^* \}$ which is accepted by some realtime REV-DQA combining and extending the constructions of Examples 7 and 9. However, the intersection of L_4 with the regular language $\{ a^m b^n \$ \{a, b\}^i \# \{a, b\}^j \mid m, n, i, j \geq 0 \}$ is again L_{rww}. This shows the non-closure under intersection with regular languages. Since the family is closed under complementation, it cannot be closed under union with regular languages.

Finally, a straightforward construction shows that

$$L_5 = \{ a^m \& b^n \$ w \# w \mid m, n \geq 0, w \in \{a, b\}^* \}$$

is accepted by some realtime REV-DQA. Let h be a homomorphism defined by $h(a) = a$, $h(b) = b$, $h(\$) = \$$, $h(\#) = \#$, and $h(\&) = a$. Then the homomorphic image $h(L_5)$ equals L_{rww}, and the non-closure under λ-free homomorphism follows.

Decidability

As for general IDQA with an unbounded number of turns, realtime REV-DQA can accept a suitable encoding of accepting computations of linear bounded automata (see [38] for details of the encodings and of how the encodings are accepted). In this way the emptiness problem for linear bounded automata reduces to the emptiness problem of realtime REV-DQA.

Theorem 35 *Emptiness is not semidecidable for realtime REV-DQA.*

Further non-semidecidability results can be derived similarly as for Theorem 18.

Theorem 36 *Finiteness, infiniteness, universality, inclusion, equivalence, regularity, and context-freeness are not semidecidable for realtime REV-DQA.*

7 Queue Heads that May Dive

The difference between pushdown stores and stack stores is that in the latter model the head may move from the top of the pushdown store into the pushdown in read-only mode. Push and pop operations are only possible if the head is at the top again. This extension has also been considered for queue automata [7], in particular with the additional constraints of quasi-realtime and finite-turn boundedness. As for reversible queue automata the transition functions considered here depend also on the symbol at the end of the queue.

In analogy to stack automata, the two queue heads from the front and from the end of the queue may move into the queue independently without altering the contents. Additionally, the automata can neither append symbols to the end of the queue nor remove symbols from the front of the queue while at least one head is inside the queue. In this way, it is possible to read but not to change the stored information. Such devices are called *diving queue automata*. To enable this behavior, the symbol at the front of the queue is marked by index $\triangleright$, the symbol at the end of the queue by index $\triangleleft$, and a single symbol in the queue by index $\bowtie$.

In order to define the transition function of a *deterministic diving queue automaton* (DDQA) with queue symbols from a set Γ, for $x \in \{\triangleright, \triangleleft, \bowtie\}$, the set Γ_x is defined to be a marked copy of Γ. That is, $\Gamma_x = \{a_x \mid a \in \Gamma\}$. Let $\bar{\Gamma}$ be the union $\Gamma \cup \Gamma_{\triangleright} \cup \Gamma_{\triangleleft} \cup \Gamma_{\bowtie}$. Then the possibly partial *transition function* δ maps from $Q \times \Sigma_\lambda \times ((\bar{\Gamma} \times \bar{\Gamma}) \cup (\bot, \bot))$ to $Q \times \{\texttt{remove}, -1, 0, +1\} \times (\Gamma \cup \{-1, 0, +1\})$, where $\texttt{remove}$ means to remove the symbol at the front of the queue, Γ is a symbol to be entered at the end of the queue, -1 means to move the head one square to the left, 0 means to keep it at the current square, and $+1$ means to move it one square to the right. Whenever $\delta(p, x, z_1, z_2) = (q, \alpha_1, \alpha_2)$ is defined, then

1. $\alpha_1 \in \{\texttt{remove}\}$ only if $(z_1, z_2) \in (\Gamma_{\triangleright} \times \Gamma_{\triangleleft}) \cup (\Gamma_{\bowtie} \times \Gamma_{\bowtie})$,
2. $\alpha_2 \in \Gamma$ only if $(z_1, z_2) \in (\Gamma_{\triangleright} \times \Gamma_{\triangleleft}) \cup (\Gamma_{\bowtie} \times \Gamma_{\bowtie}) \cup (\{\bot\} \times \{\bot\})$.

In particular, there must never be a choice of using an input symbol or of using λ input. So, it is required that for all p in Q and z_1, z_2 in $\bar{\Gamma} \cup \{\bot\}$: if $\delta(p, \lambda, z_1, z_2)$ is defined, then $\delta(p, a, z_1, z_2)$ is undefined for all a in Σ. Note that for easier writing the ordering of the components of the result of the transition function has been changed. Now the second component refers to the action at the front of the queue while the third component refers to the action at the end of the queue (see Fig. 4).

A *configuration* of a DDQA is a tuple (u, q, v, s, p_1, p_2), where $q \in Q$ is the current state, $u \in \Sigma^*$ is the already read and $v \in \Sigma^*$ is the unread part of the input, and $s \in \Gamma_{\triangleright} \Gamma^* \Gamma_{\triangleleft} \cup \Gamma_{\bowtie} \cup \{\lambda\}$ is the current queue content, where the leftmost symbol is at the front, and $0 \leq p_1, p_2 \leq |s| + 1$ are the current positions of the queue heads, where it is understood that a head at position 0 or $|s| + 1$ is out of the queue and scans nothing.

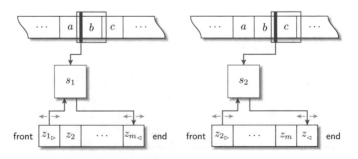

Fig. 4 Successive configurations of a DDQA, where $\delta(s_1, b, z_{1_{\triangleright}}, z_{m_{\triangleleft}}) = (s_2, \texttt{remove}, z)$

In order to implement the marking of the queue contents a mapping φ is used, that maps a string of two or more symbols from $\bar{\Gamma}^+$ to a string from $\Gamma_\triangleright \Gamma^* \Gamma_\triangleleft$ by keeping the symbols and adjusting the indices appropriately. A string of length one is mapped to the corresponding symbol from $\Gamma_\bowtie$.

The *initial configuration* for input w is set to $(\lambda, q_0, w, \lambda, 0, 0)$. Let $q, q' \in Q$, $a \in \Sigma_\lambda$, $u, v \in \Sigma^*$, $z \in \Gamma_\bowtie$, $z_1 z_2 \cdots z_m \in \Gamma_\triangleright \Gamma^* \Gamma_\triangleleft$, for $m \geq 2$, $1 \leq p_1, p_2 \leq m$, $z' \in \Gamma$, and $d_1, d_2 \in \{-1, 0, 1\}$. We set

1. $(u, q, av, \lambda, 0, 0) \vdash (ua, q', v, \lambda, 0, 0)$, if $\delta(q, a, \perp, \perp) = (q', 0, 0)$,
2. $(u, q, av, \lambda, 0, 0) \vdash (ua, q', v, z'_\bowtie, 1, 1)$, if $\delta(q, a, \perp, \perp) = (q', 1, z')$,
3. $(u, q, av, z, 1, 1) \vdash (ua, q', v, z'_\bowtie, 1, 1)$, if $\delta(q, a, z, z) = (q', \texttt{remove}, z')$,
4. $(u, q, av, z, 1, 1) \vdash (ua, q', v, \lambda, 0, 0)$, if $\delta(q, a, z, z) = (q', \texttt{remove}, 0)$,
5. $(u, q, av, z, 1, 1) \vdash (ua, q', v, \varphi(zz'), 1 + d_1, 2)$, if $\delta(q, a, z, z) = (q', d_1, z')$,
6. $(u, q, av, z, 1, 1) \vdash (ua, q', v, z, 1, 1)$, if $\delta(q, a, z, z) = (q', 0, 0)$,
7. $(u, q, av, z_1 z_2 \cdots z_m, 1, m) \vdash (ua, q', v, \varphi(z_2 z_3 \cdots z_m z'), 1, m)$,
   $\qquad$ if $\delta(q, a, z_1, z_m) = (q', \texttt{remove}, z')$,
8. $(u, q, av, z_1 z_2 \cdots z_m, 1, m) \vdash (ua, q', v, \varphi(z_2 z_3 \cdots z_m), 1, m - 1 + d_2)$,
   $\qquad$ if $\delta(q, a, z_1, z_m) = (q', \texttt{remove}, d_2)$,
9. $(u, q, av, z_1 z_2 \cdots z_m, 1, m) \vdash (ua, q', v, \varphi(z_1 z_2 \cdots z_m z'), 1 + d_1, m + 1)$,
   $\qquad$ if $\delta(q, a, z_1, z_m) = (q', d_1, z')$,
10. $(u, q, av, z_1 z_2 \cdots z_m, p_1, p_2) \vdash (ua, q', v, z_1 z_2 \cdots z_m, p_1 + d_1, p_2 + d_2)$,
    $\qquad$ if $\delta(q, a, z_{p_1}, z_{p_2}) = (q', d_1, d_2)$.

For all other situations, $\vdash$ is undefined and the DDQA *halts*.

In [7] two further modes of moving the queue heads into the queue are distinguished. If only the head from the front of the queue may move inside, the devices are called *deterministic front diving queue automata* (DfDQA), and if only the head from the tail of the queue may move inside, we call the devices *deterministic tail diving queue automata* (DtDQA). It is straightforward to see that both types of devices have the same computational capacity. For example, a DfDQA can be simulated by a DtDQA as follows. Whenever the DfDQA moves its front queue head into the queue, the DtDQA moves its tail queue head through the queue to the opposite end, and subsequently simulates the DfDQA until the head is moved to the front of the queue again. Then the DtDQA moves the head back through the entire queue and the simulation continues. Similarly, a DtDQA can be simulated by a DfDQA. So, we may use the abbreviation D1DQA for DfDQA as well as for DtDQA.

For DDQA, the maximal number of λ-steps on each square of the input tape that can be performed *when the queue heads are not inside the queue* can be limited. A DDQA is said to be *weakly quasi-realtime* if there is a constant that bounds this number for all computations. The DDQA is said to be *weakly realtime* if this constant is 0, that is, if there are no such λ-steps at all.

The results reported in this section are from [7].

Example 11 The Gladkij language $= \{ w \$ w^R \$ w \mid w \in \{a, b\}^* \}$ is accepted by a deterministic tail diving queue automaton M_1 with no turns. M_1 reads the input up to the first $\$$ and enqueues every symbol read. Thus w is stored in the queue. Then M_1

reads the second part of the input w^R while it checks with its tail head from the tail of the queue to the front symbolwise whether the current symbol of the input is the same as the symbol stored in the queue. If there is a mismatch, M_1 rejects. Otherwise, M_1 reads the second \$ and then moves its tail head from the front of the queue back to the tail while checking in the same way that the last part of the input equals w. The Gladkij language is known to be not growing context-sensitive [14]. and, hence, is also not a Church–Rosser language [47].   ∎

Example 12 The language $\{(ba^n)^n \mid n \geq 1\}$ is accepted by a deterministic diving queue automaton M_2 using no turn. Note that this language is not an indexed language [1, 18] and, thus, is not accepted by any nested stack automaton [2].   ∎

Example 13 The non-semilinear language $\{a^{n^2} \mid n \geq 0\}$ can be accepted by a deterministic tail diving queue automaton M_3 with no turns. The idea of the construction is based on the fact that n^2 can be represented as $n^2 = 1 + 3 + 5 + \cdots + (2n - 1)$. The principal idea is that M_3 starts with one A in the queue and then iteratively performs the following phases: M_3 uses its tail head to read for every A in the queue an a in the input, the tail head returns to the tail of the queue, and two additional A's are enqueued. In this way, M_3 reads one a in the first phase, a^3 in the second phase, and in general a^{2i-1} in the ith phase. Altogether, $1 + 3 + 5 + \cdots + (2i - 1) = i^2$ many a's have been read after the ith phase.   ∎

Two-Party Diving is Better Than Lonesome Diving

Here we examine the computational capacity of diving queue automata induced by the number of heads that may dive. In particular, a hierarchy on the number of heads is obtained using the unary witness languages $\{a^{n^2} \mid n \geq 0\}$ and $\{a^{n^3} \mid n \geq 0\}$. It turns out that the first language shows that one head is better than no head, whereas the latter language gives that two heads are more powerful than one head.

The next technical result settles the role played by λ-steps when the queue heads are *not* inside the queue. It is the same as for quasi-realtime and realtime computations.

Theorem 37 *For every weakly quasi-realtime diving queue automaton an equivalent weakly realtime diving queue automaton can effectively be constructed.*

The next result concerns finite-turn D1DQA that accept unary languages. By Example 13 the non-semilinear unary language $\{a^{n^2} \mid n \geq 1\}$ is accepted by some D1DQA$_0$. On the other hand, it is shown in [36] that finite-turn DQA can only accept semilinear languages.

Theorem 38 *The family of languages accepted by finite-turn DQA is properly included in the family of languages accepted by finite-turn D1DQA.*

In order to derive an upper bound for the computational power of D1DQA accepting unary languages, these devices are simulated by *deterministic one-way stack automata* (DSA) (details of the simulation can be found in [7]). Basically, a one-way stack automaton is a conventional pushdown automaton with the additional ability to move some head starting from the top inside the pushdown store and to read the contents of the pushdown store. As it is the case for diving automata, the contents of the pushdown store cannot be changed when the head is inside the store. Details on this model are given in [20]. One variant of DSA are deterministic *non-erasing* stack automata (DNESA) which can never pop symbols from the pushdown store. We will later consider also their nondeterministic variants abbreviated by NSA and NNESA.

Theorem 39 *Let $k \geq 0$ and M be a $D1DQA_k$ accepting a unary language. Then an equivalent DNESA can effectively be constructed.*

In fact, when the number of turns for D1DQA is reduced to zero, the $D1DQA_0$ can be simulated by some DNESA even for languages over arbitrary alphabets. Even better, the converse is also true.

Theorem 40 *The families of languages accepted by DNESA and $D1DQA_0$ coincide.*

So, in particular for unary languages the $D1DQA_k$ can be simulated by DNESA which in turn can be simulated by $D1DQA_0$. This implies that $D1DQA_k$ and $D1DQA_0$ are equally powerful on unary languages, and we derive that there is no turn hierarchy for unary D1DQA, that is, no turn is as good as any finite number of turns.

Corollary 3 *Let $k \geq 0$ be a constant. For any $D1DQA_k$ accepting a unary language an equivalent $D1DQA_0$ can effectively be constructed.*

By results in [51] it follows that the language $L = \{ a^{n^3} \mid n \geq 0 \}$ is not accepted by any DSA. By Theorem 39 we know that L is not accepted by any finite-turn D1DQA either. The next theorem is proved by showing that L is accepted by some $DDQA_0$.

Theorem 41 *The family of languages accepted by finite-turn D1DQA is a proper subset of the family of languages accepted by finite-turn DDQA.*

Computational Complexity of Diving Queue Automata

Now we discuss the computational complexity of the membership problem for families of languages accepted by diving queue automata in more detail. We have already seen that finite-turn DDQA can accept languages that are neither accepted by stack automata nor by nested stack automata.

The construction in the proof of Theorem 40 shows the mutual simulations of DNESA and $D1DQA_0$. However, it reveals even more, the mutual simulations are possible step by step. So, any time constraint obeyed by the DNESA is obeyed by the $D1DQA_0$ and vice versa. In [40] it is shown how a weakly quasi-realtime DNESA accepts some encodings of the P-complete *monotone circuit value problem*. So, by Theorems 40 and 37 it follows that a P-complete language is accepted by some weakly realtime $D1DQA_0$.

Theorem 42 *The membership problem for the family of languages accepted by weakly realtime $D1DQA_0$ is P-hard.*

In order to derive an upper bound for the complexity of the membership problem, the question arises to what extent the conditions of working in weakly realtime or to be turn bounded can be relaxed such that the problem is still in P.

Theorem 43 *The family of languages accepted by weakly quasi-realtime DDQA is included in the complexity class P.*

So, the condition of being turn bounded can be relaxed as long as the devices work in weakly realtime. In particular, this is even true for weakly quasi-realtime DDQA. The next theorem shows that the realtime condition can be relaxed for D1DQA as long as they are turn bounded.

Theorem 44 *Let $k \geq 0$ be a constant. The family of languages accepted by $D1DQA_k$ is included in the complexity class P.*

Clearly, a further relaxation of conditions for D1DQA leads to a language family no longer included in P, since even DQA are capable of universal compuations. Dropping the realtime conditions for DDQA and imposing a turn bound at the same time seems to lead out of P as well. The next example shows that a $DDQA_0$ can halt after performing an exponential number of steps on unary input.

Example 14 Let M be the $DDQA_0$ with unary input alphabet $\{a\}$, whose behavior is sketched as follows. Basically, M reads one input symbol and then runs through a phase of λ-steps. In this phase, it uses its two queue heads to simulate a deterministic two-way 2-head finite automaton. These devices can accept the unary language $L = \{a^{2^n} \mid n \geq 0\}$ (see, for example, [31]). In detail, M enters a symbol to the queue and, subsequently simulates a deterministic two-way 2-head finite automaton accepting L. If the length of the queue is a power of two (and, thus, the simulation ends accepting), M finishes the phase. Otherwise, it enters another symbol to the queue and simulates the 2-head finite automaton again, and so on until the queue length is a power of two. Then the behavior is repeated with the next input symbol. So, when M halts since the input a^n has been read entirely, the length of the queue is 2^n. We conclude that M performs an exponential number of steps before halting. ∎

Theorems 42, 43, and 44 imply that the membership problems for the families of languages accepted by $D1DQA_k$, as well as the family of languages accepted by weakly realtime DDQA are P-complete.

Diving and Decidability

Finally, we consider decidability questions for finite-turn diving queue automata. One of the main results is that in case of two queue heads and no turns the question of emptiness is undecidable and, thus, almost nothing is decidable. On the other hand, if only one head is available it is possible to identify some subclasses for which at least the questions of emptiness and finiteness are decidable.

We know from above that the commonly studied decidability questions such as emptiness, finiteness, universality, inclusion, and equivalence are not semidecidable for DQA, whereas the questions of emptiness and finiteness are decidable for finite-turn DQA. Since the language family accepted by DQA (with an unbounded number of turns) is contained in those accepted by D1DQA and DDQA, we derive that all above-mentioned decidability questions are not semidecidable for D1DQA and DDQA as well.

Corollary 4 *Emptiness, finiteness, infiniteness, universality, inclusion, equivalence, regularity, and context-freeness are not semidecidable for D1DQA as well as for DDQA.*

Thus, we now consider the decidability status for the remaining models, namely $DDQA_k$ and $D1DQA_k$. However, all above-mentioned decidability questions are not semidecidable for $DDQA_k$, whereas it turns out that the problems of emptiness and finiteness are decidable for several subclasses of $D1DQA_k$.

Again, the valid computations of Turing machines are the method of choice to prove the undecidability results. It is shown in [27] how a one-way two-head deterministic finite automaton M accepting the valid computations of a given Turing machine can effectively be constructed. Now, a $DDQA_0$ M' can accept the same valid computations as follows. First, the input is stored in the queue. Second, the 2-head automaton M is simulated by M' with the help of its two queue heads. Finally, M' accepts if M accepts and rejects otherwise. Clearly, M' performs no turn and thus is a $DDQA_0$.

Lemma 14 *Let T be a Turing machine. Then a $DDQA_0$ accepting the valid computations of T can effectively be constructed.*

This lemma yields the following non-semidecidability results.

Theorem 45 *Let $k \geq 0$ be a constant, M be a $DDQA_k$, and M' be a DDQA. Then emptiness, finiteness, universality, inclusion, and equivalence are not semidecidable for M. Moreover, it is not semidecidable whether or not M' is a $DDQA_k$.*

Next, we report positive decidability results for subclasses of finite-turn D1DQA, since it turns out to be possible to simulate these by nondeterministic one-way stack automata for which it is known that emptiness and finiteness are decidable. It is currently an open problem whether these positive results can be extended to the whole class of finite-turn D1DQA.

By Theorem 39, $D1DQA_k$ accepting unary languages can be simulated by non-erasing stack automata. Thus, we immediately obtain the decidability of emptiness and finiteness for such $D1DQA_k$, since both questions are decidable for one-way nondeterministic stack automata [20, 51].

Theorem 46 *Let $k \geq 0$ be a constant. Then emptiness and finiteness are decidable for $D1DQA_k$ accepting a unary language.*

We will now introduce two additional subclasses for which emptiness and finiteness are decidable as well. A $D1DQA_k$ processes its input in $2k + 1$ phases alternating between enqueuing and dequeuing of symbols. We will call such phases *enqueuing* phases and *dequeuing* phases, respectively. Within such a phase, the queue head may arbitrarily often enter the queue and move inside the queue, hence performing arbitrarily many queue head reversals, before it returns to either end of the queue and another queue operation takes place. Let $\ell \geq 0$ be a constant. Then we call a $D1DQA_k$ (∞, ℓ)-bounded $((\infty, \ell)-D1DQA_k)$, if the number of queue head reversals in every enqueuing phase may be unbounded, but the number of queue head reversals in every dequeuing phase is bounded by ℓ. Similarly, a $D1DQA_k$ is called (ℓ, ∞)-bounded $((\ell, \infty)-D1DQA_k)$, if the number of queue head reversals in every dequeuing phase may be unbounded, but the number of queue head reversals in every enqueuing phase is bounded by ℓ. Both classes of $D1DQA_k$ can be simulated by nondeterministic stack automata and hence share their decidable emptiness and finiteness problems.

Theorem 47 *Let $k, \ell \geq 0$ be constants. Then the family of languages accepted by $(\infty, \ell)-D1DQA_k$ is included in the family of languages accepted by nondeterministic non-erasing stack automata.*

Corollary 5 *Let $k, \ell \geq 0$ be constants and M be a $(\infty, \ell)-D1DQA_k$. Then the emptiness and finiteness of $L(M)$ are decidable.*

A similar simulation result can be obtained for $(\ell, \infty)-D1DQA_k$. However, the property of the simulating stack automaton to be non-erasing gets lost.

Theorem 48 *Let $k, \ell \geq 0$ be constants. Then the family of languages accepted by $(\ell, \infty)-D1DQA_k$ is included in the family of languages accepted by nondeterministic stack automata.*

Corollary 6 *Let $k, \ell \geq 0$ be constants and M be a $(\ell, \infty)-D1DQA_k$. Then the emptiness and finiteness of $L(M)$ are decidable.*

Finally, we have that the two restrictions of $D1DQA_k$ which let to decidable emptiness and finiteness problems are decidable themselves. For example, for a given $D1DQA_k$ M and some fixed $\ell \geq 0$, we construct a $D1DQA_k$ M' that accepts all inputs that cause M to perform more than ℓ queue head reversals in some dequeuing phase (regardless of whether M accepts or rejects). This can be realized by simulating M while counting in an additional component of its state set the number of queue head reversals executed in the current phase so far. Automaton M' accepts an input as soon as it would require more than ℓ queue head reversals in one phase. Otherwise, the input is rejected. Clearly, M' is a $(\infty, \ell)-D1DQA_k$ and accepts the empty set if and only if M is a $(\infty, \ell)-D1DQA_k$ as well. Now, the decidability follows from the decidability of emptiness for $(\infty, \ell)-D1DQA_k$.

Theorem 49 *Let $k, \ell \geq 0$ be constants and M be a $D1DQA_k$. Then it is decidable whether M is a $(\infty, \ell)-D1DQA_k$ and whether M is an $(\ell, \infty)-D1DQA_k$.*

References

1. Aho, A.V.: Indexed grammars - an extension of context-free grammars. J. ACM **15**, 647–671 (1968)
2. Aho, A.V.: Nested stack automata. J. ACM **16**, 383–406 (1969)
3. Alur, R., Madhusudan, P.: Visibly pushdown languages. In: Symposium on Theory of Computing (STOC 2004), pp. 202–211. ACM (2004)
4. Alur, R., Madhusudan, P.: Adding nesting structure to words. J. ACM **56**, 1–43 (2009)
5. Angluin, D.: Inference of reversible languages. J. ACM **29**, 741–765 (1982)
6. Autebert, J.M., Berstel, J., Boasson, L.: Context-free languages and pushdown automata. In: Handbook of Formal Languages, vol. 1, chap. 3, pp. 111–174. Springer (1997)
7. Beier, S., Kutrib, M., Malcher, A., Wendlandt, M.: Diving into the queue. In: Bordihn, H., Freund, R., Nagy, B., Vaszil, G. (eds.) Workshop on Non-Classical Models of Automata and Applications (NCMA 2016), books@ocg.at. Austrian Computer Society 2016, vol. 321, pp. 89–104 (2016)
8. Bennett, C.H.: Logical reversibility of computation. IBM J. Res. Dev. **17**, 525–532 (1973)
9. Book, R.V., Greibach, S.A., Wrathall, C.: Reset machines. J. Comput. Syst. Sci. **19**, 256–276 (1979)
10. Brandenburg, F.: Multiple equality sets and Post machines. J. Comput. Syst. Sci. **21**, 292–316 (1980)
11. Brandenburg, F.J.: Intersections of some families of languages. In: International Colloquium on Automata, Languages and Programming (ICALP 1986). LNCS, vol. 226, pp. 60–68. Springer (1986)
12. Brandenburg, F.J.: On the intersection of stacks and queues. Theor. Comput. Sci. **58**, 69–80 (1988)
13. Breveglieri, L., Cherubini, A., Crespi-Reghizzi, S.: Real-time scheduling by queue automata. In: Formal Techniques in Real-Time and Fault-Tolerant Systems. LNCS, vol. 571, pp. 131–147. Springer (1992)
14. Buntrock, G., Otto, F.: Growing context-sensitive languages and Church-Rosser languages. Inform. Comput. **141**, 1–36 (1998)
15. Cherubini, A., Citrini, C., Crespi-Reghizzi, S., Mandrioli, D.: QRT FIFO automata, breadth-first grammars and their relations. Theor. Comput. Sci. **85**, 171–203 (1991)
16. Chomsky, N.: On certain formal properties of grammars. Inform. Control **2**, 137–167 (1959)
17. Dymond, P.W.: Input-driven languages are in log n depth. Inform. Process. Lett. **26**, 247–250 (1988)
18. Gilman, R.H.: A shrinking lemma for indexed languages. Theor. Comput. Sci. **163**, 277–281 (1996)
19. Ginsburg, S., Spanier, E.H.: Finite-turn pushdown automata. SIAM J. Contr. **4**, 429–453 (1966)
20. Ginsburg, S., Greibach, S.A., Harrison, M.A.: One-way stack automata. J. ACM **14**, 389–418 (1967)
21. Ginsburg, S., Greibach, S.A., Harrison, M.A.: Stack automata and compiling. J. ACM **14**, 172–201 (1967)
22. Greibach, S.A.: An infinite hierarchy of context-free languages. J. ACM **16**, 91–106 (1969)
23. Hartmanis, J.: Context-free languages and Turing machine computations. Proc. Symp. Appl. Math. **19**, 42–51 (1967)
24. Holzer, M., Kutrib, M.: Flip-pushdown automata: $k + 1$ pushdown reversals are better than k. In: International Colloquium on Automata, Languages and Programming (ICALP 2003). LNCS, vol. 2719, pp. 490–501. Springer (2003)
25. Holzer, M., Kutrib, M.: Flip-pushdown automata: nondeterminism is better than determinism. In: Developments in Language Theory (DLT 2003). LNCS, vol. 2710, pp. 361–372. Springer (2003)
26. Holzer, M., Kutrib, M.: Descriptional complexity – an introductory survey. In: Scientific Applications of Language Methods, pp. 1–58. Imperial College Press (2010)

27. Holzer, M., Kutrib, M., Malcher, A.: Complexity of multi-head finite automata: origins and directions. Theor. Comput. Sci. **412**, 83–96 (2011)
28. Hopcroft, J.E., Ullman, J.D.: Introduction to Automata Theory, Languages, and Computation. Addison-Wesley, Reading (1979)
29. Ibarra, O.H.: Reversal-bounded multicounter machines and their decision problems. J. ACM **25**, 116–133 (1978)
30. Jakobi, S., Meckel, K., Mereghetti, C., Palano, B.: Queue automata of constant length. In: Descriptional Complexity of Formal Systems (DCFS 2013). LNCS, vol. 8031, pp. 124–135. Springer (2013)
31. Kutrib, M.: On the descriptional power of heads, counters, and pebbles. Theor. Comput. Sci. **330**, 311–324 (2005)
32. Kutrib, M.: Aspects of reversibility for classical automata. In: Computing with New Resources. LNCS, vol. 8808, pp. 83–98. Springer (2014)
33. Kutrib, M., Malcher, A.: Fast reversible language recognition using cellular automata. Inform. Comput. **206**, 1142–1151 (2008)
34. Kutrib, M., Malcher, A.: Reversible pushdown automata. J. Comput. Syst. Sci. **78**, 1814–1827 (2012)
35. Kutrib, M., Malcher, A.: One-way reversible multi-head finite automata. Theor. Comput. Sci **682**, 149 (2017)
36. Kutrib, M., Malcher, A., Mereghetti, C., Palano, B., Wendlandt, M.: Deterministic input-driven queue automata: Finite turns, decidability, and closure properties. Theor. Comput. Sci. **578**, 58–71 (2015)
37. Kutrib, M., Malcher, A., Wendlandt, M.: Input-driven queue automata with internal transductions. In: Language and Automata Theory and Applications (LATA 2016). LNCS, vol. 9618, pp. 156–167. Springer (2016)
38. Kutrib, M., Malcher, A., Wendlandt, M.: Reversible queue automata. Fund. Inform. (2017)
39. Landauer, R.: Irreversibility and heat generation in the computing process. IBM J. Res. Dev. **5**, 183–191 (1961)
40. Lange, K.J.: A note on the P-completeness of deterministic one-way stack language. J. UCS **16**, 795–799 (2010)
41. Lange, K.J., McKenzie, P., Tapp, A.: Reversible space equals deterministic space. J. Comput. Syst. Sci. **60**, 354–367 (2000)
42. Li, M.: Simulating two pushdown stores by one tape in $O(n^{1.5}\sqrt{rmlogn})$ time. J. Comput. Syst. Sci. **37**, 101–116 (1988)
43. Li, M., Vitányi, P.M.B.: Tape versus queue and stacks: the lower bounds. Inform. Comput. **78**, 56–85 (1988)
44. Li, M., Vitányi, P.M.B.: An Introduction to Kolmogorov Complexity and Its Applications. Springer, Berlin (1993)
45. Li, M., Longpré, L., Vitányi, P.M.B.: The power of the queue. SIAM J. Comput. **21**, 697–712 (1992)
46. Manna, Z.: Mathematical Theory of Computation. McGraw-Hill, New York (1974)
47. McNaughton, R., Narendran, P., Otto, F.: Church-Rosser Thue systems and formal languages. J. ACM **35**, 324–344 (1988)
48. Mehlhorn, K.: Pebbling moutain ranges and its application of DCFL-recognition. In: International Colloquium on Automata, Languages and Programming (ICALP 1980). LNCS, vol. 85, pp. 422–435. Springer (1980)
49. Morita, K.: Reversible computing and cellular automata - a survey. Theor. Comput. Sci. **395**, 101–131 (2008)
50. Morita, K.: Two-way reversible multi-head finite automata. Fund. Inform. **110**, 241–254 (2011)
51. Ogden, W.F.: Intercalation theorems for stack languages. In: Symposium on Theory of Computing (STOC 1969), pp. 31–42. ACM Press (1969)
52. Petersen, H.: Gegenseitige Simulation von Datenstrukturen. Habilitationsschrift (in German), Fakultät Informatik, Elektrotechnik und Informationstechnik, Universität Stuttgart (2002)

53. Petersen, H., Robson, J.M.: Efficient simulations by queue machines. SIAM J. Comput. **35**, 1059–1069 (2006)
54. Pin, J.E.: On reversible automata. In: Latin 1992: Theoretical Informatics. LNCS, vol. 583, pp. 401–416. Springer (1992)
55. Post, E.L.: Formal reductions of the classical combinatorial decision problem. Am. J. Math. **65**, 197–215 (1943)
56. Sénizergues, G.: $L(A) = L(B)$? decidability results from complete formal systems. Theor. Comput. Sci. **251**, 1–166 (2001)
57. Sudborough, I.H.: Computation by multihead writing finite automata. Ph.D. thesis, Pennsylvania State University (1971)
58. Sudborough, I.H.: One-way multihead writing finite automata. Inform. Control **30**, 1–20 (1976)
59. Valiant, L.G.: The equivalence problem for deterministic finite-turn pushdown automata. Inform. Control **25**, 123–133 (1974)
60. Vollmar, R.: Über einen Automaten mit Pufferspeicherung. Computing **5**, 57–70 (1970)
61. von Braunmühl, B., Verbeek, R.: Input-driven languages are recognized in log n space. In: Topics in the Theory of Computation. Mathematics Studies, vol. 102, pp. 1–19. North-Holland (1985)

Small Universal Reversible Counter Machines

Artiom Alhazov, Sergey Verlan and Rudolf Freund

Abstract A k-counter machine (CM(k)) is an automaton with k counters as an auxiliary memory. It is known that CM(k) are universal for $k \geq 2$. As shown by Morita reversible CM(2) are universal. Based on results from Korec we construct four small universal reversible counter machines highlighting different trade-offs: (10, 109, 129), (11, 227, 270), (9, 97, 116) and (2, 1097, 1568), where in parentheses we indicated the number of counters, states and instructions, respectively. Since counter machines are used in many areas, our results can be the starting point for corresponding reversible universal constructions.

1 Introduction

Universality is a fundamental concept in the theory of computation. The question of finding a universal computing device in the class of Turing machines was originally proposed by A. Turing himself in [17]. A universal Turing machine would be capable of simulating any other Turing machine $\mathcal{T}$: given a description of $\mathcal{T}$ and the encoding of the input tape contents, the universal machine would halt with tape contents which would correspond to the encoding of the output of $\mathcal{T}$ for the supplied input.

In a more general setting of an arbitrary class $\mathfrak{C}$ of computing devices, the universality problem consists in finding such a *fixed* element $\mathcal{M}_0$ which would be able to simulate any other element $\mathcal{M} \in \mathfrak{C}$. More formally, if the result of running $\mathcal{M}$ with

A. Alhazov
Institute of Mathematics and Computer Science, Academy of Sciences
of Moldova, Str. Academiei 5, 2028 Chicsinău, Moldova
e-mail: artiom@math.md

S. Verlan (✉)
Université Paris Est, LACL (EA 4219), UPEC, 94010 Créteil, France
e-mail: verlan@u-pec.fr

R. Freund
Faculty of Informatics, Vienna University of Technology,
Favoritenstr. 9, 1040 Vienna, Austria
e-mail: rudi@emcc.at

© Springer International Publishing AG 2018
A. Adamatzky (ed.), *Reversibility and Universality*, Emergence, Complexity
and Computation 30, https://doi.org/10.1007/978-3-319-73216-9_20

433

the input x is y (usually written as $\mathcal{M}(x) = y$), then $y = f(\mathcal{M}_0(g(\mathcal{M}'), h(x))$, where g is the function enumerating $\mathfrak{C}$, while f and h are the decoding and encoding functions respectively. We remark that in some cases it can be possible to have $\mathcal{M}_0 \in \mathfrak{C}$; then the input is the couple encoding of $g(\mathcal{M}')$ and x (e.g. using the Cantor pairing function). It is generally agreed that f and h should not be "too" complicated. Since it is relatively common to rely on exponential coding when working with devices computing numbers, the functions $f(x) = \log_a(x)$ and $h(x) = b^x$, for some $a, b \in \mathbb{N}$ are often used (cf. [9, 19]).

In this paper, we will adhere to the terminology established by I. Korec in [7] and call the element $\mathcal{M}_0$ defined as above *weakly* universal (or just universal). In case the functions f and h are additionally required to be identities, $\mathcal{M}_0$ will be referred to as *strongly* universal. Hence, the strong universality permits to capture the situations when the encoding does not alter the power of the device. For example, 2-register machines are weakly universal [9], but they cannot be strongly universal as they cannot compute even the square function [3, 15].

As a further development on the question of universality, C. Shannon [16] considered finding the *smallest* possible universal Turing machine, where the size is essentially given by the sizes of the alphabets of symbols and states. A series of important results concerning this direction were obtained [8, 14, 18]. For an overview of the recent results the reader is referred to [13]. Small universal devices are of considerable theoretical importance since they indicate the minimal choice ingredients sufficient for achieving computational completeness.

A k-counter machine $(CM(k))$ is an automaton with k counters that can hold non-negative values. In one step, the finite-state control of $CM(k)$ can increment or decrement the contents by one or test whether it is zero or not. A related model is the register machine [9], having eventual restrictions on the form of the control and a richer set of potential instructions. However, the common variant of register machines is almost identical to CMs. Register machines and hence CMs were shown to be universal and it is known that already three registers/counters suffice for strong universality and two for the weak one [6, 9].

In [10] K. Morita studied reversible CMs and showed the universality of reversible $CM(2)$. These machines are backward deterministic, i.e., each configuration has at most one predecessor. This research lines up in the study of other reversible systems such as reversible Turing machines, reversible cellular automata and reversible logic gates, see [5, 11] for a general survey. We remark that in case of reversible machines, the notion of universality is slightly different than in the classical case [2].

In 1996, I. Korec described a number of universal register machines with considerably fewer instructions than were known to be needed for universality before [7]. Based on this result small 2- and 3-register machines were constructed [1].

In this paper we consider the construction of small universal reversible counter machines. As in [1, 7, 12] we are mainly interested in the number of instructions as well as in the trade-offs between this number an the number of counters. We construct four small universal reversible counter machines highlighting different trade-offs: (10, 109, 129), (11, 227, 270), (9, 97, 116) and (2, 1097, 1568), where in parentheses we indicated the number of counters, the number of states and the number of instructions, respectively.

2 Definitions

We now recall the formal definition of CM given in [10]. We denote by $\mathbb{N}$ the set of all non-negative integers.

Definition 1 A k-counter machine (CM(k)) is the 5-tuple $M = (k, Q, \delta, q_0, q_f)$, where k is the number of counters, Q is a nonempty finite set of states, $q_0 \in Q$ is the initial state, $q_f \in Q$ is the final (halting) state and δ is the move relation, which is a subset of $Q \times \{1, \ldots, k\} \times \{Z, P\} \times Q \cup Q \times \{1, \ldots, k\} \times \{-, 0, +\} \times Q$.

We will use the notation Ri to denote the counter i.

Definition 2 An instantaneous description of a CM(k) $M = (k, Q, \delta, q_0, q_f)$ is a $k + 1$-tuple $(q, n_1, \ldots, n_k) \in Q \times \mathbb{N}^k$.

The transition relation $\vdash$ is defined as follows:

$$(q, n_1, \ldots, n_i, \ldots, n_k) \vdash (q, n_1, \ldots, n_i', \ldots, n_k)$$

iff one of the following conditions is satisfied:

1. $[q, i, Z, q'] \in \delta$ and $n_i = n_i' = 0$ (the zero test instruction).
2. $[q, i, P, q'] \in \delta$ and $n_i = n_i' > 0$ (the non-zero instruction).
3. $[q, i, -, q'] \in \delta$ and $n_i - 1 = n_i'$ (the minus instruction).
4. $[q, i, 0, q'] \in \delta$ and $n_i = n_i'$ (the jump instruction).
5. $[q, i, +, q'] \in \delta$ and $n_i + 1 = n_i'$ (the plus instruction).

In order to define the computation of the counter machine we need to consider input and output counters. Without losing the generality, we may assume that the input counters are numbered from 1 to i and the output ones are numbered from j to l. Then the result of the computation of M on the vector $(n_1, \ldots, n_i)$ can be defined as follows:

$$M(n_1, \ldots, n_i) = \{(n_j', \ldots, n_l') \mid (q_0, n_1, \ldots, n_i, 0, \ldots, 0) \vdash^* (q_f, n_1', \ldots, n_k')\}.$$

According to Korec [7], a machine $\mathcal{M}$ is (weakly) universal if there exist recursive functions h and g such that for any machine M we have

$$M(x) = f(\mathcal{M}(\#_M, h(x))), \text{ where } \#_M \text{ is the number of } M \text{ in some enumeration.}$$

A machine is said to be strongly universal if f and h are identities.

In [7] it was shown that there exist a strongly universal register machine U_{32} with 32 instructions and a weakly universal register machine with 29 instructions. While the model of register machine used in [7] is slightly different from Definition 1, there is no difficulty in translating it to/from the form used in this paper. Moreover, this translation keeps the same number of states and basically the same number of instructions (with the remark that zero-test instructions from register machines correspond to two instructions in the counter machine). We give below the corresponding list of instructions. Note that this translation adds two additional states that are not considered in [7] for technical reasons: q_0 and q_f corresponding to the start and the end state respectively. Hence, the resulting machine U_{34} has 46 instructions. Notice also that U_{32} is numbering registers from 0 to 7. We recall that the code of the simulated machine is initially stored in $R1$, the initial value in $R2$ and the result is obtained in $R0$. At the end of the computation all values of counters $R3$–$R7$ are bounded.

$[q_1, 1, P, q_2]$	$[q_1, 1, Z, q_6]$	$[q_2, 1, -, q_3]$	$[q_3, 7, +, q_1]$
$[q_4, 5, P, q_5]$	$[q_4, 5, Z, q_7]$	$[q_5, 5, -, q_6]$	$[q_6, 6, +, q_4]$
$[q_7, 6, P, q_8]$	$[q_7, 6, Z, q_4]$	$[q_8, 6, -, q_9]$	$[q_9, 5, +, q_{10}]$
$[q_{10}, 7, P, q_{11}]$	$[q_{10}, 7, Z, q_{13}]$	$[q_{11}, 7, -, q_{12}]$	$[q_{12}, 1, +, q_7]$
$[q_{13}, 6, P, q_{14}]$	$[q_{13}, 6, Z, q_1]$	$[q_{14}, 4, P, q_{15}]$	$[q_{14}, 4, Z, q_{16}]$
$[q_{15}, 4, -, q_1]$	$[q_{16}, 5, P, q_{17}]$	$[q_{16}, 5, Z, q_{23}]$	$[q_{17}, 5, -, q_{18}]$
$[q_{18}, 5, P, q_{19}]$	$[q_{18}, 5, Z, q_{27}]$	$[q_{19}, 5, -, q_{20}]$	$[q_0, 1, 0, q_1]$
$[q_{20}, 5, P, q_{21}]$	$[q_{20}, 5, Z, q_{30}]$	$[q_{21}, 5, -, q_{22}]$	$[q_{22}, 4, +, q_{16}]$
$[q_{23}, 2, P, q_{24}]$	$[q_{23}, 2, Z, q_{25}]$	$[q_{24}, 2, -, q_{32}]$	
$[q_{25}, 0, P, q_{26}]$	$[q_{25}, 0, Z, q_{32}]$	$[q_{26}, 0, -, q_1]$	
$[q_{27}, 3, P, q_{28}]$	$[q_{27}, 3, Z, q_{29}]$	$[q_{28}, 3, -, q_{32}]$	$[q_{29}, 0, +, q_1]$
$[q_{30}, 2, +, q_{31}]$	$[q_{31}, 3, +, q_{32}]$	$[q_{32}, 4, P, q_{15}]$	$[q_{32}, 4, Z, q_f]$

As for register machines, counter machines can be represented in a graphical manner as a graph whose nodes are labeled by elements from Q and having a directed edge going from q to q' labeled by iX if $[q, i, X, q'] \in \delta$, see e.g. Fig. 1.

A CM M is said to be deterministic if for any pair of instructions $[p, i, X, p']$ and $[q, j, Y, q']$ from δ it holds

$$p \neq q \lor (i = j \land X \neq Y \land X, Y \notin \{-, 0, +\}).$$

A CM M is said to be reversible if if for any pair of instructions $[p, i, X, p']$ and $[q, j, X, q']$ from δ it holds

$$p' \neq q' \lor (i = j \land X \neq Y \land X, Y \notin \{-, 0, +\}).$$

In the graphical form the deterministic property implies that each node has at most two outgoing arcs. In the case of two arcs, both of them should correspond to the zero and non-zero test of the same counter. Since U_{32} is deterministic, it is not surprising that U_{34} is deterministic too.

Similarly, the reversible property implies that each node has at most two incoming arcs. As in the deterministic case, when there are two incoming arcs then both of them should correspond to the zero and non-zero test of the same counter. We will call *non-reversible* a node (state) that is not fulfilling this property.

We also recall the following results from [10]:

Theorem 1 ([10], Theorem 3.1) *For any deterministic CM(k) $M = (k, Q, \delta, q_0, q_f)$, there is a deterministic reversible CM(k + 2) $M' = (k + 2, Q', \delta', q_0, q_f)$ such that*

$$(q_0, m_1, \ldots, m_k) \vdash_M^* (q_f, n_1, \ldots, n_k) \quad \text{iff}$$
$$\exists h \in \mathbb{N} \quad (q_0, m_1, \ldots, m_k, 0, 0) \vdash_{M'}^* (q_f, n_1, \ldots, n_k, h, 0)$$

holds for all $m_1, \ldots, m_k, n_1, \ldots, n_k \in \mathbb{N}$.

The proof of above theorem produces the history of the computation, which is recorded in the value h using a method from [4]. Obviously, this value is not known in advance and it is not bounded. The next theorem shows that using k additional counters it is possible to bound it, hence obtaining each time a "clean" computation where only the value of the output is not bounded in advance. We should call such machines *garbage-less*.

Theorem 2 ([10], Theorem 3.2) *For any deterministic CM(k) $M = (k, Q, \delta, q_0, q_f)$, there is a deterministic reversible CM(2k + 2) $M' = (2k + 2, Q', \delta', q_0, q_f')$ such that*

$$(q_0, m_1, \ldots, m_k) \vdash_M^* (q_f, n_1, \ldots, n_k) \quad \text{iff}$$
$$(q_0, m_1, \ldots, m_k, 0, \ldots, 0) \vdash_{M'}^* (q_f', m_k, \ldots, m_k, 0, 0, n_1, \ldots, n_k)$$

holds for all $m_1, \ldots, m_k, n_1, \ldots, n_k \in \mathbb{N}$.

Next theorem shows that as in [9] any number of counters can be packed into two in a reversible manner.

Theorem 3 ([10], Theorem 4.1) *For any deterministic CM(k) $M = (k, Q, \delta, q_0, q_f)$, there is a deterministic reversible CM(2) $M' = (2, Q', \delta', q_0, q_f)$ such that*

$$(q_0, m_1, \ldots, m_k) \vdash_M^* (q_f, n_1, \ldots, n_k) \quad \text{iff}$$
$$(q_0, p_1^{m_1} \cdots p_k^{m_k}, 0) \vdash_{M'}^* (q_f, p_1^{n_1} \cdots p_k^{n_k}, 0)$$

holds for all $m_1, \ldots, m_k, n_1, \ldots, n_k \in \mathbb{N}$, where p_i denotes the ith prime number.

3 Strong Universality

In this section we will construct several small universal reversible counter machines. We start by analyzing the machine U_{34}. It can be easily seen that U_{34} has the following non-reversible states:

$$q_1, q_4, q_6, q_7, q_{15}, q_{16}, q_{32}.$$

States q_1 and q_{32} are non-reversible because of multiple incoming arcs (6 and 4 respectively). The other states are non-reversible because the incoming arcs do not correspond to opposite checks of the same counter.

We start by reducing the number of incoming arcs to each state to at most four. This is performed by adding two additional states as in [10], Lemma 3.1. We remark that added states are non-reversible. This yields the following machine U_{36}, see also Fig. 1 (we emphasized in bold the differences with respect to the U_{34} machine):

$[q_1, 1, P, q_2]$	$[q_1, 1, Z, q_6]$	$[q_2, 1, -, q_3]$	$[q_3, 7, +, \mathbf{q_{34}}]$
$[q_4, 5, P, q_5]$	$[q_4, 5, Z, q_7]$	$[q_5, 5, -, q_6]$	$[q_6, 6, +, q_4]$
$[q_7, 6, P, q_8]$	$[q_7, 6, Z, q_4]$	$[q_8, 6, -, q_9]$	$[q_9, 5, +, q_{10}]$
$[q_{10}, 7, P, q_{11}]$	$[q_{10}, 7, Z, q_{13}]$	$[q_{11}, 7, -, q_{12}]$	$[q_{12}, 1, +, q_7]$
$[q_{13}, 6, P, q_{14}]$	$[q_{13}, 6, Z, \mathbf{q_{33}}]$	$[q_{14}, 4, P, q_{15}]$	$[q_{14}, 4, Z, q_{16}]$
$[q_{15}, 4, -, \mathbf{q_{33}}]$	$[q_{16}, 5, P, q_{17}]$	$[q_{16}, 5, Z, q_{23}]$	$[q_{17}, 5, -, q_{18}]$
$[q_{18}, 5, P, q_{19}]$	$[q_{18}, 5, Z, q_{27}]$	$[q_{19}, 5, -, q_{20}]$	$[q_0, 1, 0, q_1]$
$[q_{20}, 5, P, q_{21}]$	$[q_{20}, 5, Z, q_{30}]$	$[q_{21}, 5, -, q_{22}]$	$[q_{22}, 4, +, q_{16}]$
$[q_{23}, 2, P, q_{24}]$	$[q_{23}, 2, Z, q_{25}]$	$[q_{24}, 2, -, q_{32}]$	$[\mathbf{q_{34}, 1, 0, q_1}]$
$[q_{25}, 0, P, q_{26}]$	$[q_{25}, 0, Z, q_{32}]$	$[q_{26}, 0, -, \mathbf{q_{33}}]$	$[\mathbf{q_{33}, 1, 0, q_{34}}]$
$[q_{27}, 3, P, q_{28}]$	$[q_{27}, 3, Z, q_{29}]$	$[q_{28}, 3, -, q_{32}]$	$[q_{29}, 0, +, \mathbf{q_{33}}]$
$[q_{30}, 2, +, q_{31}]$	$[q_{31}, 3, +, q_{32}]$	$[q_{32}, 4, P, q_{15}]$	$[q_{32}, 4, Z, q_f]$

Note that it was possible to reduce in the same manner the number of incoming arcs to two. Then it is possible to use the construction from Theorem 1 in order to construct an equivalent reversible machine. In this case each pair of arcs (instructions) leading to a non-reversible state are replaced by 21 instructions and 16 additional states. A quick computation shows that using this method a strongly universal reversible machine with 273 instructions and 235 states is obtained.

We show below that for U_{36} a more efficient construction can be used. We will use a modified version of the technique from Theorem 1. We recall that the core of this proof is that in order to make non-reversible states reversible an additional counter is used to keep track of the computation history. More precisely, this counter stores a number whose bits keep track of which of the two incoming arcs was used to reach

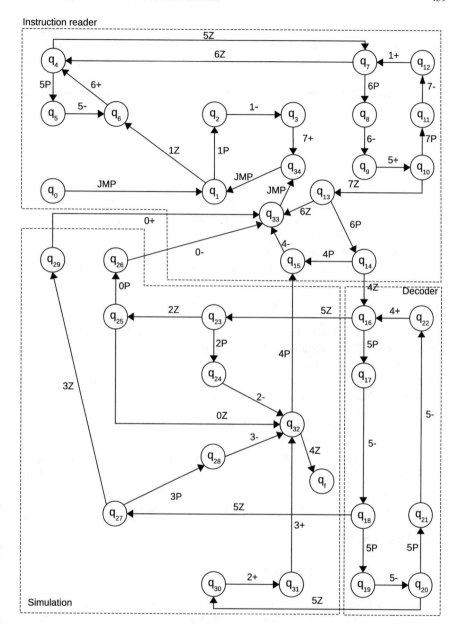

Fig. 1 Strongly universal counter machine U_{36}

a non-reversible state. Another additional counter is needed for technical reasons. With the goal of minimization of the number of instructions and states we allow up to four incoming arcs to a node. Hence, our history will keep a base-4 representation of the used choice.

We start by the observation that in the case of state q_4 the incoming arcs are labeled by $6Z$ and $6+$. We introduce a new state q_4' and we replace the instruction $[q_6, 6, +, q_4]$ by two instructions:

$$[q_6, 6, +, q_4'], \ [q_4', 6, P, q_4]$$

A similar transformation is done for the state q_{16} (using counter $R4$).

Next, we observe that the two jump instructions at state q_1 can be replaced by the test of counter $R8$ that is supposed to store the history of the computation (recall that the construction from Theorem 1 adds two additional counters $R8$ and $R9$). Since at the beginning of the computation the history is empty (equal to zero) and after the first cycle returning to q_1 it is not empty, it is possible to correctly discriminate both cases:

$$[q_0, 8, Z, q_1], \ [q_{34}, 8, P, q_1]$$

Now we concentrate on the state q_{34}. On one branch the counter $R7$ is positive (because of the $R7+$ instruction). On the other branch counter $R7$ is always zero, because the last operation on $R7$ that is performed in order to reach q_{34} is the instruction $[q_{10}, 7, Z, q_{13}]$ that ensures that $R7$ is empty. Hence, it is possible to use the following instructions to make q_{34} reversible:

$$[q_3, 7, +, q_3'], \ [q_3', 7, P, q_{34}], \ [q_{33}, 7, Z, q_{34}]$$

Consider now the state q_{15}. Using the flowchart depicted at Fig. 1 it can be easily verified that the value of counter $R5$ can discriminate the two branches. Indeed, the instruction $[q_9, 5, +, q_{10}]$ ensures that $R5$ is positive when going to q_{15} from q_{14}. On the other hand, in order to reach q_{32} one has to pass through q_{23}, q_{27} or q_{30}. But this implies a zero test on $R5$. Hence, we can use following instructions to obtain the reversible behavior of q_{15}:

$$[q_{32}, 4, P, q_{32}'], \ [q_{32}', 5, Z, q_{15}], \ [q_{14}, 4, P, q_{14}'], \ [q_{14}', 5, P, q_{15}]$$

Thus, only 4 states (q_6, q_7, q_{32}, q_{33}) need to be made reversible. We describe below the procedure that allows to replace any node with at most four incoming edges by an equivalent reversible construction.

Consider a state q_t that has 4 incoming arcs. Let the corresponding instructions be $[q_{s_j}, i_{s_j}, X_j, q_t]$, $1 \le j \le 4$. Then consider the following rules (depicted also on Fig. 2):

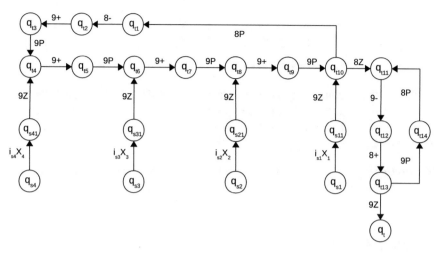

Fig. 2 Reversible junction of 4 instructions in q_t

$[q_{s_j}, i_{s_j}, X_j, q_{s_{j1}}]$, $1 \leq j \leq 4$

$[q_{s_{11}}, 9, Z, q_{t_{10}}]$	$[q_{s_{21}}, 9, Z, q_{t_8}]$	$[q_{s_{31}}, 9, Z, q_{t_6}]$	$[q_{s_{41}}, 9, Z, q_{t_4}]$
$[q_{t_1}, 8, -, q_{t_2}]$	$[q_{t_2}, 9, +, q_{t_3}]$	$[q_{t_3}, 9, P, q_{t_4}]$	$[q_{t_4}, 9, +, q_{t_5}]$
$[q_{t_5}, 9, P, q_{t_6}]$	$[q_{t_6}, 9, +, q_{t_7}]$	$[q_{t_7}, 9, P, q_{t_8}]$	$[q_{t_8}, 9, +, q_{t_9}]$
$[q_{t_9}, 9, P, q_{t_{10}}]$	$[q_{t_{10}}, 8, P, q_{t_1}]$	$[q_{t_{10}}, 8, Z, q_{t_{11}}]$	$[q_{t_{11}}, 9, -, q_{t_{12}}]$
$[q_{t_{12}}, 8, +, q_{t_{13}}]$	$[q_{t_{13}}, 9, P, q_{t_{14}}]$	$[q_{t_{13}}, 9, Z, q_t]$	$[q_{t_{14}}, 8, P, q_{t_{11}}]$

Clearly, the above rules allow to reach q_t in a reversible manner. We also observe that if a node has less than four incoming arcs, then for reversibility it is sufficient to delete unused q_{sj1} nodes.

Now considering all above constructions together it is possible to construct a strong universal reversible counter machine. Such a machine has 109 states (36 states from U_{36}, plus 5 additional states used for the reversibility of q_1, q_4, q_{15}, q_{16} and q_{34}, plus 2×18 states used for the reversibility of q_{32} and q_{33}, plus 2×16 states used for the reversibility of q_6 and q_7) and 129 instructions ($48 + 5 + 2 \times 20 + 2 \times 18$).

Theorem 4 *There exists a strongly universal reversible counter machine U_{109} with 10 counters, 109 states and 129 instructions.*

Now we will show how to bound the final value of non-output counters, i.e. obtain a garbage-less CM. We will use the construction from Theorem 2. This construction works as follows. Machine U_{109} is run leaving the history in $R8$. Next the following copy procedure is executed transferring the resulting value from $R0$ to $R10$. This is done by copying the value of $R0$ to $R9$ and then copying it back from $R9$ to $R0$ and $R10$.

$$[q_f, 9, Z, q_{c_1}] \qquad [q_{c_1}, 0, Z, q_{c_5}] \qquad [q_{c_1}, 0, P, q_{c_2}] \qquad [q_{c_2}, 0, -, q_{c_3}]$$

$$[q_{c_3}, 9, +, q_{c_4}] \qquad [q_{c_4}, 9, P, q_{c_1}] \qquad [q_{c_5}, 9, Z, p_f] \qquad [q_{c_5}, 9, P, q_{c_6}]$$

$$[q_{c_6}, 9, -, q_{c_7}] \qquad [q_{c_7}, 0, +, q_{c_8}] \qquad [q_{c_8}, 10, +, q_{c_9}] \qquad [q_{c_9}, 0, P, q_{c_5}]$$

Finally, the machine is run in a reverse manner (technically a copy of all rules with reverse operations should be provided, working on states where q replaced by p). This gives a total number of $109 \times 2 + 9 = 227$ states and $129 \times 2 + 12 = 270$ instructions.

Theorem 5 *There exists a strongly universal garbage-less reversible counter machine U_{227} with 11 counters, 227 states and 270 instructions.*

4   Weak Universality

Now consider the weakly universal machine U_{31} (based on U_{29} from [7]). Below we give the list of rules of this machine. In fact, the only modification in this machine with respect to U_{31} concerns the simulation block, which is also depicted on Fig. 3. We remark that states q_{25}, q_{26} and q_{29} are absent. Below we bold emphasized corresponding changes.

$$[q_1, 1, P, q_2] \qquad [q_1, 1, Z, q_6] \qquad [q_2, 1, -, q_3] \qquad [q_3, 7, +, q_{38}]$$

$$[q_4, 5, P, q_5] \qquad [q_4, 5, Z, q_7] \qquad [q_5, 5, -, q_6] \qquad [q_6, 6, +, q_4]$$

$$[q_7, 6, P, q_8] \qquad [q_7, 6, Z, q_4] \qquad [q_8, 6, -, q_9] \qquad [q_9, 5, +, q_{10}]$$

$$[q_{10}, 7, P, q_{11}] \qquad [q_{10}, 7, Z, q_{13}] \qquad [q_{11}, 7, -, q_{12}] \qquad [q_{12}, 1, +, q_7]$$

$$[q_{13}, 6, P, q_{14}] \qquad [q_{13}, 6, Z, q_{37}] \qquad [q_{14}, 4, P, q_{15}] \qquad [q_{14}, 4, Z, q_{16}]$$

$$[q_{15}, 4, -, q_{36}] \qquad [q_{16}, 5, P, q_{17}] \qquad [q_{16}, 5, Z, q_{23}] \qquad [q_{17}, 5, -, q_{18}]$$

$$[q_{18}, 5, P, q_{19}] \qquad [q_{18}, 5, Z, q_{27}] \qquad [q_{19}, 5, -, q_{20}] \qquad [q_0, 1, 0, q_1]$$

$$[q_{20}, 5, P, q_{21}] \qquad [q_{20}, 5, Z, q_{30}] \qquad [q_{21}, 5, -, q_{22}] \qquad [q_{22}, 4, +, q_{16}]$$

$$[q_{23}, \mathbf{0}, P, \mathbf{q_{24}}] \qquad [q_{23}, \mathbf{0}, Z, \mathbf{q_1}] \qquad [q_{24}, \mathbf{0}, -, q_{32}]$$

$$[q_{27}, \mathbf{2}, P, q_{28}] \qquad [q_{27}, \mathbf{2}, Z, \mathbf{q_1}] \qquad [q_{28}, \mathbf{2}, -, q_{32}]$$

$$[q_{30}, \mathbf{0}, +, q_{31}] \qquad [q_{31}, \mathbf{2}, +, q_{32}] \qquad [q_{32}, 4, P, q_{15}] \qquad [q_{32}, 4, Z, q_f]$$

We observe that there are only 3 incoming arcs to q_{32}. In order to minimize the number of nodes and arcs we will construct a machine with at most 3 incoming arcs to each node. First add states q_{33} and q_{34} like in the case of U_{36}, see Fig. 1. Next, consider a new state q_{35} and add following rules:

$$[q_{33}, 6, P, q_{35}] \qquad [q_{13}, 6, Z, q_{35}] \qquad [q_{35}, 1, 0, q_{34}]$$

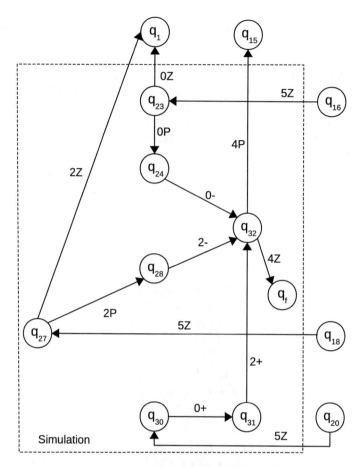

Fig. 3 Weakly universal counter machine U_{31}: simulation block

We remark that on the branch yielding to q_{33} the value of $R6$ is positive, because of the rule $[q_{13}, 6, P, q_{14}]$. Hence, q_{35} is reversible.

Now we remark that in order to make reversible a state with at most three incoming arcs we can use a slightly modified version of the flowchart from Fig. 2. In fact, states $q_{s_{41}}$, q_{t_3} and q_{t_4} should be removed and the arrow leading to q_{t_3} should lead now to q_{t_5}. This allows to store the incoming arc number in the history, interpreted as a base-3 number.

Hence we showed how it is possible to construct a weak universal reversible counter machine. This machine has 97 states ($33 + 6 + 2 \times 15 + 2 \times 14$) and 116 instructions ($44 + 6 + 2 \times 17 + 2 \times 16$).

Theorem 6 *There exists a weakly universal reversible counter machine U_{97} with 9 counters, 97 states and 116 instructions.*

Now we can use the construction from Theorem 3 to obtain a weakly universal reversible counter machine U with 2 counters. We refer to [10] for more details. In order to compute the parameters of U we recall that each jump instruction from U_{95} is performed by one instruction in U, each plus or minus instruction of Ri is performed by $p_i + 10$ instructions and $p_i + 7$ new states (p_i being the ith prime number). Each couple of zero and non-zero check on Ri instructions is performed by $3p_i + 3$ instructions using $2p_i + 2$ new states. A single zero check instruction is simulated using $3p_i + 2$ instructions and $2p_i + 1$ states, while a single non-zero check instruction is simulated using $2p_i + 4$ instructions and $p_i + 2$ states. To minimize the number of instructions we will use a different sequence of prime numbers associated to counters (trying to use smaller primes for more frequently used counters).

The table below gives the number of instructions of each type in U_{95}, as well as the chosen prime number.

i	$+$	$-$	P&Z	Z	P	p_i
0	1	1	1	0	0	23
1	1	1	1	0	0	19
2	1	1	1	0	0	17
4	1	1	2	0	1	11
5	1	4	4	1	1	5
6	1	1	2	0	2	7
7	1	1	1	1	1	13
8	4	4	4	1	5	3
9	10	4	4	12	12	2

Theorem 7 *There exists a weakly universal reversible counter machine U_{1097} with 2 counters, 1097 states and 1568 instructions.*

5 Conclusions

The table below summarizes the parameters of universal reversible counter machines constructed in this paper.

Counters	states	instructions	universality	garbage-less
10	109	129	strong	no
11	227	270	strong	yes
9	97	116	weak	no
2	1097	1568	weak	no

We remark that as noted by [2] there is no universal reversible counter machine in the strict sense, because any reversible machine computes an injective function and, obviously, the universal machine as defined above is not injective. In [2, 4] it is highlighted that corresponding constructions provide a machine for reversibly simulating any irreversible machine. Another solution provided in the paper above is to define the universal machine $\mathcal{M}$ as follows $\mathcal{M}(\#_M, x) = (\#_M, M(x))$, i.e., the result of the computation of the universal machine on the pair consisting of the code of the simulated machine M and its input x is equal to the pair of the code of the simulated machine M and the result of its computation on the given input x. Since U_{32} keeps a copy of the code of the simulated machine in $R1$, we can deduce that the simulations that we provide are universal in the above sense.

There are several possible directions for further research. One of them is the investigation of different trade-offs between the number of counters, states and instructions. It can be particularly interesting to see if as in [7] the increase of the number of counters can lead to the decrease of the number of instructions. Another interesting direction is the investigation of the universality type and independency for different number of counters.

References

1. Alhazov, A., Ivanov, S., Pelz, E., Verlan, S.: Small universal deterministic Petri nets with inhibitor arcs. J. Autom. Lang. Comb. **21**(1–2), 7–26 (2016)
2. Axelsen, H.B., Glück, R.: What do reversible programs compute? In: Hofmann, M. (ed.) Foundations of Software Science and Computational Structures - 14th International Conference, FOSSACS 2011, Held as Part of the Joint European Conferences on Theory and Practice of Software, ETAPS 2011, Saarbrücken, Germany, March 26-April 3, 2011. Proceedings, Lecture Notes in Computer Science, vol. 6604, pp. 42–56. Springer (2011)
3. Barzdin, I.M.: Ob odnom klasse machin Turinga (machiny Minskogo), russian. Algebra i Logika **1**, 42–51 (1963)
4. Bennett, C.: Logical reversibility of computation. IBM J. Res. Dev. **17**, 525–532 (1973)
5. Bennett, C.H.: Notes on the history of reversible computation. IBM J. Res. Dev. **44**(1), 270–278 (2000)
6. Ivanov, S., Pelz, E., Verlan, S.: Small universal non-deterministic Petri nets with inhibitor arcs. In: Jürgensen, H., Karhumäki, J., Okhotin, A. (eds.) Descriptional Complexity of Formal Systems - 16th International Workshop, DCFS 2014, Turku, Finland, 5–8 Aug 2014. Proceedings, Lecture Notes in Computer Science, vol. 8614, pp. 186–197. Springer (2014)
7. Korec, I.: Small universal register machines. Theor. Comput. Sci. **168**(2), 267–301 (1996)
8. Minsky, M.: Size and structure of universal Turing machines using tag systems. In: Recursive Function Theory: Proceedings, Symposium in Pure Mathematics, Provelence, vol. 5, pp. 229–238 (1962)
9. Minsky, M.: Computations: Finite and Infinite Machines. Prentice Hall, Englewood Cliffts (1967)
10. Morita, K.: Universality of a reversible two-counter machine. Theor. Comput. Sci. **168**(2), 303–320 (1996)
11. Morita, K.: Reversible cellular automata. In: Rozenberg, G., Bäck, T., Kok, J.N. (eds.) Handbook of Natural Computing, pp. 231–257. Springer, Berlin (2012)
12. Morita, K.: Universal reversible Turing machines with a small number of tape symbols. Fundam. Inform. **138**(1–2), 17–29 (2015)
13. Neary, T., Woods, D.: The complexity of small universal Turing machines: a survey. In: Bieliková, M., Friedrich, G., Gottlob, G., Katzenbeisser, S., Turán, G. (eds.) SOFSEM 2012:

38th Conference on Current Trends in Theory and Practice of Computer Science, Lecture Notes in Computer Science, vol. 7147, pp. 385–405. Springer (2012)
14. Rogozhin, Y.: Small universal Turing machines. Theor. Comput. Sci. **168**(2), 215–240 (1996)
15. Schroeppel, R.: A two counter machine cannot calculate 2N. AI Memos. MIT AI Lab (1972)
16. Shannon, C.E.: A universal Turing machine with two internal states. Autom. Stud. Ann. Math. Stud. **34**, 157–165 (1956)
17. Turing, A.M.: On computable numbers, with an application to the Entscheidungsproblem. Proc. Lond. Math. Soc. **42**(2), 230–265 (1936)
18. Watanabe, S.: 5-symbol 8-state and 5-symbol 6-state universal Turing machines. J. ACM **8**(4), 476–483 (1961)
19. Woods, D., Neary, T.: The complexity of small universal Turing machines: a survey. Theor. Comput. Sci. **410**(4–5), 443–450 (2009)

Improving the Success Probability for Shor's Factorization Algorithm

Guoliang Xu, Daowen Qiu, Xiangfu Zou and Jozef Gruska

Abstract In Shor's factorization algorithm (SFA), the task is to find a non-trivial factor of a given composite integer N. Briefly said, SFA works as follows. It chooses randomly an integer $y < N$ and checks whether y and N are co-primes. If y is co-prime with N, then SFA runs a special quantum subroutine to obtain the order $2r$ of N with a certain probability ($2r$ is here an integer). In the original SFA and all previous SFAs, if $2r$ was an even integer and $y^r \not\equiv -1 \pmod{N}$, then SFA used y and $2r$ to get a non-trivial factor of N. However, if the result r' obtained by the quantum order finding subroutine was not $2r$, or $2r$ was not an even integer, or $y^r \equiv -1 \pmod{N}$, then the quantum subroutine had to be run again (and perhaps again and again). In this paper, we show that the three constraints are strong and the success probability for the quantum subroutine can be improved. In general, if a non-trivial factor of N can be got, we can call the result r' an *available* result. Naturally, two issues arise: (1) If one of these constraints does not hold, whether these results r' can also be used to make SFA succeed sometimes? (2) If there exist some other *available* results, then what is the success probability when these results are considered? This paper proves that some factorization results are still *available* or possible even if not all of the above constraints are met, and, in addition, that a new success probability can be bigger than those of the previous SFAs. Finally, in order to demonstrate a potential of our approach, we consider factorizationof those

G. Xu · D. Qiu (✉)
Institute of Computer Science Theory, School of Data and Computer Science,
Sun Yat-sen University, Guangzhou 510006, China
e-mail: issqdw@mail.sysu.edu.cn

D. Qiu
SQIG–Instituto de Telecomunicações, Departamento de Matemática,
Instituto Superior Técnico, Av. Rovisco Pais, 1049-001 Lisbon, Portugal

X. Zou
School of Mathematics and Computational Science, Wuyi University,
Jiangmen 529020, China

J. Gruska
Masaryk University, Botanicka 68a, 60200 Brno, Czech Republic

© Springer International Publishing AG 2018
A. Adamatzky (ed.), *Reversibility and Universality*, Emergence, Complexity
and Computation 30, https://doi.org/10.1007/978-3-319-73216-9_21

447

integers N that are used as moduli for RSA, that is those N that are products of two safe primes, and we show that in this case the fault probability can be reduced to $O(1/N)$ with our method.

1 Introduction

It is already well known that quantum computers can solve certain problems more efficiently than classical computers. Shor's factoring algorithm (SFA)—a probabilistic algorithm—is one of the examples where quantum algorithms can outperform even the most efficient known classical algorithms [1]. SFA finds a non-trivial factor of a big composite integer N, exponentially faster than the best known classical algorithms [1].

1.1 The State of the Art

SFA uses a special order-finding algorithm (OFA), which needs a quantum oracle, as a subroutine [1, 2] in order to factorize an N. In particular, SFA first chooses randomly a co-prime $y < N$. Then SFA calls OFA to obtain the order $2r$ of y with a known probability ($2r$ is an integer). Finally, if $2r$ is an even and $y^r \not\equiv -1 (\text{mod } N)$, one can use y and r to obtain a non-trivial factor of N [1, 2].

In the previous approaches [1–6], in order to make SFA to succeed, a chosen *good* co-prime y should satisfy the following conditions:

$$\text{The order } 2r \text{ of } y \text{ is even, i.e., } r \text{ is an integer;} \tag{1}$$

$$y^r \not\equiv -1 (\text{mod } N). \tag{2}$$

If a *not good* enough y is selected, OFA needs to choose another co-prime and repeat computation again. In the worst case, the probability for y being *good* could be raised from $1/2$, as specified in Refs. [1, 2], to $3/4$ by choosing y having Jacobi-Symbol -1 (Leander'2002)[5]. The reason behind is that numbers of this form always satisfy Eq. (1). In particular, such an improvement of SFA implies that the success probability will be 1 for integers used as moduli in the RSA crypto protocol [7] (The reason being that in such a case the integer to be factorized is known to be the product of two safe primes).

In 2015, Lawson [4] showed that a factor of a given N can be found from a square co-prime, providing it could have been found from its (non-square) root (If $y = a^b$ where a, b are integers and a is a non-square integer, then a is called the root of y), for which it is not necessary to consider odd orders. In addition, their derivations and calculations showed that avoiding non-square co-primes raises the probability of success of SFA lower bounded by $\left(1 - \frac{1}{4\sqrt{N}}\right)^{-1} - 1 = \frac{1}{4\sqrt{N}-1} \in O\left(\frac{1}{\sqrt{N}}\right)$ [4].

Using y as the input, OFA succeeds if

$$r' \in \{r, 2r\}. \tag{3}$$

Note that $r' = r$ is considered as an *available* result, in that sense that the order we wanted to find is even and then its factor 2 may be reduced. Furthermore, OFA needs to call the oracle $O((\log N)^3)$ times and therefore succeeds with the probability $O(1)$ [2]. This result implies that the success probability of OFA is $O\left(\frac{1}{(\log N)^3}\right)$.

As a probabilistic algorithm, the success probability depends on the integer N to be factorized, on the co-prime y chosen and on the result of the quantum order finding subroutine. Since N is an input, it is the choice of a co-prime y that determines the success probability of SFA. Some ways how to choose good co-primes from the above point of view, are specified in the Refs. [4, 5]. This paper makes a full use of these results, and this way raises the success probability of SFA.

Recently, Refs. [3, 6, 8] analysed various simulations and the experimental realizations of SFA. Their results showed that SFA has indeed potential for practical applications and the quantum subroutine of SFA still is the most complex part of it. That also implies that to reduce the fault probability of SFA is not only interesting, but also practically important task.

1.2 Our Contributions

In order to raise the success probability of SFAs, our idea is to loose Eqs. (2), (3) as much as possible. In order to do that we will consider the following two problems:

- (P1) If Eq. (2) or Eq. (3) does not hold, what other results can be used to make SFA to succeed?
- (P2) If the answer of (P1) is yes, then what is the probability that needed results are obtained?

To our best knowledge, no one dealt with such problems so far. Moreover, a proper answer to (P1) is clearly not obvious, and the key problem is therefore to find some other *useful* results. In this paper we can do that, and as a consequence, we can improve efficiency of SFAs and to give a lower bound for the raising success probability of SFA.

In this paper we show that the SFA may get a non-trivial factor of N also from the more general results r'. Namely, even if one of the conditions Eq. (2) or Eq. (3) does not hold, SFA may still succeed. Then, we apply the result obtained to improve SFA and we will analyze the raising probability. In particular, for the widely-used moduli N for the RSA protocol [7], being the product of two safe primes, the fault probability of which is $O(1/\sqrt{N})$ in Leander' 2002 can be reduced to $O(1/N)$ with our result.

1.3   An Overview

The rest of this paper goes as follows. We first introduce some necessary notations, concepts and results needed in the rest of the paper, in Sect. 2. Afterwards, a generalization of the use of the usual quantum order-finding subroutine in SFA is provided in Sect. 3. As the next, in Sect. 4, we improve efficiency, namely success probability, of SFA using the results from Sect. 3. Finally, we apply our results to factorization of moduli of RSA protocols in Sect. 5, see [7]. Finally, some conclusions are discussed in Sect. 6.

2   Preliminaries

In this section, we first provide some notations, concepts and basic related results from discrete mathematics. Afterwards, we present the main results of this paper.

2.1   Basic Notations, Concepts and Algebraic Results

Notations needed:

(1)   Any odd, composite natural integer N has a unique prime factorization

$$N = \prod_{i=1}^{k} p_i^{\alpha_i} \tag{4}$$

where $k > 1$ (If $k = 1$ and $\alpha_1 = 1$, then N is prime; If $k = 1$ and $\alpha_1 > 1$, integer factorization can be done using a classical algorithm [2]), $2 < p_1 < p_2 < \ldots < p_k$ and $\alpha_i \geq 1$ for all $i \leq k$.

(2) $\text{ord}_N(y) = 2r$ denotes the order of y in $\mathbb{Z}_N^*$, i.e., the smallest positive integer $2r$ satisfying $y^{2r} \equiv 1 \pmod{N}$. We can know that $\text{ord}_N(y) = 2\text{ord}_N(y^2)$ if r is an integer. Unless explicitly stated, the order of y means the order of y in $\mathbb{Z}_N^*$.

(3) r_i is the order of y in $\mathbb{Z}_{p_i^{\alpha_i}}^*$. $\tilde{r}$ is the smaller integer of the set $\{r, r/2\}$ where r is an integer. Namely,

$$\tilde{r} = \begin{cases} r, & \text{if } r \text{ is odd;} \\ r/2, & \text{if } r \text{ is even.} \end{cases} \tag{5}$$

r' denotes a result of the quantum subroutine in SFA. That is, r' is the denominator of the fraction obtained by the continued fractions (CF) algorithm [2]. When the fraction is 0, let $r' = 1$.

(4) The greatest common divisor of integers a and b is denoted by $\gcd(a, b)$. The least common multiple of a and b is denoted by $\mathrm{lcm}\{a, b\}$.
(5) $a \mid b$ means a divides b, and $a \nmid b$ means a does not divide b. $a \mid b \mid c$ means that a divides b and b divides c.
(6) $\varphi(N)$ is Euler phi function and is defined to be the number of non-negative integers less than N which are co-prime to N. In particular, let $\varphi(1) = 1$.
(7) $R = \{r' \mid \text{There exists an index } i \text{ such that } \tilde{r}_i | r' | 2r\}$, and $R_1 = \{r' \mid r' \in R \text{ and } r' < r\}$. We can see that $R_1 = R - \{r, 2r\}$.

Basic results needed:

(1) Denote
$$\mathbb{Z}^*_{p_i^{\alpha_i}} = \{g_{p_i^{\alpha_i}}^{s_i} \pmod{p_i^{\alpha_i}} | 1 \le s_i \le \varphi(p_i^{\alpha_i})\}, \tag{6}$$

where $g_{p_i^{\alpha_i}}$ is a generator of $\mathbb{Z}^*_{p_i^{\alpha_i}}$.

(2) Suppose that y is chosen randomly such that $\gcd(y, N) = 1$. From that it follows that $\gcd(y, p_i^{\alpha_i}) = 1$. In addition,

$$\mathrm{ord}_N(y) = 2r = \mathrm{lcm}\{r_i \mid 1 \le i \le k\}. \tag{7}$$

Moreover, if r is an integer, Eq. (7) implies that at least one of r_i is even.

(3) Let $y \equiv g_{p_i^{\alpha_i}}^{s_i} \pmod{p_i^{\alpha_i}} \in \mathbb{Z}^*_{p_i^{\alpha_i}}$ and $\mathrm{ord}_{p_i^{\alpha_i}}(y) = r_i = \mathrm{ord}_{p_i^{\alpha_i}}(g_{p_i^{\alpha_i}}^{s_i})$. Then, $\varphi(p_i^{\alpha_i}) | s_i r_i$, and, we have

$$r_i = \frac{\varphi(p_i^{\alpha_i})}{\gcd(\varphi(p_i^{\alpha_i}), s_i)}. \tag{8}$$

Thus $\mathrm{ord}_{p_i^{\alpha_i}}(y^2) = \tilde{r}_i$. If r is an integer,

$$r = \mathrm{lcm}\{\tilde{r}_i \mid 1 \le i \le k\}. \tag{9}$$

Comment: In this paper, we mainly improve the following three versions of SFA [1, 2, 5]. We will call them as Algorithms 1, 2 and 3, respectively. For a more thorough introduction and analysis to SFA, see Refs. [1, 2, 5].

2.2 Our Main Results

The answers to problems (P1) and (P2) given in this paper go as follows.

- (A1) We generalize Eqs. (2), (3) to

$$y^r \not\equiv -1 \pmod{N} \tag{10}$$

and

$$r' \in R, \tag{11}$$

or

$$y^r \equiv -1 \pmod{N} \tag{12}$$

and

$$r' \in R_1. \tag{13}$$

- (A2) The raising success probability of each quantum order-finding subroutines will then be

$$O\left(\sum_{r' \in R_1} \frac{\varphi(r')}{2r}\right) \tag{14}$$

This is to be compared with Algorithm 3.

Finally, as already mentioned, we show that when factorization is the task for those integers N that are products of two safe primes, then the fault probability can be reduced to $O(1/N)$.

Example 1 Let $N = 77$, $y = 2$ with order $2r = 30$. Almost all possible results $r' \in \{1, 2, 3, 5, 6, 10, 15, 30\}$. Then, the order of 2 in $\mathbb{Z}_7^*$ $r_1 = 3$, and the order of 2 in $\mathbb{Z}_{11}^*$ $r_2 = 10$. In the previous approaches [1–6], $r' \in \{15, 30\}$ is usable. With our method, $r' \in R = \{3, 5, 6, 10, 15, 30\}$ is usable. If $y = 24$ with the order $2r = 30$ is chosen. Almost all possible result $r' \in \{1, 2, 3, 5, 6, 10, 15, 30\}$. Then, the order of 24 in $\mathbb{Z}_7^*$ $r_1 = 6$, and the order of 24 in $\mathbb{Z}_{11}^*$ $r_2 = 10$. In the previous approaches [1–6], no r' is usable, and SFA fails. With our method, $r' \in R = \{3, 5, 6, 10\}$ is usable. By that the example is finished.   □

3 An Analysis of Possible Impacts of Possible Outcomes of Quantum Subroutines in SFA

In this section we analyze the possible impacts of possible outcomes of quantum subroutines in SFA. At first we prove a theorem (and later analyze its impact) which deals with specific *availability* results of the quantum subroutine. Afterwards, we test the approach used for the case that chosen co-prime is an integer with Jacobi-Symbol -1. This is of importance since in such a case the even order, i.e., Eq. (1), can be ensured [5]. Furthermore, the result r' being a factor of the even order can be also ensured with a high probability as desired [2].

3.1 Outcomes of the Order-Finding Subroutine that Are Good Enough

In this subsection, we will make clear that even some more general results of the quantum order-finding subroutine in SFA will be good enough for factorization in some cases.

If $y^{2r} \equiv 1 \pmod{N}$, then $N \mid (y^{2r} - 1)$ implies

$$\prod_{i \in S_1} p_i^{\alpha_i} \mid (y^r - 1) \tag{15}$$

and

$$\prod_{i \in S_2} p_i^{\alpha_i} \mid (y^r + 1). \tag{16}$$

The sets S_0 and S_1 can be uniquely determined as follows. $S_1 \cap S_2 = \varnothing$ because $(y^r + 1) - (y^r - 1) = 2$ implies $\gcd(y^r + 1, y^r - 1) \leq 2 < p_i$. Furthermore, $S_1 \cup S_2 = \{i \mid 1 \leq i \leq k\}$ and $S_2 \neq \varnothing$, since $S_2 = \varnothing$ implies $S_1 = \{i \mid 1 \leq i \leq k\}$ and $N \mid (y^r - 1)$, in contradiction to $\operatorname{ord}_N(y) = 2r$. SFA then tells that it would yield factors from $\gcd(y^r \pm 1, N)$ if $S_1 \neq \varnothing$ and that it will fail if not.

Alternatively, we explore a possibility to obtain a factor of N using a factor of the order. Obviously, an immediate benefit is that we do not need unnecessary repetitions of the order finding subroutine. Moreover, that implies that SFA may succeed even if $S_1 = \varnothing$. In order to show that the following lemma will be needed.

Lemma 1 If $r_i = \operatorname{ord}_{p_i^{\alpha_i}}(y)$, where $p_i > 2$ is a prime, then $y^{\tilde{r}_i} \equiv \pm 1 \pmod{p_i^{\alpha_i}}$.

Proof If r_i is odd, the result is straightforward. If r_i is even, then

$$p_i^{\alpha_i} \mid (y^{r_i} - 1) = (y^{r_i/2} - 1)(y^{r_i/2} + 1). \tag{17}$$

Note that $p_i^{\alpha_i} \nmid (y^{r_i/2} - 1)$, because $p_i^{\alpha_i} \mid (y^{r_i/2} - 1)$ implies that $\operatorname{ord}_{p_i^{\alpha_i}}(y)$ divides $r_i/2$, in contradiction to $r_i = \operatorname{ord}_{p_i^{\alpha_i}}(y)$. If $p_i^{\alpha_i} \nmid (y^{r_i/2} + 1)$, then $p_i \mid \gcd(y^{r_i/2} - 1, y^{r_i/2} + 1)$ implies that

$$2 < p_i \leq \gcd(y^{r_i/2} - 1, y^{r_i/2} + 1) \leq (y^{r_i/2} + 1) - (y^{r_i/2} - 1) = 2. \tag{18}$$

However, $2 < 2$ is a contradiction and therefore the proof is finished. □

Using Lemma 1, we can get the following theorem.

Theorem 1 If $r' \leq r$ and $r' \in R$, then a non-trivial factor of N can be obtained from $\gcd(y^{r'} \pm 1, N)$, with the exception of the case that $S_1 = \varnothing$ and $r' = r$.

Proof According to Lemma 1, $\tilde{r}_i \mid r' \mid 2r$ implies

$$y^{r'} = (y^{\tilde{r}_i})^{r'/\tilde{r}_i} \equiv (\pm 1)^{r'/\tilde{r}_i} \equiv \pm 1 (\mathrm{mod}\ p_i^{\alpha_i}), \tag{19}$$

i.e.,

$$p_i^{\alpha_i} \mid (y^{r'} - 1) \ \text{or}\ p_i^{\alpha_i} \mid (y^{r'} + 1). \tag{20}$$

Note that $S_1 = \varnothing$ and the order of y is $2r$ implies that r is the least value such that $y^r \equiv -1 (\mathrm{mod}\ N)$. In fact, if there exists a $\hat{r}$ with $\hat{r} < r$, such that $y^{\hat{r}} \equiv -1 (\mathrm{mod}\ N)$, $y^{2\hat{r}} \equiv 1 (\mathrm{mod}\ N)$ and $2\hat{r} < 2r$, in contradiction to $\mathrm{ord}_N(y) = 2r$. Thus, $S_1 = \varnothing$ and $r' < r$ leads to

$$N \nmid (y^{r'} - 1) \ \text{and}\ N \nmid (y^{r'} + 1). \tag{21}$$

At the same time, $S_1 \neq \varnothing$ and $r' \leq r$ also implies Eq. (21).

According to Eqs. (20), (21), at least one non-trivial factor of N can be got from $\gcd(y^{r'} \pm 1, N)$ except the case that it holds $S_1 = \varnothing$ and $r' = r$. By that the proof is finished. $\qquad\square$

In order to understand consequences of Theorem 1 better, we provide and analyze two examples. Example 2 is the case that $y^r \not\equiv -1 (\mathrm{mod}\ N)$, and Example 3 is the case that $y^r \equiv -1 (\mathrm{mod}\ N)$. Observe that the Jacobi-Symbol of the co-prime y in Example 3 is not -1, but it is enough to illustrate usefulness of Theorem 1.

Example 2 Let $N = 77$, $y = 2$ with order $2r = 30$. Consider the result $r' \in \{1, 2, 3, 5, 6, 10, 15, 30\}$. Then, Algorithm 1 yields the order 30 from $\{15, 30\}$, and gets two non-trivial factors of 77 from $\gcd(2^{15} \pm 1, 77)$. However, according to Theorem 1, we can get a non-trivial factor of 77 from the set $\{3, 5, 6, 10, 15, 30\}$ by $\gcd(2^{r'} \pm 1, 77)$. That is because the order 3 of 2 in $\mathbb{Z}_7^*$ or half the order 10 of 2 in $\mathbb{Z}_{11}^*$ divides $r' \in \{3, 5, 6, 10, 15, 30\}$. This way the example is finished. $\qquad\square$

Example 3 Choose $y = 24$ with the order $2r = 30$ in Example 2. Consider the result $r' \in \{1, 2, 3, 5, 6, 10, 15, 30\}$. Then, Algorithm 1 yields the order 30 from $\{15, 30\}$ and fails, since $24^{15} \equiv -1 (\mathrm{mod}\ 77)$ which does not satisfy Eq. (2). However, according to Theorem 1, SFA can be used to get a non-trivial factor of 77 from the set $\{3, 5, 6, 10\}$ by $\gcd(2^3 \pm 1, 77)$ or $\gcd(2^5 \pm 1, 77)$. That is because the half of the order 6 of 24 in $\mathbb{Z}_7^*$ or half the order 10 of 24 in $\mathbb{Z}_{11}^*$ divides $r' \in \{3, 5, 6, 10\}$ and $r' \neq r$. By that the example is finished. $\qquad\square$

3.2 A Test and Its Efficiency

In this subsection, we give a Test algorithm which can help to see whether SFA succeeds. Then, we analyze its efficiency in SFA. The Test algorithm is as follows.

Algorithm 4 *A Test Algorithm*

Inputs: (1) A big integer N; (2) A co-prime y; (3) An integer r' (to be a result of the order finding subroutine)

Outputs: Either an integer to be a non-trivial factor of N or "A bad result" or "A bad co-prime".

- (T.1) If r' is even, compute $\gcd(y^{r'/2} \pm 1, N)$ and $\gcd(y^{r'} + 1, N)$; If r' is odd, compute $\gcd(y^{r'} \pm 1, N)$. If one of these is a non-trivial factor, return that factor.
- (T.2) If $y^{r'} \not\equiv 1 \pmod N$, output "A bad result".
- (T.3) Output "A bad co-prime". □

By Theorem 1, a *good* result r' is the multiple of a $\tilde{r}_i$ and a factor of $2r$. Note that $\{x | x \text{ is a factor of } 2r\} = \{x | x \text{ is a factor of } r\} \cup \{2x | x \text{ is a factor of } r\}$. Because $r = \text{lcm}\{\tilde{r}_i | 1 \le i \le k\} = \text{lcm}\{\text{ord}_{p_i^{\alpha_i}}(y^2) | 1 \le i \le k\} = \text{ord}_N(y^2)$, $\{x | x \text{ is a factor of } r\}$ is enough to get the non-trivial factor of N. In fact, that a factor can be obtained by $\gcd(y^{r'} \pm 1, N)$ implies that it can be obtained by $\gcd(y^{\tilde{r}'} \pm 1, N)$ or $\gcd(y^{2\tilde{r}'} + 1, N)$. It can therefore be seen that $\tilde{r}'$ corresponds to r'_{y^2} which denotes the result of a single quantum process of SFA with a black box $U_{y^2, N}$ which performs the transformation $|j\rangle|k\rangle \rightarrow |j\rangle|y^{2j}k \pmod N\rangle$. Consequently, $\{x | x \text{ is a factor of } 2r\}$ can be reduced to $\{x | x \text{ is a factor of } r\}$. This suggests that to exploit factors of $\text{ord}_N(y^2) = r$ is more convenient than to make use of the fact $\text{ord}_N(y) = 2r$.

Now, we start our analysis.

First, according to Refs. [1, 2], a single quantum subroutine of SFA succeeds if Eqs. (1)–(3) hold. This is also the meaning of r' being *available* in Refs. [1, 2].

According to Eqs. (1)–(3), the probability of r' being *available* is

$$\Pr(y^r \not\equiv -1 \pmod N) \sum_{r' \in \{r, 2r\}} \Pr(r'). \tag{22}$$

Here,

$$\Pr(r') \ge \sum_{s \in \mathbb{Z}^*_{r'}} \Pr\left(|\frac{s}{r'} - \frac{j}{2^t}| \le \frac{1}{2^{t+1}}\right), \tag{23}$$

and we know that

$$\Pr\left(|\frac{s}{r'} - \frac{j}{2^t}| \le \frac{1}{2^{t+1}}\right) \in O\left(\frac{1}{2r}\right) \tag{24}$$

where $s \in \mathbb{Z}^*_{r'}$ [1, 2]. Therefore, in Eq. (22),

$$\sum_{r' \in \{r, 2r\}} \Pr(r') \in O\left(\sum_{r' \in \{r, 2r\}} \frac{\varphi(r')}{2r}\right). \tag{25}$$

Test algorithm focuses on Eq. (22) to improve SFA further. As we have seen, Eqs. (2), (3) are generalized to Eqs. (10), (11) or Eqs. (12), (13). So far, we have adapted Eq. (22) to the sum of

$$\Pr(y' \not\equiv -1(\text{mod } N)) \sum_{r' \in R} \Pr(r') \tag{26}$$

and

$$\Pr(y' \equiv -1(\text{mod } N)) \sum_{r' \in R_1} \Pr(r'). \tag{27}$$

This sum minus Eq. (22) implies

$$\sum_{r' \in R_1} \Pr(r') \in O\left(\sum_{r' \in R_1} \frac{\varphi(r')}{2r}\right). \tag{28}$$

As a result, Eq. (28) implies that the raising success probability increases in SFA does not decrease in SFA when Algorithm 4 is used.

4 The Raising Success Probability

In this section, we estimate the efficiency of using the Test algorithm in SFA.

The success probability of each quantum process is the probability of r' being *available* that is good enough. It consists of three parts: $|2rj \pmod{2^t}| \le r(0 \le j \le 2^t - 1)$, as well as constraints on y and r'. We present the success probability of the single quantum process of SFA for Algorithms 1, 2, 3 and Algorithm 3 with the Test algorithm, respectively. Here, δ_1 and δ_2 are possible increments which depend on a specific N, and $\delta_1 = \delta_2 = 0$ is the lower bound of the corresponding probability. Moreover, ε can be set up as required in advance.

The success probabilities of each quantum process for Algorithms 1, 2, 3 and Algorithm 3 with Test algorithm are determined as follows.

- Algorithm 1 [1]. The probability of $|2rj \pmod{2^t}| \le r(0 \le j \le 2^t - 1)$ is asymptotically bounded from below by $4/\pi^2$ [1]. The probability of y being *good*—satisfying Eqs. (1), (2)—is bigger than $1/2$ and the probability of $r' \in \{r, 2r\}$ is $\varphi(r)/r$ [1]. Thus, the success probability is bounded from below by

$$\frac{4}{\pi^2} \times \left(\frac{1}{2} + \delta_1\right) \times \frac{\varphi(r) + \varphi(2r)}{2r}, 0 \le \delta_1 \le \frac{1}{2}. \tag{29}$$

Remark 1 Note that

$$\varphi(r) + \varphi(2r) = \sum_{r' \in \{r, 2r\}} \varphi(r') = \sum_{r' \in R - R_1} \varphi(r'). \tag{30}$$

Furthermore, if $\gcd(2, r) = 1$, then $\varphi(2r) = \varphi(2)\varphi(r) = \varphi(r)$ [2]. If $\gcd(2, r) = 2$, we know that

$$\{x|\gcd(x, 2r) = 1\} = \{x|\gcd(x, r) = 1\} \cup \{r + x|\gcd(x, r) = 1\}, \quad (31)$$

where $1 \leq x \leq r - 1$; Therefore, $\varphi(2r) = 2\varphi(r)$. As a result,

$$\varphi(2r) = \frac{r}{r} \times \varphi(r). \quad (32)$$

- Algorithm 2 [2]. By adding some additional qubits to the register, Ref. [2] raises the probability of $|2rj \pmod{2^t}| \leq r(0 \leq j \leq 2^t - 1)$ to $1 - \varepsilon$. Thus,

$$(1 - \varepsilon) \times \left(\frac{1}{2} + \delta_1\right) \times \frac{\varphi(r) + \varphi(2r)}{2r}, 0 \leq \delta_1 \leq \frac{1}{2}. \quad (33)$$

- Algorithm 3 [5]. By choosing the integer with Jacobi-Symbol -1, Ref. [5] raises the lower bound of the probability of y being *good* to $3/4$. Therefore,

$$(1 - \varepsilon) \times \left(\frac{3}{4} + \delta_2\right) \times \frac{\varphi(r) + \varphi(2r)}{2r}, 0 \leq \delta_2 \leq \frac{1}{4}. \quad (34)$$

- Algorithm 3 with the Test algorithm. According to Theorem 1, the success probability for this algorithm is

$$\frac{1 - \varepsilon}{2r} \left[(\frac{3}{4} + \delta_2)(\sum_{r' \in R} \varphi(r')) + (\frac{1}{4} - \delta_2)(\sum_{r' \in R_1} \varphi(r')) \right]. \quad (35)$$

When compared with the Algorithm 3 [5], the raising probability is

$$\frac{1 - \varepsilon}{2r}(\sum_{r' \in R_1} \varphi(r')) \quad (36)$$

what follows from Eq. (35) minus Eq. (34). Thus, the total raising probability for SFA is

$$\sum_{2r} \Pr(\text{ord}_N(y) = 2r) \frac{1 - \varepsilon}{2r} \left(\sum_{r' \in R_1} \varphi(r')\right) \quad (37)$$

Consequently, we can see that by loosing the constraints Eqs. (1)–(3) we can obtain an *available* result and make a full use of that result.

5 An Example

In this section, we show how much can our results be used to improve success probability when the RSA moduli [7] are factorized. In doing that we make use the fact that such moduli are products of two safe primes.

Example 4 Let $N = p_1 p_2$ where $p_i - 1 = 2q_i > 10^{100}$, $\log_2(p_1/p_2)$ is small, $|p_1 - p_2|$ is big, p_i and q_i are odd safe primes and $i = 1, 2$. Numbers of this form are likely candidates for moduli of the RSA [7] protocols.

In this case, we will make use of the following remark.

Remark 2 When Algorithm 3 [5] is used in Example 4, Eq. (34) becomes

$$(1 - \varepsilon) \times \frac{\varphi(r)}{r}. \tag{38}$$

Proof First,

$$\left(\frac{y}{N}\right) = -1 = \left(\frac{y}{p_1}\right)\left(\frac{y}{p_2}\right) \tag{39}$$

implies that

$$\left(\frac{y}{p_1}\right) = 1 \text{ and } \left(\frac{y}{p_2}\right) = -1 \tag{40}$$

or

$$\left(\frac{y}{p_1}\right) = -1 \text{ and } \left(\frac{y}{p_2}\right) = 1. \tag{41}$$

Without loss of generality, we can assume that Eq. (40) is true. Combining with Eqs. (6), (8) and (40) implies that s_1 is even and s_2 is odd, and

$$r_1 = \frac{q_1}{\gcd(q_1, s_1/2)} \text{ is odd} \tag{42}$$

and

$$r_2 = \frac{2q_2}{\gcd(2q_2, s_2)} \text{ is even.} \tag{43}$$

According to Eq. (7), $\text{ord}_N(y) = 2r = \text{lcm}\{r_1, r_2\}$. Combined with Eqs. (42), (43), we have $r = \text{lcm}\{r_1, r_2/2\}$. Then, $r_1 \mid r$ implies that $y^r \equiv 1 \pmod{p_1}$. However, $y^r \equiv -1 \pmod{N}$ implies that $y^r \equiv -1 \pmod{p_1}$. Thus, $\left(\frac{y}{N}\right) = -1$ implies that Eq. (2) always holds and $\delta_2 = \frac{1}{4}$ in Eq. (34).

Then, applying Eq. (8),

$$r_i = \frac{\varphi(p_i)}{\gcd(\varphi(p_i), s_i)} = \frac{2q_i}{\gcd(2q_i, s_i)}, \quad 1 \le s_i \le 2q_i. \tag{44}$$

Thus, r_i must be the factor of $2q_i$, i.e., $1, 2, q_i$ and $2q_i$. According to Eqs. (40), (41), (r_1, r_2) is one of $(1, 2), (2, 1), (1, 2q_2), (2q_1, 1), (2, q_2), (q_1, 2), (q_1, 2q_2)$, and $(2q_1, q_2)$. We have

$$\text{ord}_N(y) = \text{lcm}\{r_1, r_2\} \in \{2, 2q_1, 2q_2, 2q_1q_2\}. \tag{45}$$

Combining with Remark 2 the Eq. (38) can be obtained. By that the proof is finished. □

Now, we can start to determine the improvement success probability in the following remark.

Remark 3 Compared with Algorithm 3 [5], the raising probability of Algorithm 3 with the Test algorithm for RSA [7] is greater than $2/\sqrt{N}$ in each quantum process.

Proof First, note that the raising probability in the Algorithm 3 is

$$P_1 = \sum_t \Pr(r' = t, \frac{t}{2} \mid 2r = t)\Pr(2r = t) \tag{46}$$

where t can take any value in $\{2, 2q_1, 2q_2, 2q_1q_2\}$. Equivalently, it can be written as

$$P_1 = \sum_t \Pr(r'_{y^2} = t \mid r = t)\Pr(r = t) \tag{47}$$

where $t \in \{1, q_1, q_2, q_1q_2\}$.

In Algorithm 3 with the Test algorithm, according to Eq. (44), $\tilde{r}_1$ and $\tilde{r}_2$ is one of $(1, 1), (q_1, 1), (1, q_2)$ and (q_1, q_2). Thus, $\text{ord}_N(y^2) = r = \text{lcm}\{\tilde{r}_1, \tilde{r}_2\}$ is one of $1, q_1, q_2$ and q_1q_2. According to the discussion in Sect. 3.1, the success probability of Algorithm 3 with the Test algorithm is

$$P_2 = \sum_t \Pr(\tilde{r}_i \text{ divides } 2r'_{y^2} \mid r = t)\Pr(r = t) \tag{48}$$

where $t \in \{1, q_1, q_2, q_1q_2\}$. Using Eqs. (47), (48), the raising probability is:

$$P_2 - P_1 \geq \sum_{j=1}^{2} \Pr(r'_{y^2} = q_j \mid r = q_1q_2)\Pr(r = q_1q_2). \tag{49}$$

According to Eqs. (44), (45), we have

$$\Pr(r = q_1q_2) = \prod_{j=1}^{2} \Pr(\tilde{r}_j = q_j) = \frac{(q_1 - 1)(q_2 - 1)}{q_1q_2}. \tag{50}$$

Because $r'_{y^2} = r/\gcd(r, s)$ where $1 \leq s \leq r$, we get

$$\sum_{j=1}^{2} \Pr(r'_{y^2} = q_j \mid r = q_1 q_2) = \frac{q_1 + q_2 - 2}{q_1 q_2}. \tag{51}$$

Computing the product of Eqs. (50), (51), we have

$$P_2 - P_1 \geq \frac{(q_1 - 1)(q_2 - 1)}{q_1 q_2} \frac{q_1 + q_2 - 2}{q_1 q_2} \tag{52}$$

$$> \frac{1}{q_1} + \frac{1}{q_2} - \frac{1}{q_1^2} - \frac{1}{q_2^2} - \frac{5}{q_1 q_2} \tag{53}$$

$$> \frac{1}{q_1} + \frac{1}{q_2} - \frac{7}{\min^2\{q_1, q_2\}} \tag{54}$$

$$> \frac{1}{\min\{q_1, q_2\}} \tag{55}$$

$$> \frac{2}{\min\{p_1, p_2\}} \tag{56}$$

$$> \frac{2}{\sqrt{N}}. \tag{57}$$

where $\min^2\{q_1, q_2\} > 7\max\{q_1, q_2\}$ is used in that $\log_2(p_1/p_2)$ is small and $p_i - 1 = 2q_i > 10^{100}$. By that the proof is finished. □

Meanwhile, according to Eqs. (50), (51), we can see that the success probability is close to 1 even if SFA does not have to its disposal the ideal value of the order from the order determining algorithm. However, Algorithm 3 with the Test algorithm is behind the possibility to increase success probability to 1. In fact, only for $r'_{y^2} = 1$ and $r = q_1 q_2$, we can not obtain a factor in the improved version of the SFA. But in such a case the raising probability is

$$\Pr(r'_{y^2} = 1 \mid r = q_1 q_2) \Pr(r = q_1 q_2). \tag{58}$$

It is not difficult to verify that Eq. (58) is $O(1/N)$.

Furthermore, using a similar argument with Eq. (38), we can show that the fault probability in Ref. [5] is $O(1/\sqrt{N})$. Thereby, we have reduced the fault probability to $O(1/N)$ in Example 3. This way the example is finished. □

As a consequence, Algorithm 3 with the Test algorithm can use the *available* result to obtain a non-trivial factor of large integers and can improve the success probability in related quantum processes.

Example 5 Let $N = 77 = 7 \times 11$. $\varphi(77) = \varphi(7) \times \varphi(11) = 6 \times 10 = 60$. In these numbers, the number of numbers satisfying Eq. (39) is 30, and the number of numbers satisfying Eq. (40) is 15. According to Eqs. (42), (43), $r_1 \in \{1, 3\}$ and $r_2 \in \{2, 10\}$. Thus, $2r = \mathrm{lcm}\{r_1, r_2\} \in \{2, 6, 10, 30\}$. Numbers in $\mathbb{Z}_7^*$ with $r_1 = 1$ is 1, and with $r_1 = 3$ is in the set $\{2, 4\}$. Meanwhile, numbers in $\mathbb{Z}_{11}^*$ with $r_2 = 2$ is 10, and with

$r_2 = 10$ is in the set $\{2, 6, 7, 8\}$. Here, we list these numbers of the order $2r \in \{2, 6, 10, 30\}$, respectively.

- (1) The numbers with $2r = 2$: 43. According to Theorem 1, all factors of the order is usable.
- (2) The numbers with $2r = 6$: 32, 65. According to Theorem 1, all factors of the order is usable.
- (3) The numbers with $2r = 10$: 8, 29, 50, 57. According to Theorem 1, all factors of the order is usable.
- (4) The numbers with $2r = 30$: 2, 18, 30, 39, 46, 51, 72, 74. According to Theorem 1, $r' \in \{3, 5, 6, 10, 15, 30\}$ is usable, and only $r' \in \{1, 2\}$ is unusable. Note that we can get $r' \in \{1, 2\}$ if the numerator of the fraction obtained by the continued fractions (CF) algorithm [2] is 15 and 30. Therefore, this probability is $1/15$. The raising probability is the case that $r' \in \{3, 5, 6, 10\}$ and the probability is $2/5$. And, we choose the numbers with $2r = 30$ with a probability $8/15$. As a result, the fault probability is reduced to $8/225$.

This way the example is finished.   □

6 Conclusion

This paper showed that we can get a non-trivial factor even from some "unuseful" results of the order finding subroutines in previous Shor's factoring algorithms [1, 2, 5]. This discovery shows that the previous constraints are only a sufficient but not necessary condition for SFA being successful in factorization.

A natural research challenge is therefore now to extend results obtained in this paper and to show some other unideal values of the results of the order finding algorithm that can be actually used to factorize with sufficiently large probability.

Acknowledgements This work is supported in part by the National Natural Science Foundation of China (Nos. 61572532, 61272058), the National Natural Science Foundation of Guangdong Province of China (No. 2017B030311011), and the Fundamental Research Funds for the Central Universities of China (No. 17lgjc24). Qiu is also Funded by FCT project UID/EEA/50008/2013.

References

1. Shor, P.W.: Polynomial-time algorithms for prime factorization and discrete logarithms on a quantum computer. SIAM J. Comput. **26**(5), 1484–1509 (1997)
2. Nielson, M.A., Chuang, I.L.: Quantum Computation and Quantum Information. Cambridge University Press, Cambridge (2000)
3. Bocharov, A., Roetteler, M., Svore, K.M.: Factoring with qutrits: shor's algorithm on ternary and metaplectic quantum architectures. arXiv:1605.02756v3 [quant-ph] (2016)
4. Lawson, T.: Odd orders in Shor's factoring algorithm. Quantum Inf. Process. **14**(3), 831–838 (2015)

5. Leander, G.: Improving the success probability for shor's factoring algorithm. arXiv: 0208183 [quant-ph] (2002)
6. Nagaich, S., Goswami, Y.C.: Shor's algorithm for quantum numbers using matlab simulator. In: International Conference on Advanced Computing Communication Technologies, Panipat, 165–168 (2015)
7. Cramer, R., Shoup, V.: Signature schemes based on the strong RSA assumption. ACM Trans. Inf. Syst. Secur. 3(3), 46–51 (1999)
8. Davies, J.T., Rickerd, C.J., Grimes, M.A., Guney, D.O.: An n-bit general implementation of shor's quantum period-finding algorithm. arXiv:1612.07424v1 [quant-ph] (2016)

Simple Block-Substitution Rule Exhibits Interesting Patterns

Rolf Hoffmann

Abstract The goal is to find all patterns (space-time evolutions) that can be generated by a simple binary 1D block substitution rule (BCA), the Twin-Toggle Rule. The 1D space is partitioned into even or odd cell pairs. The even and odd partition are alternated in time. If the two bits in a pair are equal, then they are substituted by their logical inverses. Firstly, all possible initial configurations are reduced to a set of representatives taking into account 0/1-inversion, cyclic shift and mirroring. Secondly, the BCA was simulated and all patterns were compared to each other for similarities (black/white-inversion, shift, rotation, mirroring). Only a small amount of them was stored, representing all possible patterns. Most of the patterns are *single* patterns (the pixel structure is the same for black and white pixels). For $N = 8$, 12 cells interesting *dual* patterns were discovered, which show a different structure for black and white pixels.

1 Introduction

When we are thinking of Cellular Automata (CA) we associate that they are able to create useful, interesting, or beautiful patterns [1–3]. A challenge is often to find a combination of a simple CA rule together with a special initial configuration that results in the desired effect. In principal we may use any kind of CA, like classical CA, partitioned CA, lattice gas CA, block substitution CA, or PSA (Parallel Substitution Algorithm). PSA [4] is a powerful generalization of the CA model. And we may use any updating scheme, like synchronous, block-synchronous, or asynchronous. We are also able to define multi-agent-systems (MAS) on top of these models, using specific CA rules. CA-MAS are systems with moving agents/particles which act according to a certain 'intelligence' which is modeled within the chosen CA model variant, often as a finite state machine [5].

R. Hoffmann (✉)
Computer Science Department, Technical University Darmstadt,
Hochschulstr. 10, 64289 Darmstadt, Germany
e-mail: hoffmann@rbg.informatik.tu-darmstadt.de

© Springer International Publishing AG 2018
A. Adamatzky (ed.), *Reversibility and Universality*, Emergence, Complexity
and Computation 30, https://doi.org/10.1007/978-3-319-73216-9_22

463

The goal is to find all different patterns (space-time evolutions) that can be generated by a simple binary 1D block substitution rule (1D-BCA), starting from any initial configuration. BCA were already studied by Margolus 1984 [6] in the context of reversibility [7–9]. One question is: are there special initial configurations that lead to interesting patterns. We will see later, that seldom so called 'dual' patterns can be discovered.

Our generated patterns are not only appealing from the aesthetic point of view. They may be used for cloth, carpets, wallpapers and facades. Moreover, they could be used to achieve certain properties in mechanical, physical or chemical systems built from two components.

1.1 1D Block Substitution Cellular Automata

The Block Substitution CA (BCA) that we will consider is formed by N cells $c_i \in \{0, 1\}$ which are cyclically connected, and N is assumed to be even:

$$c = (c_0, c_1, \ldots, c_{N-1}).$$

If a cell with an index i outside this range is addressed, then it is mapped by *mod* N to the valid range. The cell space is partitioned into the *even partition* or the *odd partition*:

$$\pi_{even} = \{(c_0, c_1), (c_2, c_3), \ldots (c_{N-2}, c_{N-1})\},$$
$$\pi_{odd} = \{(c_1, c_2), (c_3, c_4), \ldots (c_{N-1}, c_0)\}.$$

A substitution rule is a mapping

$$f: \{0, 1\} \times \{0, 1\} \rightarrow \{0, 1\} \times \{0, 1\}.$$

When all cell pairs (c_i, c_{i+1}) of π_{even} are substituted by $f(c_i, c_{i+1})$, then the global transition function is denoted by $F_{even}(c)$. When all cell pairs of π_{odd} are substituted then the global transition function is denoted by $F_{odd}(c)$. The substitutions can be performed in any order (including synchronously simultaneous), because the cell pairs do not overlap in each of the partitions.

The even and odd computations F_{even} and F_{odd} are alternated in time. If the condition Φ_{even} (even phase) is true, then F_{even} is executed, and if the condition Φ_{odd} (odd phase) is true, then F_{odd} is executed. If the condition $StartOdd = 0$, then the whole computation starts with F_{even}. If the condition $StartOdd = 1$, then the whole computation starts with F_{odd}. The initial generation is $c^{t=0}$. The following generations are $c^{t=1,2,\cdots}$. The following algorithm describes the simulation of the BCA.

input(c^0), **input**($Start\,Odd$)
output($c = c^0$)
for $t = 1$ **to** t_{max} **do**
        $\Phi_{even} = (t + Start\,Odd) \bmod 2 = 1, \Phi_{odd} = \mathbf{not}\ \Phi_{even}$
        if Φ_{even} **then for** $i = 0, 2, 4, \ldots, N - 2$ **do** $(c_i, c_{i+1}) := f(c_i, c_{i+1})$

            if Φ_{odd} **then for** $i = 1, 3, 5, \ldots, N - 1$ **do** $(c_i, c_{i+1}) := f(c_i, c_{i+1})$
            output$(c = c^t)$
  end for t

In our simulations with the rule described in the next subsection we stop the computation when c^{N-1} is computed ($t_{max} = N - 1$) because then the outputs are repeating ($c^{t \bmod N} = c^t$).

1.2   The Twin-Toggle Rule

A very simple rule was selected for pattern generation, called the *Twin-Toggle Rule* (TTR).

$$f(a, b) = \begin{cases} (1, 1) & \text{if } (a, b) = (0, 0) \\ (0, 0) & \text{if } (a, b) = (1, 1) \\ (a, b) & \text{otherwise no change} \end{cases} \tag{1}$$

The rule can also be described by a list of substitutions:

S0: $(0, 0) \to (1, 1)$,
S1: $(1, 1) \to (0, 0)$.

If zero and one are interchanged, the substitutions are interchanged, too, and the whole rule remains the same. Thus the rule is invariant under 0/1 interchange, or 0/1-symmetric. The function (Eq. 1) is also its inverse (self-inverse), i.e. $f(f(a, b)) = (a, b)$.

We can also easily find (e.g. by writing down a truth table) that

$$f(a, b) = (\bar{b}, \bar{a}), \tag{2}$$

where $\bar{a}$ denotes the logical *NOT(a)* operation.

Now the aim is to discover all different patterns which can be generated by this rule. In order to reduce the number of unnecessary simulations which yield the same pattern (bw-similar, mirror symmetric), the whole set of initial configurations will first be reduced to a smaller set of representatives (Sect. 2), which are still able to produce all desired patterns. Then the generated patterns will also be reduced to their representatives (Sects. 3 and 4), because it turned out, that the 'same' pattern (similar under shift, 0/1-inversion, rotation, mirroring) can be generated by different initial configurations.

2   All Initial Configurations

Any possible initial configurations will be considered for this 1D BCA with width
$N = 2, 4, 6, 8, 10$, and 12 cells.

Width N = 2. There are the configurations $c = 00, 01, 10, 11$. When starting with
00, the time evolution is cyclic ($00 \leftrightarrow 11$) with period $T = 2$. When starting with
11, the time evolution is also cyclic ($11 \leftrightarrow 00$) with period $T = 2$. When starting
with 01 or 10, the configuration remains constant.

```
t=3            11     00     01     10
t=2            00     11     01     10
t=1            11     00     01     10
t=0            00     11     01     10


space
index x =  01     01     01     01
```

We can notice that we do not need to study all initial configurations because the rule is
0/1-symmetric. It is sufficient to consider 00 as the representative for 00 and 11, and
01 as the representative for 01 and 10. When starting with an inverse configuration
$\bar{c}$ at $t = 0$ we will obtain a time evolution of configurations which is also inverse.
Note that for this particular trivial case ($N = 2$), it does not matter whether the
computation starts on the even or odd partition.

In order to perceive more distinctly the generated patterns, the space-time area
will be displayed doubled in space ($x = 0, 1, \ldots, 2N - 1$) and doubled in time
($t = 0, 1, \ldots, 2N - 1$). Then the displayed area (also called *space-time-layout* or
picture) is square of size $2N \times 2N$. For $N = 2$ the four possible initial configurations
the space-time-layouts are as follows:

```
t=3           1111    0000    0101    1010
t=2           0000    1111    0101    1010
t=1           1111    0000    0101    1010
t=0           0000    1111    0101    1010


space
index x =  0101    0101    0101    0101
```

Now we can easily observe that horizontal or vertical line patterns alternating in
black and white are generated. The pictures in Sect. 4 are also displayed in this way.

Width N = 4. There are 16 possible initial configurations (Table 1a). 12 of them
are redundant because they are equivalent under cyclic shift or 0/1-inversion. The
remaining representatives are $d(C) = 0, 1, 3, 5$ (Table 1b).

Table 1 **a** All possible initial configurations for $N = 4$ cells. **b** Only four non-redundant configurations remain if equivalents under cyclic shift and 0/1-inversion are eliminated

(a)

init config, dual number C	decimal number d(C)	cyclic shift equivalent	cyclic shift count	bw inverse
0 0 0 0	0			
0 0 0 1	1			
0 0 1 0	2	1	1	
0 1 0 0	4	1	2	
1 0 0 0	8	1	3	
0 0 1 1	3			
0 1 0 1	5			
0 1 1 0	6	3	1	
1 0 0 1	9	3	3	6
1 0 1 0	10	5	1	5
1 1 0 0	12	3	2	3
0 1 1 1	7			8
1 0 1 1	11	7	3	4
1 1 0 1	13	7	2	2
1 1 1 0	14	7	1	1
1 1 1 1	15			0

(b)

init config, dual number C	decimal number d(C)	config. index i
0 0 0 0	0	0
0 0 0 1	1	1
0 0 1 1	3	2
0 1 0 1	5	3

For example, the configuration $C = 0010$ ($d = 2$) becomes 0001 ($d = 1$) when it is shifted one position to the right: $sh(c = 0010, count = 1) = 0001$. The cyclic shift operator $\boldsymbol{sh}(c, count)$ shifts a bit string c (or a bit matrix) $count$ positions cyclically to the right if $count > 0$. If $count < 0$ then the shift goes to the left.

The bit string $c^{shifted}$ is cyclically equivalent to c if there exists a shift count such that

$$c^{shifted} = sh(c, count) \Leftrightarrow \forall i : c_i^{shifted} = c_{(i-count) \bmod N} . \tag{3}$$

Width N = 6. There 64 possible configurations. If the cyclic and inverse equivalents are removed, nine are remaining (Table 2a). Now we observe another equivalence, the *mirror symmetry*:

- Configuration $d(c) = 11$, shifted 2 positions to the left and then mirrored in the middle (positions i and $N - 1 - i$ interchanged, $i = 0 \dots N/2 - 1$) yields configuration 13.

Mirror-symmetrical configurations will also be removed from the set of *non-redundant initial configurations* because we consider symmetrical time-evolutions as equivalent. After removing configuration 13 (the only symmetrical one for $N = 6$) only 8 representatives remain. Note that because of cyclic shift equivalence there may be up to N equivalent mirrors.

Table 2 a Case $N = 6$ cells: Only 8 non-redundant initial configurations remain if the mirror-symmetrical one is also eliminated. **b** Case $N = 8$ cells: Only 19 non-redundant initial configurations exist

(a)

init config. / dual number C	decimal number d(C)	mirror-symmetric to	config. index i
000000	0		0
000001	1		1
000011	3		2
000101	5		3
001001	9		4
000111	7		5
001011	11		6
001101	13	11	
010101	21		7

(b)

init config. / dual number C	decimal number d(C)	config. index i
00000000	0	0
00000001	1	1
00000011	3	2
00000101	5	3
00001001	9	4
00010001	17	5
00000111	7	6
00001011	11	7
00010011	19	8
00010101	21	9
00100101	37	10
00001111	15	11
00010111	23	12
00011011	27	13
00100111	39	14
00101011	43	15
00101101	45	16
00110011	51	17
01010101	85	18

The mirror of c, $mirror(c) = mirsh(c, 0)$ is obtained by reflecting the bits on a mirror which is placed in the middle of the bit string:

$$c^{mirror} = mirror(c) \Leftrightarrow \forall i : c_i^{mirror} = c_{N-1-i} , \quad i = 0 \ldots N - 1. \quad (4)$$

For each shift count $\in \{1, \ldots, N - 1\}$ there exists a cyclically shifted mirror:

$$mirsh(c, count) = mirror(sh(c, count)) = sh(mirror(c), -count). \quad (5)$$

There is a special mirror **mir** that will be used later in Sect. 3.3 where it will be applied to each row c^t of a space-time matrix C.

$$mir(C) = mirsh(C, 1). \quad (6)$$

Here is an example for the mirror and its equivalents under shift:

```
    c       mirsh(c,0) mirsh(c,-1) mirsh(c,-2) mirsh(c,-3)mirsh(c,-4)mirsh(c,-5)
            mirsh(c,0) mirsh(c,5)  mirsh(c,4)  mirsh(c,3) mirsh(c,2) mirsh(c,1)
  001011    110100     101001      010011      100110     001101     011010
i=012345    543210     432105      321054      210543     105432     054321
imir=       012345     012345      012345      012345     012345     012345
i=imir                   x  x                    x  x                  x  x
```

Table 3 The 47 representatives for the different initial configuration classes with $N = 10$ cells

init config. / dual number C	decimal number d(C)	config. index i	init config. / dual number C	decimal number d(C)	config index i
0 0 0 0 0 0 0 0 0 0	0	0	0 0 0 1 0 0 1 1 0 1	77	24
0 0 0 0 0 0 0 0 0 1	1	1	0 0 0 1 0 1 0 0 1 1	83	25
0 0 0 0 0 0 0 0 1 1	3	2	0 0 0 1 0 1 0 1 0 1	85	26
0 0 0 0 0 0 0 1 0 1	5	3	0 0 0 1 1 0 0 0 1 1	99	27
0 0 0 0 0 0 1 0 0 1	9	4	0 0 1 0 0 1 0 0 1 1	147	28
0 0 0 0 0 1 0 0 0 1	17	5	0 0 1 0 0 1 0 1 0 1	149	29
0 0 0 0 1 0 0 0 0 1	33	6	0 0 1 0 1 0 0 1 0 1	165	30
0 0 0 0 0 0 0 1 1 1	7	7	0 0 0 0 0 1 1 1 1 1	31	31
0 0 0 0 0 0 1 0 1 1	11	8	0 0 0 0 1 0 1 1 1 1	47	32
0 0 0 0 0 1 0 0 1 1	19	9	0 0 0 0 1 1 0 1 1 1	55	33
0 0 0 0 0 1 0 1 0 1	21	10	0 0 0 1 0 0 1 1 1 1	79	34
0 0 0 0 1 0 0 0 1 1	35	11	0 0 0 1 0 1 0 1 1 1	87	35
0 0 0 0 1 0 0 1 0 1	37	12	0 0 0 1 0 1 1 0 1 1	91	36
0 0 0 1 0 0 0 1 0 1	69	13	0 0 0 1 0 1 1 1 0 1	93	37
0 0 0 1 0 0 1 0 0 1	73	14	0 0 0 1 1 0 0 1 1 1	103	38
0 0 0 0 0 0 1 1 1 1	15	15	0 0 0 1 1 0 1 0 1 1	107	39
0 0 0 0 0 1 0 1 1 1	23	16	0 0 1 0 0 1 0 1 1 1	151	40
0 0 0 0 0 1 1 0 1 1	27	17	0 0 1 0 0 1 1 0 1 1	155	41
0 0 0 0 1 0 0 1 1 1	39	18	0 0 1 0 1 0 0 1 1 1	167	42
0 0 0 0 1 0 1 0 1 1	43	19	0 0 1 0 1 0 1 0 1 1	171	43
0 0 0 0 1 0 1 1 0 1	45	20	0 0 1 0 1 0 1 1 0 1	173	44
0 0 0 0 1 1 0 0 1 1	51	21	0 0 1 0 1 1 0 0 1 1	179	45
0 0 0 1 0 0 0 1 1 1	71	22	0 1 0 1 0 1 0 1 0 1	341	46
0 0 0 1 0 0 1 0 1 1	75	23			

Width $N = 8$. Altogether there are 2^8 possible initial configurations. 23 remain if cyclically and 0/1-equivalents are eliminated, and at last 19 remain after eliminating mirror-symmetrical ones (Table 2b).

Width $N = 10$. Altogether there are 2^{10} possible initial configurations. 67 remain if cyclically and 0/1-equivalents are eliminated, and then 47 remain after eliminating mirror-symmetrically ones (Table 3).

Width $N = 12$. There are 2^{12} possible initial configurations. 216 remain if cyclically and 0/1-equivalents are eliminated, and then 137 remain after eliminating mirror-symmetrically ones (Table 3).

Width $N = 14$. There are 2^{14} possible initial configurations. 714 remain if cyclically and 0/1-equivalents are eliminated, and then 410 remain after eliminating mirror-symmetrically ones.

Fig. 1 The logarithm of the
number of representatives R
versus the number of cells

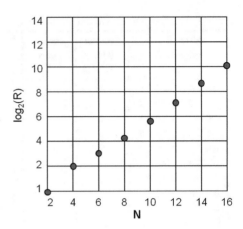

Width N = 16. There are 2^{16} possible initial configurations. 2463 remain if cyclically
and bw equivalents are eliminated, and then 1345 remain after eliminating mirror-
symmetrically ones.

The number of representatives. The number R(N) of representatives for $N =$
2, 4, 6, 8, 10, 12, 14, 16 is

$$R(N) = 2, 4, 8, 19, 47, 137, 410, 1345, \tag{7}$$

and the percentage is $R/2^N = 50, 25, 12.5, 7.42, 4.59, 3.34, 2.50, 2.05\%$.

This dependency is depicted in Fig. 1. Thus the number of initial configurations
to be investigated is significantly lower than the whole set.

3 Finding All Possible Patterns

Now the representative initial configurations are used to generate all possible patterns by simulation. The space-time evolution was simulated for N cells and
$N-1$ time steps. This yields a square area/matrix C (Fig. 2), with space coordinate $x = 0 \ldots N - 1$ and time coordinate $t = 0 \ldots N - 1$. Observing the generated
$N \times N$ space-time patterns it turned out that many of them a symmetric by rotation,
by mirroring (reflection), or by 0/1-inversion (black/white inversion). The goal is
to reduce the whole set of generated patterns to a set of non-redundant representatives (also called *essential patterns*). Then any possible TTR-pattern can be derived
from the essential patterns by applying rotation, mirroring or 0/1-inversion in any
combination.

The $N \times N$ space-time area is then displayed by doubling it in space and in time
(4 areas altogether) in order to highlight the inherent pattern structure. We call this
fourfold area *"picture"* (marked in pattern A1 of Fig. 3). We will use the general term
"pattern" for the square $N \times N$ simulation area or for the $2N \times 2N$ picture.

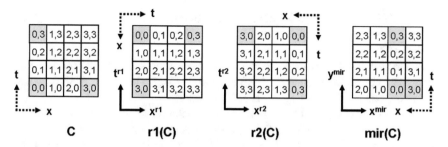

Fig. 2 The elements of the original matrix C are rotated $i \times 90°$ clockwise by $ri(C)$, and vertical mirrored by $mir(C)$, keeping $(x, t) = (N - 1, 0)$ at its original position

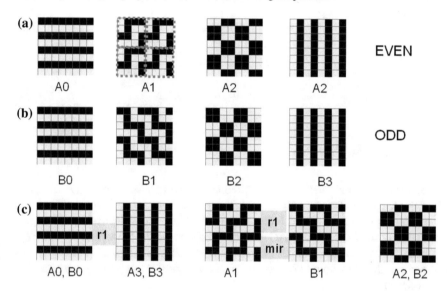

Fig. 3 Generated patterns for a 4×4 space-time area. The area is displayed as an 8×8 area, the area 4×4 area is doubled in time and space in order to visualize the pattern more distinctly. **a** The computation starts with even substitutions. **b** The computation starts with odd substitutions. **c** There are only 3 pattern classes, with equivalence under rotation r1 (rotate 90 degrees to the right)

In order to reduce the whole set of patterns to the essential ones the following symmetry relations were used.

3.1 Symmetries

We will associate the color **white** with the cell value 0 and the color **black** with the value 1 throughout this presentation. We define the black/white inversion **bw(c)** of a 0/1 bit string c (or a bit matrix C) by inverting every bit c_i of c.

- **Black/White Inversion: bw(C)** exchanges black (1) pixels with white (0) pixels, also called 0/1-inversion.

  $\forall t, i : C_i^t \leftarrow \overline{C_i^t}$

- **Mirror: mir(C)** mirrors C (including one shift right before mirroring in the middle – or including one shift left after mirroring). The included shift ensures that the cell at position $C_{x=N-1}^t$ keeps it position.

  This operator is especially useful to compare patterns computed that start with the even resp. odd partition first. This operator was already defined (Eq. 6).

- **Rotation: r1(C)** rotates C one quarter (90 degrees) to the right (clock-wise), **r2(C)** rotates C two quarters to the right, and **r3(C)** rotates C three quarters to the right. Formally the rotation operators are defined as follows, where indexes $x, t \in \{0, \ldots, N-1\}$

$$\forall x, t : [r1(C)]_x^t = C_t^{N-1-x} \tag{8}$$

$$\forall x, t : [r2(C)]_x^t = C_{N-1-x}^{N-1-t} \tag{9}$$

$$\forall x, t : [r3(C)]_x^t = C_{N-1-t}^{x} . \tag{10}$$

3.2 Tested Similarities

In order to test for similar patterns we define a two-dimensional $shift(P, \Delta x, \Delta t)$. It shifts a pattern P cyclically Δx positions in x direction and Δt positions in t direction. Two patterns Q and P are **sh-similar** if there exists a shift count $(\Delta x, \Delta t)$, such that $Q = shift(P, \Delta x, \Delta t)$. For this predicate we use the notion $P \sim sh(Q)$. A pattern P can be self-similar (or **repetitive**) under shift $P \sim sh(P)$. If we want to specify the shift counts for repetition we say $rep(P, \Delta x, \Delta t)$ is true, or P is $rep(\Delta x, \Delta t)$.

Two patterns Q and P are **bw-similar** if there exists a shift count $(\Delta x, \Delta t)$, such that $Q = bw(shift(P, \Delta x, \Delta t))$. For this predicate we use the notion $P \sim bw(Q)$.

A pattern P is called **bw-self-similar** if $P \sim bw(P)$ is true. Most of the generated patterns are bw-self-similar, i.e. the structure given by the black pixels is the same (under shift) as the structure given by the white pixels, or the layer of black pixels looks as the layer of white pixels. For the human perception one color appears dominant, this is supposed to be black here. So what we mainly perceive is a black structure consisting of black lines, pixels, and compounds of them. For example all of the essential patterns for N = 4, 6, 10 (Sects. 4.1, 4.2 and 4.4) are bw-self-similar. We call a bw-similar pattern *single pattern*. Patterns which are not bw-self-similar are attractive and of special interest, because they are composed of two different structures. Such patterns will be called *double patterns*. For instance for N = 8 there exist two double patterns, A6 and A13 (Sect. 4.3).

A pattern Q is **r1-similar** to P if there exist a shift count $(\Delta x, \Delta t)$, such that $Q = r1(shift(P, \Delta x, \Delta t))$. For this predicate we use the notion $Q \sim r1(P)$.

A pattern Q is *r2-similar* to P if there exist a shift count $(\Delta x, \Delta t)$, such that $Q = r2(shift(P, \Delta x, \Delta t))$. For this predicate we use the notion $Q \sim r2(P)$.

A pattern Q is *r3-similar* to P if there exist a shift count $(\Delta x, \Delta t)$, such that $Q = r3(shift(P, \Delta x, \Delta t))$. For this predicate we use the notion $Q \sim r3(P)$.

A pattern Q is *mir-similar* to P if there exist a shift count $(\Delta x, \Delta t)$, such that $Q = mir(shift(P, \Delta x, \Delta t))$. For this predicate we use the notion $Q \sim mir(P)$.

3.3 Observed General Properties

At first we define some notations:

- A *configuration* at time t is denoted as
  $c^t = (c_0^t, c_1^t, \ldots, c_{N-1}^t)$.
- A *space-time area* is the concatenation of the configurations from time 0 to time $N - 1$:
  $$C = \begin{bmatrix} c^{t=N-1} \\ \cdots \\ c^{t=1} \\ c^{t=0} \end{bmatrix}$$
- $\#b(C)$ denotes the number of black cells in a configuration c^t or in a space-time area C.
- $\#w(C)$ denotes the number of white cells in a configuration c^t or in a space-time area C.

The observed properties are

- Any pattern C ($N \times N$ space-time area) generated by the TT Rule is repetitive with offset N in space and in time. Thus $rep(C, N, 0), rep(C, 0, N)$, and $rep(C, N, N)$ are always true.
- The number of black and white pixels of a pattern C is equal:
  $\#b(C) = \#w(C) = N^2/2$.
- The number of black cells of two consecutive configurations is N:
  $\#b(c^t) + \#b(c^{t+1}) = N$.
  The same assertion is true for the white cells.
- The frequency of black and white cells is the same for configurations at time-steps t and $t + 2$:
  $\#b(c^t) = \#b(c^{t+2})$,
  $\#w(c^t) = \#w(c^{t+2})$.
- The computation can be reversed:
  $c^t = F_{odd}(F_{odd}(c^t))$,
  $c^t = F_{even}(F_{even}(c^t))$,
  $c^t = F_{even}(F_{odd}(F_{odd}(F_{even}(c^t))))$,
  $c^t = F_{odd}(F_{even}(F_{even}(F_{odd}(c^t))))$.

- After two generations, when starting with the even/odd computation, the even/odd bit C_i is shifted two positions to the right
  $$C_{i+2}^{t+2} = C_i^t,$$
  and the odd/even bit C_k two positions to the left.
  $$C_{k-2}^{t+2} = C_k^t.$$
  This means that the cell bit information is either traveling to the right or to the left.
- When starting with the even computation, the generated pattern is denoted as A, otherwise as B. Then B is the mirror of A (Eq. 6)
  $$B = mir(A).$$

4 The Representative Patterns

For every initial configuration i in the set of representatives a simulation was performed and the space-time picture was captured. Two simulations were performed, one starting with the even partition and one starting with the odd partition. The captured pictures are denoted as Ai/Bi, starting with the even resp. odd partition. The configuration index is $i \in \{0, \ldots, R(N) - 1\}$, where R is given by Eq. 7.

The generated patterns were compared to each other under an arbitrary combination of rotation, mirror, black/white inversion, and cyclic shift (as defined in Sect. 3.2). Each pattern Q was compared to any other pattern P according to the following relations.

$Q = P^*, bw(P^*), r1(P^*), r2(P^*), r3(P^*), r1(bw(P^*)), r2(bw(P^*)), r3(bw(P^*))$,

$mir(P^*), mir(bw(P^*)), mir(r1(P^*)), mir(r2(P^*)), mir(r3(P^*)), mir(r1(bw(P^*)))$,

$mir(r2(bw(P^*))), mir(r3(bw(P^*)))$.

P^* means that every possible shift of P, out of the set $\{sh(P, \Delta x, \Delta t)\}$ was used for testing. Only one representative for each pattern class (under similarity) was finally kept. Patterns which are not bw-self-similar (*double patterns*) were kept twice, as Ai and $bw(Ai)$.

4.1 Width 4

For $N = 4$ there are four non-redundant initial configurations (representatives): 000, 001, 0011, 0101 (Table 1). Starting with even substitutions, the pictures A1, A2, A3, A4 are generated, and starting with odd substitutions, the pictures B1, B2, B3, B4 are generated. We observe that A1 $\sim$ B1, A3 $\sim$ B4, A4 $\sim$ r1(A4), B2 $\sim$ r1(A2) $\sim$ mir(A2), A3 $\sim$ B3. There are only three essential patterns A1, A2, A3, because all others can be derived from them by rotation (r1). Other transformations (r2, r3, mir, bw, or combinations of them) are not necessary to derive all other patterns for $N = 4$. All theses patterns are *single patterns* (single layer structured, bw-self-similar).

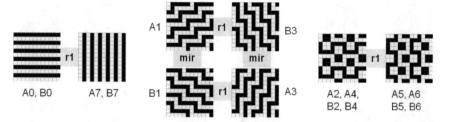

Fig. 4 There are three different essential patterns: A0, A1, A2

The following repetitions are true: $rep(A9, any, 2)$, $rep(A3, 2, any)$, $rep(A1, -2, 2)$, $rep(A2, -2, 2)$, $rep(A1, 4, 0)$, $rep(A1, 0, 4)$, $rep(A3, 4, 0)$, $rep(A3, 0, 4)$.

4.2 Width 6

There are only three essential patterns, A0, A1, and A2. All of them are single patterns. Figure 4 shows the relations between the generated patterns A0...A7, B0...B7, for example A7 ∼ r1(A0), B3 ∼ r1(A1), A1 ∼ r1(B3), B1 ∼ mir(A1), A3 ∼ r1(mir(A1)). Note that the operators r1 and mir can here be applied forwards and backwards.

The following repetitions are true: $rep(A0, any, 2)$, $rep(A7, 2, any)$, $rep(A1, -2, 2)$.

Compared to the case $N = 4$ the number of essential patterns is the same, and we can find the simple stripe pattern A0 in both cases.

4.3 Width 8

There are 10 essential patterns, 8 of them are single patterns (A0, A1, A2, A3, A5, A8, A11, A17), and two of them are dual patterns (A6, A13).

The patterns bw(A6) ∼ B7 shows a different black structure than A6, and the same is true for A13, because A6 and A13 are not bw-self-similar. Thus the total number of different structures (layers of pixels with the same color) is $12 = 10 + 2$.

Figure 5 shows relations between the generated patterns A0...A17, B0...B17, for example A12 ∼ r3(A2), A2 ∼ r1(A12), A4 ∼ r2(A2), A15 ∼ r3(A4) ∼ r3(r2(A2)) ∼ r1(A2). Note that the forward and the backward transformation between A2 and A12 are not the same, as for A4 and A15.

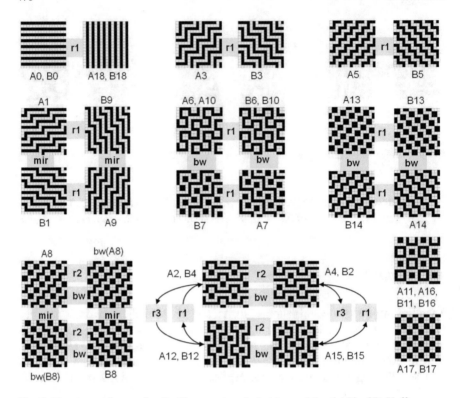

Fig. 5 The patterns for $n = 8$ cells. There are two dual-patterns, A6 and A13. ©R. Hoffmann

4.4 Width 10

There are only 10 essential patterns (Fig. 6), and all of them are single patterns (bw-self-similar).

4.5 Width 12

There are 37 essential patterns, 27 of them are single patterns and 10 are dual patterns (Fig. 7). Altogether there are $47 = 27 + 2 \times 10$ different pattern structures.

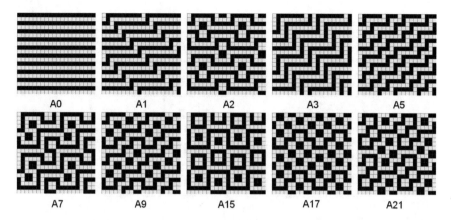

Fig. 6 The 10 essential patterns. All of them are single patterns. ©R. Hoffmann

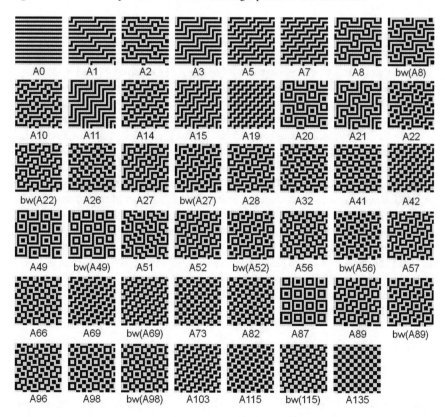

Fig. 7 The 37 essential patterns for N = 12. 27 of them are single patterns, and 10 of them are double patterns. ©R. Hoffmann

Table 4 The number of different patterns for a CA with N cells

N	2	4	6	8	10	12
single patterns	1	3	3	8	10	27
dual patterns	0	0	0	2	0	10
essential patterns	1	3	3	10	10	37
structures	1	3	3	12	10	47

5 Summary

The goal was to find all different patterns (space-time evolutions) that are generated by the Twin-Toggle Rule, a simple 1D block substitution rule, for an even number of cells up to 12. This rule is self-inverse and its time-evolution is reversible. Firstly, all possible initial configurations (binary bit strings) were reduced to a smaller set of representatives by deleting strings which are similar under 0/1-inversion, cyclic shift or mirroring. Secondly, the BCA was simulated twice for all representative initial configurations, once starting with the even partition, and another time starting with the odd partition. All simulated patterns were captured and compared to each other for similarities. Patterns were defined to be similar if black/white-inversion, shift, rotation and mirroring are performed in any combination. The number of different patterns appeared is summarized in Table 4. All patterns contain the same number of black and white pixels. Most of the patterns are *single* patterns (the pixel structure is the same for black and white pixels). For $N = 8$, 12 *dual* patterns appeared, where the black pixel structure is unequal the white pixel structure. In further work, the patterns, structures and its quantities could be discovered and studied for $N > 12$, theoretically and by simulation.

References

1. Prusinkiewicz, P.: Modeling and visualization of biological structures. In: Proceeding of Graphics Interface, pp. 128–137. Toronto (1993)
2. Pickover, C.A.: The Pattern Book: Fractals, Art and Nature. World Scientific Publishing Co., Inc., River Edge (1995)
3. Adamatzky, A., Martnez, J. (eds.): Designing Beauty: The Art of Cellular Automata. Emergence, Complexity and Computation, vol. 20. Springer, Berlin (2016)

4. Achasova, S., Bandman, O., Markova, V., Piskunov, S.: Parallel Substitution Algorithm: Theory and Application. World Scientific, Singapore (1994)
5. Hoffmann, R.: How agents can form a specific pattern. In: ACRI Conference on 2014. LNCS, vol. 8751, pp. 660–669 (2014)
6. Margolus, N.: Physics-like models of computation. Phys. D **10**, 81–95 (1984)
7. Morita, K., Harao, M.: Computation Universality of one dimensional reversible (injective) cellular automata. Trans. IEICE Jpn. **E72**, 758–762 (1989)
8. Morita, K.: Reversible simulation of one-dimensional irreversible cellular automata. Theor. Comput. Sci. **148**, 157–163 (1995)
9. Kari, J.: Representation of reversible cellular automata with block permutations. Theory Comput. Syst. **29**, 47–61 (1996). Springer

Index

© Springer International Publishing AG 2018
A. Adamatzky (ed.), *Reversibility and Universality*, Emergence, Complexity
and Computation 30, https://doi.org/10.1007/978-3-319-73216-9

Printed in the United States
By Bookmasters